Hospitals, **2** (42), 38; **4** (20), 103; **5** (57), 156; **5** (64), 157; **6** (12), 171; **7** (29), 247; **7** (51), 256; **9** (86), 352; **10** (19), 374; **12** (6), 478
Housing, **3** (41), 78; **7** (47), 255

Insurance, **3** (32), 69; **5** (1), 116; **5** (45), 154
I.Q., **2** (2), 12; **4** (45), 111; **5** (59), 156; **5** (66), 158; **8** (55), 292; **11** (26), 444
Integration, **3** (43), 79

Jogging, **2** (48), 39; **2** (49), 39
Juvenile delinquency, **6** (64), 220; **9** (37), 321

Linguistics, **3** (11), 57; **6** (30), 190; **7** (15), 234

Marketing, **2** (3), 13; **2** (7), 14; **2** (54), 41; **2** (55), 41; **3** (20), 63; **4** (30), 106; **6** (5), 165; **6** (14), 174; **6** (29), 190; **6** (54), 218; **9** (74), 350; **9** (78), 350; **12** (28), 504; **12** (44), 510
Marriage, **11** (26), 444
Media, **2** (4), 13; **2** (36), 35; **5** (41), 150; **5** (43), 151; **12** (38), 508; **12** (42), 509
Medicine, **2** (1), 12; **2** (56), 41; **4** (30), 106; **8** (63), 293; **9** (36), 321; **9** (47), 325; **9** (50), 344; **9** (53), 344; **9** (60), 345; **9** (79), 350; **9** (81), 351; **10** (23), 385; **10** (46), 399; **10** (65), 406; **11** (20), 434; **11** (26), 444; **11** (31), 449; **12** (5), 477; **12** (6), 478; **12** (7), 481; **12** (8), 482; **12** (20), 490; **12** (27), 495
Meteorology, **4** (28), 106; **5** (46), 154; **10** (27), 386; **12** (10), 482

Nuclear power, **2** (23), 26; **4** (42), 111; **5** (29), 137; **5** (42), 151; **9** (73), 350

Parapsychology, **9** (31), 314; **9** (91), 352
Physics, **2** (23), 26; **7** (34), 248; **12** (32), 504
Physiology, **9** (50), 344
Population, **2** (45), 39; **3** (16), 60; **3** (58), 81; **5** (56), 156; **5** (62), 157; **7** (42), 252; **11** (60), 472; **12** (12), 483
Pricing, **3** (15), 60; **3** (25), 63
Production, **2** (8), 14; **6** (37), 204; **6** (59), 219; **9** (61), 220; **9** (62), 220; **12** (23), 494; **12** (29), 504

Psychiatry, **6** (12), 171; **11** (16), 433; **12** (7), 481; **12** (8), 482
Psychology, **2** (35), 35; **2** (43), 39; **2** (49), 39; **2** (52), 40; **2** (58), 42; **2** (60), 42; **3** (12), 57; **3** (30), 69; **3** (55), 80; **4** (11), 91; **4** (12), 92; **4** (21), 103; **4** (24), 104; **4** (26), 105; **4** (37), 110; **4** (45), 111; **5** (15), 124; **5** (39), 150; **5** (66), 158; **6** (11), 170; **6** (13), 171; **6** (28), 190; **6** (41), 204; **6** (50), 217; **6** (64), 220; **6** (66), 221; **6** (67), 221; **7** (43), 255; **8** (32), 279; **8** (37), 280; **8** (50), 290; **8** (55), 292; **8** (62), 293; **8** (72), 294; **8** (75), 295; **9** (30), 314; **9** (45), 325; **9** (52), 344; **9** (55), 344; **9** (62), 346; **9** (80), 351; **9** (81), 351; **9** (85), 352; **10** (17), 373; **10** (28), 386; **10** (38), 393; **10** (63), 405; **11** (16), 433; **11** (19), 434; **12** (32), 504; **12** (33), 505; **12** (36), 508

Radiation, **3** (61), 81; **5** (42), 151
Real estate, **5** (50), 154
Rehabilitation, **8** (66), 294; **10** (32), 387
Resorts, **8** (64), 293
Restaurants, **6** (5), 165

Safety, **2** (23), 26; **2** (40), 38; **4** (44), 111; **5** (2), 116; **5** (20), 129; **5** (22), 130; **5** (42), 151; **5** (53), 155; **6** (42), 204; **6** (49), 217; **6** (58), 219; **6** (63), 220; **6** (69), 222; **7** (16), 234; **7** (48), 255; **7** (52), 256; **8** (28), 276; **8** (71), 294; **8** (74), 295; **9** (24), 309; **10** (20), 374; **12** (31), 504
Sales, **3** (20), 63; **6** (14), 174; **12** (28), 504
Sanitation, **4** (46), 112
Schools, **2** (38), 37; **2** (41), 38; **2** (49), 39; **2** (53), 41; **2** (57), 42; **3** (10), 57; **3** (21), 63; **3** (24), 63; **3** (43), 79; **3** (44), 79; **3** (51), 80; **3** (56), 81; **5** (4), 117; **5** (34), 144; **5** (61), 157; **7** (46), 255; **7** (55), 256; **9** (67), 349; **9** (89), 352; **10** (57), 404
Sex education, **2** (53), 41; **8** (34), 279
Social psychology, **5** (28), 137; **5** (29), 137; **5** (39), 150; **10** (62), 405; **10** (67), 406; **12** (34), 505
Social sciences, **6** (48), 214; **12** (20), 490
Sociology, **2** (59), 42; **4** (39), 110; **6** (20), 184; **6** (27), 190; **6** (39), 204; **7** (30), 247; **7** (47), 255; **8** (38), 280; **10** (12), 369; **10** (48), 399; **10** (49), 399; **10** (61), 405; **11** (16), 433; **11** (49), 470; **11** (60), 472; **12** (12), 483

Solar heat, **3** (38), 75; **3** (49), 80
Sports, **2** (49), 39; **3** (53), 80; **4** (22), 104; **4** (31), 106; **6** (40), 204; **6** (61), 220; **7** (57), 257; **8** (64), 293; **10** (47), 399; **12** (36), 508
Stocks, **2** (25), 26
Subsidies, **4** (23), 104
Suicide, **11** (16), 433
Surveys, **2** (50), 40; **2** (53), 41; **3** (46), 79; **3** (64), 82; **4** (33), 110; **7** (7), 228; **7** (12), 234; **7** (13), 234; **7** (16), 234; **7** (22), 237; **7** (33), 247; **7** (47), 255; **8** (32), 279; **8** (34), 279; **8** (38), 280; **8** (44), 286; **8** (66), 294; **9** (51), 344; **10** (54), 404; **9** (66), 349; **10** (7), 368; **10** (9), 368; **10** (11), 368; **10** (56), 404; **10** (61), 405; **10** (64), 406; **12** (45), 510; **12** (46), 511

Television, **3** (57), 81; **9** (47), 325; **12** (38), 508
Transportation, **2** (17), 21; **2** (26), 27; **2** (28), 27; **2** (40), 38; **2** (50), 40; **3** (30), 69; **3** (32), 69; **3** (52), 80; **3** (54), 80; **3** (62), 82; **4** (1), 87; **4** (13), 92; **4** (14), 92; **4** (48), 112; **5** (20), 129; **5** (31), 143; **5** (58), 156; **6** (15), 175; **6** (21), 184; **6** (31), 190; **6** (42), 204; **6** (49), 217; **6** (55), 218; **6** (65), 221; **7** (16), 234; **8** (28), 276; **8** (57), 293; **8** (67), 294; **8** (68), 294; **9** (12), 301; **9** (24), 309; **9** (58), 345; **9** (59), 345; **10** (40), 393; **10** (56), 404; **10** (60), 405; **11** (33), 449; **11** (62), 472; **12** (18), 489; **12** (37), 508; **12** (47), 511; **12** (48), 511

Unemployment, **11** (16), 433; **11** (49), 470; **11** (54), 471
Utilities, **3** (42), 78; **11** (51), 470

Vision, **11** (31), 449
Voting, **3** (60), 81; **8** (10), 267; **8** (46), 290; **9** (38), 321

Weapons, **4** (38), 110; **4** (42), 111
Welfare, **4** (36), 110; **4** (41), 111; **5** (35), 144; **6** (60), 219; **8** (69), 294; **9** (29), 314; **9** (69), 349

X-rays, **2** (39), 38; **3** (61), 81; **5** (57), 156

APPLIED ELEMENTARY STATISTICS

RICHARD I. LEVIN

DAVID S. RUBIN

both of the University of North Carolina, Chapel Hill

PRENTICE-HALL, INC. *Englewood Cliffs, New Jersey 07632*

APPLIED ELEMENTARY STATISTICS

Library of Congress Cataloging in Publication Data

LEVIN, RICHARD I
    Applied elementary statistics.

    Bibliography: p.
    Includes index.
    1. Statistics. I. Rubin, David S., joint author.
    II. Title
QA276.12.L48 1980   519.5   79-20520
ISBN 0-13-040113-7

10 9 8 7 6 5 4 3 2 1

This is a Special Projects book, designed and edited by
Maurine Lewis, *director*
Ray Keating, *manufacturing buyer*
Page layouts: Judith A. Matz
Cover: Olympia Shahbaz

Prentice-Hall International, Inc., *London*
Prentice-Hall of Australia Pty. Limited, *Sydney*
Prentice-Hall of Canada, Ltd., *Toronto*
Prentice Hall of India Private Limited, *New Delhi*
Prentice-Hall of Japan, Inc., *Tokyo*
Prentice-Hall of Southeast Asia Pte. Ltd., *Singapore*
Whitehall Books Limited, *Wellington, New Zealand*

To the student

This is a short Preface, because most of what is usually said in the Preface is in the first chapter: the goals of the book, its methods, and a sneak preview of the chapters. We've both taught beginning courses in statistics for many years, and we've listened to complaints from students about "imponderable" textbooks. We've answered thousands of questions beginning, "What do the authors mean here?" and we've sympathized with those for whom most textbooks on this subject bring on an anxiety attack.

So we wrote our own statistics book. We started off with the idea that we would try to explain concepts well enough so that the student could lean on the book instead of the teacher. Many of our explanations may be long-winded, and sometimes we use three examples to make a point when perhaps one would do, but neither of us believes it possible to overexplain statistics. So if you'll bear with our surplus examples and copious explanations, we think we can learn statistics together and enjoy doing it!

Our approach uses intuitive explanations, extensions of something you already know, instead of complicated statistical proofs. When something cannot be explained in this commonsense manner, we leave it out.

You will find nearly a thousand text problems and exercises in our book. They cover the widest possible range of topic areas, all the way from abortion to X-rays—nearly one hundred different facets of our lives. If you want to get a quick idea of these different areas, turn to the front endpapers and look at the Index of Applications. Many entries, such as biology, economics, education, government, medicine, and psychology, are fields we encounter or read about every day. But some are newer concerns, such as communication, computers, consumer protection, ecology, energy, environment, housing, nuclear power, population, and solar heat, where a lot of the action is today.

Lots of people made inputs into our book, and a few deserve a special vote of thanks: Carolyn Ezzell, whose creativity concerning problems is everywhere; Harry Gaines, our editor, whose keen insights pick up what students and instructors really want in a book; Lisa Levin, whose indexing abilities are truly amazing, and Maurine Lewis, whose editorial and artistic ideas are always exciting.

We are grateful, too, to our reviewers, each of whom leaves a special imprint on the book: Professor George Casella, Rutgers—The State University; Professor Susan S. Lenker, Central Michigan University; Professor Albert Liberi, Westchester Community College; Professor Norman Locksley, Prince George's Community College; Professor Richard C. Orr, State University of New York at Oswego; Professor J. Roger Teller, Georgetown University, and Professor Stephen B. Vardeman, Purdue University. We are also grateful to the literary executor of the late Sir Ronald A. Fisher, F.R.S., to Dr. Frank Yates, F.R.S., and to Longman Group Ltd., London, for permission to reprint Tables III and IV from their book, *Statistical Tables for Biological, Agricultural, and Medical Research* (6th edition, 1974).

We hope you like what we've put together, and we hope it helps make your study of statistics interesting, nearly painless, and even a little fun!

Chapel Hill, N.C.   DICK LEVIN

                                                                 DAVE RUBIN

Contents

1 Introduction 1

1. Definitions • 2. History • 3. Subdivisions within statistics • 4. Chapter preview • 5. Strategy, assumptions, and approach

2 Arranging data to convey meaning: tables and graphs 8

1. How can we arrange data? • 2. Examples of raw data • 3. Arranging data using the data array and the frequency distribution • 4. Constructing a frequency distribution • 5. Graphing frequency distributions • 6. Terms introduced • 7. Equations introduced • 8. Chapter review exercises • 9. Chapter concepts test

3 Summary measures of frequency distributions 44

1. Beyond tables and graphs: descriptive measures of frequency distributions • 2. A measure of central tendency: the arithmetic mean • 3. A second measure of central tendency: the weighted mean • 4. A third measure of central tendency: the geometric mean • 5. A fourth measure of central tendency: the median • 6. A final measure of central tendency: the mode • 7. Comparing the mean, median, and mode • 8. Terms introduced • 9. Equations introduced • 10. Chapter review exercises • 11. Chapter concepts test

4 Measuring variability 84

1. Measures of dispersion • 2. Dispersion: distance measures • 3. Dispersion: average deviation measures • 4. Relative dispersion: the coefficient of variation • 5. Terms introduced • 6. Equations introduced • 7. Chapter review exercises • 8. Chapter concepts test

5 Probability I: introductory ideas 114

1. History and relevance of probability theory • 2. Some basic concepts in probability • 3. Three types of probability • 4. Probability rules • 5. Probabilities under conditions of statistical independence • 6. Probabilities under conditions of statistical dependence • 7. Revising prior estimates of probabilities: Bayes' theorem • 8. Terms introduced • 9. Equations introduced • 10. Chapter review exercises • 11. Chapter concepts test

6 Probability II: distributions 160

1. Introduction to probability distributions • 2. Random variables • 3. Use of expected value in decision making • 4. The binomial distribution • 5. The Poisson distribution • 6. The normal distribution: a distribution of a continuous random variable • 7. Expected value decision making with continuous distributions • 8. Choosing the correct probability distribution • 9. Terms introduced • 10. Equations introduced • 11. Chapter review exercises • 12. Chapter concepts test

7 Sampling and sampling distributions 224

1. Introduction to sampling • 2. Random sampling • 3. Introduction to sampling distributions • 4. Sampling distributions in more detail • 5. An operational consideration in sampling: the relationship between sample size and standard error • 6. Terms introduced • 7. Equations introduced • 8. Chapter review exercises • 9. Chapter concepts test

8 Estimation 258

1. Introduction • 2. Point estimates • 3. Interval estimates: basic concepts • 4. Interval estimates and confidence intervals • 5. Calculating interval estimates of the mean from large samples • 6. Calculating interval estimates of the proportion—large samples • 7. Interval estimates using the t distribution • 8. Determining the sample size in estimation • 9. Terms introduced • 10. Equations introduced • 11. Chapter review exercises • 12. Chapter concepts test

9 Testing hypotheses 296

1. Introduction. • 2. Concepts basic to the hypothesis testing procedure • 3. Testing hypotheses • 4. Hypothesis testing of means—samples with population standard deviations known • 5. Measuring the power of a hypothesis test • 6. Hypothesis testing of

proportions—large samples • 7. Hypothesis testing of means under different conditions • 8. Hypothesis testing for differences between means and proportions • 9. Terms introduced • 10. Equations introduced • 11. Chapter review exercises • 12. Chapter concepts test

10 Chi-square and analysis of variance 354

1. Introduction • 2. Chi-square as a test of independence • 3. Chi-square as a test of goodness of fit • 4. Analysis of variance • 5. Inferences about a population variance • 6. Inferences about two population variances • 7. Terms introduced • 8. Equations introduced • 9. Chapter review exercises • 10. Chapter concepts test

11 Regression and correlation analysis 408

1. Introduction • 2. Estimation using the regression line • 3. Correlation analysis • 4. Making inferences about population parameters • 5. Using regression and correlation analysis: limitations, errors, and caveats • 6. Multiple regression and correlation analysis • 7. The computer and multiple regression • 8. Terms introduced • 9. Equations introduced • 10. Chapter review exercises • 11. Chapter concepts test

12 Nonparametric methods 474

1. Introduction to nonparametric statistics • 2. The sign test for paired data • 3. A rank sum test: the Mann-Whitney U test • 4. One-sample runs tests • 5. Rank correlation • 6. Terms introduced • 7. Equations introduced • 8. Chapter review exercises • 9. Chapter concepts test

Afterword 513

1. Falling off the true path • 2. Design of experiments • 3. Time series • 4. Index numbers

Answers to selected even-numbered exercises 529

Answers to chapter concepts tests 547

Appendix tables 548

1. Areas under the standard normal probability distribution • 2. Areas in both tails combined for student's t distribution • 3. The cumulative binomial distribution • 4. Values of $e^{-\lambda}$ for computing Poisson probabilities • 5. Area in the right tail of a chi-square distribution • 6. Values of F for F distributions (.01 and .05) • 7. Values for Spearman's rank correlation • 8. Square roots

Bibliography 571

1. Introduction to statistics for the layman • 2. General statistics • 3. Probability • 4. Sampling theory and techniques • 5. Nonparametric statistics • 6. Statistical decision theory • 7. Special statistical topics • 8. Sources of statistical data • 9. Statistical tables • 10. Dictionaries and general reference works

Index 573

ABOUT THE AUTHORS

RICHARD I. LEVIN graduated in industrial enginering from North Carolina State University, then worked as an industrial engineer for the Glen Alden Corporation. He returned to school, however, for his M.S. in engineering and his Ph.D. in economics. His interests are in systems approaches to quantitative decision making and in turning statistics into a less fearsome course for students at Chapel Hill than it was for him. Family statistics: one married daughter, a son who is a reporter, and a daughter who is a student in special education. Dr. Levin loves to fish and to do carpentry work.

DAVID S. RUBIN did his undergraduate work in mathematics at Harvard, his graduate work in operations research at the University of Chicago. After a stint at Bell Laboratories in New Jersey, he joined the faculty of the University of North Carolina at Chapel Hill, where he has been since 1971. His research is in mathematical programming, with particular interests in discrete optimization problems. Dr. Rubin continues to teach elementary statistics and courses in operations research. His interests include jogging, gardening, and observing and (he hopes) helping his three children grow up.

Introduction

1. Definitions, 2

2. History, 2

3. Subdivisions within statistics, 3

4. Chapter preview, 4

5. Strategy, assumptions, and approach, 6

This book was written for students taking statistics for the first time. A glance at this chapter should convince any concerned citizen and future decision maker that a working knowledge of basic statistics will be quite useful in coping with the complex problems of our society. Your first look will also convince you that this book is dedicated to helping you acquire that knowledge with virtually no previous formal mathematical training and with no pain at all.

Different meanings of statistics depending on use

1 DEFINITIONS The word *statistics* means different things to different people. To a football fan, statistics are the information about rushing yardage, passing yardage, and first downs, given at halftime. To an administrator in the Environmental Protection Agency, statistics may be information about the quantity of pollutants being released into the atmosphere. To a school principal, statistics are information on absenteeism, test scores, and teacher salaries. To a medical researcher investigating the effects of a new drug, statistics are evidence of the success of his research efforts. And to a college student, statistics are the grades made on all the quizzes in a course this semester.

Each of these people is using the word *statistics* correctly, yet each uses it in a slightly different way and for a somewhat different purpose. *Statistics* is a word that can refer to quantitative data (such as wheat yield per acre) or to a field of study (you can, for example, major in statistics).

Today, statistics and statistical analysis are used in nearly every profession. For decision makers, in particular, statistics have become a most valuable tool.

Origin of the word

2 HISTORY The word *statistik* comes from the Italian word *statista* (meaning "statesman"). It was first used by Gottfried Achenwall (1719–1772), a professor at Marlborough and Göttingen. Dr. E. A. W. Zimmerman introduced the word *statistics* into England. Its use was popularized by Sir John Sinclair in his work, *Statistical Account of Scotland 1791–1799.* Long before the eighteenth century, however, people had been recording and using data.

Early government records

Official government statistics are as old as recorded history. The Old Testament contains several accounts of census taking. The governments of ancient Babylonia, Egypt, and Rome gathered detailed records of populations and resources. In the Middle Ages, governments began to register the ownership of land. In A.D. 762, Charlemagne asked for detailed descriptions of church-owned properties. Early in the ninth century, he completed a statistical enumeration of the serfs attached to the land. About 1086, William the Conqueror ordered the writing of the *Domesday Book,* a record of the ownership, extent, and value of the lands of England. This work was England's first statistical abstract.

Because of Henry VII's fear of the plague, England began to register its dead in 1532. About this same time, French law required the clergy to register baptisms, deaths, and marriages. During an outbreak of the plague in the late 1500s, the English government started publishing weekly death statistics. This practice continued, and by 1632 these *Bills of Mortality* listed births and deaths by sex. In 1662, Captain John Graunt used thirty years of these Bills to make predictions about the number of persons who would die from various diseases and the proportion of male and female births that could be expected. Summarized in his work, *Natural and Political Observations . . . Made Upon the Bills of Mortality,* Graunt's study was a pioneer effort in statistical analysis. For his achievement using past records to predict future events, Graunt was made a member of the original Royal Society.

An early prediction from statistics

The history of the development of statistical theory and practice is a lengthy one. We have only begun to list the people who have made significant contributions to this field. Later we will encounter others whose names are now attached to specific laws and methods. Many people have brought to the study of statistics refinements or innovations that, taken together, form the theoretical basis of what we will study in this book.

3 SUBDIVISIONS WITHIN STATISTICS   Decision makers apply some statistical technique to virtually every branch of public and private enterprise. These techniques are so diverse that statisticians commonly separate them into two broad categories: *descriptive statistics* and *inferential statistics.* Let's use examples to understand the difference between the two.

Suppose a professor computes an average grade for one history class. Since statistics describe the performance of that one class but do not make a generalization about several classes, we can say that the professor is using *descriptive* statistics. Graphs, tables, and charts that display data so that they are easier to understand are all examples of descriptive statistics.

Descriptive statistics

Suppose now that the history professor decides to use the average grade achieved by one history class to estimate the average grade achieved in all ten sections of the same history course. The process of estimating this average grade would be a problem in *inferential* statistics. Statisticians also refer to this category as *statistical inference.* Obviously, any conclusion the professor makes about the ten sections of the course will be based on a generalization that goes far beyond the data for the original history class; and the generalization may not be completely valid, so the professor must state how likely it is to be true. Similarly, statistical inference involves generalizations and statements about the *probability* of their validity.

Inferential statistics

The methods and techniques of statistical inference can also be used in a branch of statistics called *decision theory.* Knowledge of decision theory is very helpful for decision makers because it is used to make decisions

Decision theory

under conditions of uncertainty—when, for example, a state government social service agency cannot specify precisely the demand for its services or when the chairman of the English department at your school must schedule faculty teaching assignments without knowing precisely the student enrollment for next fall.

Chapter 2

4 CHAPTER PREVIEW Chapters 2, 3, and 4 will introduce the concepts and techniques of descriptive statistics. Chapter 2 examines two methods for describing a collection of items: tables and graphs. If you have ever heard someone give a long-winded economic report (such as how many dues are owed by all eighty club members) and wished for a quick graphic display to ease the pain, you already have an appreciation of what's to come in Chapter 2.

Chapter 3

Chapter 3 focuses on special ways to describe a collection of items, particularly the way observations tend to cluster or bunch up. Here, we shall encounter some familiar terms, such as the concept of an average. If the basketball coach at your university says the average height of the members of his team is 6'11", he is really saying that there is a tendency for the heights of the players to bunch up around 6'11". For a basketball team with this much height, you intuitively know that the chances of a winning season are quite good—even before you formally study statistics. In Chapter 3, we'll also study the mean, the median, and the mode—all ways of measuring and locating data.

Chapter 4

Chapter 4 finishes our study of descriptive statistics by looking at methods that enable us to measure the tendency for a group of data to spread out, or disperse. Suppose an airline requires that its pilots average 6' in height, and you recruit a 4' person and an 8' person to apply for jobs. You would not get much praise for your efforts even though these two unusual persons do average 6'. Instead, the airline is likely to reject both of your candidates because their heights are too far from the desired average. In this situation the 6' average is an inadequate summary description of your two candidates. Chapter 4 provides better descriptions of variability.

Chapter 5

Chapter 5 introduces the basic concepts of *probability* (or chance) and, with Chapter 6, gives us a foundation for our study of statistical inference, which will follow in later chapters. Here, we shall examine methods of calculating and using probabilities under various conditions. If you are one of 200 students in a class and the professor seems to call on you each time the class meets, you might accuse that professor of not calling on students at random. If, on the other hand, you are one student in a class of 8 and you never prepare for class, assuming that the professor will not get around to you, then you may be the one who needs to examine probability ideas a bit more.

In Chapter 6, we are concerned with probability *distributions,* that is, Chapter 6 with the various ways in which data array themselves when we graph them. Here again we are laying the foundation for later work in statistical inference. You may already have a notion about probability distributions if you have dealt with the concept of the bell-shaped curve in psychology or mathematics courses. If this term means nothing to you, but you are a male who wears a size 16 EE shoe or a female who wears a size 3 AAAA shoe, you may have an intuitive idea about probability distributions, too. Each time you try to get fitted and can't, you probably wish the shoe store manager would order a larger distribution of shoes. If the manager thinks in terms of correct probability distributions, however, he will not order such unusual sizes and won't be able to accommodate people with very large or very small feet.

Most consequential decisions in the public and private sectors are made under conditions of uncertainty because decision makers seldom have complete information about what the future will bring. Also introduced in Chapter 6 is *statistical decision theory,* those methods that are useful when we must decide among alternatives despite uncertain conditions.

Statistical sampling, the subject of Chapter 7, is a systematic approach Chapter 7 to selecting a few elements (a *sample*) from an entire collection of data (a *population*) in order to make some inferences about the total collection. Here we shall learn methods that help to ensure that the samples we collect are actually representative of the entire collection. If you have ever examined a peach on the top of a basket, bought the whole basket on the basis of that peach's condition, and then found the bottom of the basket filled with overripe fruit, you already have a good (if somewhat expensive) understanding of statistical sampling and the need for better sampling methods.

Chapters 8 and 9 deal with statistical inference. In Chapter 8, we shall Chapter 8 learn to *estimate* the characteristics of a population by observing the characteristics of a sample. Two characteristics of special interest will be how a population tends to "bunch up" and how it spreads out.

The subject of Chapter 9 is *hypothesis testing.* Here we are trying to Chapter 9 determine when it is reasonable to conclude, from analysis of a sample, that the entire population possesses a certain property and when it is not reasonable to reach such a conclusion. Suppose a student purchases a $500 second-hand car from a dealer who advertises "our cars are the finest, most dependable automobiles in town." If the car's repair bills during the first month are $600, that one-car sample may cause the student to conclude that the dealer's population of used cars is probably not as advertised. Chapter 9 will allow us to test and evaluate larger samples than those available to the buyer of the used car.

Chapter 10 discusses two statistical techniques: chi-square tests and Chapter 10 analysis of variance. *Chi-square tests* are useful in analyzing more than two populations. They can be helpful in political situations to determine whether, in fact, there is a difference between the proportions of the population supporting two opposing candidates in data taken from different states or geographic regions. Chi-square tests also enable us to determine whether a group of data that we think could be described by the bell-shaped curve actually does conform to that graphic pattern. *Analysis of variance,* the second subject of Chapter 10, is used to test the difference between several sample means. It is a method the Environmental Protection Agency might use when evaluating five series of tests on the same model car. This method can help the agency answer the question, "Are the miles per gallon results really the same, or do they only appear to be?"

Chapter 11

If your university used your high school grade point average and your college board scores to predict your college grade point average, it may have used the technique of *regression analysis,* one of the subjects of Chapter 11. And if you have heard the statement that there is a high correlation between smoking and lung cancer, then the word *correlation* (another topic in Chapter 11) is no stranger to you. *Correlation analysis* is used to measure the degree of association between two variables.

Chapter 12

In Chapters 7 to 11, then, we shall learn how statisticians take samples from populations and attempt to reach conclusions from those samples. But how can we handle cases in which we do not know what kind of population we are sampling (that is, when we do not know the shape of the population distribution)? In these cases, we can often apply the techniques of *nonparametric statistics* presented in Chapter 12.

For students, not statisticians

5 STRATEGY, ASSUMPTIONS, AND APPROACH   This book is designed to help you get the feel of statistics—what it is, how and when to apply statistical techniques to decision making situations, and how to interpret the results you get. Since we are not writing for professional statisticians, this book is tailored to the backgrounds and needs of college students who, as future citizens, probably accept the fact that statistics can be of considerable help to them in their future occupations but are, very likely, apprehensive about studying the subject.

Symbols are simple and explained

In this book, we discard mathematical proofs in favor of intuitive ones. You will be guided through the learning process by reminders of what you already know, by examples with which you can identify, and by a step-by-step process instead of statements like "it can be shown" or "it therefore follows."

As you thumb through this book and compare it with other basic statistics textbooks, you will notice a minimum of mathematical notation. In the past, the complexity of the notation has intimidated many students who got lost in the symbols, even though they were motivated and intellectually capable of understanding the ideas. Each symbol and formula that is used is explained in detail, not only at the point at which it is introduced but also in a section at the end of the chapter.

If you felt reasonably comfortable when you finished your college algebra course, you have enough background to understand *everything* in this book. Nothing beyond basic algebra is either assumed or used. This book's goals are for you to be comfortable as you learn and for you to get a good intuitive grasp of statistical concepts and techniques. As a member of either a private or a public organization, you will need to know when statistics can help your decision process and which tools to use. If you do need statistical help, you can find a statistical expert to handle the details.

No math beyond simple algebra required

The problems used to introduce material in the chapters and the exercises at the end of each section within the chapter are drawn from a wide variety of situations you are already familiar with or are likely to confront quite soon. You will see problems involving public education, social services, health systems, environmental protection, consumer advocacy, history, psychology, biology, anthropology, sociology, political science, economics, and the private sector.

Text problems cover a wide variety of situations

In each problem situation, a decision maker is attempting to use statistics productively. Helping you become comfortable doing exactly that is the goal of this book.

Arranging data to convey meaning: tables and graphs

1. How can we arrange data? 10

2. Examples of raw data, 13

3. Arranging data using the data array and the frequency distribution, 14

4. Constructing a frequency distribution, 21

5. Graphing frequency distributions, 28

6. Terms introduced, 36

7. Equations introduced, 37

8. Chapter review exercises, 37

9. Chapter concepts test, 43

The water quality control engineer of the Orange Water and Sewer Authority is responsible for the chlorination level of the water. It must be close to the level required by the department of health. To keep watch on the chlorine, without checking the content of every gallon leaving the water filtration and treatment plant, the engineer samples several gallons every day, measures their chlorine content, and draws a conclusion about the average chlorination level of water treated that day. The table below shows the chlorine levels of the 30 gallons selected as one day's sample. These levels are the raw data from which the engineer can draw conclusions about the entire population of that day's treatment.

Chlorine levels in parts per million (ppm) in 30 gallons of treated water

16.2	15.4	16.0	16.6	15.9	15.8	16.0	16.8	16.9	16.8
15.7	16.4	15.2	15.8	15.9	16.1	15.6	15.9	15.6	16.0
16.4	15.8	15.7	16.2	15.6	15.9	16.3	16.3	16.0	16.3

Using the methods introduced in this chapter, we can help the water quality control engineer draw the proper conclusions.

Data are collections of any number of related observations. We can collect the number of telephones installed on a given day by several workers or the number of telephones installed per day over a period of several days by one worker and call the results our data. A collection of data is called a *data set,* and a single observation a *data point.*

1 HOW CAN WE ARRANGE DATA?

For data to be useful, our observations need to be organized so that we can pick out trends and come to logical conclusions. This chapter introduces the techniques of arranging data in tabular and graphical forms. Chapter 3 will show how to use numbers to describe data.

Collecting data

Statisticians select their observations so that all relevant groups are represented in the data. To determine the potential market for a new product, for example, analysts might study 100 consumers in a certain geographical area. The analysts must be certain that this group contains a variety of people representing variables such as income level, race, education, and neighborhood.

Data can come from actual observations or from records that are kept for ·normal purposes. A hospital, for example, will record the number of patients using the X-ray facilities both for billing purposes and for doctors' reports. But this information can also be organized to produce data that statisticians can describe and interpret.

Use data
about the past
to make decisions
about the future

Data can assist decision makers in making educated guesses about the *causes* and therefore the probable *effects* of certain characteristics in given situations. Also, knowledge of trends from past experience can enable concerned citizens to be aware of potential outcomes and plan in advance. Our marketing survey may reveal that the product is preferred by black housewives of suburban communities, average incomes, and average educations. The product's advertising copy should address this target audience. And if hospital records show that more patients used the X-ray facilities in June than in January, the hospital Personnel Division should determine if this was accidental to this year or an indication of a trend, and perhaps it should adjust its hiring and vacation practices accordingly.

When data are arranged in compact, useable form, decision makers can take reliable information from the environment and use it to make intelligent decisions. Today, computers allow statisticians to collect enormous amounts of observations and compress them instantly into tables, graphs, and numbers. These are all compact, useable forms—but are they reliable? Remember that the data that come out of a computer are only as accurate as the data that go in. As computer programmers say, "GIGO!" or "Garbage In, Garbage Out!" Managers must be very careful to be sure that the data they are using are based on correct assumptions and interpretations. Before

relying on any interpreted data, from a computer or not, test the data by
asking these questions:

1. Where did the data come from? Is the source biased; that is, is it likely to
have an interest in supplying data points that will lead to one conclusion
rather than another?
2. Do the data support or contradict other evidence we have?
3. Is evidence missing that might cause us to come to a different conclusion?
4. How many observations do we have? Do they represent all the groups we
wish to study?
5. Is the conclusion logical? Have we made conclusions that the data do not
support?

Study your answers to these questions. Are the data worth using? Or should
we wait and collect more information before acting? If the hospital was
caught short-handed because it hired too few nurses to staff the X-ray room,
its administration relied on insufficient data. If the advertising agency tar-
geted its copy only toward black suburban housewives when it could have
tripled its sales by appealing to white suburban housewives too, it relied on
insufficient data also. In both cases, testing available data would have
helped managers make better decisions.

Difference between samples and populations

Statisticians gather data from a sample. They use this information to
make inferences about the population that sample represents. Thus, *sample*
and *population* are relative terms. A population is a whole, and a sample is
a fraction or segment of that whole. A cherry pie, for example, is a population consisting of cherry filling and pie crust. A wedge cut from that pie is a
fraction, or one sample of that pie. The sample contains both filling and pie
crust, just as the population does.

Sample
and population
defined

We will study samples in order to be able to describe populations. Our
hospital may study a small, not unrepresentative group of X-ray records
rather than examine each record for the last fifty years. The Gallup Poll may
interview a sample of only 2,500 adult Americans in order to predict the
opinion of all adults living in the United States. Studying samples is obviously easier than studying whole populations, and it is reliable if carefully
and properly done.

Function of samples

A *population* is a collection of all the elements we are studying and
about which we are trying to draw conclusions. We must define this population so that it is clear whether or not an element is a member of the population. The population for our marketing study may be all women within a
15-mile radius of center-city Cincinnati whose annual family incomes lie
between $10,000 and $25,000 and who have completed at least 11 years of
school. A woman living in downtown Cincinnati with a family income of
$15,000 and a college degree would be a part of this population. A woman

Function
of populations

living in San Francisco or with a family income of $7,000 or with 5 years of schooling would not qualify as a member of this population.

A *sample* is a collection of some, but not all, of the elements of the population. The population of our marketing survey is *all* women who meet the qualifications listed above. Any group of women who meet these qualifications can be a sample, as long as the group is only a fraction of the whole population. A large helping of cherry filling with only a few crumbs of crust is a sample of a pie, but it is not a representative sample because the proportions of the ingredients are not the same in the sample as they are in the whole.

Need for a representative sample

A *representative sample* contains the relevant characteristics of the population *in the same proportion* as they are included in that population. If our population of women is $1/3$ black, then a sample of the population that is representative in terms of race will also be $1/3$ black. Specific methods for sampling will be covered in detail in Chapter 7.

Finding a meaningful pattern in the data

Data come in a variety of forms

There are many ways to sort data. We can simply collect it and keep it in order. Or if the observations are measured in numbers, we can list the data points from the lowest to the highest in numerical value. But if the data are skilled workers (such as carpenters, masons, and iron workers) required at construction sites, or the different types of automobiles manufactured by all automakers, or the various colors of sweaters manufactured by a given firm, we will need to organize them differently. We will need to present the data points in alphabetical order or by some other organizing principle. One useful way to organize data is to divide them into similar categories or classes and then count the number of observations that fall into each category. This method produces a *frequency distribution* and is discussed later in this chapter.

Why should we arrange data?

The purpose of organizing data is to enable us to see quickly all the possible characteristics in the data we have collected. We look for things such as the range (the largest and smallest values), apparent trends, what values the data may tend to group around, what values appear most often, and so on. The more information of this kind that we can learn from our sample, the better we can understand the population from which it came and the better we can make decisions.

EXERCISES

1. "Three out of four doctors recommend aspirin." Is this conclusion drawn from a sample or a population? Explain.

2. "Third graders' IQ's have declined in the past 5 years. A survey of 30 third grade students from 5 states showed a mean decrease of 4.5 points on a popular IQ form." Comment on this statement from the point of view of populations and samples.

3. An electronics firm recently introduced a new amplifier, and warranty cards indicate that 10,000 of these have been sold so far. The president of the firm, very upset after reading three letters of complaint about the new amplifiers, informed the production manager that costly control measures would be implemented immediately to ensure that the defects would not appear again. Comment on the president's reaction from the standpoint of the five tests for data given on page 11.

4. "Dewey Beats Truman," announced the newspaper headlines the morning following the 1948 election. For weeks, the pollsters had been predicting a Dewey landslide. Everyone was so confident Dewey would win that some newspapers had preset the headline type and printed the morning papers without waiting for full returns. Truman, however, was elected. Give some possible reasons for the pollsters' incorrect predictions.

2 EXAMPLES OF RAW DATA

Information before it is arranged and analyzed is called *raw data*. It is "raw" because it is unprocessed by statistical methods.

The chlorine data in the chapter opening problem was one example of raw data. Let's consider a second example of raw data. Suppose that the admissions staff of a university, concerned with the success of the students it selects for admission, wishes to compare the students' college performances with other achievements, such as high school grades, test scores, and extracurricular activities. Rather than study every student from every year, the staff can draw a sample of the population of all the students in a given time period and study only that group to conclude what characteristics appear to predict success. The staff can, for example, compare high school grades with college grade point average (GPA) for students in the sample. The staff can assign each grade a numerical value. Then it can add the grades and divide by the total number of grades to get an average for each student. Table 2·1 shows a sample of this raw data in tabular form: 20 pairs of average grades in high school and college.

Problem facing admissions staff

H.S.	College	H.S.	College	H.S.	College	H.S.	College
3.6	2.5	3.5	3.6	3.4	3.6	2.2	2.8
2.6	2.7	3.5	3.8	2.9	3.0	3.4	3.4
2.7	2.2	2.2	3.5	3.9	4.0	3.6	3.0
3.7	3.2	3.9	3.7	3.2	3.5	2.6	1.9
4.0	3.8	4.0	3.9	2.1	2.5	2.4	3.2

TABLE 2·1
High school and college grade point averages of 20 college seniors

When designing a bridge, engineers are concerned with the stress that a given material, such as concrete, will withstand. Rather than test every cubic inch of the concrete to determine its stress capacity, the engineers can take a sample of the concrete, test it, and conclude how much stress,

Bridge-building problem

on the average, that kind of concrete can withstand. Table 2·2 summarizes the raw data gathered from a sample of 40 batches of concrete that will be used in constructing a bridge.

TABLE 2·2
Pounds of pressure per square inch that concrete can withstand

2500.2	2497.8	2496.9	2500.8	2491.6	2503.7	2501.3	2500.0
2500.8	2502.5	2503.2	2496.9	2495.3	2497.1	2499.7	2505.0
2490.5	2504.1	2508.2	2500.8	2502.2	2508.1	2493.8	2497.8
2499.2	2498.3	2496.7	2490.4	2493.4	2500.7	2502.0	2502.5
2506.4	2499.9	2508.4	2502.3	2491.3	2509.5	2498.4	2498.1

EXERCISES

5. Discuss the data given in the chapter opening problem in terms of the 5 tests for data.

6. Look at the data in Table 2·1. Why do these data need further arranging? Can you form any conclusions from the data as they exist now?

7. The marketing manager of a large company receives a report each month on the sales activity of one of the company's products. The report is a listing of the sales of the product by state during the previous month. Is this an example of raw data?

8. The production manager in a large company receives a report each month from the quality control section. The report gives the reject rate for the production line (the number of rejects per hundred units produced), the machine causing the greatest number of rejects, and the average cost of repairing the rejected units. Is this an example of raw data?

3 ARRANGING DATA USING THE DATA ARRAY AND THE FREQUENCY DISTRIBUTION

Data array defined

The data array is one of the simplest ways to present data. It arranges values in ascending or descending order. Table 2·3 repeats the chlorine data from our chapter opening problem, and Table 2·4 rearranges these numbers in a data array in ascending order.

TABLE 2·3
Chlorine levels in ppm of 30 gallons of treated water

16.2	15.8	15.8	15.8	16.3	15.6
15.7	16.0	16.2	16.1	16.8	16.0
16.4	15.2	15.9	15.9	15.9	16.8
15.4	15.7	15.9	16.0	16.3	16.0
16.4	16.6	15.6	15.6	16.9	16.3

TABLE 2·4
Data array of chlorine levels in ppm of 30 gallons of treated water

15.2	15.7	15.9	16.0	16.2	16.4
15.4	15.7	15.9	16.0	16.3	16.6
15.6	15.8	15.9	16.0	16.3	16.8
15.6	15.8	15.9	16.1	16.3	16.8
15.6	15.8	16.0	16.2	16.4	16.9

14

Data arrays offer several advantages over raw data:

1. We can quickly notice the lowest and highest values in the data. In our chlorination example, the range is from 15.2 ppm to 16.9 ppm.
2. We can easily divide the data into sections. In Table 2·4, the first 15 values (the lower half of the data) are between 15.2 and 16.0 ppm, and the last 15 values (the upper half) are between 16.0 and 16.9 ppm. Similarly, the lowest third of the values range from 15.2 to 15.8 ppm, the middle third from 15.9 to 16.2 ppm, and the upper third from 16.2 to 16.9 ppm.
3. We can see whether any values appear more than once in the array. Equal values appear together. Table 2·4 shows that nine levels occurred more than once when the sample of 30 gallons of water was tested.
4. We can observe the distance between succeeding values in the data. In Table 2·4, 16.6 and 16.8 are succeeding values. The distance between them is .2 ppm, (16.8 − 16.6).

In spite of these advantages, sometimes a data array isn't helpful. Since it lists every observation, it is a cumbersome form for displaying large quantities of data. We need to compress the information and still be able to use it for interpretation and decision making. How can we do this?

A better way to arrange data: the frequency distribution

One way we can compress data is to use a *frequency table* or a *frequency distribution.* To understand the difference between this and an array, take as an example the average staff sizes for members of 20 state legislatures.

In Tables 2·5 and 2·6, we have taken identical data concerning the average staff sizes in these legislatures and displayed them first as an array in ascending order and then as a frequency distribution. To obtain Table

Advantages of
data arrays

Disadvantages
of data arrays

Frequency
distributions
handle more data

They lose
some information

TABLE 2·5

**Data array
of average staff
sizes for members
of 20 state
legislatures**

TABLE 2·6

**Frequency
distribution
of average staff
sizes for members
of 20 state
legislatures**

| 2.0 | 3.4 | 3.8 | 4.1 | 4.1 | 4.3 | 4.7 | 4.9 | 5.5 | 5.5 |
| 3.4 | 3.8 | 4.0 | 4.1 | 4.2 | 4.7 | 4.8 | 4.9 | 5.5 | 5.5 |

Class (group of similar values of data points)	Frequency (number of observations in each class)
2.0 to 2.5	1
2.6 to 3.1	0
3.2 to 3.7	2
3.8 to 4.3	8
4.4 to 4.9	5
5.0 to 5.5	4

2·6, we had to divide the data into groups of similar values. Then we recorded the number of data points that fell into each group. Notice that we lose some information in constructing the frequency distribution. We no longer know, for example, that the value 5.5 appears three times or that the value 5.1 does not appear at all. Yet we gain information concerning the *pattern* of staff sizes. We can see from Table 2·6 that average staff size falls most often in the range from 3.8 to 4.3 people. It is unusual to find an average staff size in the range from 2.0 to 2.5 people or from 2.6 to 3.1 people. Sizes in the ranges of 4.4 to 4.9 people and 5.0 to 5.5 people are not prevalent but occur more frequently than some others. Thus, frequency distributions sacrifice some detail but offer us new insights into patterns of data.

But they gain other information

A frequency distribution is a table that organizes data into *classes,* that is, into groups of values describing one characteristic of the data. "The average staff size for members" is one characteristic of the 20 state legislatures. In Table 2·5, this characteristic has 11 different values. But this same data could be divided into any number of classes. Table 2·6, for example, uses 6. We could compress the data even further and use only the 2 classes "less than 3.8" and "greater than or equal to 3.8." Or we could increase the number of classes by using smaller intervals, such as we have done in Table 2·7.

Function of classes in a frequency distribution

**TABLE 2·7
Frequency distribution of legislative staff sizes**

Classes	Frequencies	Classes	Frequencies
2.0 to 2.2	1	3.8 to 4.0	3
2.3 to 2.5	0	4.1 to 4.3	5
2.6 to 2.8	0	4.4 to 4.6	0
2.9 to 3.1	0	4.7 to 4.9	5
3.2 to 3.4	2	5.0 to 5.2	0
3.5 to 3.7	0	5.3 to 5.5	4

Why it is called a "frequency" distribution

A frequency distribution shows **the number of observations from the data set that fall into each of the classes.** If you can determine the frequency with which values occur in each class of a data set, you can construct a frequency distribution.

Characteristics of relative frequency distributions

Relative frequency distribution defined

So far, we have expressed the frequency with which values occur in each class as the total number of data points that fall within that class. We can also express the frequency of each value as a *fraction* or a *percentage* of the total number of observations. The frequency of an average staff size of 4.4 to 4.9 people for example, is "5" in Table 2·6 but ".25" in Table 2·8. To get this value of .25, we divided the frequency for that class (5) by the total number of observations in the data set (20). The answer can be expressed as a fraction ($\frac{5}{20}$), a decimal (.25), or a percentage (25%). A *relative frequency distribution* presents frequencies in terms of fractions or percentages.

Classes	Frequencies	Relative frequencies: fraction of observations in each class	
2.0 to 2.5	1	.05	
2.6 to 3.1	0	.00	
3.2 to 3.7	2	.10	
3.8 to 4.3	8	.40	
4.4 to 4.9	5	.25	
5.0 to 5.5	4	.20	
	20	**1.00**	sum of the relative frequencies of all classes

Notice in Table 2·8 that the sum of all the relative frequencies equals 1.00, or 100 percent. This is true because a relative frequency distribution pairs each class with its appropriate fraction or percentage of the total data. Therefore, the classes in any relative or simple frequency distribution are *all-inclusive.* All the data fit into one category or another. Also notice that the classes in Table 2·8 are *mutually exclusive;* that is, no data point falls into more than one category. Table 2·9 illustrates this concept by comparing mutually exclusive classes with ones that overlap. In frequency distributions, there are no overlapping classes.

Classes are all-inclusive

They are mutually exclusive

TABLE 2·9

Mutually exclusive and overlapping classes

Mutually exclusive	1 to 4	5 to 8	10 to 13	14 to 17
Not mutually exclusive	1 to 4	3 to 6	5 to 8	7 to 10

To this point, our classes have consisted of numbers and have described some quantitative attribute of the items sampled. We can also classify information according to qualitative characteristics, such as race, religion, and sex, which do not fall naturally into numerical categories. Like classes of quantitative attributes, these classes must be all-inclusive and mutually exclusive. Table 2·10 shows how to construct both simple and

Classes of qualitative data

TABLE 2·10

Occupations of sample of 100 graduates of Central College

Occupational class	Frequency distribution (1)	Relative frequency distribution (1) ÷ 100
Actor	5	.05
Banker	8	.08
Businessman	22	.22
Chemist	7	.07
Doctor	10	.10
Insurance representative	6	.06
Journalist	2	.02
Lawyer	14	.14
Teacher	9	.09
Other	17	.17
	100	**1.00**

relative frequency distributions using the qualitative attribute of occupations.

Although Table 2·10 does not list every occupation held by the graduates of Central College, it is still all-inclusive. Why? The class "other" covers all the observations that fail to fit one of the enumerated categories. We will use a word like this whenever our list does not specifically list all the possibilities. If, for example, our characteristic can occur in any month of the year, a complete list would include 12 categories. But if we wish to list only the 8 months from January to August, we can use the term "other" to account for our observations during the four months of September, October, November, and December. Although our list does not specifically list all the possibilities, it is all-inclusive. This "other" is called an *open-ended class* when it allows either the upper or the lower end of a quantitative classification scheme to be limitless. The last class in Table 2·11 ("72 and older") is open-ended.

Open-ended classes
for lists that are
not exhaustive

TABLE 2·11
Ages of Bunder
County residents

Class: age (1)	Frequency (2)	Relative frequency (2) ÷ 89,592
Birth to 7	8,873	.0990
8 to 15	9,246	.1032
16 to 23	12,060	.1346
24 to 31	11,949	.1334
32 to 39	9,853	.1100
40 to 47	8,439	.0942
48 to 55	8,267	.0923
56 to 63	7,430	.0829
64 to 71	7,283	.0813
72 and older	6,192	.0691
	89,592	**1.0000**

Discrete classes

Classification schemes can be either quantitative or qualitative *and* either discrete or continuous. *Discrete* classes are separate entities that do not progress from one class to the next without a break. Such classes as the number of children in each family, the number of trucks owned by moving companies, or the occupations of Central College graduates are discrete. Discrete data are data that can take on only a limited number of values. Central College graduates can be classified as either doctors or chemists but not something in between. The closing price of AT&T stock can be 56¾, or your basketball team can have a center who is 7 feet 1½ inches tall.

Continuous classes

Continuous data do progress from one class to the next without a break. They involve numerical measurement such as the weights of cans of tomatoes, the pounds of pressure on concrete, or the high school GPA's of college seniors. Continuous data can be expressed in either fractions or whole numbers.

9. Arrange the data below in a data array from lowest to highest.

```
708  541  528  546  631  541  622  592  534  663
546  641  603  650  502  592  618  631  599  637
578  483  578  619  586  567  644  641  622  547
644  689  557  612  644  531  536  695  645  578
```

a) What are the highest and lowest data values?

b) Between what values do the lowest ¼ of the data fall? The highest ¼ of the data?

c) How many values appear more than once in the data set, and what are they?

10. For the data set in 9, above, determine how many observations fall between 450.0 and 499.9, between 500.0 and 549.9, between 550.0 and 599.9, between 600.0 and 649.9, between 650.0 and 699.9, and between 700.0 and 749.9.

11. Construct a frequency distribution with intervals of .5 from the following set of measurements.

```
3.9  4.9  5.9  3.7  6.9  4.5  3.6  3.9  3.9
4.0  5.2  4.9  4.0  5.4  3.7  6.1  4.0  4.4
5.6  4.8  5.4  4.0  4.1  3.9  4.8  3.5  4.7
5.1  3.9  5.0  3.9  3.7  3.8  5.2  5.0  4.5
4.2  5.4  3.7  5.5  3.8  6.2  3.2  5.4  4.2
```

12. Given the following data set, construct a relative frequency distribution using (a) 7 equal intervals and (b) 13 equal intervals.

```
80  52  07  59  60  79  62  55  52  90
04  87  65  64  50  71  72  64  71  67
40  56  74  69  97  67  81  77  77  57
35  86  71  99  88  43  54  48  68  77
99  70  84  78  68  60  47  56  60  57
```

13. Arrange the data in Table 2·2 on page 14 in an array from highest to lowest.

a) Suppose that state law requires bridge concrete to withstand at least 2,500 lbs./sq. in. How many samples would fail this test?

b) How many samples could withstand a pressure of at least 2,497 lbs./sq. in. but could not withstand a pressure greater than 2,504 lbs./sq. in.?

c) As you examine the array, you should notice that some samples can withstand identical amounts of pressure. List these pressures and the number of samples that can withstand each amount.

14. Using Table 2·1 on page 13, arrange the data in an array from highest to lowest high school GPA. Now, arrange the data into an array from highest to lowest college GPA.

15. The Environmental Protection Agency took water samples from 10 different rivers and streams that feed into Lake Erie. These samples were tested in the EPA laboratory and rated as to the amount of solid pollution suspended in each sample. The results of the testing are given in the following table:

Sample	1	2	3	4	5	6	7	8	9	10
Pollution rating (ppm)	27.2	38.7	64.3	52.8	47.6	23.4	33.9	45.0	56.7	41.1

19

a) Arrange the data into an array from highest to lowest.

b) Determine the number of samples having a pollution content between 20.0 and 29.9, 30.0 and 39.9, 40.0 and 49.9, 50.0 and 59.9, 60.0 and 69.9.

c) If 40.0 is the number used by the EPA to indicate excessive pollution, how many samples would be rated as having excessive pollution?

d) What is the largest distance between any two consecutive samples?

16. Suppose that the admissions staff mentioned in the discussion of Table 2·1 on page 13 wishes to examine the relationship between a student's differential on the college SAT examination (the difference between actual and expected score based on the student's high school GPA) and the spread between the student's high school and college GPA (the difference between the high school and college GPA). The admissions staff will use the following data:

H.S. GPA	College GPA	SAT score	H.S. GPA	College GPA	SAT score
3.6	2.5	1090	3.4	3.6	1170
2.6	2.7	955	2.9	3.0	1025
2.7	2.2	940	3.9	4.0	1315
3.7	3.2	1170	3.2	3.5	1160
4.0	3.8	1330	2.1	2.5	925
3.5	3.6	1190	2.2	2.8	975
3.5	3.8	1240	3.4	3.4	1160
2.2	3.5	1050	3.6	3.0	1110
3.9	3.7	1300	2.6	1.9	850
4.0	3.9	1345	2.4	3.2	1080

In addition, the admissions staff has received the following information from the Educational Testing Service.

H.S. GPA	Avg. SAT score	H.S. GPA	Avg. SAT score
4.0	1340	2.9	1020
3.9	1310	2.8	1000
3.8	1280	2.7	980
3.7	1250	2.6	960
3.6	1220	2.5	940
3.5	1190	2.4	920
3.4	1160	2.3	910
3.3	1130	2.2	900
3.2	1100	2.1	880
3.1	1070	2.0	860
3.0	1040		

a) Arrange these data into an array of spreads from highest to lowest. (Consider an increase in college GPA over high school GPA as positive and a decrease in college GPA below high school GPA as negative.) Include with each spread the appropriate SAT differential. (Consider an SAT score below expected as negative and above expected as positive.)

b) What is the most common spread?

c) For this spread in part b, what is the most common SAT differential?

17. Construct a frequency distribution with intervals of 7 days from the following data obtained from shipping records of a mail order firm.

Time from receipt of order to delivery (in days)

3	11	7	13	10	5	5	12	14	10
12	22	6	23	9	14	22	8	25	5

18. Refer to Table 2·2 on page 14 and construct a relative frequency distribution using intervals of 4.0 lbs./sq. in.

4 CONSTRUCTING A FREQUENCY DISTRIBUTION

Now that we have learned how to divide a sample into classes, we can take raw data and actually construct a frequency distribution. To solve the clorination problem on the first page of the chapter, follow these three steps:

1. Decide on the type and number of classes for dividing the data. In this case, we have already chosen to classify the data by the quantitative measure of the number of ppm of chlorine in treated water rather than by a qualitative attribute like the color or odor of the water. Next, we need to decide how many different classes to use and the range (from where to where) each class should cover. The range must be divided by *equal* classes; that is, the width of the interval from the beginning of one class to the beginning of the next class needs to be the same for every class. If we choose a width of .5 ppm for each class in our water example, the classes will be those shown in Table 2·12.

Classify the data

Divide the range by equal classes

Class in ppm	Frequency
15.1–15.5	2
15.6–16.0	16
16.1–16.5	8
16.6–17.0	4
	30

TABLE 2·12
Chlorine levels in samples of treated water with .5 ppm class intervals

If the classes were unequal and the width of the intervals differed among the classes, then we would have a distribution that is much more difficult to interpret than one with equal intervals. Imagine how hard it would be to interpret the data presented in Table 2·13!

Problems with unequal classes

Class	Width of class intervals	Frequency
15.1–15.5	15.6 − 15.1 = .5	2
15.6–15.8	15.9 − 15.6 = .3	8
15.9–16.1	16.2 − 15.9 = .3	9
16.2–16.5	16.6 − 16.2 = .4	7
16.6–16.9	17.0 − 16.6 = .4	4
		30

TABLE 2·13
Chlorine levels in samples of treated water using unequal class intervals

The number of classes depends on the number of data points and the range of the data collected. The more data points or the wider the range of the data, the more classes it takes to divide the data. Of course, if we have only 10 data points, it is senseless to have as many as 10 classes. As a rule, statisticians rarely use fewer than 6 or more than 15 classes.

Because we need to make the class intervals of equal size, the number of classes determines the width of each class. To find the intervals, we can use this equation:

$$\frac{\text{Width of}}{\text{class intervals}} = \frac{\begin{array}{c}\text{Next unit value after} \\ \text{largest value in data} - \text{Smallest value in data}\end{array}}{\text{Total number of class intervals}} \quad [2 \cdot 1]$$

We must use the *next value of the same units* because we are measuring the *interval* between the first value of one class and the first value of the next class. In our water study, the last value is 16.9, so 17.0 is the next value. Since we are using 6 classes in this example, the width of each class will be:

$$\frac{\text{Next unit value after largest value in data} - \text{Smallest value in data}}{\text{Total number of class intervals}} \quad [2 \cdot 1]$$

$$= \frac{17.0 - 15.2}{6}$$

$$= \frac{1.8}{6}$$

$$= .3 \text{ ounces} \leftarrow \text{width of class intervals}$$

Step 1 is now complete. We have decided to classify the data by the quantitative measure of how many ppm of chlorine are in the treated water. We have chosen 6 classes to cover the range of 15.2 to 16.9 and, as a result, will use .3 ppm as the width of our class intervals.

2. Sort the data points into classes and count the number of points in each class. This we have done in Table 2·14. Every data point fits into at least one class, and no data point fits into more than one class. Therefore, our classes are all-inclusive and mutually exclusive. Notice that the lower boundary of the first class corresponds with the smallest data point in our

TABLE 2·14
Chlorine levels in samples of treated water with .3 ppm class intervals

Class	Frequency
15.2–15.4	2
15.5–15.7	5
15.8–16.0	11
16.1–16.3	6
16.4–16.6	3
16.7–16.9	3
	30

sample, and the upper boundary of the last class corresponds with the largest data point.

    3. **Illustrate the data in a chart.** (See Fig. 2·1).

    These three steps enable us to arrange the data in both tabular and graphic form. In this case, our information is displayed in Table 2·14 and in Fig. 2·1. These two frequency distributions omit some of the detail contained in the raw data of Table 2·3, but they make it easier for us to notice trends in the data. One obvious characteristic, for example, is that the class 15.8–16.0 contains the most elements; class 15.2–15.4, the fewest.

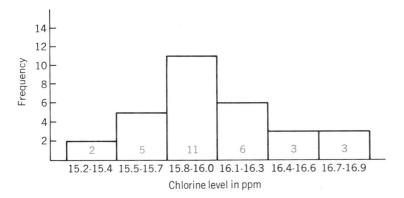

FIG. 2·1
Frequency distribution of chlorine levels in samples of treated water using .3 ppm class intervals

    Notice in Fig. 2·1 that the frequencies in the classes of .3 ppm widths follow a regular progression: the number of data points begins with 2 for the first class, builds to 5, reaches 11 in the third class, falls to 6, and tumbles to 3 in the fifth and sixth classes. We will find that the larger the width of the class intervals, the smoother this progression will be. However, if the classes are too wide, we lose so much information that the chart is almost meaningless. If, for example, we collapse Fig. 2·1 into only two categories, we obscure the trend. This is evident in Fig. 2·2.

Notice any trends

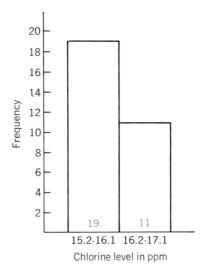

FIG. 2·2
Frequency distribution of chlorine levels in samples of treated water using 1 ppm class intervals

Class limits and class marks

Stated class limits
vs.
real class limits

Statisticians use the term *class limits* to refer to the smallest and largest values that go into any given class. But two different kinds of class limits exist. In Table 2·14 the lower and upper class limits for the first class are 15.2 and 15.4, respectively. These are the *stated* limits. The *real* limits, however, are 15.15 and 15.45. This is true because we round the values 15.15, 15.16, 15.17, 15.18, and 15.19 *up* to 15.2. As a result, all these values fall into the class with the lower limit equal to 15.2.

The real limits of the next class are 15.45 and 15.75. Notice that the upper limit of one class is the lower limit of the succeeding class. Therefore, 15.45 is the upper limit of the class with stated limits of 15.2 and 15.4 *and* the lower limit of the class with stated limits of 15.5 and 15.7.

Distinction between
real and stated
class limits

The distinction between real limits and stated limits is made only with continuous variables; we make this distinction because continuous variables are rounded. However, for discrete variables, which are usually counted (rather than measured), we do not make any distinction between the two kinds of limits. For instance, if we have a frequency distribution of weekly car sales by salespeople, and if one of the frequency classes is 0–5, then zero is *both* the real and the stated lower limit, and 5 is *both* the real and the stated upper limit.

Statisticians use another term, *class marks,* to describe the midpoints of the classes. To calculate the class mark, we simply average the lower and upper limits by applying Equation 2·2:

$$\text{Class mark} = \frac{\text{Stated lower limit} + \text{Stated upper limit}}{2} \qquad [2·2]$$

Class marks
or midpoints

Using this equation and Table 2·14, we can find the class mark for the class 15.2–15.4 like this:

$$\frac{\text{Stated lower limit} + \text{Stated upper limit}}{2} = \frac{15.2 + 15.4}{2} \qquad [2·2]$$

$$= \frac{30.6}{2}$$

$$= 15.3 \text{ ppm} \leftarrow \text{class mark}$$

Now suppose that we are classifying a hi-fi store's accounts receivable, and our first two classes are $0–$49.99 and $50–$99.99. Then the class mark for the first class would be:

$$\frac{\text{Stated lower limit} + \text{Stated upper limit}}{2} = \frac{\$0 + \$49.99}{2} \qquad [2·2]$$

$$= \$24.995 \leftarrow \text{class mark}$$

In this case, we would round up so that $25.00 would be the class mark. This rounding enables us to work with much more convenient values, and is usually done whenever we deal with discrete variables (cents, in this case) and with wide intervals (5,000 cents, in this case). In such situations we use a modified version of Equation 2·2:

$$\text{Class mark} = \frac{\text{Stated lower limit} + \text{Stated lower limit of the next class}}{2} \quad [2\cdot3]$$

So, in our accounts-receivable problem, we would find the class mark like this:

$$\frac{\text{Stated lower limit} + \text{Stated lower limit of the next class}}{2} = \frac{\$0 + \$50}{2} \quad [2\cdot3]$$
$$= \$25 \leftarrow \text{class mark}$$

A word of advice

A hint
when constructing
class intervals

If possible, try to construct class intervals so that values *cluster* around the values of class marks. To do this, examine the raw data and look for values around which data points are concentrated. Look at Table 2·15, which illustrates the raw data from a sample of 20 weeks of penny production at the Philadelphia mint. The director of the mint wants to know how many pennies are stamped each week.

TABLE 2·15

Philadelphia mint weekly penny production (in millions)

4	4	6	7	7	7	7	9	9	10
4	5	7	7	7	7	8	9	9	10

Notice that in Table 2·15 the values cluster around 4, 7, and 9. In constructing class intervals for the frequency distribution, then, we should attempt to have these three values as class marks. We have done this in Table 2·16. Of course, values do not always cluster so neatly. The important thing is to choose intervals so that as many values as possible are close to the values of the class marks.

TABLE 2·16

Frequency distribution of Philadelphia mint weekly penny production

Class in millions	Frequency	Class mark
3 to 5	4	4
6 to 8	10	7
9 to 11	6	10

19. For the following data, construct
    a) a 6-category closed classification.
    b) a 5-category open-ended classification.
    c) relative frequency distributions to go with the above frequency distributions.

34.1	39.0	38.3	41.6	36.4	43.9	33.2	56.4	33.9	34.5
46.4	42.1	41.8	49.4	42.2	51.7	42.4	44.5	46.7	40.6
45.7	50.7	37.6	36.0	34.9	38.9	44.6	49.0	51.4	48.3

20. For the frequency distribution constructed above in 19c, give the
    a) real class limits.
    b) stated class limits.
    c) class marks for the intervals used.

21. Determine the class marks for the intervals of the following frequency distribution:

Class		
17.50–19.99	25.00–27.49	32.50–34.99
20.00–22.49	27.50–29.99	35.00–37.49
22.50–24.99	30.00–32.49	37.50–39.99

22. Given the following class marks for the intervals of a frequency distribution, determine the real and stated class limits of the intervals.

Class mark			
8.50	14.50	20.50	26.50
11.50	17.50	23.50	29.50

23. Mr. Franks, a safety engineer for the Mars Point Nuclear Power Generating Station, has charted the peak reactor temperature each day for the past year and has prepared the following frequency distribution:

Temperatures in °C	Frequency
Below 500°	4
501–510	7
511–520	32
521–530	59
530–540	82
550–560	65
561–570	33
571–580	28
580–590	27
591–600	23
Total	**360**

    List and explain any errors you can find in Mr. Franks's distribution.

24. Construct a discrete, closed classification for the possible responses to the "marital status" portion of an employment application. Also, construct a three-category, discrete, open-ended classification for the same responses.

25. Listings for a stock exchange usually contain the company name, high and low bids, closing price, and the change from the previous day's closing price. For example:

Name	High bid	Low bid	Closing	Change
Jefferson Pilot	28½	27¾	28¼	+1¼

Is a distribution of all

a) stocks on the New York Stock Exchange by industry

b) closing prices on a given day

c) changes in prices on a given day

1) quantitative or qualitative? 2) continuous or discrete? 3) open-ended or closed?

Would your answer to part c be different if the change were expressed simply as "higher," "lower," or "unchanged"?

26. The noise level in decibels of aircraft taking off from JFK Airport in New York City were rounded to the nearest tenth of a decibel and grouped in a table having the following class marks: 102.45, 107.45, 112.45, 117.45, 122.45, 127.45, 132.45, and 137.45. What are the stated and real class limits?

27. Anthropometric data were collected from a group of infants in a Guatemalan village. One of the variables of interest was cranial circumference, recorded in inches and measured to the nearest one-thousandth of an inch. The data were compiled into a distribution with class marks for the intervals of 5.8695, 5.8895, 5.9095, 5.9295, 5.9495, 5.9695, and 5.9895. Determine the real limits and the stated limits for the intervals of the distribution.

28. The president of Ocean Airlines is trying to estimate when the Civil Aeronautics Board (CAB) is most likely to rule on the company's application for a new route between Charlotte and Nashville. Assistants to the president have assembled the following waiting times for applications filed during the past year. The data are given in days from the date of application until a CAB ruling.

```
32  38  26  29  32  41  28  31  45  36
45  35  40  30  31  40  27  33  28  30
30  41  39  38  33  35  31  36  37  32
23  45  39  37  38  36  33  35  42  38
34  22  37  43  52  32  35  30  46  36
```

a) Construct a frequency distribution using 10 closed intervals, equally spaced. Which interval occurs most often?

b) Construct a frequency distribution using 5 closed intervals, equally spaced. Which interval occurs most often?

29. A fifth-grade teacher administered a very long test to her students and recorded for each student the total number of problems answered correctly in 10 minutes. She got the following results for her 40 students:

```
 7   8   5  10   9  10   5  12   8   6
10  11   6   5  10  11  10   5   9  13
 8  12   8   8  10  15   7   6   8   8
 5   6   9   7  14   8   7   5   5  14
```

a) Based on frequency, what would be the desired class marks?

b) Construct a frequency and relative frequency distribution having as many of the marks as possible. Make your intervals evenly spaced and at least two problems wide.

5 GRAPHING FREQUENCY DISTRIBUTIONS
Figures 2·1 and 2·2 (both on page 23) are previews of what we are going to discuss now: how to present frequency distributions graphically. Graphs give data in a two-dimensional picture. On the *horizontal* axis, we can show the values of the variable (the characteristic we are measuring), such as the chlorine level in ppm. On the *vertical* axis, we mark the frequencies of the classes shown on the horizontal axis. Thus, the height of the boxes in Fig. 2·1 measures the number of observations in each of the classes marked on the horizontal axis.

Graphs of frequency distributions and relative frequency distributions are useful because they emphasize and clarify trends that are not so readily discernible in tables. They attract a reader's attention to trends in the data. Graphs can also help us do problems concerning frequency distributions. They will enable us to estimate some values at a glance and will provide us with a pictorial check on the accuracy of our solutions.

Histograms

Figures 2·1 and 2·2 on page 23 are two examples of histograms. A *histogram* is a series of rectangles, each proportional in width to the range of values within a class and proportional in height to the number of items falling in the class. If the classes we use in the frequency distribution are of equal width, then the vertical bars in the histogram are also of equal width. The height of the bar for each class corresponds to the number of items in the class. As a result, the area contained in each rectangle (width times height) is the same percentage of the area of all the rectangles as the relative frequency of that class is to all the observations made.

A histogram that uses the relative frequency of data points in each of the classes rather than the actual number of points is called a *relative frequency histogram.* The relative frequency histogram has the same shape as an absolute frequency histogram made from the same data set. This is true because in both the relative size of each rectangle is the frequency of that class compared to the total number of observations.

Recall that the relative frequency of any class is the number of observations in that class divided by the total number of observations made. The sum of all the relative frequencies for any data set is equal to 1.0. With this in mind, we can convert the histogram of Fig. 2·1 into a relative frequency histogram such as we find in Fig. 2·3. Notice that the only difference between these two is the left-hand vertical scale. Whereas the scale in Fig. 2·1 is the *absolute* number of observations in each class, the scale in Fig. 2·3 is the number of observations in each class as a *fraction* of the total number of observations.

Being able to present data in terms of the relative rather than the absolute frequency of observations in each class is useful because, while the

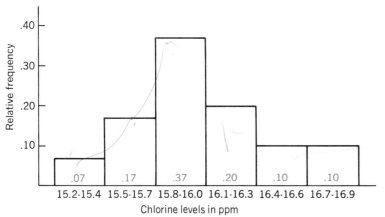

FIG. 2·3
**Relative frequency
distribution of
chlorine levels in
samples of treated
water using .3 ppm
class intervals**

absolute numbers may change (as we test more gallons of water for example), the relationship among the classes may remain stable. Twenty percent of all the gallons of water may fall in the class "16.1–16.3 ppm" whether we test 30 gallons or 300 gallons. It is easy to compare data from different sizes of samples when we use relative frequency histograms.

Frequency polygons

Although less widely used, frequency polygons are another way to portray graphically both simple and relative frequency distributions. To construct a frequency polygon, we mark the frequencies on the vertical axis and the values of the variable we are measuring on the horizontal axis, as we did with histograms. Next, we plot each class frequency by drawing a dot above its class mark, or midpoint, and connect the successive dots with a straight line to form a polygon (a many-sided figure).

Use class marks on the horizontal axis

Figure 2·4 is a frequency polygon constructed from the data in Table 2·14. If you compare this figure with Fig. 2·1, you will notice that classes have been added at *each end* of the scale of observed values. These two new classes contain zero observations but allow the polygon to reach the horizontal axis at both ends of the distribution.

Add two classes

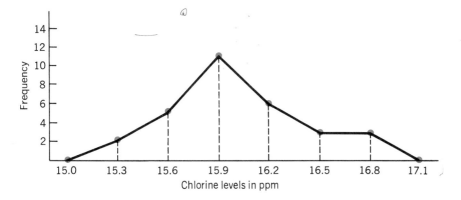

FIG. 2·4
**Frequency polygon
of chlorine levels
in samples of
treated water
using .3 ppm class
intervals**

FIG. 2·5
**Histogram drawn
from the points of
the frequency
polygon in Fig. 2·4**

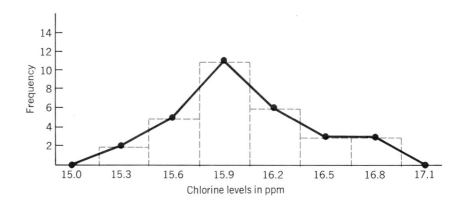

Chlorine levels in ppm

Converting a
frequency polygon
to a histogram

How can we turn a frequency polygon into a histogram? A frequency polygon is simply a line graph that connects the midpoints of all the bars in a histogram. Therefore, we can reproduce the histogram by drawing vertical lines from the bounds of the classes (as marked on the horizontal axis) and connecting them with horizontal lines at the heights of the polygon at each class mark. We have done this with dotted lines in Fig. 2·5.

Constructing a
relative frequency
polygon

A frequency polygon that uses the relative frequency of data points in each of the classes rather than the actual number of points is called a *relative frequency polygon.* The relative frequency polygon has the same shape as the frequency polygon made from the same data set but a different scale of values on the vertical axis. Rather than the absolute number of observations, the scale is the number of observations in each class as a fraction of the total number of observations.

Advantages
of histograms

Histograms and frequency polygons are similar. Why do we need both? The advantages of histograms are:

1. The rectangle clearly shows each separate class in the distribution.
2. The area of each rectangle, relative to all the other rectangles, shows the proportion of the total number of observations that occur in that class.

Advantages
of polygons

Frequency polygons, however, have certain advantages too.

1. The frequency polygon is simpler than its histogram counterpart.
2. It sketches an outline of the data pattern more clearly.
3. The polygon becomes increasingly smooth and curvelike as we increase the number of classes and the number of observations.

Creating a
frequency curve

A polygon such as the one we have just described, smoothed by added classes and data points, is called a *frequency curve.* In Fig. 2·6, we have used our water example, but we have increased the number of observations to 300 and the number of classes to 10 (the first and last dots do not represent class marks). Notice that we have connected the points with curved lines to approximate the way the polygon would look if we had an infinite number of data points and very small class intervals.

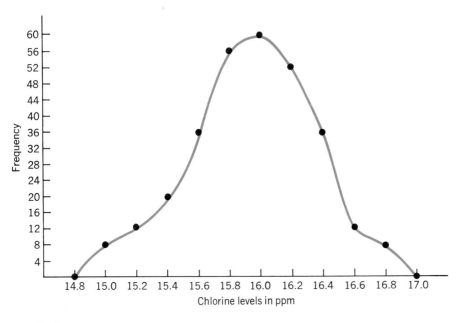

FIG. 2·6
**Frequency curve
of the chlorine
levels in 300
gallons of water
using .2 ppm
intervals**

Ogives

A cumulative frequency distribution enables us to see how many obser-vations lie above or below certain values, rather than merely recording the numbers of items within intervals. If, for example, we wish to know how many of our original 30 gallons of water contain more than 16.0 ppm of chlorine we would refer to a table of cumulative "more than" frequencies, such as Table 2·17. To know how many gallons contain less than 17.0 ppm we can use a table recording the cumulative "less than" frequencies in our sample, such as Table 2·18.

Cumulative frequency distribution defined

Tables of "more than" and "less than" frequencies

Class	Cumulative frequency
More than 15.1	30
More than 15.4	28
More than 15.7	23
More than 16.0	12
More than 16.3	6
More than 16.6	3
More than 17.0	0

TABLE 2·17
**Cumulative
frequency
distribution of
chlorine levels
in ppm**

Class	Cumulative frequency
Less than 15.2	0
Less than 15.5	2
Less than 15.8	7
Less than 16.1	18
Less than 16.4	24
Less than 16.7	27
Less than 17.1	30

TABLE 2·18
**Cumulative
frequency
distribution of
chlorine levels
in ppm**

FIG. 2·7
"More than" ogive
of the distribution
of chlorine levels
in ppm for 30
gallons of treated
water

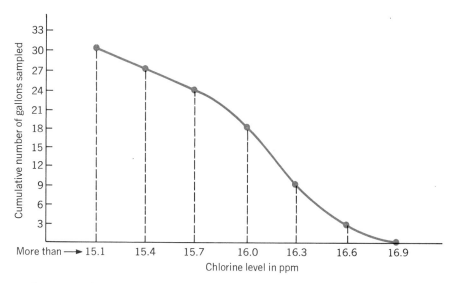

A "more than" ogive

A graph of a cumulative frequency distribution is called an *ogive* (pronounced "**oh**-jive"). The ogive for the cumulative distribution in Table 2·17 is shown in Fig. 2·7. The plotted points represent the number of gallons having more chlorine than the ppm shown on the horizontal axis. Notice that the upper bound of the classes in the table becomes the lower bound of the cumulative distribution of the ogive.

A "less than" ogive

Likewise, we can use the ogive in Fig. 2·8, which plots the cumulative distribution of Table 2·18, to find the number of gallons having less chlorine than the ppm shown on the horizontal axis. In this case, the lower bound of the classes is the upper bound of the cumulative distribution in Fig. 2·8.

Shapes of ogives

The *S-shaped* curves shown in Figs. 2·7 and 2·8 are typical of ogives. Notice that the "more than" curve slopes down and to the right. The "less than" curve slopes up and to the right.

FIG. 2·8
"Less than" ogive
of the distribution
of chlorine levels
in ppm for 30
gallons of treated
water

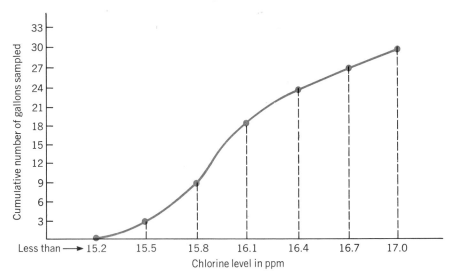

TABLE 2·19

Class	Cumulative frequency	Cumulative relative frequency
Less than 15.2	0	.00
Less than 15.5	2	.07
Less than 15.8	7	.23
Less than 16.1	18	.60
Less than 16.4	24	.80
Less than 16.7	27	.90
Less than 17.0	30	1.00

We can construct an ogive of a relative frequency distribution in the same manner in which we drew the ogives of absolute frequency distributions in Figs. 2·7 and 2·8. There will be one change—the vertical scale. As in Fig. 2·3 on page 29, this scale must mark the *fraction* of the total number of observations that fall into each class.

Ogives of relative frequencies

To construct a cumulative "less than" ogive in terms of relative frequencies, we can refer to a relative frequency distribution (like Fig. 2·3) and set up a table using the data (like Table 2·19). Then, we can convert the figures there to an ogive (as in Fig. 2·9). Notice that Figs. 2·8 and 2·9 are equivalent except for the left-hand vertical axis.

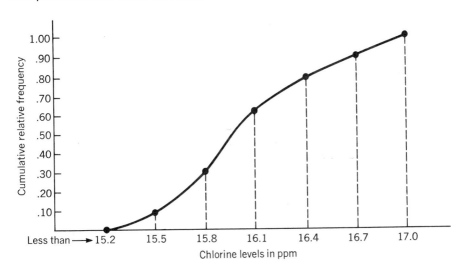

FIG. 2·9
"Less than" ogive of the distribution of chlorine levels in ppm for 30 gallons of treated water using relative frequencies

Suppose we now draw a line perpendicular to the vertical axis at the .50 mark to intersect our ogive. (We have done this in Fig. 2·10.) In this way, we can read an approximate value for the chlorine level in the 15th gallon of an array of the 30 gallons. Thus, we are back to the first data arrangement discussed in this chapter. From the data array, we can construct frequency distributions. From frequency distributions, we can construct cumulative frequency distributions. From these, we can graph an ogive. And from this ogive, we can approximate the values we had in the data array. However, we normally cannot recover the *exact* original data from any of the graphic representations we have discussed.

Approximating the data array

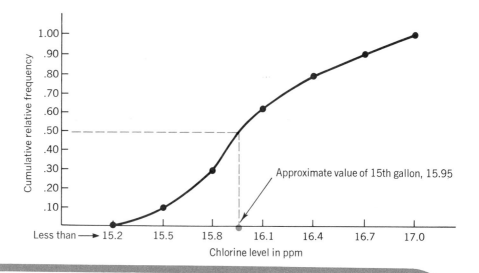

FIG. 2·10
"Less than" ogive of the distribution of chlorine levels in ppm for 30 gallons of treated water indicating approximate middle value in original data array

Approximate value of 15th gallon, 15.95

EXERCISES

30. Construct a histogram for the data in the following frequency distribution:

Class	Frequency	Class	Frequency
75–89	9	150–164	24
90–104	12	165–179	10
105–119	24	180–194	8
120–134	25	195–209	6
135–149	30	210–224	2

31. Construct a frequency distribution for the data below.

a) Using intervals 50–59, 60–69, 70–79, 80–89, and 90–99.

b) Using intervals 52–57, 58–63, 64–69, 70–75, 76–81, 82–87, 88–93, and 94–99.

c) Construct a frequency polygon for parts a and b.

```
67   68   95   79   70   82   83   84   71   93
92   67   97   72   60   57   58   86   54   68
98   80   85   84   74   66   60   56   78   79
59   81   78   69   74   90   71   82   73   65
80   72   76   88   83   70   71   91   69   83
```

32. For the following frequency distribution, construct

a) A cumulative frequency distribution for frequencies of values "less than" the interval limits.

b) An ogive based on part a.

c) A cumulative frequency distribution for frequencies of values "more than" the interval limits.

d) An ogive for part c.

Class	Frequency	Class	Frequency
3.00–3.19	1	4.00–4.19	11
3.20–3.39	4	4.20–4.39	8
3.40–3.59	11	4.40–4.59	7
3.60–3.79	15	4.60–4.79	6
3.80–3.99	12		

33. For the following frequency distribution,

   a) Construct a cumulative relative frequency ogive using frequencies of values "less than" the interval limits.

   b) Estimate the value of the middle observation in the original data set.

Class	Frequency	Class	Frequency
0.5–0.9	13	2.5–2.9	30
1.0–1.4	20	3.0–3.4	35
1.5–1.9	22	3.5–3.9	41
2.0–2.4	24	4.0–4.4	15

34. Prior to constructing a dam on the Colorado River, the U.S. Army Corps of Engineers performed a series of tests to measure the water flow past the proposed location of the dam. The results of the testing were used to construct the following frequency distribution:

River flow (thousands of gallons per minute)	Frequency
1001–1050	9
1051–1100	20
1101–1150	31
1151–1200	45
1201–1250	53
1251–1300	39
1301–1350	27
1351–1400	13
Total	237

   a) Use the data given in the table to construct a "more than" cumulative frequency distribution and ogive.

   b) Use the data given in the table to construct a "less than" cumulative frequency distribution and ogive.

35. Debbie Mayberry, a consulting psychologist, ran an experiment with a group of people to determine their indifference points in gambling with poker dice. She believed that if the subjects could win above a certain amount of money, they would be eager to make a bet, but below that amount they would be reluctant. From a group of 50 subjects, she obtained the following data:

Indifference point	Frequency
$ .90– .94	2
.95– .99	6
1.00–1.04	9
1.05–1.09	13
1.10–1.14	10
1.15–1.19	4
1.20–1.24	3
1.25–1.29	3
	50

   a) Construct a "more than" and "less than" cumulative relative frequency distribution.

   b) Graph the two distributions in part a into relative frequency ogives.

36. At a newspaper office, the time required to set the entire front page in type was recorded for 50 days. The data, to the nearest tenth of a minute, are given below.

20.8	22.8	21.9	22.0	20.7	20.9	25.0	22.2	22.8	20.1
25.3	20.7	22.5	21.2	23.8	23.3	20.9	22.9	23.5	19.5
23.7	20.3	23.6	19.0	25.1	25.0	19.5	24.1	24.2	21.8
21.3	21.5	23.1	19.9	24.2	24.1	19.8	23.9	22.8	23.9
19.7	24.2	23.8	20.7	23.8	24.3	21.1	20.9	21.6	22.7

a) Arrange the data in an array from lowest to highest.

b) Construct a frequency distribution and a "less than" cumulative frequency distribution from the data using intervals of .8 minutes.

c) Construct a frequency polygon from the data.

d) Construct a "less than" frequency ogive from the data.

6 TERMS INTRODUCED IN CHAPTER 2

- **CLASS LIMITS** The smallest and largest values that go into any given class.
- **CLASS MARK** The midpoint of a class in a frequency distribution, the average of the lower and upper limits.
- **CONTINUOUS DATA** Data that may progress from one class to the next without a break and may be expressed by either whole numbers or fractions.
- **CUMULATIVE FREQUENCY DISTRIBUTION** A tabular display of data showing how many observations lie above, or below, certain values.
- **DATA** A collection of any number of related observations on one or more variables.
- **DATA ARRAY** The arrangement of raw data by observations in either ascending or descending order.
- **DATA POINT** A single observation from a data set.
- **DATA SET** A collection of data.
- **DISCRETE DATA** Data that do not progress from one class to the next without a break; i.e., where classes represent distinct categories or counts and may be represented by whole numbers.
- **FREQUENCY CURVE** A frequency polygon smoothed by adding classes and data points to a data set.
- **FREQUENCY DISTRIBUTION** An organized display of data that shows the number of observations from the data set that fall into each of a set of mutually exclusive classes.

- **FREQUENCY POLYGON** A line graph connecting the midpoints of each class in a data set, plotted at a height corresponding to the frequency of the class.
- **HISTOGRAM** A graph of a data set, composed of a series of rectangles, each proportional in width to the range of values in a class and proportional in height to the number of items falling in the class, or the fraction of items in the class.
- **OGIVE** A graph of a cumulative frequency distribution.
- **OPEN-ENDED CLASS** A class that allows either the upper or lower end of a quantitative classification scheme to be limitless.
- **POPULATION** A collection of all the elements we are studying and about which we are trying to draw conclusions.
- **RAW DATA** Information before it is arranged or analyzed by statistical methods.
- **RELATIVE FREQUENCY DISTRIBUTION** The display of a data set that shows the fraction or percentage of the total data set that falls into each of a set of mutually exclusive classes.
- **REPRESENTATIVE SAMPLE** A sample which contains the relevant characteristics of the population in the same proportion as they are included in that population.
- **SAMPLE** A collection of some, but not all, of the elements of the population under study, used to describe the population.

p. 22:

$$\text{Width of class intervals} = \frac{\text{Next unit value after largest value in data} - \text{Smallest value in data}}{\text{Total number of class intervals}}$$

[2·1]

To arrange raw data, decide the number of classes in which you will divide the data (normally, between 6 and 15), and then use Equation 2·1 to determine the *width of class intervals of equal size.* This formula uses the next value of the same units because it measures the interval between the first value of one class and the first value of the next class.

p. 24:

$$\text{Class mark} = \frac{\text{Stated lower limit} + \text{Stated upper limit}}{2}$$

[2·2]

The midpoint of a class, that is, its *class mark,* is calculated by averaging the lower and upper limits of that class.

p. 25:

$$\text{Class mark} = \frac{\text{Stated lower limit} + \text{Stated lower limit of the next class}}{2}$$

[2·3]

When we are dealing with discrete variables and wide intervals, the midpoint of a class (its class mark) is calculated by using a slight modification of Equation 2·2. This enables us to work with more convenient values.

8   *CHAPTER REVIEW EXERCISES*

37. The following set of raw data gives income and education level for a sample of individuals. Would rearranging the data help us to draw some conclusions? Rearrange the data in a way that makes it more meaningful.

Income	Education	Income	Education	Income	Education
$ 8,500	High school	$10,600	B.S.	$8,600	2 yrs. college
10,400	B.S.	14,000	B.S.	9,800	B.A.
13,500	M.A.	15,100	High school	18,100	M.S.
35,000	M.D.	11,200	2 yrs. college	7,200	1 yr. college
14,500	Ph.D.	50,000	M.D.	9,200	2 yrs. college
7,200	10th grade	38,000	Law degree	17,200	B.A.
9,500	High school	22,000	Ph.D.	13,000	High school
11,600	M.A.	8,800	11th grade	26,000	Law degree
15,200	High school	12,900	High school	32,000	Ph. D.
12,800	B.A.	10,100	1 yr. college	16,400	B.S.

38. The state department of education receives the average attendance per day for each of the nine weeks in a grading period from all the state's counties, plus information on the percentage of full attendance in each county. Is this an example of raw data? Why or why not?

39. The lengths of X-ray treatments given at a local hospital were recorded in milliseconds (1/1,000 of a second) and are given below.

$$
\begin{array}{cccccccccc}
.3 & 1.8 & 1.4 & .8 & .2 & 1.5 & .3 & 1.3 & 1.1 & .7 \\
.8 & .9 & .7 & .7 & .9 & 1.6 & .8 & 1.2 & 1.2 & 1.5 \\
1.2 & 1.0 & 1.1 & .9 & .8 & .7 & .1 & .7 & 1.8 & 1.4 \\
.1 & 1.5 & 1.3 & 1.7 & 1.0 & .6 & .5 & .5 & 1.1 & 1.0
\end{array}
$$

a) Arrange the data in an array from highest to lowest.

b) Construct a relative frequency distribution and a "more than" cumulative relative frequency distribution using intervals of .25 milliseconds.

c) Construct a histogram from the data.

d) Construct a relative frequency "more than" ogive from the data.

e) Verify that the 20th data point in the array is close to the intersection of a horizontal line drawn from .50 on the vertical axis to the ogive curve.

40. The National Safety Council randomly sampled the tread depth of 60 right front tires on passenger vehicles stopped at a rest area on an interstate highway. From their data, they constructed the following frequency distribution:

Tread depth (inches)	Frequency	Tread depth (inches)	Frequency
16/32 (new tire)	5	4/32–6/32	7
13/32–15/32	10	1/32–3/32	4
10/32–12/32	20	0/32 (bald)	2
7/32–9/32	12		

Approximately what was the tread depth of the 30th tire in the data array?

41. The state department of education keeps records on the number of absences from school in its 15 county school systems. These absences are then classified according to whether they were in elementary or secondary schools in order to examine the relative numbers of absences in each. Records show that the following number of absences were recorded in the state for the last month of 20 school days:

Total number of students absent for the month

9,897	10,052	10,028	9,722	9,908
10,098	10,587	9,872	9,956	9,928
10,123	10,507	9,910	9,992	10,237

Construct both a frequency distribution and a relative frequency distribution of absences recorded per day, using intervals of 5 absences per day.

42. The administrator of a hospital has ordered a study of the amount of time a patient must wait before being treated by emergency room personnel. The following data were collected during a typical day:

Waiting time (min.)

15	13	19	23	22	5	15	12	28	20
28	2	8	17	24	7	20	26	13	9

a) Arrange the data in an array, lowest to highest.

b) Construct a frequency distribution, using 6 equal intervals.

c) Construct a frequency distribution, using 10 equal intervals.

43. For her Master's thesis in psychology, Anne Jones presented controversial words on a tachistoscope to a group of subjects. The data were recognition times, measured to the nearest tenth of a millisecond. They were recorded in a distribution with 15.25, 15.85, 16.45, 16.65, 16.85, 17.35, 17.95, and 18.55 as the real class limits. Determine the stated class limits and class marks for each interval of the distribution.

44. Below are the measurements on an entire population of 100 elements.

   a) Select two samples: one sample of the first 10 elements and another sample of the largest 10 elements.

   b) Are the two samples equally representative of the population? If not, which sample is more representative and why?

226	198	210	233	222	175	215	191	201	175
264	204	193	244	180	185	190	216	178	190
174	183	201	238	232	257	236	222	213	207
233	205	180	267	236	186	192	245	218	193
189	180	175	184	234	234	180	252	201	187
155	175	196	172	248	198	226	185	180	175
217	190	212	198	212	228	184	219	196	212
220	213	191	170	258	192	194	180	243	230
180	135	243	180	209	202	242	259	238	227
207	218	230	224	228	188	210	205	197	169

45. In the population under study, there are 2,000 women and 8,000 men. If we are to select a sample of 250 individuals from this population, how many should be women, to make our sample considered strictly representative?

46. The U.S. Department of Labor publishes several classifications of the unemployment rate, as well as the rate itself. Recently, the unemployment rate was 7.3 percent. The department reported the following educational categories:

Level of education	Relative frequency (% of those unemployed)
Did not complete high school	.38
Received high school diploma	.29
Attended college but did not receive a degree	.17
Received a college degree	.08
Attended graduate school but did not receive a degree	.05
Received a graduate degree	.03
Total	**1.00**

Using these data, construct a relative frequency histogram.

47. Using the relative frequency distribution given in problem 55 on page 41, construct a relative frequency histogram and polygon. For the purposes of the present problem, assume that the upper limit of the last class is $51.00.

48. Using the frequency distribution given in problem 49 for miles per week of jogging, construct a frequency histogram and polygon. For the purpose of the present problem, assume that the upper limit of the last class is 5.39 miles.

49. A sports psychologist studying the effect of jogging on college students' grades collected data from a group of college joggers. Along with some other variables, he recorded the average number of miles run per week. He compiled his results into the following distribution:

Miles per day	Frequency
1.00–1.39	32
1.40–1.79	43
1.80–2.19	81
2.20–2.59	122
2.60–2.99	131
3.00–3.39	130
3.40–3.79	111
3.80–4.19	95
4.20–4.59	82
4.60–4.99	47
5.00 and up	53
	927

Determine the stated limits, real limits, and class marks for the intervals of the distribution.

50. City engineers made a study of the average time (in hours) cars remained parked at a new city parking lot. The data were rounded to the nearest tenth of an hour and grouped in a table whose classes have the following real limits: .05, .35, .65, .95, 1.25, 1.55, 1.85, 2.15, 2.45, 2.75, 3.05, and no limit for the last interval. Determine the stated limits and class marks for each interval.

51. If the following age groups are included in the proportions indicated, how many of each age group should be included in a sample of 3,000 individuals to make the sample representative?

Age group	Relative proportion in population
12–17	.15
18–23	.33
24–29	.25
30–35	.17
36+	.10
	1.00

52. State University has 3 campuses, each with its own psychology department. Last year, State's psychology professors published numerous articles in prestigious professional journals, and the board of regents counted these articles as a measure of the productivity of each department.

Journal number	Number of publications	Campus	Journal number	Number of publications	Campus
9	3	North	14	20	South
12	6	North	10	18	South
3	12	South	3	12	West
15	8	West	5	6	North
2	9	West	7	5	North
5	15	South	7	15	West
1	2	North	6	2	North
15	5	West	2	3	West
12	3	North	9	1	North
11	4	North	11	8	North
7	9	North	14	10	West
6	10	West	8	17	South

a) Construct a frequency distribution and a relative frequency distribution by journal.

b) Construct a frequency distribution and a relative frequency distribution by university branch.

c) Construct a frequency distribution and a relative frequency distribution by number of publications (using intervals of 3).

53. A questionnaire on attitudes about sex education in the schools is sent out to a random sample of 2,000 people; 880 are completed and returned to the researcher. Comment on the data available from these questionnaires in terms of the five tests for data.

54. With each appliance that Central Electric produces, the company includes a warranty card for the purchaser. In addition to validating the warranty and furnishing the company with the purchaser's name and address, the card also asks for certain other information that is used for marketing studies.

Name_____	Marital Status___③_____
Address_____	Where was appliance purchased?
City_____ State_____	___④___
Zip Code_____	Why was appliance purchased?
Age__①__ Yearly Income___②___	___⑤___

For each of the numbered blanks on the card, determine the most likely characteristics of the categories that would be used by the company to record the information. In particular, would they be: (1) quantitative or qualitative? (2) continuous or discrete? (3) open-ended or closed? Briefly state the reasoning behind your answers.

55. The following relative frequency distribution resulted from a study of the dollar amounts spent per visit by customers at a supermarket:

Amount spent	Relative frequency
$ 0–$ 5.99	1%
6.00–$10.99	3
11.00–$15.99	4
16.00–$20.99	6
21.00–$25.99	7
26.00–$30.99	9
31.00–$35.99	11
36.00–$40.99	19
41.00–$45.99	32
46.00 and above	8
Total	**100%**

Determine the class marks for each of the intervals.

56. The following responses were given by two groups of hospital patients: one receiving a new treatment, the other receiving a standard treatment for an illness. The question asked was, "What degree of discomfort are you experiencing?"

Group 1			Group 2		
Mild	Moderate	Severe	Moderate	Mild	Severe
None	Severe	Mild	Severe	None	Moderate
Moderate	Mild	Mild	Mild	Moderate	Moderate
Mild	Moderate	None	Moderate	Mild	Severe
Moderate	Mild	Mild	Severe	Moderate	Moderate
None	Moderate	Severe	Severe	Mild	Moderate

Suggest a better way to display these data. Explain why it is better.

41

57. A teacher posts final grade averages for her class, based on homework, tests, and a paper. Is this an example of raw data? Why or why not? If not, what would it be in this situation?

58. The head of a very large psychology department wanted to classify the specialties of its 67 members. He asked Jack Manning, a Ph.D. candidate, to get the information from the faculty members' publications. Jack compiled the following:

Specialty	Faculty members publishing
Social psychology only	1
Clinical psychology only	5
Experimental psychology only	4
Developmental psychology only	2
Social and clinical psychology	7
Social and experimental psychology	6
Developmental and social psychology	3
Developmental and clinical psychology	8
Experimental and developmental psychology	9
Clinical and experimental psychology	21
No publications	1
	67

Construct a relative frequency distribution for the types of specialties. (Hint: the categories of your distribution will be mutually exclusive, but any individual faculty member may fall into several categories.)

59. A sociologist has been studying the residents of a high rise apartment building in a lower-middle class section of New York City. She is interested in whether people of similar income levels tend to live in the same parts of the building. Her income data suggest that nobody who lives in the apartment makes over $20,000, and several people appear to have no income at all. In a preliminary analysis, she decides to construct both a frequency and a relative frequency distribution for income, and she wants to have $2,000 intervals.

a) Develop a continuous, closed distribution that meets her requirements.

b) Develop a continuous distribution with 9 categories that meets her requirements and that is open at both ends. You may relax the requirement for $2,000 intervals for the open-ended category.

60. Twenty rats were placed in a maze, and the total time it took each rat to find the goal box was recorded in seconds. Following are a frequency distribution and a relative frequency distribution of these times. Fill in all the missing data.

Classes	Frequency	Relative frequency
0–10 sec.	?	.05
11–20	0	?
21–30	1	?
31–40	?	?
41–50	?	.15
51–60	?	.20
61–70	2	?
71–80	?	.00
81–90	3	?
91–100	?	.00
Total	?	?

Answer true *or* false. *Answers are in the back of the book.*

1. In comparison to a data array, the frequency distribution has the advantage of representing data in compressed form.

2. The smallest and largest values that go into any given class of a frequency distribution are referred to as the class limits.

3. A histogram is a series of rectangles, each proportional in width to the number of items falling within a specific class of data.

4. A single observation is called a data point, whereas a collection of data is known as a tabular.

5. The classes in any relative frequency distribution are all-inclusive and mutually exclusive.

6. When a sample contains the relevant characteristics of a certain population in the same proportion as they are included in that population, the sample is said to be a representative sample.

7. The distinction between real class limits and stated class limits is made only when we are dealing with continuous variables.

8. If we were to connect the midpoints of the consecutive bars of a frequency histogram with a series of lines, we would be graphing a frequency polygon.

9. Before information is arranged and analyzed, using statistical methods, it is known as pre-processed data.

10. One disadvantage of the data array is that it does not allow us easily to find the highest and lowest values in the data set.

3

Summary measures of frequency distributions

1. Beyond tables and graphs: descriptive measures of frequency distributions, 46

2. A measure of central tendency: the arithmetic mean, 50

3. A second measure of central tendency: the weighted mean, 58

4. A third measure of central tendency: the geometric mean, 61

5. A fourth measure of central tendency: the median, 64

6. A final measure of central tendency: the mode, 69

7. Comparing the mean, median, and mode, 75

8. Terms introduced, 76

9. Equations introduced, 77

10. Chapter review exercises, 78

11. Chapter concepts test, 83

The transportation director of the Chapel Hill bus system needs some measure of the time her ten buses are out of service. With this information, she can plan supporting service, rearrange routes, or otherwise adjust schedules. This table represents data from last year for each of the ten buses.

Bus number	1	2	3	4	5	6	7	8	9	10
Days out of service	7	23	4	8	2	12	6	13	9	4

The director would like some single measure of days out of service for all of her buses, so she could use that figure in her planning. This chapter introduces several measures that would be useful to her and to others who must make similar plans.

1 BEYOND TABLES AND GRAPHS: DESCRIPTIVE MEASURES OF FREQUENCY DISTRIBUTIONS   In Chapter 2, we learned to construct tables and graphs using raw data. The resulting ''pictures'' of frequency distributions enabled us to discern trends and patterns in the data. But what if we need more exact measures of a data set? In that case, we can use single numbers, called *summary statistics,* to describe certain characteristics of a data set. From these, we can gain a more precise understanding of the data than we can from our tables and graphs. And these numbers will enable us to make quicker and better decisions because we will not need to consult our original observations.

Summary statistics describe the characteristics of a data set

Four of these characteristics are particularly important:

Middle of a data set

1. **Measures of central tendency.**   Like averages, measures of central tendency tell us what we can expect a typical or middle data point to be. They are also called *measures of location.* In Fig. 3·1, the central location of curve B lies to the right of those of curve A and curve C. Notice that the central location of curve A is equal to that of curve C.

**FIG. 3·1
Comparison of central location of three curves**

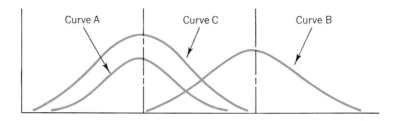

Range of a data set

2. **Measures of dispersion.**   Dispersion refers to the spread of the data, that is, the extent to which the observations are scattered. In Chapter 2, we studied a measure of dispersion called the range. The range indicates how far it is from the lowest data point to the highest. Notice that curve A in Fig. 3·2 has a wider spread, or dispersion, than curve B.

**FIG. 3·2
Comparison of dispersion of two curves**

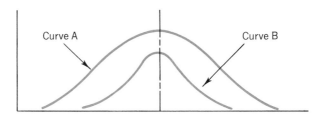

Symmetry of a data set

3. **Measures of skewness.**   Curves representing the data points in the data set may be either symmetrical or skewed. *Symmetrical* curves, like the one in Fig. 3·3, are such that a vertical line drawn from the peak of the curve to the horizontal axis will divide the area of the curve into two equal parts. Each part is the mirror image of the other.

FIG. 3·3
Symmetrical curve

Curves A and B in Fig. 3·4 are *skewed* curves. They are skewed because values in their frequency distributions are concentrated at either the low end or the high end of the measuring scale on the horizontal axis. The values are not equally distributed. Curve A is skewed to the right (or *positively* skewed), because it tails off toward the high end of the scale. Curve B is just the opposite. It is skewed to the left (*negatively* skewed), because it tails off toward the low end of the scale.

Skewness of a data set

Curve A might represent the frequency distribution of the number of days' supply on hand in the perishable grocery business. The curve would be skewed to the right with many values at the low end and few at the high because the inventory must turn over rapidly. Similarly, curve B could represent the frequency of the number of days a real-estate broker requires to sell a house. It would be skewed to the left with many values at the high end and few at the low because the inventory of houses turns over very slowly.

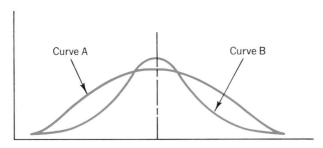

FIG. 3·4
Comparison of two skewed curves

4. Measures of kurtosis.   When we measure the *kurtosis* of a distribution, we are measuring its peakedness. In Fig. 3·5, for example, curves A and B differ only by the fact that one is more peaked than the other. They have the same central location and dispersion, and both are symmetrical. Statisticians say that the two curves have different degrees of kurtosis.

Peakedness of a data set

FIG. 3·5
Two curves with the same central location but different kurtosis

Three degrees of peakedness

There are many different degrees of kurtosis, but statisticians commonly use three broad classes. A curve such as the one in Fig. 3·6 is called *mesokurtic;* a curve that is more peaked like the one in Fig. 3·7 is called *leptokurtic;* and a curve that is less peaked, as in Fig. 3·8, is called *platykurtic.*

Now that we have briefly described these characteristics of frequency distributions, we can discuss in greater detail three common *measures of central tendency:* the *mean,* the *median,* and the *mode.*

FIG. 3·6
Mesokurtic curve (*meso-*, meaning "intermediate")

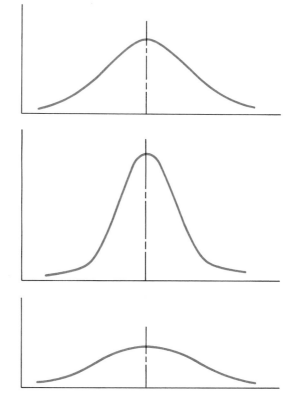

FIG. 3·7
Leptokurtic curve (*lepto-*, meaning "slender")

FIG. 3·8
Platykurtic curve (*platy-*, meaning "broad" or "flat")

EXERCISES

1. Draw examples of the following distributions, A and B:
   a) A: Symmetrical, mesokurtic, range from −1.0 to +1.0, central tendency of 0.0.
      B: Symmetrical, platykurtic, range from −1.5 to +1.5, central tendency of 0.0.
   b) A: Skewed left, mesokurtic, range from −1.0 to +1.0, peak at +0.5.
      B: Symmetrical, mesokurtic, range from −1.0 to +1.0, central tendency of 0.0.
   c) A: Symmetrical, leptokurtic, range from −0.5 to +0.5, central tendency of 0.0.
      B: Skewed right, leptokurtic, range from −1.0 to +1.0, peak at −0.5.

2. Draw three curves, all symmetrical and with the same dispersion, but with the following central locations:

   a) 0.0

   b) 1.0

   c) −1.0

3. Drawn below are four distribution curves. For each, indicate its peak, its degree of kurtosis, and whether it is symmetrical, positively skewed, or negatively skewed.

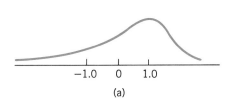

(a)

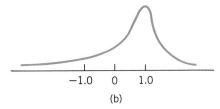

(b)

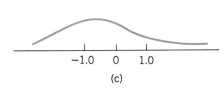

(c)

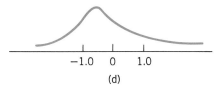

(d)

4. For the following distribution, indicate which distribution

   a) has the larger average value.

   b) is more likely to produce a small value than a large value.

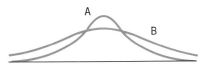

   For the next two distributions, indicate which distribution, if any

   c) has values most evenly distributed across the range of possible values.

   d) is more likely to produce a value near 0.

   e) has a greater likelihood of producing large values than small values.

5. If the following two curves represent the distribution of scores for a group of students on two tests, which test appears to be more difficult for the students, A or B? Explain.

TABLE 3·1
**Downtime of
Chapel Hill's
buses**

The arithmetic mean
is an average

2 A MEASURE OF CENTRAL TENDENCY: THE
ARITHMETIC MEAN   Most of the time when we refer to the "average" of something, we are talking about the arithmetic mean. This is true in cases such as the average winter temperature of New York City, the average life-time of a flashlight battery, and the average corn yield from an acre of land.

Bus number	1	2	3	4	5	6	7	8	9	10
Days out of service	7	23	4	8	2	12	6	13	9	4

Table 3·1 repeats the data from our chapter opening example. Data in the table represent the number of days the buses were out of operation, owing to regular maintenance or some malfunction. To find the arithmetic mean, we sum the values and divide by the number of observations:

$$\text{Arithmetic mean} = \frac{7 + 23 + 4 + 8 + 2 + 12 + 6 + 13 + 9 + 4}{10}$$

$$= \frac{88}{10}$$

$$= 8.8 \text{ days}$$

In this one year period the buses were out of operation for an average of 8.8 days. With this figure, the transportation director has a reasonable single measure of the behavior of *all* her buses.

Conventional symbols

Characteristics
of a sample:
statistics

To write equations for these measures of frequency distributions, we need to learn the mathematical notations used by statisticians. A *sample* of a population consists of n observations (a lower case n) with a mean of $\bar{x}$ (read *x-bar*). Remember that the measures we compute for a sample are called *statistics.*

Characteristics
of a population:
parameters

The notation is different when we are computing measures for the entire *population;* that is, for the group containing every element we are describing. The mean of a population is symbolized by μ, which is the Greek letter *mu.* The number of elements in a population is denoted by the capital letter N. Generally in statistics, we use Roman letters to symbolize sample information and Greek letters to symbolize population information.

Calculating the mean from ungrouped data

Finding
the population
and sample means

In the bus example, the average of 8.8 days would be μ (the population mean) if the population of buses is exactly 10. It would $\bar{x}$ (the sample mean) if the ten buses are a sample drawn from a larger population of buses. To write the formulas for these two means, we combine our mathematical symbols and the steps we used to determine the arithmetic mean. If we add the values of the observations and divide this sum by the number of observations, we will get:

Population mean   Sum of values of all observations

$$\mu = \frac{\Sigma x}{N}$$ [3·1]

Number of elements in the
population

and:

Sample mean   Sum of values of all observations

$$\bar{x} = \frac{\Sigma x}{n}$$ [3·2]

Number of elements in the
sample

Since μ is the *population arithmetic mean,* we use N to indicate that we divide by the number of observations or elements in the population. Similarly, $\bar{x}$ is the *sample arithmetic mean,* and n is the number of observations in the sample. The greek letter *sigma,* Σ, indicates that all the values of x are summed together.

TABLE 3·2
**Percentile
increase in S.A.T
verbal scores**

Student	1	2	3	4	5	6	7
Increase	9	7	7	6	4	4	2

Another example: Table 3·2 lists the percentile increase in S.A.T verbal scores shown by 7 different students taking an S.A.T preparatory course. The data are arrayed in descending order. We assume that there are too many students in the course to survey each one. Therefore, we use our sample and compute the mean as follows:

$$\bar{x} = \frac{\Sigma x}{n}$$ [3·2]

$$= \frac{9 + 7 + 7 + 6 + 4 + 4 + 2}{7}$$

$$= \frac{39}{7}$$

$$= 5.6 \text{ points per student} \longleftarrow \text{sample mean}$$

Notice that to calculate this mean, we added every observation separately, in no special order. Statisticians call this *ungrouped* data. The computations were not difficult because our sample size was small. But suppose we are dealing with the weight of 5,000 head of cattle and prefer not to add each of our data points separately. Or suppose we have access to only the

Dealing with
ungrouped data

frequency distribution of the data, not to every individual observation. In these cases, we will need a different way to calculate the arithmetic mean.

Calculating the mean from grouped data

A frequency distribution consists of data that are grouped by classes. Each value of an observation falls somewhere in one of the classes. Unlike the S.A.T example, we do not know the separate values of every observation. Suppose we have a frequency distribution (illustrated in Table 3·3) of average monthly checking account balances of 600 customers at a branch bank. From the information in this table, we can easily compute an *estimate* of the value of the mean of this grouped data. It is an estimate because we do not use all 600 data points in the sample. Had we used the original, ungrouped data, we could have calculated the actual value of the mean—but only after we had averaged the 600 separate values. For ease of calculation we must give up accuracy.

TABLE 3·3
Average monthly balances of 600 customers

Class (dollars)	Frequency
0– 49.99	78
50.00– 99.99	123
100.00–149.99	187
150.00–199.99	82
200.00–249.99	51
250.00–299.99	47
300.00–349.99	13
350.00–399.99	9
400.00–449.99	6
450.00–499.99	4
	600

To find the arithmetic mean of grouped data, we first calculate the midpoint of each class (the class mark) using the modified form of Equation 2·2. Then we multiply each class mark by the frequency of observations in that class, sum all these results, and divide the sum by the total number of observations in the sample. The formula looks like this:

$$\overline{x} = \frac{\Sigma(f \times x)}{n}$$

[3·3]

where:
$\overline{x}$ is the sample mean
Σ is the symbol meaning "the sum of"
f is the frequency (number of observations) in each class
x represents the class mark for each class in the sample
n is the number of observations in the sample

Table 3·4 illustrates how to calculate the arithmetic mean from our grouped data, using Equation 3·3.

Class (dollars) (1)	Class marks (x) (2)		Frequency (f) (3)		f × x (3) × (2)		TABLE 3·4
0– 49.99	25.00	×	78	=	1,950		**Calculation of arithmetic sample mean from grouped data in Table 3·3**
50.00– 99.99	75.00	×	123	=	9,225		
100.00–149.99	125.00	×	187	=	23,375		
150.00–199.99	175.00	×	82	=	14,350		
200.00–249.99	225.00	×	51	=	11,475		
250.00–299.99	275.00	×	47	=	12,925		
300.00–349.99	325.00	×	13	=	4,225		
350.00–399.99	375.00	×	9	=	3,375		
400.00–449.99	425.00	×	6	=	2,550		
450.00–499.99	475.00	×	4	=	1,900		

$$\Sigma f = n = 600 \qquad 85,350 \leftarrow \Sigma(f \times x)$$

$$\overline{x} = \frac{\Sigma(f \times x)}{n} \qquad [3 \cdot 3]$$

$$= \frac{85,350}{600}$$

$$= 142.25 \leftarrow \text{sample mean (dollars)}$$

In our sample of 600 customers, the average monthly checking account balance is $142.25. This is our approximation from the frequency distribution. Notice that since we did not know every data point in the sample, we assumed that every value in a class was equal to its class mark. Our results, then, can only approximate the actual average monthly balance.

Let's compare an approximate mean calculated from grouped data with an actual mean compiled from ungrouped data. Consider the example presented in Tables 3·5 and 3·6 recording the annual snowfall (in inches) over

Comparing the estimated mean with the actual mean

TABLE 3·5 **Annual snowfall in Harlan, Kentucky**

Year	Snowfall (inches)	Year	Snowfall (inches)
1960	23	1970	12
1961	8	1971	28
1962	14	1972	8
1963	31	1973	36
1964	5	1974	16
1965	26	1975	9
1966	11	1976	42
1967	27	1977	30
1968	32	1978	7
1969	46	1979	22

$$433 \leftarrow \text{total snowfall}$$

$$\overline{x} = \frac{\Sigma x}{n} \qquad [3 \cdot 2]$$

$$= \frac{433}{20}$$

$$= 21.65 \leftarrow \text{average annual snowfall}$$

the past 20 years in Harlan, Kentucky. If we use ungrouped data, the average annual snowfall is 21.65 inches. If we use grouped data, the estimated average is 21.5. The difference is small. And when the number of observations is, large, you will appreciate the convenience offered by using grouped data.

TABLE 3·6 **Annual snowfall in Harlan, Kentucky**

Class (grouped data) (1)	Class mark (x) (2)		Frequency (f) (3)		f × x (3) × (2)
0– 7	3.5	×	2	=	7.0
8–15	11.5	×	6	=	69.0
16–23	19.5	×	3	=	58.5
24–31	27.5	×	5	=	137.5
32–39	35.5	×	2	=	71.0
40–47	43.5	×	2	=	87.0
					430.0 ← $\Sigma(f \times x)$

$$\overline{x} = \frac{\Sigma(f \times x)}{n} \qquad [3 \cdot 3]$$

$$= \frac{430}{20}$$

$$= 21.5 \leftarrow \text{average annual snowfall}$$

Coding

Giving codes to the class marks

We can further simplify our calculation of the mean from grouped data. Using a technique called coding, we eliminate the problem of large or inconvenient class marks. Instead of using the actual class marks to perform our calculations, we can assign small-value consecutive integers (whole numbers) called *codes* to each of the class marks. The integer zero can be assigned anywhere, but to keep the integers small, we will assign zero to the class mark in the *middle* (or the one nearest to the middle) of the frequency distribution. Then we can assign negative integers to values smaller than that class mark and positive integers to those larger, as follows:

Class	1–5	6–10	11–15	16–20	21–25	26–30	31–35	36–40	41–45
Code (u)	−4	−3	−2	−1	0	1	2	3	4
					↑ x_0				

Calculating the mean from grouped data, using codes

Symbolically, statisticians use x_0 to represent the class mark that is assigned the code 0 and u for the coded class marks. The following formula is used to determine the sample mean using codes:

$$\overline{x} = x_0 + w \frac{\Sigma(u \times f)}{n} \qquad [3 \cdot 4]$$

where:

$\bar{x}$ = mean of the sample
x_0 = value of the class mark assigned the code 0
w = numerical width of the class interval
u = code assigned to each class
f = frequency or number of observations in each class
n = total number of observations in the sample

Keep in mind that $\Sigma(u \times f)$ simply means that we (1) multiply u by f for every class in the frequency distribution and (2) sum all of these products. Table 3·7 illustrates how to code the class marks and find the sample mean. The result is the same as it was when we calculated the mean from grouped data without coding (illustrated in Table 3·6).

TABLE 3·7 **Annual snowfall in Harlan, Kentucky**

Class (1)	Class mark (x) (2)	Code (u) (3)	Frequency (f) (4)	u × f (3) × (4)
0– 7	3.5	−2	× 2	= −4
8–15	11.5	−1	× 6	= −6
16–23	19.5	0	× 3	= 0 ← x_0
24–31	27.5	1	× 5	= 5
32–39	35.5	2	× 2	= 4
40–47	43.5	3	× 2	= 6

$$\Sigma f = n = 20 \qquad\qquad 5 \leftarrow \Sigma(u \times f)$$

$$\bar{x} = x_0 + w\,\frac{\Sigma(u \times f)}{n} \qquad [3\cdot4]$$

$$= 19.5 + (8)\left(\frac{5}{20}\right)$$

$$= 19.5 + 2$$

$$= 21.5 \leftarrow \text{average annual snowfall}$$

Advantages and disadvantages of the arithmetic mean

The arithmetic mean, as a single number representing a whole data set, has important advantages. First, its concept is familiar to most people and intuitively clear. Second, every data set has a mean. It is a measure that can be calculated, and it is unique because every data set has one and only one mean. Finally, the mean is useful for performing statistical procedures such as comparing the means from several data sets (a procedure we will carry out in Chapter 9).

Advantages of the mean

Yet, like any statistical measure, the arithmetic mean has disadvantages of which we must be aware. **First**, while the mean is reliable in that it reflects all the values in the data set, it may also be affected by extreme values that

Disadvantages of the mean

TABLE 3·8
**Times for track
team members in
a one-mile race**

Member	1	2	3	4	5	6	7
Time in minutes	4.2	4.3	4.7	4.8	5.0	5.1	9.0

are not representative of the rest of the data. Notice that if the seven members of a track team have times in a mile race shown in Table 3·8, the mean time is:

$$\mu = \frac{\Sigma x}{N} \qquad\qquad [3 \cdot 1]$$

$$= \frac{4.2 + 4.3 + 4.7 + 4.8 + 5.0 + 5.1 + 9.0}{7}$$

$$= \frac{37.1}{7}$$

$$= 5.3 \text{ minutes} \qquad \text{population mean}$$

If we compute a mean time for the first six members, however, and exclude the 9.0 value, the answer is about 4.7 minutes. The one *extreme value* of 9.0 distorts the value we get for the mean. It would be more representative to calculate the mean without including such an extreme value.

A **second** problem with the mean is the same one we encountered with our 600 checking account balances: it is tedious to compute the mean because we *do* use every data point in our calculation (unless, of course, we take the short-cut method of using grouped data to approximate the mean).

The **third** disadvantage is that we are unable to compute the mean for a data set that has open-ended classes at either the high or low end of the scale. Suppose the data in Table 3·8 had been arranged in the frequency distribution shown in Table 3·9. We could not compute a mean value for this data because of the open-ended class of "5.4 and above." We have no way of knowing whether the value is 5.4, near to 5.4, or far above 5.4.

TABLE 3·9
**Times for track
team members in
a one-mile race**

Class in minutes	4.2–4.5	4.6–4.9	5.0–5.3	5.4 and above
Frequency	2	2	2	1

EXERCISES

6. Compute the sample mean for the following sets of data:

   a) 10, 15, 16, 11, 18, 15, 13, 12,

   b) 1.472, 1.341, 1.403, 1.459, 1.299, 1.391, 1.430

   c) 314, 237, 557, 425, 518, 473, 490, 316, 375, 341, 423, 479

   d) 43.0, 48.7, 58.4, 40.9, 44.2, 43.6, 52.7, 48.6, 53.4, 46.5

7. Compute the sample mean for the following sets of grouped data using the class mark method.

a)

Class	200–224	225–249	250–274	275–299	300–324	325–349
Frequency	6	21	32	26	10	5

b)

Class	.95–1.04	1.05–1.14	1.15–1.24	1.25–1.34	1.35–1.44	1.45–1.54	1.55–1.64
Frequency	2	6	7	10	9	5	1

8. Using the following set of data:

a) Construct a frequency distribution using intervals 35–44, etc.

b) Compute the sample mean from the raw data.

c) Compute the sample mean from the frequency distribution.

d) Compare b and c.

```
95  76  72  67  69  48  37  76  74   60
78  80  48  86  59  68  73  77  51   82
94  95  48  58  75  69  55  51  89   91
89  93  69  81  68  49  86  74  79  100
```

9. From the frequency distribution below:

a) Compute the sample mean, using the class mark method.

b) Compute the sample mean, using the coding method and assigning 0 to the fourth class.

c) Compute the sample mean, using the coding method and assigning 0 to the sixth class.

d) Verify that a, b, and c are equal.

Class	Frequency	Class	Frequency
10.0–10.9	2	15.0–15.9	10
11.0–11.9	3	16.0–16.9	9
12.0–12.9	5	17.0–17.9	6
13.0–13.9	7	18.0–18.9	8
14.0–14.9	12	19.0–19.9	2

10. The twelve counties in the state showed the following public school enrollments last year:

Bradford	75,800	Dade	36,700
Duval	70,100	Kent	38,200
Meecham	45,500	Burton	31,500
Carter	45,500	Easton	33,000
Simpson	45,500	Fulcrum	30,750
Northwood	35,800	Pope	28,800

What was the mean enrollment for the counties last year?

11. A subject in a linguistics study was run on 18 trials of a task involving judgments of sentence ambiguity. The response times (in seconds) for the trials were as follows:

```
4.01  3.90  3.91  3.85  3.89  3.98  3.94  3.95  3.92
4.00  3.98  3.94  3.91  3.93  3.97  3.98  4.00  3.96
```

What is the mean response time?

12. A psychologist studying reading disabilities wishes to determine the average time that adults with no reading difficulties require to read aloud a short passage. He will compare

this average to that for adults who have reading disabilities. Using a stopwatch, the psychologist collects the following times in seconds:

20.3	19.9	22.1	23.7	21.2	25.0	21.1	22.8	28.1	24.2
21.9	24.6	25.6	24.8	22.6	24.3	24.2	23.5	23.1	20.9

What should he conclude is the average time to read the passage?

13. The price of the pound sterling on the Paris money market, measured in U.S. dollars, was recorded at the close of each day for two weeks. What was the average (mean) value for the pound during (a) the first week? (b) the second week? (c) the two-weeks' period?

Week 1	$1.973	$1.970	$1.972	$1.975	$1.976
Week 2	1.969	1.892	1.893	1.887	1.895

3 A SECOND MEASURE OF CENTRAL TENDENCY: THE WEIGHTED MEAN

A weighted average

The weighted mean enables us to calculate an average that takes into account the importance of each value to the overall total. Consider, for example, the company in Table 3·10, which uses three grades of labor—unskilled, semiskilled, and skilled—to produce two end products. The company wants to know the average cost of labor per hour for each of the products.

TABLE 3·10
Labor input in manufacturing process

		Labor hours per unit of output	
Grade of labor	Hourly wage (x)	Product 1	Product 2
Unskilled	$3.00	1	4
Semiskilled	5.00	2	3
Skilled	7.00	5	3

A simple arithmetic average of the labor wage rates would be:

$$\bar{x} = \frac{\Sigma x}{n} \qquad\qquad [3\cdot 2]$$

$$= \frac{\$3 + \$5 + \$7}{3}$$

$$= \frac{\$15}{3}$$

$$= \$5.00/\text{hour}$$

In this case, the arithmetic mean is incorrect

Using this average rate, we would compute the labor cost of one unit of product 1 to be $5(1 + 2 + 5) = $40, and of one unit of product 2 to be $5(4 + 3 + 3) = $50. But these answers are incorrect.

To be correct, the answers must take into account the fact that different amounts of each grade of labor are used. We can determine the correct answers in the following manner. For product 1, the total labor cost per unit is ($3 × 1) + ($5 × 2) + ($7 × 5) = $48, and, since there are 8 hours of labor input, the average labor cost per hour is $48/8 = $6.00 per hour. For product 2, the total labor cost per unit is ($3 × 4) + ($5 × 3) + ($7 × 3) = $48, for an average labor cost per hour of $48/10 or $4.80 per hour.

Another way to calculate the correct average cost per hour for the two products is to take a *weighted average* of the cost of the three grades of labor. To do this, we weight the hourly wage for each grade by its proportion of the total labor required to produce the product. One unit of product 1, for example, requires 8 hours of labor. Unskilled labor uses ⅛ of this time, semi-skilled labor uses ⅔ of this time, and skilled labor requires ⅝ of this time. If we use these fractions as our weights, then one hour of labor for product 1 costs an average of:

The correct answer is the weighted mean

$$\left(\frac{1}{8} \times \$3\right) + \left(\frac{2}{8} \times \$5\right) + \left(\frac{5}{8} \times \$7\right) = \$6.00/\text{hour}$$

Similarly, a unit of product 2 requires 10 labor hours, of which ⁴⁄₁₀ is used for unskilled labor, ³⁄₁₀ for semiskilled labor, and ³⁄₁₀ for skilled labor. Using these fractions as weights, one hour of labor for product 2 costs:

$$\left(\frac{4}{10} \times \$3\right) + \left(\frac{3}{10} \times \$5\right) + \left(\frac{3}{10} \times \$7\right) = \$4.80/\text{hour}$$

Thus, we see that the weighted averages give the correct values for the average hourly labor costs of the two products because they take into account the fact that different amounts of each grade of labor are used in the products.

Calculating the weighted mean

Symbolically, the formula for calculating the weighted average is:

$$\overline{x}_w = \frac{\Sigma(w \times x)}{\Sigma w} \qquad [3\cdot5]$$

where:
$\overline{x}_w$ = the symbol for the weighted mean*
  w = weight assigned to each observation (⅛, ⅔, and ⅝ for product 1 in our example).
$\Sigma(w \times x)$ = sum of the weight of each element times that element
  Σw = sum of all of the weights

---

*The symbol $\overline{x}_w$ is read *x-bar sub w*. The lower-case *w* is called a subscript and is a reminder that this is not an ordinary mean but one that is weighted according to the relative importance of the values of *x*.

If we apply Equation 3·5 to product 1 in our labor cost example, we find:

$$\overline{x}_w = \frac{\Sigma(w \times x)}{\Sigma w} \qquad [3 \cdot 5]$$

$$= \frac{(\frac{1}{8} \times \$3) + (\frac{2}{8} \times \$5) + (\frac{5}{8} \times \$7)}{\frac{1}{8} + \frac{2}{8} + \frac{5}{8}}$$

$$= \frac{\$6}{1}$$

$$= \$6.00/\text{hour}$$

The arithmetic mean of grouped data: the weighted mean

Notice that Equation 3·5 states more formally something we have done previously. When we calculated the arithmetic mean from grouped data (page 52), we actually found a weighted mean, using the class marks for the *x* values and the frequencies of each class as the weights. We divided this product by the sum of all the frequencies, which is the same as dividing by the sum of all the weights.

In like manner, *any* mean computed from all the values in a data set according to Equation 3·1 or 3·2 is really a weighted average of the components of the data set. What those components are, of course, determines what the mean measures. In a factory, for example, we could determine the weighted mean of all the wages (skilled, semiskilled, and unskilled), or of the wages of men workers, women workers, or union and nonunion members.

EXERCISES

14. A professor has decided to use a weighted average in figuring final grades for his seminar students. The homework average will count for 30 percent of a student's grade; the midterm, 20 percent; the final, 25 percent; the term paper, 15 percent; and quizzes, 10 percent. From the data below, compute the final average for the five students in the seminar.

Student	Homework	Quizzes	Paper	Midterm	Final
1	85	89	94	87	90
2	78	84	88	91	92
3	94	88	93	86	89
4	82	79	88	84	93
5	95	90	92	82	88

15. Given the following prices and the number of each item sold, find the average price of the items sold.

Price	$1.29	$2.95	$3.49	$5.00	$7.50	$10.95
Number sold	7	9	12	8	6	3

16. From the following, find the average number of children per family in a certain city:

Number of children in family	0	1	2	3	4	5	6
Frequency	998	983	1,417	727	294	236	210

17. The state commission in charge of educational budgeting is expanding the budget for the coming year in the following areas: Research and Evaluation, with a current budget of

$57.5 million, will increase by 7.15 percent; Personnel, with a budget of $193.8 million, will increase by 7.25 percent; and Curriculum Development, with a current budget of $79.3 million, will increase its budget by 8.20 percent. What is the average rate of increase among these three areas of the educational budget?

18. The U.S. Postal Service handles 7 basic types of letters and cards: third class, second class, first class, air mail, special delivery, registered, and certified. The mail volume during the past year is given in the following table:

Type of mailing	Ounces delivered (in millions)	Price per ounce
Third class	15,500	$.05
Second class	23,900	.08
First class	79,100	.13
Air mail	1,800	.17
Special delivery	1,200	.35
Registered	800	.40
Certified	700	.45

What was the average revenue per ounce for these services during the year?

4 A THIRD MEASURE OF CENTRAL TENDENCY: THE GEOMETRIC MEAN

Finding the growth rate: the geometric mean

Sometimes when we are dealing with quantities that change over a period of time, we need to know an average rate of change, such as an average growth rate over a period of several years. In such cases, the simple arithmetic mean is inappropriate because it gives the wrong answers. What we need to find is the *geometric mean,* called simply the G.M.

Consider, for example, the growth of a savings account. Suppose we deposit $100 initially and let it accrue interest at varying rates for five years. The growth is summarized in Table 3·11.

Year	Interest rate	Growth factor	Savings at end of year
1	7%	1.07	$107.00
2	8	1.08	115.56
3	10	1.10	127.12
4	12	1.12	142.37
5	18	1.18	168.00

TABLE 3·11 Growth of $100 deposit in a savings account

The entry labeled "growth factor" is equal to:

$$1 + \frac{\text{interest rate}}{100}$$

The growth factor is the amount by which we multiply the savings at the beginning of the year to get the savings at the end of the year. The simple arithmetic mean growth factor would be (1.07 + 1.08 + 1.10 + 1.12 + 1.18)/ 5 = 1.11, which corresponds to an average interest rate of 11 percent per

In this case, the arithmetic mean growth rate is incorrect

year. If the bank gives interest at a constant rate of 11 percent per year, however, a $100 deposit would grow in five years to:

$$\$100 \times 1.11 \times 1.11 \times 1.11 \times 1.11 \times 1.11 = \$168.51$$

The correct answer

Table 3·11 shows that the actual figure is only $168.00. Thus, the correct average growth factor must be slightly less than 1.11.

Calculating
the geometric mean

To find the correct average growth factor, we can multiply together the five years' growth factors and then take the fifth root of the product—the number that, when multiplied by itself four times, is equal to the product we started with. The result is the *geometric mean growth rate,* which is the appropriate average to use here. The formula for finding the geometric mean of a series of numbers is:

Number of x values

$$\text{G.M.} = \sqrt[n]{\text{Product of all the } x \text{ values}} \qquad [3·6]$$

If we apply this equation to our savings account problem, we can determine that 1.1093 is the correct average growth factor.

$$
\begin{aligned}
\text{G.M.} &= \sqrt[n]{\text{Product of all the } x \text{ values}} \qquad [3·6]\\
&= \sqrt[5]{1.07 \times 1.08 \times 1.10 \times 1.12 \times 1.18}\\
&= \sqrt[5]{1.679965}\\
&= 1.1093 \leftarrow \text{average growth factor}
\end{aligned}
$$

Warning: use the
appropriate mean

Notice that the correct average interest rate of 10.93 percent per year obtained with the geometric mean is very close to the incorrect average rate of 11 percent obtained with the arithmetic mean. This happens because the interest rates are relatively small. Be careful, however, not to be tempted to use the arithmetic mean instead of the more complicated geometric mean. The following example demonstrates why.

In highly inflationary economies, banks must pay high interest rates to attract savings. Suppose that over five years in an unbelievably inflationary economy, banks pay interest at annual rates of 100, 200, 250, 300, and 400 percent, which correspond to growth factors of 2, 3, 3.5, 4, and 5. (We've calculated these growth factors just as we did in Table 3·11).

In five years, an initial deposit of $100 would grow to $100 × 2 × 3 × 3.5 × 4 × 5 = $42,000. The arithmetic mean growth factor is (2 + 3 + 3.5 + 4 + 5)/5, or 3.5. This corresponds to an average interest rate of 250 percent. Yet if the banks actually give interest at a constant rate of 250 percent per year, then $100 would grow to $52,521.88 in five years:

$$\$100 \times 3.5 \times 3.5 \times 3.5 \times 3.5 \times 3.5 = \$52,521.88$$

This answer exceeds the actual $42,000 by more than $10,500, a sizeable error.

Let's use the formula for finding the geometric mean of a series of numbers to determine the correct growth factor:

$$\text{G.M.} = \sqrt[n]{\text{Product of all the } x \text{ values}} \qquad [3 \cdot 6]$$
$$= \sqrt[5]{2 \times 3 \times 3.5 \times 4 \times 5}$$
$$= \sqrt[5]{420}$$
$$= 3.347 \leftarrow \text{average growth factor}$$

This growth factor corresponds to an average interest rate of 235 percent per year. In this case the use of the appropriate mean does make a significant difference.

EXERCISES

19. From the information below, calculate the average percentage increase in employee pay over the last 5 years.

1973	1974	1975	1976	1977
5%	10.5%	9.0%	6.0%	7.5%

20. The growth of sales for a small store over the last several years is given below. Calculate the average growth rate over this time period.

1972	1973	1974	1975	1976	1977	1978
.11	.09	.075	.08	.095	.108	.120

21. Given the following enrollments over a number of years at a university, calculate the average increase in enrollment over the time span (in terms of percent increase).

1969	1970	1971	1972	1973
12,500	13,250	14,310	15,741	17,630

22. If the geometric mean of a set of 6 values is 1.24, and 5 of the values are 1.18, 1.32, 1.27, 1.15, and 1.22, find the last of the 6 values.

23. Over a three-months' period, a home owner purchased $60 worth of fertilizer for his garden in 3 equal purchases of $20 each. The first batch of fertilizer was $1.00 per pound; the second, $1.10; the third, $1.15. What was the average price per pound paid for all the fertilizer?

24. Mary Carter is with a consulting firm that conducts testing and evaluation programs for schools. She has recently been asked to head a project involving the testing of three schools that implemented special reading programs last year. They have agreed to charge 10 percent over the average cost of testing each classroom. Each of three agency employees was assigned to a school and given $1,000 for materials and time. The average classroom costs involved in conducting the testing at the three schools were $100.00, $111.11, and $125.00, differing as a result of classroom sizes and other variables. What price per classroom should the firm charge?

25. The U.S. Public Health Service calculates that costs of health care have increased from $43.00 to $46.50 to $49.80 to $53.65 per patient during the last 4 years; yet the budget for the service has remained the same during each of these years. What has been the average cost per patient over the 4-year period?

5 A FOURTH MEASURE OF CENTRAL
TENDENCY: THE MEDIAN

The median is a measure of central tendency different from any of the means we have discussed so far. The median is a single value from the data set that measures the central item in the data. This single item is the *middlemost* or *most central* item in the set of numbers. Half of the items lie above this point, and the other half lie below it.

Calculating the median from ungrouped data

To find the median of a data set, first array the data in ascending or descending order. If the data set contains an *odd* number of items, the middle item of the array is the median. If there is an *even* number of items, the median is the average of the two middle items. In formal language, the median is:

Number of items in the array

$$\text{Median} = \text{the } \left(\frac{n + 1}{2} \right) \text{th item in a data array} \qquad [3 \cdot 7]$$

Suppose we wish to find the median of seven items in a data array. According to Equation 3·7, the median is the $(7 + 1)/2 = $ 4th item in the array. If we apply this to our previous example of the times for seven members of a track team, we discover that the fourth element in the array is 4.8 minutes. This is the median time for the track team. Notice that unlike the arithmetic mean we calculated earlier, the median we calculated in Table 3·12 was *not* distorted by the presence of the last value (9.0). This value could have been 15.0 or even 45.0 minutes, and the median would have been the same!

TABLE 3·12
Times for track
team members

Item in data array	1	2	3	4	5	6	7
Time in minutes	4.2	4.3	4.7	4.8	5.0	5.1	9.0

median

Now let's calculate the median for an array with an even number of items. Consider the data shown in Table 3·13 concerning the number of patients treated daily in the emergency room of a hospital. The data are arrayed in descending order. The median of this data set would be:

$$\text{Median} = \text{the } \left(\frac{n + 1}{2} \right) \text{th item in a data array} \qquad [3 \cdot 7]$$

$$= \frac{8 + 1}{2}$$

$$= 4.5\text{th item}$$

TABLE 3·13
Patients treated in emergency room on 8 consecutive days

Item in data array	1	2	3	4	5	6	7	8
Number of patients	86	52	49	43	35	31	30	11

↑
median of 39

Since the median is the 4.5th element in the array, we need to average the fourth and fifth elements. The fourth element in Table 3·13 is 43, and the fifth is 35. The average of these two elements is equal to (43 + 35)/2, or 39. Therefore, 39 is the median number of patients treated in the emergency room per day during the eight-day period.

Calculating the median from grouped data

Often, we have access to data only after it has been grouped in a frequency distribution. We do not, for example, know every observation that led to the construction of Table 3·14, the data on 600 bank customers originally introduced earlier. Instead, we have ten class intervals and a record of the frequency with which the observations appear in each of the intervals.

Finding the median of grouped data

TABLE 3·14
Average monthly balances for 600 customers

Class in dollars	Frequency
0– 49.99	78
50.00– 99.99	123
100.00–149.99	187← median class
150.00–199.99	82
200.00–249.99	51
250.00–299.99	47
300.00–349.99	13
350.00–399.99	9
400.00–449.99	6
450.00–499.99	4
	600

Nevertheless, we can compute the median checking account balance of these 600 customers by determining which of the ten class intervals *contains* the median. To do this, we must add the frequencies in the frequency column in Table 3·14 until we reach the $(n + 1)/2$th item. Since there are 600 accounts, the value for $(n + 1)/2$ is 300.5 (the average of the 300th and 301st items). The problem is to find the class intervals containing the 300th and 301st elements. The cumulative frequency for the first two classes is only $78 + 123 = 201$. But when we move to the third class interval, 187 elements are added to 201 for a total of 388. Therefore, the 300th and 301st observations must be located in this third class (the interval from $100.00–$149.99).

Locate the median class

The *median class* for this data set contains 187 items. If we assume that these 187 items begin at $100.00 and are *evenly spaced over the entire class*

Interpolate to find the median

65

interval from $100.00 to $149.99, then we can interpolate and find values for the 300th and 301st items. First, we determine that the 300th item is the 99th element in the median class:

$$300 - 201 \text{ [items in the first two classes]} = 99$$

and that the 301st item is the 100th element in the median class:

$$301 - 201 = 100$$

Then we can calculate the *width* of the 187 equal steps from $100.00 to $149.99, as follows:

First item of next class   **First item of median class**

$$\frac{\$150.00 - \$100.00}{187} = \$.267 \text{ in width}$$

Now, if there are 187 steps of $.267 each and if 98 steps will take us to the 99th item, then the 99th item is:

$$(\$.267 \times 98) + \$100 = \$126.17$$

and the 100th item is one additional step:

$$\$126.17 + \$.267 = \$126.44$$

Therefore, we can use $126.17 and $126.44 as the values of the 300th and 301st items, respectively.

The actual median for this data set is the value of the 300.5th item, that is, the average of the 300th and 301st items. This average is:

$$\frac{\$126.17 + \$126.44}{2} = \$126.30$$

This figure ($126.30) is the median monthly checking account balance, as estimated from the grouped data in Table 3·14.

Steps for finding
the median
of grouped data

In summary, we can calculate the median of grouped data as follows:

1. Use Equation 3·7 to determine which element in the distribution is center-most (in this case, the average of the 300th and 301st items).
2. Add the frequencies in each class to find the class that contains that center-most element (the third class, or $100.00 − $149.99).
3. Determine the number of elements in the class (187) and the location in the class of the median element (item 300 was the 99th element; item 301, the 100th element).

4. Learn the width of each step in the median class by dividing the class interval by the number of elements in the class (width = $.267).

5. Determine the number of steps from the lower bound of the median class to the appropriate item for the median (98 steps for the 99th element: 99 steps for the 100th element).

6. Calculate the estimated value of the median element by multiplying the number of steps to the median element times the width of each step and by adding the result to the lower bound of the median class ($100 + 98 × $.267 = $126.17; $126.17 + $.267 = $126.44).

7. If, as in our example, there is an even number of elements in the distribution, average the values of the median element calculated in step #6 ($126.30).

To shorten this procedure, statisticians use an equation to determine the median of grouped data. For a sample, this equation would be:

An easier method

Sample median

$$\tilde{m} = \left(\frac{(n + 1)/2 - (F + 1)}{f_m}\right) w + L_m \qquad [3\cdot8]$$

where:
$\tilde{m}$ = sample median
n = total number of items in the distribution
F = sum of all the class frequencies *up to*, but *not including*, the median class
f_m = frequency of the median class
w = class interval width
L_m = lower limit of the median class interval

If we use Equation 3·8 to compute the median of our sample of checking account balances, then $n = 600$, $F = 201$, $f_m = 187$, $w = 50, and $L_m = 100.

$$\tilde{m} = \left(\frac{(n + 1)/2 - (F + 1)}{f_m}\right) w + L_m \qquad [3\cdot8]$$

$$= \left(\frac{601/2 - 202}{187}\right) $50 + $100$$

$$= \left(\frac{300.5 - 202}{187}\right) $50 + $100$$

$$= \left(\frac{98.5}{187}\right) $50 + $100$$

$$= (.527)($50) + $100$$

$$= $126.35 \leftarrow \text{estimated sample median}$$

The slight difference between this answer and our answer calculated the long way is due to rounding.

Advantages and disadvantages of the median

The median has several advantages over the mean. The most important, demonstrated in our track team example in Table 3·12 on page 64, is that extreme values do not affect the median as strongly as they do the mean. The median is easy to understand and can be calculated from any kind of data—even for grouped data with open-ended classes such as the frequency distribution in Table 3·9—*unless* the median falls into an open-ended class.

Advantages of the median

We can find the median even when our data are qualitative descriptions like color or sharpness, rather than numbers. Suppose, for example, we have five runs of a printing press, the results from which must be rated according to sharpness of the image. We can array the results from best to worst: extremely sharp, very sharp, sharp, slightly blurred, and very blurred. The median of the five ratings is the (5 + 1)/2, or third rating (sharp).

Disadvantages of the median

The median has some disadvantages as well. Certain statistical procedures that use the median are more complex than those that use the mean. Also, because the median is an average of position, we must array the data before we can perform any calculations. This is time-consuming for any data set with a large number of elements. Therefore, if we want to use a sample statistic as an estimate of a population location parameter, the mean is easier to use than the median. Chapter 8 will discuss estimation in detail.

EXERCISES

26. Calculate the median of the following data set:

810	444	748	593	762	729	528
1,190	843	824	858	802	579	485
622	734	432	500	524	733	676
733	604	555	484	740	831	673
485	605	803	881	720	657	609

27. Find the median of the following set of data:

1.08	.98	.97	1.10	1.03	1.13	1.07	1.24	.99	1.13
.99	1.43	1.18	1.02	1.12	1.17	.98	1.28	.98	1.09

28. For the frequency distribution below, determine

a) which is the median class

b) which number item represents the median item

c) the width of the equal steps in the median class

d) the estimated value of the median for these data

Class	Frequency	Class	Frequency
100–149.5	12	300–349.5	72
150–199.5	14	350–399.5	63
200–249.5	27	400–449.5	36
250–299.5	58	450–499.5	18

29. For the following data, calculate an estimate of the median using Equation 3·8:

Class	Frequency	Class	Frequency
0–24.9	6	75–99.9	16
25–49.9	11	100–124.9	13
50–74.9	14	125–149.9	10

30. Calculate the median for the data given in the following table.

Minutes for bus ride from O'Hare Airport to John Hancock Center

15	15	16	16	17	17	17	18	18	18
18	18	18	18	19	20	20	21	22	25
26	27	27	27	27	27	28	28	29	29
30	30	31	31	33	33	33	34	34	34
34	34	34	34	35	35	35	35	35	36
37	38	40	43	49	50	51	53	58	64

31. An anthropologist studying some remains from an ancient civilization found a number of bones resembling the radius (one of the forearm bones) in humans today. The bones were of varying lengths, probably as a result of differences in the ages of the individuals at the time of death. Calculate the median bone length (in inches) from the following data:

$5\frac{1}{4}$	$6\frac{1}{4}$	6	$7\frac{7}{8}$	$9\frac{1}{4}$	$9\frac{1}{2}$	$10\frac{1}{2}$
$5\frac{3}{8}$	6	$6\frac{1}{4}$	8	$9\frac{1}{2}$	$9\frac{7}{8}$	$10\frac{1}{4}$
$5\frac{1}{2}$	$5\frac{7}{8}$	$6\frac{1}{2}$	$8\frac{1}{4}$	$9\frac{3}{8}$	$10\frac{1}{4}$	$10\frac{1}{8}$
$5\frac{7}{8}$	$5\frac{3}{4}$	7	$8\frac{1}{2}$	$9\frac{1}{8}$	$10\frac{1}{2}$	$10\frac{1}{8}$
6	$5\frac{7}{8}$	$7\frac{1}{2}$	9	$9\frac{1}{4}$	$10\frac{7}{8}$	10

32. If insurance claims for automobile accidents follow the distribution given below, determine the median using the method outlined on page 66 in this chapter. Verify that you get the same answer using Equation 3·8.

Amount of claim ($)	Frequency	Amount of claim ($)	Frequency
less than 150	52	350–399.99	816
150–199.99	108	400–449.99	993
200–249.99	230	450–499.99	825
250–299.99	528	500 and above	650
300–349.99	663		

6 A FINAL MEASURE OF CENTRAL TENDENCY: THE MODE

The mode is a measure of central tendency that is different from the mean but somewhat like the median because it is not actually calculated by the ordinary processes of arithmetic. The mode is *that value that is repeated most often in the data set.*

Mode defined

As in every other aspect of life, chance can play a role in the arrangement of data. Sometimes chance causes a single unrepresentative item to be repeated often enough to be the most frequent value in the data set. For

Limited use of mode of ungrouped data

this reason, we rarely use the mode of ungrouped data as a measure of central tendency. Table 3·15, for example, shows the number of billing errors per day made by an office of a local hospital. The modal value is 15 because it occurs more often than any other value (three times). A mode of 15 implies that the hospital's performance record on billing errors is poorer than 6.7 (6.7 is the answer we'd get if we calculated the mean). The mode tells us that 15 is the most frequent number of errors, but it fails to let us know that most of the values are under 10.

TABLE 3·15
Hospital billing errors per day in one 20-day period

Errors arrayed in ascending order				
0	2	5	7	15
0	2	5	7	15 ← mode
1	4	6	8	15
1	4	6	12	19

Finding the modal class of grouped data

Now let's group this data into a frequency distribution, as we have done in Table 3·16. If we select the class with the most observations, which we can call the *modal class,* we would choose "4–7" errors. This class is more representative of the accuracy of the billing department than is the mode of 15 errors per day. For this reason, whenever we use the mode as a measure of the central tendency of a data set, we should calculate the mode from grouped data.

TABLE 3·16
Billing errors per day

Class in number of errors	0–3	4–7	8–11	12 and more
Frequency	6	8	1	5

↑
modal class

The mode in symmetrical and skewed distributions

Let's study Figs. 3·9 through 3·11, each of which shows a frequency distribution. Figure 3·9 is symmetrical; Fig. 3·10 is skewed to the right; and Fig. 3·11 is skewed to the left.

Location of the mode

In Fig. 3·9, where the distribution is symmetrical and there is only one mode, the three measures of central tendency—the mode, median, and mean—coincide with the highest point on the graph. In Fig. 3·10, the data

FIG. 3·9

Symmetrical distribution, showing that the mean, median, and mode coincide

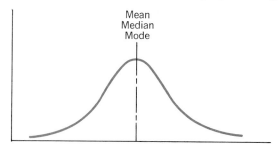

Mean
Median
Mode

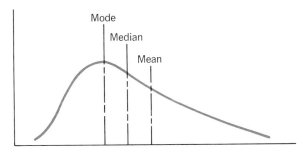

FIG. 3·10
Distribution is skewed to the right

set is skewed to the right. Here, the mode is still at the highest point on the graph, but the median lies to the right of this point and the mean falls to the right of the median. When the distribution is skewed to the left as in Fig. 3·11, the mode is at the highest point on the graph, the median lies to the left of the mode, and the mean falls to the left of the median. No matter what the shape of the curve, the mode is always located at the highest point.

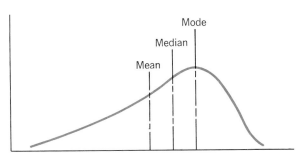

FIG. 3·11
Distribution is skewed to the left

Calculating the mode from grouped data

When our data are already grouped in a frequency distribution, we must assume that the mode is located in the class with the most items, that is, with the highest frequency. But how can we determine a single value for the mode from this modal class? Two methods are available to us. The first enables us to estimate the mode from a graph. The second method uses an equation.

Finding the mode in the modal class

To demonstrate these two ways of finding the mode in grouped data, let's use the data in Table 3·14 on page 65 (our example of the checking account balances). First, we can construct a histogram of the data as shown in Fig. 3·12. Then, since the modal class is the tallest rectangle, we can locate the mode in it by:

1. Drawing a line from the top right corner of the tallest rectangle to the top right corner of the rectangle to its immediate left.
2. Drawing a second line from the top left corner of the tallest rectangle to the top left corner of the rectangle to its immediate right.
3. Drawing a line perpendicular to the horizontal axis through the point where the lines drawn in steps 1 and 2 cross.

A graphical solution

FIG. 3·12
Calculation of the
mode from
grouped data
using graph
method

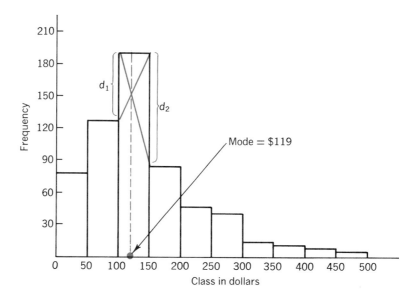

Mode = $119

The value on the horizontal axis marked by the line drawn in step 3 will approximate the modal value. In this case, the mode is about 119.

A mathematical solution

A second way of finding the mode in grouped data is to use Equation 3·9:

Mode

$$Mo = L_{Mo} + \frac{d_1}{d_1 + d_2} w \qquad [3·9]$$

where:

L_{Mo} = lower limit of the modal class

d_1 = frequency of the modal class minus the frequency of the class *directly below it*

d_2 = frequency of the modal class minus the frequency of the class *directly above it*

w = width of the modal class interval

If we use Equation 3·9 to compute the mode of our checking account balances, then L_{Mo} = $100, d_1 = 187 − 123 = 64, d_2 = 187 − 82 = 105, and w = $50.

$$Mo = L_{Mo} + \frac{d_1}{d_1 + d_2} w \qquad [3·9]$$

$$= \$100 + \frac{64}{64 + 105} \$50$$

$$= \$100 + (.38)(\$50)$$

$$= \$100 + \$19$$

$$= \$119.00 \leftarrow \text{Mode}$$

Our answer of $119 is the estimate of the mode using either the graphic or the mathematical method of calculation.

Multimodal distributions

What happens when we have two different values that *each* appear the greatest number of times of any values in the data set? Table 3·17 shows the billing errors for a 20-day period in another hospital office. Notice that both 1 and 4 appear the greatest number of times in the data set. They each appear three times. This distribution, then, has two modes and is called a *bimodal distribution.*

Bimodal distributions

Errors arrayed in ascending order			
0	2	6	9
0	4 ⎫	6	9
1 ⎫	4 ⎬ ←mode	7	10
1 ⎬ ←mode	4 ⎭	8	12
1 ⎭	5	8	12

TABLE 3·17
Billing errors per day in 20-day period

In Fig. 3·13, we have graphed the data in Table 3·17. Notice that there are *two* highest points on the graph. They occur at the values 1 and 4 billing errors. The distribution in Fig. 3·14 is also called bimodal, even though the two highest points are not equal. Clearly these points stand out above the neighboring values in the frequency with which they are observed.

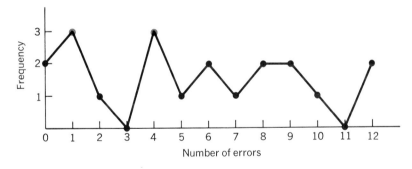

FIG. 3·13
Data in Table 3·17, showing bimodal distribution

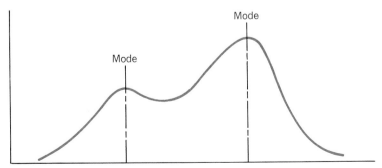

FIG. 3·14
Bimodal distribution with two unequal modes

Advantages and disadvantages of the mode

Advantages
of the mode

The mode, like the median, can be used as a central location for qualitative as well as quantitative data. If a printing press turns out five impressions, which we rate "very sharp," "sharp," "sharp," "sharp," and "blurred," then the modal value is "sharp." Similarly, we can talk about modal styles when, for example, furniture customers prefer Early American furniture to other styles.

Also like the median **the mode is not unduly affected by extreme values.** Even if the high values are very high and the low values very low, we choose the most frequent value of the data set to be the modal value. We can use the mode no matter how large, how small, or how spread out the values in the data set happen to be.

A third advantage of the mode is that we can use it even when one or more of the classes are open-ended. Notice, for example, that Table 3·16 on page 70 contains the open-ended class "12 errors and more."

Disadvantages
of the mode

Despite these advantages, the mode is not used as often to measure central tendency as are the mean and median. Too often, there is no modal value because the data set contains no values that are repeated more than once. Other times, every value is the mode because every value occurs the same number of times. Clearly, the mode is a useless measure in these cases. Another disadvantage is that when data sets contain two, three, or many modes, they are difficult to interpret and compare.

EXERCISES

33. Find:

   a) The modal class for the following data set

Class	20–23	24–27	28–31	32–35	36–39	40–43	44–47
Frequency	3	4	7	15	12	6	2

   b) The mode of the following sample: 5, 8, 11, 9, 8, 6, 8, 7, 12, 8, 7, 7, 11, 8, 6, 10, 13, 7, 8

34. Estimate the modal value of the following distribution by:

   a) The graphical method   b) Formula 3·9

Class	Frequency	Class	Frequency
48–51.9	2	64–67.9	56
52–55.9	8	68–71.9	28
56–59.9	20	72–75.9	4
60–63.9	32		

35. What are the modal values for the following distributions?

a) *Hair color*	*Frequency*	b) *Blood type*	*Frequency*
Black	11	AB	4
Brunette	24	O	12
Redhead	6	A	35
Blonde	18	B	16

c)

Day of birth	Frequency	Day of birth	Frequency
Monday	22	Friday	13
Tuesday	10	Saturday	32
Wednesday	32	Sunday	14
Thursday	17		

36. For the following data:

   a) Find the mode of the data.

   b) Construct a frequency distribution with intervals 10–14.9, 15–19.9, etc.

   c) Estimate the modal value using equation 3·9.

   d) Compare a and c.

```
19  15  14  11  20  13  17  24  12  20
13  19  25  15  19  20  15  12  13  16
18  16  15  26  21  11  19  20  11  24
16  16  17  18  16  11  10  27  18  13
```

37. Estimate the mode for the distribution given in problem 32.

38. There are many different types of solar heating systems available to the public. Depending on the type of system, heat can be stored for varying lengths of time while the sun is not shining. The following frequency distribution gives the heat storage capacity in days for 20 systems that were tested.

Days	0–0.99	1–1.99	2–2.99	3–3.99	4–4.99	5–5.99	6 or more
Frequency	2	5	2	3	5	2	1

Estimate the modes for the distribution.

7 COMPARING THE MEAN, MEDIAN, AND MODE

Mean, median, and mode are identical in symmetrical distribution

When we work statistical problems, we must decide whether to use the mean, the median, or the mode as the measure of central tendency. Symmetrical distributions that contain only one mode always have the same value for the mean, the median, and the mode, as we illustrated in Fig. 3·9 on page 70. In these cases, we need not choose the measure of central tendency because the choice has been made for us.

In a positively skewed distribution (one skewed to the right, such as the one in Fig. 3·10), the values are concentrated at the left end of the horizontal axis. Here, the mode is at the highest point of the distribution; the median is to the right of that; and the mean is to the right of both the mode and the median. In a negatively skewed distribution, such as in Fig. 3·11, the values are concentrated at the right end of the horizontal axis. The mode is at the highest point of the distribution, and the median is to the left of that. The mean is to the left of both the mode and the median.

The median may be best in skewed distributions

When the population is skewed negatively or positively, the median is often the best measure of location because it is always between the mean and the mode. The median is not as highly influenced by the frequency of occurrence of a single value as is the mode, nor is it pulled by extreme values as is the mean.

Otherwise, there are no universal guidelines for applying the mean, median, or mode as the measure of central tendency for different populations. Each case must be judged independently, according to the guidelines we have discussed.

EXERCISES

39. When the distribution of data is symmetrical and bell-shaped, selection of a measure of location is considerably simplified. Why?

40. For which type of distribution (positively skewed, negatively skewed, or symmetric) is

    a) the mean less than the median?

    b) the mode less than the mean?

    c) the median less than the mode?

8 TERMS INTRODUCED IN CHAPTER 3

- **BIMODAL DISTRIBUTION** A distribution of data points in which two values occur more frequently than the rest of the values in the data set.

- **CODING** A method of calculating the mean for grouped data by recoding values of class marks to more simple values.

- **GEOMETRIC MEAN** A measure of central tendency used to measure the average rate of change or growth for some quantity, computed by taking the nth root of the product of n values representing change.

- **KURTOSIS** The degree of peakedness of a distribution of points.

- **LEPTOKURTIC** A strongly peaked distribution.

- **MEAN**   A central tendency measure representing the arithmetic average of a set of observations.

- **MEASURE OF CENTRAL TENDENCY**   A measure indicating the value to be expected of a typical or middle data point.

- **MEASURE OF DISPERSION** A measure describing how scattered or spread out the observations in a data set are.

- **MEDIAN** The middle point of a data set, a measure of location that divides the data set into halves.

- **MEDIAN CLASS** The class in a frequency distribution that contains the median value for a data set.

- **MESOKURTIC** A moderately peaked distribution.

- **MODE** The value most often repeated in the data set. It is represented by the highest point in the distribution curve of a data set.

- **PARAMETERS** Numerical values that describe the characteristics of a whole population, commonly represented by Greek letters.

- **PLATYKURTIC**   A slightly peaked distribution.

- **SKEWNESS** The extent to which a distribution of data points is concentrated at one end or the other; the lack of symmetry.

- **STATISTICS** Numerical measures describing the characteristics of a sample.

- **SUMMARY STATISTICS**   Single numbers that describe certain characteristics of a data set.

- **SYMMETRICAL**   A characteristic of a distribution in which each half is the mirror image of the other half.

- **WEIGHTED MEAN** An average calculated to take into account the importance of each value to the overall total; i.e., an average in which each observation value is weighted by some index of its importance.

p. 51:

$$\mu = \frac{\Sigma x}{N}$$

[3·1]

The *population arithmetic mean* is equal to the sum of the values of all the elements in the population (Σx) divided by the number of elements in the population (N).

p. 51:

$$\bar{x} = \frac{\Sigma x}{n}$$

[3·2]

To derive the *sample arithmetic mean,* sum the values of all the elements in the sample (Σx) and divide by the number of elements in the sample (n).

p. 52:

$$\bar{x} = \frac{\Sigma(f \times x)}{n}$$

[3·3]

To find the *sample arithmetic mean of grouped data,* calculate the class marks (x) for each class in the sample. Then multiply each class mark by the frequency (f) of observations in that class, sum (Σ) all these results, and divide by the total number of observations in the sample (n).

p. 54:

$$\bar{x} = x_0 + w\frac{\Sigma(u \times f)}{n}$$

[3·4]

This formula enables us to calculate the *sample arithmetic mean of grouped data* using codes to eliminate dealing with large or inconvenient class marks. Assign these codes (u) as follows: give the value of zero to the middle class mark (called x_0), positive consecutive integers to class marks larger than x_0, and negative consecutive integers to smaller class marks. Then, multiply the code assigned to each class (u) by the frequency (f) of observations in each class and sum (Σ) all of these products. Divide this result by the total number of observations in the sample (n), multiply by the numerical width of the class interval (w), and add the value of the class mark assigned the code zero (x_0).

p. 59:

$$\bar{x}_w = \frac{\Sigma(w \times x)}{\Sigma w}$$

[3·5]

The *weighted mean,* $\bar{x}_w$, is an average that takes into account how important each value is to the overall total. We can calculate this average by multiplying the weight, or proportion, of each element (w) times that element (x), summing the results (Σ), and dividing this amount by the sum of all the weights (Σw).

p. 62:

$$G.M. = \sqrt[n]{\text{Product of all the } x \text{ values}}$$

[3·6]

The *geometric mean,* or G.M., is appropriate to use whenever we need to measure the average rate of change (the growth rate) over a period of time. In this equation, n is equal to the number of x values dealt with in the problem.

p. 64:
$$\text{Median} = \text{the } \left(\frac{n + 1}{2}\right) \text{ th item in a data array}$$

[3·7]

where: n = the number of items in the data array

The *median* is a single value that measures the central item in the data set. Half the items lie above the median, half below it. If the data set contains an odd number of items, the middle item of the array is the median. For an even number of items, the median is the average of the two middle items. Use this formula when the data is ungrouped.

p. 67:
$$\tilde{m} = \left(\frac{(n + 1)/2 - (F + 1)}{f_m}\right) w + L_m$$

[3·8]

This formula enables us to find the *sample median of grouped data.* In it, n equals the total number of items in the distribution; F equals the sum of all the class frequencies up to, but not including, the median class; f_m is the frequency of observations in the median class; w is the class interval width; and L_m is the lower limit of the median class interval.

p. 72:
$$Mo = L_{Mo} + \frac{d_1}{d_1 + d_2} w$$

[3·9]

The *mode* is that value most often repeated in the data set. To find the *mode of grouped data* (symbolized *Mo*), use this formula and let L_{Mo} = the lower limit of the modal class; d_1 = the frequency of the modal class minus the frequency of the class directly below it; d_2 = the frequency of the modal class minus the frequency of the class directly above it; and w = the width of the modal class interval.

10   CHAPTER REVIEW EXERCISES

41. The table below gives the relative distribution of apartment sizes by number of bedrooms for all apartments constructed during last year.

Number of bedrooms	Efficiency	1	2	3	4	5 or more
Relative frequency	.21	.18	.38	.19	.03	.01

What is the mode of the distribution?

42. A utility company made a study of its coal usage per day for a 4-week period, with the following results:

Coal usage in kilotons

Monday	Tuesday	Wednesday	Thursday	Friday	Saturday	Sunday
22.3	20.8	25.0	20.9	21.5	27.0	26.2
25.1	24.8	23.1	20.5	21.2	24.1	23.7
22.2	22.9	24.2	24.7	20.9	22.9	25.0
21.1	20.9	26.2	25.7	24.8	24.6	23.2

a) What is the modal day according to total usage over the four weeks?

b) What are the modal class and modal value for the data arranged by one-kiloton classes?

43. The Board of Education in a major city is concerned over the increasing percentage of black students attending schools in the urban areas of the city. The board has the following data for all urban schools for the year just ended:

Black students (%)	0–9.9	10–19.9	20–29.9	30–39.9	40–49.9	50–59.9	60–69.9	70–79.9
Frequency	1	3	2	7	6	10	12	9

If the Department of Justice has established 50 percent black enrollment as the maximum average for all urban schools in any system, is this city in violation?

44. There are 4 private secondary schools in the city. The schools are all similar in terms of tuition costs and academic curriculum. During the past year, the average enrollment capacity per school was 420 children. This year, by adding faculty and facilities, two of the schools have increased their capacity by 10 percent, which will then make all four schools' enrollment capacities equal to each other. What capacities did the 4 schools have before the increase?

45. For a skewed distribution, the best measure of central tendency to report is (choose one)

a) the mean

b) the median

c) the geometric mean

d) depends upon the direction of skewness

e) the mode

Why is this the case?

46. A firm that makes interviewing surveys for businesses and other organizations in New York City finds that the type of sampling its clients desire affects the cost of the interviews. If a client wants a sample by area, more work and time are involved, since more areas must be reached. The firm has divided the city into sectors and has arrived at the following schedule of charges for interviewing, depending on the number of people to be surveyed in a sector:

Dozens of interviews/sector	1–10	11–15	16–20	21–30	31–40	41–50	Over 50
Cost (per dozen)	$12.00	$11.40	$10.80	$10.50	$10.20	$9.90	$9.60

If a particular organization using a sampling plan wants 660 (or 55 dozen) persons interviewed in one of the sectors, what is the average cost per interview?

47. Using the frequency distribution given in problem 62:

a) Construct a frequency histogram.

b) Graphically determine the mode.

48. Referring to problem 31:

a) What is the mode?

b) Construct a frequency distribution using 1-inch-wide classes. What is the modal class?

c) Estimate the mode from the distribution in part b using Equation 3·9. Compare your answer to that obtained by using a graphical calculation.

49. Suppose that the distributions illustrated in Fig. 3·1 represent the heat output distributions from 3 types of solar furnaces.

   a) On average, which furnace produces the most heat?

   b) Is it possible for all three types to have the same heat output?

   c) Is it possible for type A to have a greater output than type B?

   d) Is it possible for type C to have a smaller output than type A?
   Briefly explain your answers.

50. Suppose that the two curves illustrated in Fig. 3·2 represent the distribution of possible winnings involved in 2 series of gambles in an experiment. Further suppose that both curves are symmetric, with their centers (highest points) representing zero winnings. Positions to the left of center represent losses, and positions to the right of center represent winnings. Intuitively, which set of gambles, A or B, would you prefer?

51. Compare and contrast the central position, skewness, and kurtosis of the distributions of student/teacher ratios for all

   a) colleges in the U.S.

   b) large state universities in the U.S.

   c) private colleges in the U.S.

52. The gross displacement of each ship utilizing the Panama Canal during a one-week period was compiled into the frequency distribution below.

Gross displacement (thousands of tons)	0–2.99	3–5.99	6–8.99	9–11.99	12–14.99	15–17.99	18–20.99
Frequency	30	47	69	32	18	3	1

   a) Compute the sample mean of this data, using the class mark method.

   b) Verify that you get the same result using the coding method.

53. From the data below, find the average age of the participants at a state sports meet:

Age	Frequency	Age	Frequency
13	25	20	40
14	48	21	63
15	62	22	32
16	81	23	24
17	105	24	22
18	75	25	19
19	54		

54. The following table gives the distribution of miles per gallon (mpg) ratings for the engines produced by one Detroit automobile manufacturer:

MPG	10–12.99	13–15.99	16–18.99	19–21.99	22–24.99	25–27.99	28–30.99
Relative frequency	.05	.10	.20	.30	.25	.05	.05

   What is the mean value for the engines tested?

55. Jean Wiggins, a psychologist concerned about learning problems, gives all children entering the first grade a perceptual-motor task, which is timed. Slow performance may indicate some perceptual problem or perceptual-motor development less advanced than that of other children the same age. Based on her screening of the children, she obtains the following results:

Time (minutes)	Frequency	Time (minutes)	Frequency
0.00–0.39	15	2.00–2.39	7
0.40–0.79	43	2.40–2.79	4
0.80–1.19	58	2.80–3.19	3
1.20–1.59	21	3.20–3.59	2
1.60–1.99	9	3.60–3.99	2

Using the coding method, show that you get the same value for the mean when you assign 0 to either the fifth or the sixth class.

56. Over the past 5 years, the number of high school seniors going on to college has shown an average increase of 5.3 percent per year. The increases for the first 4 years were 6.2, 5.8, 5.0, and 4.2 percent. What was the increase in the fifth year?

57. A survey of 20 households concerning the quality of a particular TV program yielded the following distribution of ratings, with positive numbers indicating a favorable rating and negative numbers (denoted in parentheses) indicating an unfavorable one:

Quality rating	(30.0)–(20.1)	(20.0)–(10.1)	(10.0)–(0.1)	0.0–9.9	10.0–19.9	20.0–29.9
Frequency	3	7	5	2	2	1

Use the coding method to find the average rating given the TV program.

58. Over the past several years, the growth pattern in the population of a developing industrial town has followed the pattern given in the table below.

Year	1970	1971	1972	1973	1974	1975	1976	1977
Growth	.024	.062	.043	.201	.022	.005	.427	.201

Calculate the geometric mean of the increases.

59. Which measure of central tendency would you recommend to represent the following distributions?

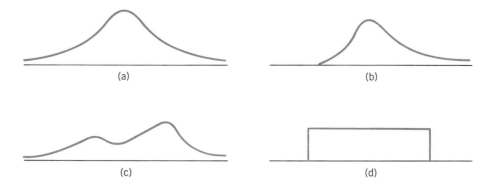

(a)

(b)

(c)

(d)

60. Dorothy Langston, supervisor of elections in Webb County, is preparing a report on the increase in voter registration over the past several years. Records show the following numbers of registered voters in the general election:

Year	1970	1972	1974	1976	1978
Voters registered	120,000	126,000	137,340	157,940	167,420

What should Langston report as the average percentage increase over the 5 voting years?

61. Scientists categorize radiation according to wavelength. Typical categories are infrared, visible, ultraviolet, X-ray, and cosmic ray. What is the median category?

62. The weights of a sample of packages shipped by air freight are given in the following distribution:

Weight (lbs)	0–9.99	10.0–19.99	20.0–29.99	30.0–39.99	40.0–49.99	50.0 and above
Frequency	28	25	14	8	4	1

What is the median?

63. On January 3rd, Don Ackerman had a better day than usual at the tracks. He placed the following bets in 5 races:

Race number	1	2	3	4	5
Number of tickets bought	10	20	10	25	10
Price per ticket	$15.40	$16.80	$5.00	$32.50	$3.20

a) What was the average price per ticket?

b) At the end of the races, Don received the following payoffs for each ticket:

Race number	1	2	3	4	5
Payoff per ticket	$1.00	$2.50	$.20	$.80	$.75

What was the average payoff on investment for these bets?

64. As part of a study of mathematical ability, a number of people were asked to perform arithmetic computations of moderate difficulty without the use of paper and pencil. Their response times (correct answers only) were recorded, and the frequency distribution of response times is given below.

Response time (secs)	Frequency	Response time (secs)	Frequency
5.00 and below	15	7.51–8.00	58
5.01–5.50	21	8.01–8.50	63
5.51–6.00	38	8.51–9.00	27
6.01–6.50	39	9.01–9.50	21
6.51–7.00	45	9.51–10.00	19
7.01–7.50	51	Above 10.00	14

Determine the median.

Answer true *or* false. *Answers are in the back of the book.*

F 1. The value of every observation in the data set is taken into account when we calculate its median.

F 2. When the population is either negatively or positively skewed, it is often preferable to use the median as the best measure of location because it always lies between the mean and the mode.

F 3. Measures of central tendency in a data set refer to the extent to which the observations are scattered.

F 4. A measure of the peakedness of a distribution curve is its skewness.

T 5. With ungrouped data, the mode is most frequently used as the measure of central tendency.

T 6. If we arrange the observations in a data set from highest to lowest, the data point lying in the middle is the median of the data set.

T 7. When working with grouped data, we may compute an approximate mean by assuming each value in a given class is equal to its class mark.

T 8. The value most often repeated in a data set is called the arithmetic mean.

F 9. If the curve of a certain distribution tails off toward the left end of the measuring scale on the horizontal axis, the distribution is said to be negatively skewed.

T 10. After grouping a set of data into a number of classes, we may identify the median class as being the one that has the largest number of observations.

Measuring variability

1. Measures of dispersion, 86

2. Dispersion: distance measures, 88

3. Dispersion: average deviation measures, 92

4. Relative dispersion: the coefficient of variation, 104

5. Terms introduced, 106

6. Equations introduced, 107

7. Chapter review exercises, 109

8. Chapter concepts test, 113

The North Carolina Commissioner of Elections is studying the workload of local election boards throughout the state. Since the amount of work depends almost entirely on the size of individual precincts, the commissioner has compiled the following frequency distribution for registrations in the 100 precincts of the Research Triangle area (Raleigh, Durham, and Chapel Hill).

Number of voters	Frequency	Number of voters	Frequency
700–799	4	1300–1399	13
800–899	7	1400–1499	10
900–999	8	1500–1599	9
1000–1099	10	1600–1699	7
1100–1199	12	1700–1799	2
1200–1299	17	1800–1899	1

The commissioner would like to compare that area with the Piedmont Crescent (Greensboro, High Point, Winston Salem, and Charlotte). To do so, he will summarize the distribution, but with an eye toward getting more information than just a measure of central tendency. This chapter discusses how he can measure the *variability* in a distribution and thus get a much better feel for the data.

1 MEASURES OF DISPERSION

In Chapter 3, we learned that two sets of data can have the same central location and yet be very different if one is more spread out than the other. This is true of the three distributions in Fig. 4·1. The mean of all three curves is the same, but curve A has less spread (or *variability*) than curve B, and curve B has less variability than curve C. If we measure only the mean of these three distributions, we will miss an important difference among the three curves. Likewise for any data, the mean, the median, and the mode tell us only part of what we need to know about the characteristics of the data. To increase our understanding of the pattern of the data, we must also measure its *dispersion*—its spread, or variability.

FIG. 4·1

Three curves with the same mean but different variabilities

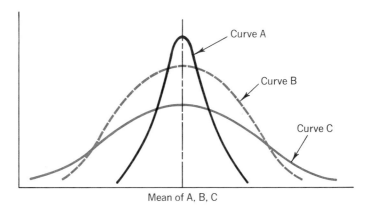

Mean of A, B, C

Uses of dispersion measures

Why is the dispersion of the distribution such an important characteristic to understand and measure? **First,** it gives us additional information that enables us to judge the reliability of our measure of the central tendency. If data are widely dispersed, such as those in curve C in Fig. 4·1, the central location is less representative of the data as a whole than it would be for data more closely centered around the mean, as in curve A. **Second,** because there are problems peculiar to widely dispersed data, we must be able to recognize that data are widely dispersed before we can tackle those problems. And **third,** we may wish to compare dispersions of various samples. If a wide spread of values away from the center is undesirable or presents an unacceptable risk, we need to be able to recognize and avoid choosing those distributions with the greatest dispersion.

Financial use

Quality control use

Financial analysts are concerned about the dispersion of a firm's earnings. Widely dispersed earnings—those varying from extremely high to low or even negative levels—indicate a higher risk to stockholders and creditors than do earnings remaining relatively stable. Similarly, quality control experts analyze the dispersion of a product's quality levels. A drug that is average in purity but ranges from very pure to highly impure may endanger lives.

1. A firm using two different methods to ship orders to its customers found the following distributions of delivery time for the two methods, based on past records. From available evidence, which shipment method would you recommend?

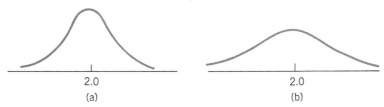

(a)   2.0   2.0 (b)

2. For which of the following distributions is the mean more representative of the data as a whole? Why?

(a)   (b)

3. To measure scholastic achievement, educators need to test students' levels of knowledge and ability. Taking students' individual differences into account, teachers may better plan their curricula. The curves below represent distributions based on previous scores of two different tests. Which would you select as the better for the teachers' purpose?

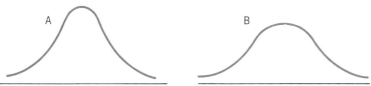

A   B

4. Which of the following is not one of the reasons for measuring the dispersion of a distribution?

   a) It provides an indication of the reliability of the central tendency measure.

   b) It enables us to compare several samples with similar averages.

   c) It uses more data in describing a distribution.

   d) It draws attention to problems associated with very small or very large variability distributions.

5. Of the three curves shown in Fig. 4·1, choose one that would best describe the distribution of values for the ages of the following groups: members of Congress, newly elected members of the House of Representatives, the chairmen of major congressional committees. In making your choices, disregard the common mean of the curves in Fig. 4·1 and consider only the variability of the distributions. Briefly state your reasons for your choices.

6. How do you think the concept of variability might apply to an investigation by the Federal Trade Commission (FTC) into possible price fixing by a group of manufacturers?

2 DISPERSION: DISTANCE MEASURES   Dispersion may be measured in terms of the difference between two values selected from the data set. In this section, we shall study three of these so-called *distance measures:* the range, the interfractile range, and the quartile deviation.

Range

Defining and
computing the range

As we said in Chapter 2, *the range is the difference between the highest and lowest observed values.* In equation form, we can say:

$$\text{Range} = \frac{\text{Value of highest}}{\text{observation}} - \frac{\text{Value of lowest}}{\text{observation}} \qquad [4 \cdot 1]$$

Using this equation, let's compare the ranges of annual payments from Blue Cross-Blue Shield received by the two hospitals illustrated in Table 4·1.

The range of annual payments to Cumberland is $1,883,000 − $863,000 = $1,020,000. For Valley Falls, the range is $690,000 − $490,000 = $200,000.

TABLE 4·1
**Annual payments
from Blue Cross-
Blue Shield
(000's omitted)**

Cumberland	863	903	957	1,041	1,138	1,204	1,354	1,624	1,698	1,745	1,802	1,883
Valley Falls	490	540	560	570	590	600	610	620	630	660	670	690

Characteristics of
the range

The range is easy to understand and to find, but its usefulness as a measure of dispersion is limited. The range considers only the highest and lowest values of a distribution and fails to take account of any other observation in the data set. As a result, it ignores the nature of the variation among all the other observations, and it is heavily influenced by extreme values. Because it measures only two values, the range is likely to change drastically from one sample to the next in a given population, even though the values that fall between the highest and lowest value may be quite similar. Keep in mind, too, that open-ended distributions have no range because no "highest" or "lowest" value exists in the open-ended class.

Interfractile range

Fractiles

In a frequency distribution, a given fraction or proportion of the data lie at or below a *fractile.* The median, for example, is the .5 fractile because half of the data set are less than or equal to this value. You will notice that fractiles are similar to percentages. In any distribution, 25 percent of the data lie at or below the .25 fractile; likewise, 25 percent of the data lie at or below the 25th percentile. The *interfractile range* is a measure of the spread between two fractiles in a frequency distribution; that is, the difference between the values of two fractiles.

Meaning of the
interfractile range

Calculating the
interfractile range

Suppose we wish to find the interfractile range between the first and second *thirds* of Cumberland's receipts from Blue Cross-Blue Shield. We begin by dividing the observations into thirds, as we have done in Table 4·2. Each third contains 4 items (⅓ of the total of 12 items). Therefore, 33⅓ percent of the items lie at $1,041,000 or below it, and 66⅔ percent are equal to or less than $1,624,000. Now we can calculate the interfractile range between the ⅓ and ⅔ fractiles by subtracting the value $1,041,000 from the value $1,624,000. This difference of $583,000 is the spread between the top of the first third of the payments and the top of the second third.

TABLE 4·2

**Annual
Blue Cross-
Blue Shield
payments to
Cumberland
Hospital
(000's omitted)**

First third	Second third	Last third
863	1,138	1,698
903	1,204	1,745
957	1,354	1,802
1,041 ← ⅓ fractile	1,624 ← ⅔ fractile	1,883

Fractiles may have special names, depending on the number of equal parts into which they divide the data. Fractiles that divide the data into 10 equal parts are called *deciles*. *Quartiles* divide the data into 4 equal parts. *Percentiles* divide the data into 100 equal parts. You've probably encountered percentiles in reported test scores. You know that if you scored in the 75th percentile, ¾ or 75 percent of all the people who took the test did no better than you did.

Special fractiles: deciles, quartiles and percentiles

Interquartile range and quartile deviation

The interquartile range measures approximately how far from the median we must go on either side before we can include one-half of the values of the data set. To compute this range, we divide our data into 4 parts, each of which contains 25 percent of the items in the distribution. The *quartiles* are then the highest values in each of these 4 parts, and the *interquartile range* is the difference between the values of the first and third quartiles:

Computing the interquartile range

$$\text{Interquartile range} = Q_3 - Q_1 \qquad [4 \cdot 2]$$

Figure 4·2 shows the concept of the interquartile range graphically. Notice in that figure that the width of the 4 quartiles need *not* be the same.

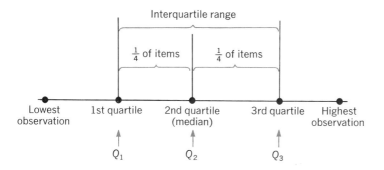

FIG. 4·2

Interquartile range

One-half of the interquartile range is a measure called the *quartile deviation:*

$$\text{Quartile deviation} = \frac{Q_3 - Q_1}{2} \qquad [4 \cdot 3]$$

Quartile deviation

The quartile deviation, then, measures the *average* range of one-fourth of the data. It is representative of the spread of all the data since it is found by taking an average of the middle half of the items rather than by choosing one of the fourths.

Illustrative problem for interquartile range and quartile deviation

Let's find the interquartile range and the quartile deviation of the annual Blue Cross-Blue Shield payments to Cumberland in Table 4·1. We begin by dividing the items into four equal parts, as we have done in Table 4·3. There, we see that the third quartile is $1,698,000 and the first quartile is $957,000. Using Equation 4·2, we find that the interquartile range is $741,000:

$$\begin{aligned} \text{Interquartile range} &= Q_3 - Q_1 \qquad [4 \cdot 2] \\ &= 1,698 - 957 \\ &= 741 \text{ thousand dollars} \end{aligned}$$

and the *quartile deviation* is $370,500:

$$\begin{aligned} \text{Quartile deviation} &= \frac{Q_3 - Q_1}{2} \qquad [4 \cdot 3] \\ &= \frac{1,698 - 957}{2} \\ &= \frac{741}{2} \\ &= 370.5 \text{ thousand dollars} \end{aligned}$$

TABLE 4·3
Annual Blue Cross- Blue Shield payments to Cumberland Hospital (000's omitted)

First fourth	Second fourth	Third fourth	Last fourth
863	1,041	1,354	1,745
903	1,138	1,624	1,802
957 ← first quartile	1,204	1,698 ← third quartile	1,883

Advantages of interquartile range and quartile deviation

Like the range, the interquartile range and the quartile deviation are based on only two values from the data set. Although they are more complicated to calculate than the range, they avoid extreme values by using only the middle half of the data. Thus, they have a distinct advantage over the range, which is affected by the extreme values.

7. For the data below, calculate the

    a) range

    b) interfractile range between the third and seventh deciles

    c) interfractile range between the fourth and sixth deciles

98	69	58	87	73	89	83	65	82	63
88	91	77	68	94	86	96	89	98	85
55	59	87	84	59	82	73	95	68	81

8. Divide the following data set into 5 fractiles and list the elements of each. Calculate the

    a) interfractile range between the second and fourth fractiles

    b) interfractile range between the second and third fractiles

    c) range

4.73	4.90	5.02	5.10	5.24	4.81	4.96	5.03	5.13	5.25
4.85	4.97	5.07	5.17	5.31	4.88	5.00	5.09	5.18	5.43

9. For the following data, compute the

    a) interquartile range

    b) quartile deviation

97	72	87	57	39	81	70	84	93	79
84	81	65	97	75	72	84	46	94	77

10. For the sample below, compute the

    a) range

    b) interfractile range between the 20th and 80th percentiles

    c) interquartile range

    d) quartile deviation

    e) interfractile range between the first and second quartiles.

    f) Compare parts *d* and *e*.

2696	2880	2575	2748	2762	2572	3233	2733	2890	2878
3100	3321	2693	2865	2784	3296	2977	2090	2905	3350

11. A psychologist studying drivers' reaction time developed a simulation film showing a car stopped in the road 300 feet ahead. A driving simulator measured the time that elapsed before the driver first touched the brake. Calculate the quartile deviation and interquartile range for the data.

    Latency of driver response (in seconds)

.10	.20	.38	.45	.50	.61	.71	.83	.88	.98	1.02	1.18
.12	.28	.40	.46	.53	.68	.73	.84	.91	1.00	1.10	1.20
.15	.32	.42	.49	.59	.70	.75	.86	.96	1.01	1.15	1.24

12. Joan Herlihy, a psychologist, has done research studying subjects' performance in tedious tasks. In a previous study, she asked subjects to read nonsense syllables aloud until they got tired, and then to stop. She found that most would read for about an hour before they stopped. She also found a surprisingly small quartile deviation of 7.2 minutes. In an attempt to replicate her results, she repeated the experiment, hoping that she might perhaps find an even smaller quartile deviation. Her data are given below. Were her hopes confirmed?

Reading time (in minutes)

52	35	48	46	46	43	40	61	49	57	58	65
72	69	38	37	55	52	50	31	41	60	45	41

13. The Department of Transportation recently concluded a study of the miles driven by a sample of 40 U.S. families over a one-year period. The results of the study are given below:

3,600	4,200	4,700	4,900	5,300	5,700	6,700	7,300
7,700	8,100	8,300	8,400	8,700	8,700	8,900	9,300
9,500	9,500	9,700	10,000	10,300	10,500	10,700	10,800
11,000	11,300	11,300	11,800	12,100	12,700	12,900	13,100
13,500	13,800	14,600	14,900	16,300	17,200	18,500	20,300

Calculate the interquartile range and quartile deviation.

14. The New Mexico State Highway Department is charged with maintaining all state roads in good condition. One measure of condition is the number of cracks present in each 100 feet of roadway. From the department's yearly survey, the following distribution was constructed.

Cracks per 100 feet

2	5	6	7	7	8	9	10	10	11
12	12	12	13	13	13	14	14	15	15
15	15	16	16	17	17	18	19	19	20

Calculate interfractile range between the 20th, 40th, 60th, and 80th percentiles.

Two measures of average deviation

3 DISPERSION: AVERAGE DEVIATION MEASURES The most comprehensive descriptions of dispersion are those that deal with the average deviation from some measure of central tendency. Two of these measures are important to our study of statistics: the *variance* and the *standard deviation*. Both of these tell us an average distance of any observation in the data set from the mean of the distribution.

Average absolute deviation

Meaning of average absolute deviation

We will have a better understanding of the variance and the standard deviation if we focus first on what statisticians call the *average absolute deviation*. To compute this, we begin by finding the mean of our sample. Then we determine the absolute value of the difference between each item in the data set and the mean. In other words, we subtract the mean from every value in the data set and ignore the sign (positive or negative), thereby taking each to be positive. Finally, we add all these differences together and divide by the total number of items in our sample.

Symbolically, the formula for finding the average absolute deviation looks like this:

$$\text{Average absolute deviation} = \frac{\Sigma|x - \mu|}{N} \text{ for a population} \qquad [4\cdot4]$$

and like this:

$$\text{Average absolute deviation} = \frac{\Sigma|x - \bar{x}|}{n} \text{ for a sample} \qquad [4\cdot5]$$

where:

x = the item or observation
μ = the population mean
N = number of items in the population
$\bar{x}$ = sample mean
n = number of items in the sample

Remember that Σ means "the sum of all the values." In this case, they are $|x - \mu|$ or $|x - \bar{x}|$. Also notice the straight lines surrounding $|x - \mu|$ and $|x - \bar{x}|$, which indicate that we want the *absolute value* of that distance (expressed in positive, not negative, numbers). This means that if the distance $x - \bar{x}$ is -10, then the absolute value is 10. The absolute value of -25 is 25.

Let's compute the average absolute deviation of the annual Blue Cross-Blue Shield payments to Cumberland in Table 4·1. First we find the mean:

Calculating the average absolute deviation

$$\bar{x} = \frac{\Sigma x}{n} \qquad [3\cdot2]$$

$$= \frac{\begin{array}{c}863 + 903 + 957 + 1041 + 1138 + 1204 + 1354 \\ + 1624 + 1698 + 1745 + 1802 + 1883\end{array}}{12}$$

$$\frac{16{,}212}{12}$$

$$= 1{,}351 \text{ thousand dollars}$$

Using the step-by-step process outlined in Table 4·4, we find the absolute deviation of every observation from this mean of $1,351,000. Now we can divide the sum of these absolute deviations by the number of items in the sample to learn the value of the average absolute deviation:

$$\text{Average absolute deviation} = \frac{\Sigma|x - \overline{x}|}{n} \qquad [4 \cdot 5]$$

$$= \frac{4,000}{12}$$

$$= 333.3 \text{ thousand dollars}$$

TABLE 4·4
Determination of the average absolute deviation of annual Blue Cross- Blue Shield payments to Cumberland Hospital (000's omitted)

Observation (x) (1)		Mean ($\overline{x}$) (2)		Deviation $(x - \overline{x})$ (1) − (2)	Absolute deviation $(|x - \overline{x}|)$ $|(1) - (2)|$
863	−	1,351	=	−488	488
903	−	1,351	=	−448	448
957	−	1,351	=	−394	394
1,041	−	1,351	=	−310	310
1,138	−	1,351	=	−213	213
1,204	−	1,351	=	−147	147
1,354	−	1,351	=	3	3
1,624	−	1,351	=	273	273
1,698	−	1,351	=	347	347
1,745	−	1,351	=	394	394
1,802	−	1,351	=	451	451
1,883	−	1,351	=	532	532
16,212 ← Σx					4,000 ← $\Sigma|x - \overline{x}|$

Characteristics of the average absolute deviation

This average absolute deviation is a better measure of dispersion than the ranges we have already calculated because it takes *every* observation into account. It weights each item equally and indicates how far, on average, each observation lies from the mean. In spite of this advantage, however, and for reasons that will be explained later, this average deviation method is rarely used.

Population variance

Variance

In the next two sections, we are going to focus on populations rather than samples. We shall begin with the fact that each population has a variance, which is symbolized by σ^2 (*sigma squared*).

Relationship of variance to average absolute deviation

The population variance is similar to an average absolute deviation computed for an entire population. But in this case, we are using the sum of the *squared* distances between the mean and each item divided by the total number of elements in the population. By squaring each distance, we automatically make every number positive and therefore have no need to take the absolute value of each deviation.

Formula for the variance

The formula for calculating the variance is similar to Equation 4·4. But this time, because we are finding the average squared distance between the mean and each item in the population, we will square each difference of $x - \mu$:

$$\sigma^2 = \frac{\Sigma(x - \mu)^2}{N} = \frac{\Sigma x^2}{N} - \mu^2 \qquad [4 \cdot 6]$$

where:

σ^2 = the population variance
x = the item or observation
μ = population mean
N = total number of items in the population
Σ = sum of all the values $(x - \mu)^2$ (or all the values x^2)

In Equation 4·6, the middle expression $\frac{\Sigma(x - \mu)^2}{N}$ is the definition of σ^2. The last expression $\frac{\Sigma x^2}{N} - \mu^2$ is *mathematically* equivalent to the definition but is often much more convenient to use if we actually must compute the value of σ^2, since it frees us from calculating the deviations from the mean. However, when the x values are large and the $x - \mu$ values are small, it may be more convenient to use the middle expression $\frac{\Sigma(x - \mu)^2}{N}$ to compute σ^2.

Before we can use this formula in an example, we need to discuss an important problem concerning the variance. In solving that problem, we will learn what the standard deviation is and how to calculate it. Then we can return to the variance itself.

Earlier, when we calculated the range, the interquartile deviation, and the average absolute deviation, the answers were expressed in the same units as the data itself. (In our examples, the units were "thousands of dollars of payments.") For the variance, however, the units are the *squares of the units* of the data, for example, "squared dollars" or "dollars squared." Squared dollars or dollars squared are not intuitively clear or easily interpreted. For this reason, we have to make a significant change in the variance to compute a useful measure of deviation, one that does not give us a problem with units of measure and thus is less confusing. This measure is called the standard deviation, and it is the square root of the variance. The square root of $100 squared is $10 because we take the square root of both the value and the units in which it is measured. The standard deviation, then, is in units that are the same as the original data.

Units in which
the variance
is expressed
cause a problem

Population standard deviation

The population standard deviation, or σ, is simply the square root of the population variance. Since the variance is the average of the squared distances of the observations from the mean, the standard deviation is the

Relationship of
standard deviation
to the variance

square root of the average of the squared distances of the observations from the mean. While the variance is expressed in the square of the units used in the data, the standard deviation is in the same units as those used in the data. The formula for the standard deviation is:

$$\sigma = \sqrt{\sigma^2} = \sqrt{\frac{\Sigma(x - \mu)^2}{N}} = \sqrt{\frac{\Sigma x^2}{N} - \mu^2} \qquad [4\cdot7]$$

where:

x = the observation
μ = the population mean
N = the total number of elements in the population
Σ = the symbol for the sum of all the $(x - \mu)^2$, or all the values x^2
σ = the population standard deviation
σ^2 = the population variance

Use the positive
square root

The square root of a positive number may be either positive or negative since $a^2 = (-a)^2$. When taking the square root of the variance to calculate the standard deviation, however, statisticians consider only the positive square root.

Computing the
standard deviation

To calculate either the variance or the standard deviation, we construct a table, using every element of the population. If we have a population of 15 vials of compound produced in one day and we test each vial to determine its purity, our data might look like Table 4·5. In Table 4·6 we show how to use this data to compute the mean (column 1 divided by N = 2.49/15), the deviation of each value from the mean (column 3), the square of the deviation of each value from the mean (column 4), and the sum of the squared deviations. From this, we can compute the variance, which is .0034 percent squared. (Table 4·6 also computes σ^2 using the second half of Equation 4·6 $\frac{\Sigma x^2}{N} - \mu^2$. Note that we get the same result but do a bit less work, since we do not have to compute the deviations from the mean.) Taking the square root of σ^2, we can compute the standard deviation, .058 percent.

TABLE 4·5
**Results of purity
test on
compounds**

Observed percent of impurity				
.04	.14	.17	.19	.22
.06	.14	.17	.21	.24
.12	.15	.18	.21	.25

TABLE 4·6 **Determination of the variance and standard deviation of percent impurity of compounds**

Observation (x) (1)	Mean $(\mu) = 2.49/15$ (2)			Deviation $(x - \mu)$ (3) = (1) − (2)	Deviation squared $(x - \mu)^2$ (4) = [(1) − (2)]²	Observation squared x^2 (5) = (1)²
.04	−	.166	=	−.126	.016	.0016
.06	−	.166	=	−.106	.011	.0036
.12	−	.166	=	−.046	.002	.0144
.14	−	.166	=	−.026	.001	.0196
.14	−	.166	=	−.026	.001	.0196
.15	−	.166	=	−.016	.000	.0225
.17	−	.166	=	.004	.000	.0289
.17	−	.166	=	.004	.000	.0289
.18	−	.166	=	.014	.000	.0324
.19	−	.166	=	.024	.001	.0361
.21	−	.166	=	.044	.002	.0441
.21	−	.166	=	.044	.002	.0441
.22	−	.166	=	.054	.003	.0484
.24	−	.166	=	.074	.005	.0576
.25	−	.166	=	.084	.007	.0625
2.49 ← Σx					.051 ← Σ$(x - \mu)^2$	.4643 ← Σx^2

$$\sigma^2 = \frac{\Sigma(x - \mu)^2}{N} \qquad [4·6]$$

$$= \frac{.051}{15}$$

$$= .0034 \text{ percent squared}$$

$$\sigma = \sqrt{\sigma^2} \qquad [4·7]$$

$$= \sqrt{.0034}$$

$$= .058 \text{ percent}$$

$$\sigma^2 = \frac{\Sigma x^2}{N} - \mu^2 \qquad [4·6]$$

$$= \frac{.4643}{15} - (.166)^2$$

$$= .0034 \text{ percent squared}$$

Uses of the standard deviation

Chebyshev's theorem

The standard deviation enables us to determine, with a great deal of accuracy, where the values of a frequency distribution are located in relation to the mean. We can do this according to a theorem devised by the Russian mathematician P.L. Chebyshev (1821–1894). Chebyshev's theorem says that no matter what the shape of the distribution, at least 75 percent of the values will fall within plus-and-minus 2 standard deviations from the mean of the distribution, and at least 89 percent of the values will lie within plus-and-minus 3 standard deviations from the mean.

We can measure with even more precision the percentage of items that fall within specific ranges under a symmetrical, bell-shaped curve like the one in Fig. 4·3. In these cases, we can say that:

1. About 68 percent of the values in the population will fall within plus-or-minus 1 standard deviation from the mean.
2. About 95 percent of the values will lie within plus-or-minus 2 standard deviations from the mean.
3. About 99 percent of the values will be in an interval ranging from 3 standard deviations below the mean to 3 standard deviations above the mean.

FIG. 4·3
Location of observations around the mean of a bell-shaped frequency distribution

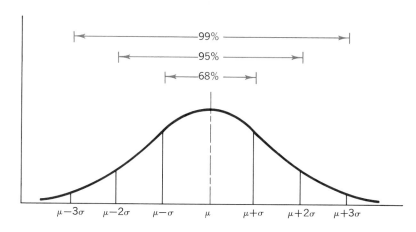

Using Chebyshev's theorem

In light of Chebyshev's theorem, let's analyze the data in Table 4·6. There, the mean impurity of the 15 vials of compound is .166 percent, and the standard deviation is .058 percent. Chebyshev's theorem tells us that at least 75 percent of the values (at least 11 of our 15 items) are between .166 − 2(.058) = .050 and .166 + 2(.058) = .282. In fact, 93 percent of the values (14 of the 15 values) are actually in that interval. Notice that the distribution is reasonably symmetrical and that 93 percent is close to the theoretical 95 percent for an interval of plus-or-minus 2 standard deviations from the mean of a bell-shaped curve.

Concept of the standard score

The standard deviation is also useful in describing how far individual items in a distribution depart from the mean of the distribution. A measure called the standard score gives us the number of standard deviations a particular observation lies below or above the mean. If we let x symbolize the observation, the standard score computed from population data is:

$$\text{Population standard score} = \frac{x - \mu}{\sigma} \qquad [4·8]$$

where:

x = the observation from the population
μ = the population mean
σ = the population standard deviation

Suppose we observe a vial of compound that is .108 percent impure. Since our population has a mean of .166 and a standard deviation of .058, an observation of .108 would have a standard score of -1:

$$\text{Standard score} = \frac{x - \mu}{\sigma} \qquad [4\cdot 8]$$

$$= \frac{.108 - .166}{.058}$$

$$= -\frac{.058}{.058}$$

$$= -1$$

Calculating the
standard score

An observed impurity of .282 percent would have a standard score of $+2$:

$$\text{Standard score} = \frac{x - \mu}{\sigma} \qquad [4\cdot 8]$$

$$= \frac{.282 - .166}{.058}$$

$$= \frac{.116}{.058}$$

$$= 2$$

The standard score indicates that an impurity of .282 percent deviates from the mean by $2(.058) = .116$ units, which is equal to $+2$ in terms of units of standard deviations away from the mean.

Interpreting the
standard score

Calculation of variance and standard deviation using grouped data

In our chapter-opening example, data on voter registration in North Carolina precincts were already grouped in a frequency distribution. With such data we can use the following formulas to calculate the variance and the standard deviation:

Calculating
the variance and
standard deviation
for grouped data

$$\sigma^2 = \frac{\Sigma f(x - \mu)^2}{N} = \frac{\Sigma fx^2}{N} - \mu^2 \qquad [4\cdot 9]$$

and:

$$\sigma = \sqrt{\sigma^2} = \sqrt{\frac{\Sigma f(x - \mu)^2}{N}} = \sqrt{\frac{\Sigma fx^2}{N} - \mu^2} \qquad [4\cdot 10]$$

where:

σ^2 = population variance
σ = population standard deviation
f = frequency of each of the classes

x = class mark for each class

μ = population mean

N = size of the population

Table 4·7 shows how to apply these equations to find the variance and standard deviation of the number of voters registered in 100 North Carolina precincts. We leave it as an exercise for the curious reader to verify that the second half of Equation 4·9, $\dfrac{\Sigma fx^2}{N} - \mu^2$, will yield the same value of σ^2.

TABLE 4·7 Determination of the variance and standard deviation of voter registrations for 100 North Carolina precincts

Class	Class mark (x) (1)	Frequency (f) (2)	f × x (3) = (2) × (1)	Mean (μ) (4)	x − μ (1) − (4)	(x − μ)² [(1) − (4)]²	f(x − μ)² (2) × [(1) − (4)]²
700– 799	750	4	3,000	1,250	−500	250,000	1,000,000
800– 899	850	7	5,950	1,250	−400	160,000	1,120,000
900–, 999	950	8	7,600	1,250	−300	90,000	720,000
1,000–1,099	1,050	10	10,500	1,250	−200	40,000	400,000
1,100–1,199	1,150	12	13,800	1,250	−100	10,000	120,000
1,200–1,299	1,250	17	21,250	1,250	0	0	0
1,300–1,399	1,350	13	17,550	1,250	100	10,000	130,000
1,400–1,499	1,450	10	14,500	1,250	200	40,000	400,000
1,500–1,599	1,550	9	13,950	1,250	300	90,000	810,000
1,600–1,699	1,650	7	11,550	1,250	400	160,000	1,120,000
1,700–1,799	1,750	2	3,500	1,250	500	250,000	500,000
1,800–1,899	1,850	1	1,850	1,250	600	360,000	360,000
		100	125,000				6,680,000

$$\overline{x} = \frac{\Sigma(f \times x)}{n} \qquad [3\cdot3]$$

$$= \frac{125,000}{100}$$

$$= 1,250 \text{ voters} \leftarrow \text{mean}$$

$$\sigma^2 = \frac{\Sigma f(x - \mu)^2}{N} \qquad [4\cdot9]$$

$$= \frac{6,680,000}{100}$$

$$= 66,800 \text{ (or 66,800 voters squared)} \leftarrow \text{variance}$$

$$\sigma = \sqrt{\sigma^2} \qquad [4\cdot10]$$

$$= \sqrt{66,800}$$

$$= 258.5 \text{ voters} \leftarrow \text{standard deviation}$$

Switching to
sample variance
and sample
standard deviation

Now we are ready to compute the sample statistics that are analogous to the population variance σ^2 and the population standard deviation σ. These are the sample variance s^2 and the sample standard deviation s. In the next section, you'll notice we are changing from Greek letters (which denote population parameters) to the Latin letters of sample statistics.

Sample standard deviation

101

Section 3
AVERAGE
DEVIATION
MEASURES

To compute the sample variance and the same standard deviation, we use the same formulas as Equations 4·6 and 4·7, replacing μ with $\bar{x}$ and N with $n - 1$. The formulas look like this:

$$s^2 = \frac{\Sigma(x - \bar{x})^2}{n - 1} = \frac{\Sigma x^2}{n - 1} - \frac{n\bar{x}^2}{n - 1} \qquad [4 \cdot 11]$$

and:

Computing
the sample
standard deviation

$$s = \sqrt{s^2} = \sqrt{\frac{\Sigma(x - \bar{x})^2}{n - 1}} = \sqrt{\frac{\Sigma x^2}{n - 1} - \frac{n\bar{x}^2}{n - 1}} \qquad [4 \cdot 12]$$

where:

s^2 = sample variance
s = sample standard deviation
x = value of each of the n observations
$\bar{x}$ = mean of the sample
$n - 1$ = number of observations in the sample minus one

Why do we use $n - 1$ as the denominator instead of n? Statisticians can prove that if we take many samples from a given population, find the sample variance (s^2) for each sample, and average each of these together, then this average tends not to equal the population variance, σ^2, unless we use $n - 1$ as the denominator. In Chapter 8, we shall learn the statistical explanation of why this is true.

Use of $n - 1$ as
the denominator

Let's use Equations 4·11 and 4·12 to find the sample variance and the sample standard deviation of the annual Blue Cross-Blue Shield payments to Cumberland Hospital discussed in Table 4·4 on page 94. We do this in Table 4·8, noting that both halves of Equation 4·11 yield the same result.

Calculating
sample variance and
standard deviation
for Cumberland
Hospital data

Just as we used the population standard deviation to derive population standard scores, we may also use the sample standard deviation to compute sample standard scores. These sample standard scores tell us how many standard deviations a particular sample observation lies below or above the sample mean. The appropriate formula is:

Computing sample
standard scores

$$\text{Sample standard score} = \frac{x - \bar{x}}{s} \qquad [4 \cdot 13]$$

where:

x = the observation from the sample
$\bar{x}$ = the sample mean
s = the sample standard deviation

TABLE 4·8 **Determination of the sample variance and standard deviation of annual Blue Cross-Blue Shield payments to Cumberland Hospital (000's omitted)**

Observation (x) (1)	Mean $(\bar{x})$ (2)	$x - \bar{x}$ (1) − (2)	$(x - \bar{x})^2$ [(1) − (2)]^2	x^2 (1)^2
863	1,351	−488	238,144	744,769
903	1,351	−448	200,704	815,409
957	1,351	−394	155,236	915,849
1,041	1,351	−310	96,100	1,083,681
1,138	1,351	−213	45,369	1,295,044
1,204	1,351	−147	21,609	1,449,616
1,354	1,351	3	9	1,833,316
1,624	1,351	273	74,529	2,637,376
1,698	1,351	347	120,409	2,883,204
1,745	1,351	394	155,236	3,045,025
1,802	1,351	451	203,401	3,247,204
1,883	1,351	532	283,024	3,545,689
			$\Sigma(x - \bar{x})^2 \to$ 1,593,770	23,496,182 $\leftarrow \Sigma x^2$

$$s^2 = \frac{\Sigma(x - \bar{x})^2}{n - 1} \qquad [4·11]$$

$$= \frac{1,593,770}{11}$$

$= 144,888$ (or \$144,888 billion squared) ← sample variance

$$s = \sqrt{s^2} \qquad [4·12]$$

$$= \sqrt{144,888}$$

$= 380.64$ (that is, \$380,640) ← sample standard deviation

$$s^2 = \frac{\Sigma x^2}{n - 1} - \frac{n\bar{x}^2}{n - 1} \qquad [4·11]$$

$$= \frac{23,496,182}{11} - \frac{12(1351)^2}{11}$$

$$= \frac{1,593,770}{11}$$

$$= 144,888$$

In the example we just did, we see that the observation 863 corresponds to a standard score of −1.28:

$$\text{Sample standard score} = \frac{x - \bar{x}}{s} \qquad [4·13]$$

$$= \frac{863 - 1351}{380.64}$$

$$= \frac{-488}{380.64}$$

$$= -1.28$$

Characteristics of the standard deviation

This section has demonstrated why the standard deviation is the measure of dispersion used most often. We can use it to compare distributions and to compute standard scores, an important element of statistical inference to be discussed later. Like the average absolute deviation, it takes into account every observation in the data set. But the standard deviation has some disadvantages too. It is not as easy to calculate as the range, and it cannot be computed from open-ended distributions. In addition, extreme values in the data set distort the value of the standard deviation, although to a lesser extent than they do the range.

15. For the following measurements, compute the

    a) average absolute deviation

    b) population variance

    c) population standard deviation

50	53	52	51	43	52	51	50	56	54
45	48	54	52	49	48	47	58	51	56

16. The following values represent a sample from a large population. Compute the

    a) average absolute deviation for the sample

    b) sample variance

    c) sample standard deviation

26	17	24	29	26	21	33	31	29	27

17. In a set of 60 observations with a mean of 39.5, a variance of 17.64, and an unknown distribution shape,

    a) between what values should at least 75 percent of the observations fall, according to Chebyshev's theorem?

    b) if the distribution is symmetrical and bell-shaped, approximately how many observations should be found in the interval 35.3 to 43.7?

    c) find the standard scores for the following observations from the distribution: 36.35, 44.54, 50.0, and 30.5.

18. Calculate the population variance for the following set of grouped data:

Class	0–199	200–399	400–599	600–799	800–999
Frequency	8	13	20	12	7

19. The Federal Reserve Board has given permission to all member banks to raise interest rates $\frac{1}{2}$ percent for all depositors. Old rates for passbook savings were $5\frac{1}{4}\%$; for certificates of deposit (CD's): 1-year CD, $5\frac{1}{2}\%$; 18-month CD, $5\frac{3}{4}\%$; 2-year CD, 6%; 3-year CD, $6\frac{1}{2}\%$; 5-year CD, 7%. The president of The First State Bank wants to know what the characteristics of the new distribution of rates will be if the full $\frac{1}{2}$ percent is added to all rates. How are the new characteristics related to the old ones?

20. The administrator of a Georgia hospital conducted a survey of the number of days patients stayed in the hospital following an operation. The data are given below:

Hospital stay (in days)	1–3	4–6	7–9	10–12	13–15	16–18	19–21	22–24
Frequency	32	108	67	28	14	7	3	1

    a) Calculate the mean and standard deviation.

    b) According to Chebyshev's theorem, how many stays should be between 0 and 15 days? How many are actually in that interval?

    c) Since this distribution is roughly bell-shaped, how many of the stays can we expect to fall between 0 and 15 days?

21. A family psychologist did a study asking parents how many sons there would be on the average in families who had "ideal" sex compositions. For each couple, she averaged the husband's and wife's responses, to get the overall couple response. She tabulated these in a frequency distribution:

Number of sons	0–.49	.50–.99	1.00–1.49	1.50–1.99	2.00–2.49	2.50–2.99
Frequency	1	12	26	8	3	2

a) Calculate the variance and the standard deviation.

b) Since the distribution is bell-shaped, how many of the observations should theoretically fall between .8 and 1.8? between .3 and 2.3? How many actually do fall in these intervals?

22. Coach Lefty Dryspell says the average attendance at his university's basketball games during the past year was 11,398 with a variance of 49,729. If the data used to compute the results were collected for 32 games, during how many games was the attendance below 11,175? Above 11,844?

23. In Minnesota, the average farm subsidy payment for the past year was $1,250 with a standard deviation of $355. In Iowa, the average was $2,580 with a standard deviation of $578; and in Alabama, the average was $533 with a standard deviation of $42. Three farmers were interviewed, one from each state. The Minnesota farmer received a payment of $1,000; the Iowa farmer received a payment of $2,300; and the Alabama farmer received a payment of $500. Which of the three had the smallest payment in relation to the mean and standard deviation of his state?

24. In a cognitive processing experiment, subjects were asked to search paragraphs of three different lengths for a pre-specified key word. Subjects responded as quickly as possible, "yes" if the word was in the paragraph or "no" if it was not. The three paragraphs were constructed and tested so that the first had an average response latency of 2 seconds with a standard deviation of .005 seconds, the second had an average latency of 3 seconds with a standard deviation of .007 seconds, and the third had an average latency of 4 seconds with a standard deviation of .10 seconds. One particular subject had the following latencies: 1.994 seconds for paragraph I, 3.007 seconds for paragraph II, and 3.990 seconds for paragraph III. On which paragraph was this subject furthest from average performance, in standard deviation units?

4 RELATIVE DISPERSION: THE COEFFICIENT OF VARIATION

The standard deviation is an *absolute* measure of dispersion that expresses variation in the same units as the original data. The annual Blue Cross-Blue Shield payments to Cumberland Hospital (Table 4·8) have a standard deviation of $380,640. The annual Blue Cross-Blue Shield payments to Valley Falls Hospital (Table 4·1) have a standard deviation (which you can compute) of $57,390. Can we compare the values of these two standard deviations? Unfortunately, no.

Shortcomings of the standard deviation

The standard deviation cannot be the sole basis for comparing two distributions. If we have a standard deviation of 10 and a mean of 5, the values vary by an amount twice as large as the mean itself. If, on the other hand, we have a standard deviation of 10 and a mean of 5,000, the variation relative to the mean is insignificant. Therefore, we cannot know the dispersion of a set of data until we know the standard deviation, the mean, *and* how the standard deviation compares with the mean.

The coefficient of variation, a relative measure

What we need is a *relative* measure that will give us a feel for the magnitude of the deviation relative to the magnitude of the mean. The *coefficient of variation* is one such relative measure of dispersion. It relates the standard deviation and the mean by expressing the standard deviation as a percentage of the mean. The unit of measure, then, is "percent" rather than the same units as the original data. For a population, the formula for the coefficient of variation is:

Standard deviation of the population

$$\text{Population coefficient of variation} = \frac{\sigma}{\mu}(100) \qquad [4 \cdot 14]$$

Mean of the population

Let's use this formula in an example. Suppose each day, laboratory technician A completes 40 analyses with a standard deviation of 5. Technician B completes 160 analyses per day with a standard deviation of 15. Which employee shows less variability?

At first glance, it appears that technician B has three times more variation in the output rate than technician A. But B completes analyses at a rate four times faster than A. To take all this information into account, let's compute the coefficient of variation for both technicians:

$$\text{Coefficient of variation} = \frac{\sigma}{\mu}(100) \qquad [4 \cdot 14] \qquad \text{Computing the coefficient of variation}$$

$$= \frac{5}{40}(100)$$

$$= 12.5\% \leftarrow \text{for technician A}$$

and:

$$\text{Coefficient of variation} = \frac{15}{160}(100)$$

$$= 9.4\% \leftarrow \text{for technician B}$$

So we find that technician B, who has more *absolute* variation in output than technician A, has less *relative* variation because the mean output for B is much greater than for A.

EXERCISES

25. In two samples of size 50 each, the mean value for the first sample was 1.16 with a standard deviation of .21; the second sample had a mean of 1.75 and a standard deviation of .35. Which sample exhibits greater relative dispersion?

26. Two groups were given a task to do after receiving different training for the task. For the first group, the average score on the task was 28.74, and the scores had a variance of 79.39. The second group showed an average of 20.5 and a variance of 54.76. Which group has less relative variability in its performance?

27. The following two samples are believed to represent the same population but were collected separately. Determine the relative dispersion of scores for the two samples and indicate which sample shows greater relative variability.

    Sample 1   20   19   27   20   18   26   30   24   25   26
    Sample 2   24   25   27   16   22   20   35   28   18   32

28. A meteorologist interested in the consistency of temperatures in three Florida cities during a traditionally big convention week collected the following data. Last year, the temperatures for the five days of the convention week in the three cities were:

    City 1   74   80   72   65   78
    City 2   82   76   69   70   84
    City 3   77   80   75   69   73

    Which city seems to have the most consistent temperature, based on these data?

29. In a gambling experiment pitting subjects against a computer, one group of subjects had an average win of $2.80 with a standard deviation of 53¢, while another group had an average win of $3.78 with a standard deviation of 48¢. If we consider greater risk to be associated with the greater relative deviation, which was the riskier group?

30. A drug company that supplies hospitals with premeasured doses of certain medications uses different machines for medications requiring different dosage amounts. One machine designed to produce doses of 100 cc has as its mean dose 100 cc with a standard deviation of 2.6 cc. Another machine produces premeasured amounts of 180 cc of medication and it has a standard deviation of 5.3 cc. Which machine has the least accuracy from the standpoint of relative dispersion?

31. The coach of the Boston University swimming team is evaluating three freshmen for the final spot on the team. The three swimmers compete in five 100-meter freestyle races with these results:

    Swimmer #1:   62.1,   61.8,   63.2,   62.9,   61.7   seconds
    Swimmer #2:   62.5,   61.9,   62.8,   63.0,   60.7   seconds
    Swimmer #3:   61.9,   61.9,   62.9,   63.7,   61.5   seconds

    The coach feels that consistency, as well as the best average, is important. Based on relative dispersion, which swimmer would make the team?

5   TERMS INTRODUCED IN CHAPTER 4

- **AVERAGE ABSOLUTE DEVIATION**   In a data set, the average distance of the observations from the mean.

- **CHEBYSHEV'S THEOREM**   No matter what the shape of a distribution, at least 75 percent of the values in the population will fall within 2 standard deviations of the mean, and at least 89 percent will fall within 3 standard deviations.

- **COEFFICIENT OF VARIATION**   A relative measure of dispersion, comparable across distributions, which expresses the standard deviation as a percentage of the mean.

- **DECILES**   Fractiles that divide the data into 10 equal parts.

- **DISPERSION**   The scatter or variability in a set of data.

- **DISTANCE MEASURE** A measure of dispersion in terms of the difference between two values in the data set.
- **FRACTILE** In a frequency distribution, the location of a value at, or above, a given fraction of the data.
- **INTERFRACTILE RANGE** A measure of the spread between two fractiles in a distribution; i.e., the difference between the values of two fractiles.
- **INTERQUARTILE RANGE** The difference between the values of the first and third quartiles, indicating the range of the middle half of the data set.
- **PERCENTILES** Fractiles that divide the data into 100 equal parts.
- **QUARTILES** Fractiles that divide the data into 4 equal parts.

- **QUARTILE DEVIATION** Half of the interquartile range; a measure of the average range of one-fourth of the data.
- **RANGE** The distance between the highest and lowest values in a data set.
- **STANDARD DEVIATION** The positive square root of the variance; a measure of dispersion in the same units as the original data, rather than in the squared units of the variance.
- **STANDARD SCORE** Expressing an observation in terms of standard deviation units above or below the mean; i.e., the transformation of an observation by subtracting the mean and dividing by the standard deviation.
- **VARIANCE** A measure of the average squared distance between the mean and each item in the population.

6 EQUATIONS INTRODUCED IN CHAPTER 4

p. 88:

$$\text{Range} = \frac{\text{Value of highest}}{\text{observation}} - \frac{\text{Value of lowest}}{\text{observation}}$$

[4·1]

The *range* is the difference between the highest and lowest observed values in a frequency distribution.

p. 89:

$$\text{Interquartile range} = Q_3 - Q_1$$

[4·2]

The *interquartile range* measures approximately how far from the median we must go on either side before we can include one-half of the values of the data set. To compute this range, divide the data into four equal parts. The *quartiles* (*Q*) are the highest values in each of these four parts. The *interquartile range* is the difference between the values of the first and third quartiles (Q_1 and Q_3).

p. 90:

$$\text{Quartile deviation} = \frac{Q_3 - Q_1}{2}$$

[4·3]

The *quartile* deviation measures the average of one-fourth of the data in a distribution. It is equal to one-half of the interquartile range.

p. 93:

$$\text{Average absolute deviation} = \frac{\Sigma|x - \mu|}{N}$$

[4·4]

This formula enables us to calculate the *average absolute deviation for a population.* Because this measure of variability deals with the absolute value of the difference between each item in the data set and the mean, it is not as useful for further calculation as is the variance, which squares each distance.

Here, x represents the item or observation, μ the population mean, N the number of items in the population, and $\Sigma|x - \mu|$ the sum of all the values of $|x - \mu|$.

p. 93:
$$\text{Average absolute deviation} = \frac{\Sigma|x - \bar{x}|}{n}$$
[4·5]

For a *sample,* use this formula to determine the average absolute deviation. Unlike Equation 4·4, this formula uses the sample mean $\bar{x}$ and the number of items in the sample n.

p. 95:
$$\sigma^2 = \frac{\Sigma(x - \mu)^2}{N} = \frac{\Sigma x^2}{N} - \mu^2$$
[4·6]

This formula enables us to calculate the *population variance,* a measure of the average *squared* distance between the mean and each item in the population. The middle expression, $\dfrac{\Sigma(x - \mu)^2}{N}$ is the definition of σ^2. The last expression $\dfrac{\Sigma x^2}{N} - \mu^2$ is mathematically equivalent to the definition but is often much more convenient to use, since it frees us from calculating the deviations from the mean.

p. 96:
$$\sigma = \sqrt{\sigma^2} = \sqrt{\frac{\Sigma(x - \mu)^2}{N}} = \sqrt{\frac{\Sigma x^2}{N} - \mu^2}$$
[4·7]

The population standard deviation, σ, is the square root of the population variance. It is a more useful parameter than the variance because it is expressed in the same units as the data itself (whereas the units of the variance are the squares of the units of the data). Notice that the standard deviation is always the *positive* square root of the variance.

p. 99:
$$\text{Population standard score} = \frac{x - \mu}{\sigma}$$
[4·8]

The *standard score* of an observation is the number of standard deviations the observation lies below or above the mean of the distribution. The standard score enables us to make comparisons between distribution items that differ in order of magnitude or in the units employed. Use Equation 4·8 to find the standard score of an item in a *population.*

p. 99:
$$\sigma^2 = \frac{\Sigma f(x - \mu)^2}{N} = \frac{\Sigma f x^2}{N} - \mu^2$$
[4·9]

This formula in either form enables us to calculate the *variance* of data already *grouped* in a frequency distribution. Here, *f* represents the frequency of the class, and *x* represents the class mark.

p. 99:

$$\sigma = \sqrt{\sigma^2} = \sqrt{\frac{\Sigma f(x - \mu)^2}{N}} = \sqrt{\frac{\Sigma fx^2}{N} - \mu^2}$$

[4·10]

Take the square root of the variance, and you have the *standard deviation using grouped data.*

p. 101:

$$s^2 = \frac{\Sigma(x - \bar{x})^2}{n - 1} = \frac{\Sigma x^2}{n - 1} - \frac{n\bar{x}^2}{n - 1}$$

[4·11]

To compute the *sample variance,* use the same formula as Equation 4·6, replacing μ with $\bar{x}$ and *N* with *n* − 1. Chapter 8 contains an explanation of why we use *n* − 1 rather than *n* to calculate the sample variance.

p. 101:

$$s = \sqrt{s^2} = \sqrt{\frac{\Sigma(x - \bar{x})^2}{n - 1}} = \sqrt{\frac{\Sigma x^2}{n - 1} - \frac{n\bar{x}^2}{n - 1}}$$

[4·12]

The *sample standard deviation* is the square root of the sample variance. It is similar to Equation 4·7, except that μ is replaced by the sample mean $\bar{x}$ and *N* is changed to *n* − 1.

p. 101:

$$\text{Sample standard score} = \frac{x - \bar{x}}{s}$$

[4·13]

Use this equation to find the standard score of an item in a *sample.*

p. 105:

$$\text{Population coefficient of variation} = \frac{\sigma}{\mu}(100)$$

[4·14]

The *coefficient of variation* is a relative measure of dispersion that enables us to compare two distributions. It relates the standard deviation and the mean by expressing the standard deviation as a percentage of the mean.

7   CHAPTER REVIEW EXERCISES

32. Calculate the standard deviation for the following data:

**Salaries of civil service employees
living in Washington, D.C.**

$36,500	$31,850	$18,885
29,750	29,900	25,510
47,000	16,275	30,000

33. An attitude survey, assessing political self-perception in 6 different communities, found the following percentages of people who categorized themselves as ultra-conservative:

5.25%, 5.5%, 5.75%, 6%, 6.5%, 7%.

Calculate the mean, variance, and standard deviation for these percentages.

34. How would you reply to the following statement? "Variability is not an important factor, because even though the outcome is more uncertain, you still have an equal chance of falling either above or below the median. Therefore, on average, the outcome will be the same."

35. Below are three general sections of one year's defense budget, each of which was allocated the same amount of funding by Congress:

a) Officer salaries (total)

b) Aircraft maintenance

c) Food purchases (total)

Considering the distribution of possible outcomes for the funds actually spent in each of these areas, match each section to one of the curves in Fig. 4· 1. Support your answers.

36. Weekly welfare checks in New York average $98.20 with a standard deviation of $15.40. Checks in California average $120.80, with a standard deviation of $21.40. Which state has the greater relative dispersion?

37. An educational psychologist has been keeping extensive records on the SAT scores of students admitted to his college over the past 20 years. He has noticed that in the last two years, although the average SAT score has remained about the same, the distribution of SAT scores has widened. Students' scores from these last two years have had significantly larger variations from the mean than students' scores from any of the previous two year periods for which he had records. What conclusions might be drawn from these observations?

38. In a study conducted by the U.S. Army, data were collected on the weekly tonnage of explosives expended on the test range at Aberdeen Proving Grounds. A sample of the data collected is given below.

Tons of explosives expended

2.65	2.89	3.02	3.51	3.78
3.98	4.02	4.15	4.39	4.59
4.88	5.01	5.90	6.01	6.95

Calculate the

a) range

b) interfractile range between the $1/3$ and $2/3$ fractiles

c) interfractile range between the $2/5$ and $3/5$ fractiles

d) interfractile range between the $3/5$ and $4/5$ fractiles

39. Two different sociologists, one in Bombay, India and one in Chicago, Illinois, are studying family size. What differences would you expect to find in the variability of their data?

40. James Wiley, the representative from the 7th district of New Mexico, claims that the military base located in his district has greater variation in personnel than any other base in the U.S. He cites a standard deviation of 1,051.5 personnel during the past year to support his conclusion. Howard Brackson, the representative from the 28th district of Idaho, has for years made the same claim. His latest figures show a standard deviation of 1,020.8 personnel for the base in his district. If the average number of personnel is 10,280 for the New Mexico base and 9,935 for the Idaho base, which representative is correct?

41. As part of a control program, samples are taken of welfare payments issued at each regional welfare office each week. The following data were collected during the final week in July at one such office:

Individual welfare payments (in dollars)

89.70	112.35	113.90	114.90	116.75
90.25	112.40	114.05	115.00	117.60
102.75	113.00	114.55	115.50	119.00

Calculate the range and the interfractile range between the $\frac{1}{3}$ and $\frac{2}{3}$ fractiles.

42. The U.S. Navy is planning to purchase two missiles for a new nuclear-powered attack ship. One missile has an average miss-distance of 700 feet with a standard deviation of 35 feet. The other missile has an average miss-distance of 300 feet and a standard deviation of 16 feet. Which missile is relatively less accurate?

43. Using the following population data, calculate the average absolute deviation, variance, and standard deviation:

Average heating fuel cost per gallon for eight states

53¢	58¢	54¢	63¢	62¢	55¢	59¢	56¢

44. Below is the average number of New York City policemen and policewomen on duty each day between 8 and 12 P.M. in the borough of Manhattan.

Monday 2,950	Wednesday 2,900	Friday 3,285	Sunday 2,975
Tuesday 2,900	Thursday 2,980	Saturday 3,430	

Calculate the variance and standard deviation of the distribution.

45. A psychologist wrote a computer program to simulate the way a person responds to a standard IQ test. To test the program, he gave the computer 15 different forms of a popular IQ test and computed its IQ from each form.

IQ values

121.85	123.50	124.75	125.15	125.15
130.05	131.00	131.75	132.50	132.95
133.10	133.50	133.75	135.50	141.40

a) Calculate the mean and standard deviation of the IQ scores.

b) According to Chebyshev's theorem, how many of the values should be between 120.11 and 140.67? How many are actually in that interval?

46. On a particular day, a city sanitation department measured the garbage weight in tons collected by the department's 40 trucks. The data were arranged in the following array:

Garbage weight (tons)

16.2	15.8	15.5	15.3	15.0	14.9	14.9	14.8	14.7	14.6
14.6	14.5	14.5	14.4	14.3	14.0	13.9	13.9	13.8	13.7
13.5	13.2	13.0	12.9	12.7	12.4	12.2	12.0	12.0	11.9
11.8	11.5	11.4	11.1	11.0	10.9	10.9	10.0	9.5	9.1

List the values in each decile.

47. Calculate the average absolute deviation, variance, and standard deviation for the following sample data:

Grade-point average of selected midshipmen at the U.S. Naval Academy

3.21	3.10	2.05	2.33	2.68	3.05	2.91	2.87	2.55	2.85

48. On two successive days, a sample was taken of the lengths of missions flown by pilots at an overseas air force base. The data are shown below:

Length of mission (hours)

Day 1	1.1	1.3	1.4	1.5	3.0
Day 2	1.2	1.4	1.6	1.9	2.2

a) Calculate the range of the two distributions.

b) Comment on using the range as a measure of dispersion for these data.

Answer true *or* false. *Answers are in the back of the book.*

1. The dispersion of a data set gives insight into the reliability of the measure of central tendency.

2. The standard deviation is equal to the square root of the variance.

3. The difference between the highest and lowest observations in a data set is called the quartile range.

4. The quartile deviation is based upon only two values taken from the data set.

5. The standard deviation is measured in the same units as the observations in the data set.

6. A fractile is a location in a frequency distribution at which a given proportion (or fraction) of the data lies at or above it.

7. The average absolute deviation, like the standard deviation, takes into account every observation in the data set.

8. The coefficient of variation is an absolute measure of dispersion.

9. The measure of dispersion most often used by statisticians is the standard deviation.

10. One of the advantages of dispersion measures is that any statistic that measures absolute variation also measures relative variation.

5

Probability I:

introductory ideas

1. History and relevance of probability theory, 116

2. Some basic concepts in probability, 117

3. Three types of probability, 119

4. Probability rules, 125

5. Probabilities under conditions of statistical independence, 130

6. Probabilities under conditions of statistical dependence, 138

7. Revising prior estimates of probabilities: Bayes' theorem, 144

8. Terms introduced, 151

9. Equations introduced, 152

10. Chapter review exercises, 154

11. Chapter concepts test, 159

Gamblers have used odds to make bets during most of recorded history. But it wasn't until the seventeenth century that a French nobleman named Antoine Gombauld (1607–1684) questioned the mathematical basis for success and failure at the dice tables. He asked the French mathematician Blaise Pascal (1623–1662), "What are the odds of rolling two sixes at least once in twenty-four rolls of a pair of dice?" Pascal solved the problem, having become as interested in the idea of probabilities as was Gombauld. They shared their ideas with the famous mathematician Pierre de Fermat (1601–1665), and the letters written by these three constitute the first academic journals in probability theory. We have no record of the degree of success enjoyed by these gentlemen at the dice tables, but we do know that their curiosity and research introduced many of the concepts we shall study in this chapter and the next.

1 HISTORY AND RELEVANCE OF PROBABILITY THEORY

Jacob Bernoulli (1654–1705), Abraham de Moivre (1667–1754), Reverend Thomas Bayes (1702–1761), and Joseph Lagrange (1736–1813) developed probability formulas and techniques. In the nineteenth century, Pierre Simon, Marquis de Laplace (1749–1827) unified all these early ideas and compiled the first general theory of probability.

Need for
probability theory

Probability theory was successfully applied at the gambling tables and, more relevant to our study, eventually to other social and economic problems. The insurance industry which emerged in the nineteenth century, required precise knowledge about the risk of loss in order to calculate premiums. Within fifty years, many learning centers were studying probability as a tool for understanding social phenomena. Today, the mathematical theory of probability is the basis for statistical applications in both social and decision-making research.

Examples of
the use of
probability theory

Probability is a part of our everyday lives. In personal and professional decisions, we face uncertainty and use probability theory whether or not we admit the use of something so sophisticated. When we hear a weather forecast of a 70 percent chance of rain, we change our plans from a picnic to a pool game. Playing bridge, we make some probability estimate before attempting a finesse. Managers who deal with inventories of highly styled women's clothing must wonder about the chances that sales will reach or exceed a certain level, and the buyer who stocks up on skateboards considers the probability of the life of this particular fad. Before Muhammad Ali's highly publicized fight with Leon Spinks, Ali was reputed to have said, "I'll give you **odds** I'm still the greatest when it's over." And when you begin to study for the inevitable quiz attached to the use of this book, you may ask yourself, "What are the chances the professor will ask us to recall something about the history of probability theory?"

We live in a world in which we are unable to forecast the future with complete certainty. Our need to cope with uncertainty leads us to the study and use of probability theory. In many instances we, as concerned citizens, will have some knowledge about the possible outcomes of a decision. By organizing this information and considering it systematically, we will be able to recognize our assumptions, communicate our reasoning to others, and make a sounder decision than we could by using a shot-in-the-dark approach.

EXERCISES

1. The insurance industry uses probability theory to calculate premium rates; but life insurers know for *certain* that every policyholder is going to die. Does this mean that probability theory does not apply to the life insurance business? Explain.

2. "Warning: the Surgeon General has determined that cigarette smoking is hazardous to your health." How might probability theory have played a part in that statement?

3. Is there really any such thing as an "uncalculated risk"? Explain.

4. A private school that has grades 1 through 8 at present decides to expand to a 12-year school, adding grades 9 through 12. In what ways do you think the decision involved probability theory?

2 SOME BASIC CONCEPTS IN PROBABILITY

In general, probability is the chance something will happen. Probabilities are expressed as fractions (¹⁄₆, ¹⁄₂, ⁸⁄₉) or as decimals (.167, .500, .889) between zero and one. Assigning a probability of zero means that something can never happen; a probability of one indicates that something will always happen.

In probability theory, an *event* is one or more of the possible outcomes of doing something. If we toss a coin, getting a tail would be an *event,* and getting a head would be another event. Similarly, if we are drawing from a deck of cards, selecting the ace of spades would be an event. An example of an event closer to your life, perhaps, is being picked from a class of 100 students to answer a question. When we hear the frightening predictions of highway traffic deaths, we hope not to be one of those events.

The activity that produces such an event is referred to in probability theory as an *experiment.* Using this formal language, we could ask the question, "In a coin toss *experiment,* what is the probability of the event *head*?" And, of course, if it is a fair coin with an equal chance of coming down on either side (and no chance of landing on its edge), we would answer, "¹⁄₂" or ".5." The set of all possible outcomes of an experiment is called the *sample space* for the experiment. In the coin toss experiment, the sample space is:

Sample space

$$S = \{\text{head, tail}\}$$

In the card drawing experiment, the sample space has 52 members: ace of hearts, deuce of hearts, and so on.

Most of us are less excited about coins or cards than we are interested in questions like, "What are the chances of making that plane connection?" or "What are my chances of getting a second job interview?" In short, we are concerned with the chances that certain events will happen.

Events are said to be *mutually exclusive* if one and only one of them can take place at a time. Consider again our example of the coin. We have two possible outcomes, heads and tails. On any toss, either heads or tails may turn up but not both. As a result, the events heads and tails on a single toss are said to be mutually exclusive. Similarly, you will either pass or fail this course or, before the course is over, you may drop it without a grade. Only one of those three outcomes can happen; they are said to be mutually exclusive events. The crucial question to ask in deciding whether events are

Mutually exclusive events

118

Chapter 5

PROBABILITY I:
INTRODUCTORY
IDEAS

A collectively
exhaustive list

really mutually exclusive is, "Can two or more of these events occur at one time?" If the answer is yes, the events are *not* mutually exclusive.

When a list of the possible events that can result from an experiment includes every possible outcome, the list is said to be *collectively exhaustive.* In our coin example, the list "head and tail" is collectively exhaustive (unless, of course, the coin stands on its edge when we toss it). In a presidential campaign, the list of outcomes "Democratic candidate and Republican candidate" is *not* a collectively exhaustive list of outcomes, since an independent candidate or the candidate of another party could conceivably win.

EXERCISES

5. Give a collectively exhaustive list of the possible outcomes of tossing two dice.

6. Which of the following are pairs of mutually exclusive events in the drawing of a single card from a standard deck of 52?

   a) A heart and a queen

   b) A club and a red card

   c) An even number and a spade

   d) An ace and an even number

   Which of the following are mutually exclusive outcomes in the rolling of two dice?

   a) A total of 5 and a 5 on one die

   b) A total of 7 and an even number of points on both dice

   c) A total of 8 and an odd number of points on both dice

   d) A total of 9 points and a 2 on one die

   e) A total of 10 points and a 4 on one die

7. Give the sample space of outcomes for the following "experiments" in terms of their sex makeup: the birth of (a) twins, (b) triplets.

8. Give the probability for each of the following totals in the rolling of two dice: 1, 2, 6, 7, 5, 10, 11.

9. In a recent meeting of campaign workers supporting Bill Remus for mayor, the campaign manager stated that the 'chances were good' that Remus would defeat the single opponent facing him in the election.

   a) What are the "events" that could take place with regard to the election?

   b) Is your list collectively exhaustive? Are the events in your list mutually exclusive?

   c) Disregarding the campaign manager's comments and knowing no other additional information, what probabilities would you assign to each of your events?

10. In the coming election, Joan Sprague, campaign manager for one of the gubernatorial candidates, is considering the distribution of campaign spending funds for promotions aimed at specific interest groups. The following table lists the only voting blocs in the state that she considers worthy of focused promotions.

Voting bloc	Cost of special campaign aimed at group
Minorities	$500,000
Big business	750,000
Women	250,000
Professionals and middle class	250,000
Labor	500,000

There is up to $1 million available for these special campaigns.

a) Are the voting blocs listed in the table collectively exhaustive? Are they mutually exclusive?

b) Make a collectively exhaustive and mutually exclusive list of the possible events of this spending decision.

c) Suppose the campaign manager has decided to spend the entire $1 million on special campaigns. Does this change your answer to part b? If so, what is your new answer?

3 THREE TYPES OF PROBABILITY

There are three basic ways of classifying probability. These three represent rather different conceptual approaches to the study of probability theory, and in fact experts disagree about which approach is the proper one to use. Let us begin by defining the

1. classical approach
2. relative frequency approach
3. subjective approach

Classical probability

Classical probability defines the probability that an event will occur as: Classical probability defined

$$\text{Probability of an event} = \frac{\text{Number of outcomes where the event occurs}}{\text{Total number of possible outcomes}} \quad [5 \cdot 1]$$

It must be emphasized that in order for Equation 5·1 to be valid, each of the possible outcomes must be equally likely. This is a rather complex way of defining something that may seem intuitively obvious to us, but we can use it to write our coin toss and dice rolling examples in symbolic form. First, we would state the question, "What is the probability of getting a head on one toss?" as:

P(Head)

Then, using formal terms, we get:

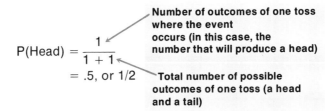

$$P(\text{Head}) = \frac{1}{1 + 1}$$
$$= .5, \text{ or } 1/2$$

Number of outcomes of one toss where the event occurs (in this case, the number that will produce a head)

Total number of possible outcomes of one toss (a head and a tail)

And for the dice rolling example:

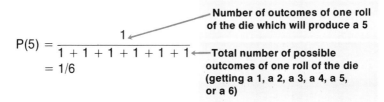

$$P(5) = \frac{1}{1 + 1 + 1 + 1 + 1 + 1}$$
$$= 1/6$$

Number of outcomes of one roll of the die which will produce a 5

Total number of possible outcomes of one roll of the die (getting a 1, a 2, a 3, a 4, a 5, or a 6)

A priori
probability

Classical probability is often called *a priori* probability because if we keep using orderly examples like fair coins, unbiased dice, and standard decks of cards, we can state the answer in advance (a priori) *without* tossing a coin, rolling a die, or drawing a card. We do not have to perform experiments to make our probability statements about fair coins, standard card decks, and unbiased dice. Instead, we can make statements based on logical reasoning before any experiments take place.

Shortcomings of the
classical approach

This approach to probability is useful when we deal with card games, dice games, coin tosses, and the like but has serious problems when we try to apply it to the less orderly decision problems we encounter in the real world. The classical approach to probability assumes a world that does not exist. It assumes away situations that are very unlikely but that could conceivably happen. Such occurrences as a coin landing on its edge, your classroom burning down during a discussion of probabilities, or your eating pizza while on a business trip at the North Pole are all extremely unlikely but not impossible. Nevertheless, the classical approach assumes them all away. Classical probability also assumes a kind of symmetry about the world, and that assumption can get us into trouble. Real-life situations, disorderly and unlikely as they often are, make it useful to define probabilities in other ways.

Relative frequency of occurrence

Suppose we begin asking ourselves complex questions such as, "What is the probability that I will live to be 85?" or, "What are the chances that I will blow one of my stereo speakers if I turn my 200-watt amplifier up to wide open?" or "What is the probability that the location of a new paper plant on the river near our town will cause a substantial fish kill?" We quickly see that we may not be able to state in advance, without experimentation, what these probabilities are. Other approaches may be more useful.

In the 1800s, British statisticians, interested in a theoretical foundation for calculating risk of losses in life insurance and commercial insurance, began defining probabilities from statistical data collected on births and deaths. Today this approach is called *relative frequency of occurrence.* It defines probability as either:

Probability redefined

1. the observed relative frequency of an event in a very large number of trials or

2. the proportion of times that an event occurs in the long run when conditions are stable

This method uses the relative frequencies of past occurrences as probabilities. We determine how often something has happened in the past and use that figure to predict the probability that it will happen again in the future. Let us look at an example. Suppose an insurance company knows from past actuarial data that of all males 40 years old, about 60 out of every 100,000 will die within a one-year period. Using this method, the company estimates the probability of death for that age group as:

Using the relative frequency of occurrence approach

$$\frac{60}{100,000} \text{ or } .0006$$

A second characteristic of probabilities established by the relative frequency of occurrence method can be shown by tossing one of our fair coins 300 times. Figure 5·1 illustrates the outcomes of these 300 tosses. Here we can see that although the proportion of heads was far from .5 in the first hundred tosses, it seemed to stabilize and approach .5 as the number of tosses increased. In statistical language, we would say that the relative frequency becomes stable as the number of tosses becomes large (if we are tossing the coin under uniform conditions). Thus, when we use the relative frequency approach to establish probabilities, our probability figure will gain accuracy as we increase the number of observations. Of course, this improved accuracy is not free; although more tosses of our coin will produce a more accurate probability of heads occurring, we must bear both the time and the cost of additional observations.

More trials, greater accuracy

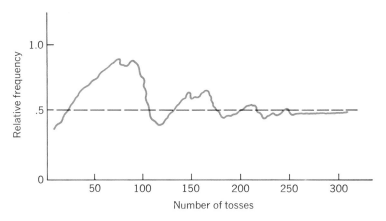

Number of tosses

FIG. 5·1
Relative frequency of occurrence of heads in 300 tosses of a fair coin

One difficulty with the relative frequency approach is that people often use it without evaluating a sufficient number of outcomes. If you heard someone say, "My aunt and uncle got the flu this year, and are both over 65, so everyone in that age bracket will probably get the flu," you would know that your friend did not base his assumptions on enough evidence. He had insufficient data for establishing a relative frequency of occurrence probability.

But what about a different kind of estimate, one that seems not to be based on statistics at all? Suppose your school's basketball team lost the first ten games of the year. You were a loyal fan, however, and bet $100 that your team would beat Indiana's in the eleventh game. To everyone's surprise, you won your bet. We would have difficulty convincing you that you were statistically incorrect. And you would be right to be skeptical about our argument. Perhaps without knowing that you did so, you may have based your bet on the statistical foundation described in the next approach to establishing probabilities.

Subjective probabilities

Subjective probabilities are based on the beliefs of the person making the probability assessment. In fact, subjective probability can be defined as the probability assigned to an event by an individual, based on whatever evidence is available. This evidence may be in the form of relative frequencies of past occurrences, or it may be just an educated guess. Probably the earliest subjective probability estimate of the likelihood of rain occurred when someone's Aunt Bess said, "My corns hurt; I think we're in for a downpour." Subjective assessments of probability permit the widest flexibility of the three concepts we have discussed. The decision maker can use whatever evidence is available and temper this with personal feelings about the situation.

Subjective probability assignments are frequently found when events occur only once or at most a very few times. Say that it is your job to interview and select a new social services caseworker. You have narrowed your choice to three people. Each has an attractive appearance, a high level of energy, abounding self-confidence, a record of past accomplishments, and a state of mind that seems to welcome challenges. What are the chances each will relate to clients successfully? Answering this question and choosing among the three will require you to assign a subjective probability to each person's potential.

Let's illustrate this kind of probability assignment with one more example. A judge is deciding whether to allow the construction of a nuclear power plant on a site where there is some evidence of a geological fault. He must ask himself the question, "What is the probability of a major nuclear accident at this location?" The fact that there is no relative frequency of occurrence evidence of previous accidents at this location does not excuse

him from making a decision. He must use his best judgment in trying to
determine the subjective probabilities of a nuclear accident.

   Since most higher level social and managerial decisions are concerned
with specific, unique situations, rather than with a long series of identical
situations, decision makers at this level make considerable use of subjective
probabilities.

   The subjective approach to assigning probabilities was introduced in
1926 by Frank Ramsey in his book, *The Foundation of Mathematics and
Other Logical Essays.* The concept was further developed by Bernard Koop-
man, Richard Good, and Leonard Savage, names that appear regularly in
advanced work in this field. Professor Savage pointed out that two reasonable people faced with the same evidence could easily come up with quite
different subjective probabilities for the same event. The two people who
made opposing bets on the outcome of the Indiana basketball game would
understand quite well what he meant.

EXERCISES

11. Below are the frequency distributions of height for groups of 25 fourteen-year-old boys and
   girls.

Height (in.)	# Boys in class	# Girls in class
48–51	2	4
52–55	4	5
56–59	3	7
60–63	8	6
64–67	7	3
68–71	1	0
	25	**25**

   a) What is the probability that a girl selected at random from the group of girls is between
      52 and 56 inches tall?

   b) What is the probability that a boy selected from the group of boys is between 60 and
      64 inches tall?

   c) Suppose we were to combine the boys and girls into one group and form a new fre-
      quency distribution. What would the probability of selecting an individual between 52
      and 56 inches tall be, based on this new distribution?

   d) What type of probability estimates are these?

12. Determine the probabilities of the following events in drawing a card from a standard deck
   of 52.

   a) A queen

   b) A club

   c) An ace in a red suit

   d) A red card

   e) A face card (king, queen, or jack)

   f) What type of probability estimates are these?

13. Below is a frequency distribution of annual incomes from a survey of 250 households. Based on this information, what is the probability that a household has an income

a) between $8,000 and $12,000

b) less than $8,000

c) more than $24,000

d) between $12,000 and $16,000

Income	Frequency
$ 0– 3,999	5
4,000– 7,999	15
8,000–11,999	40
12,000–15,999	90
16,000–19,999	30
20,000–23,999	25
24,000+	20

14. A campaign worker is trying to predict the number of votes his candidate will receive in the primary election. He has limited his possible predictions to 200,000; 250,000; 300,000; 350,000 or 400,000 votes. He feels unable to decide whether 300,000 or 350,000 votes are more likely. But he feels that 350,000 votes are twice as likely as 400,000 and that 300,000 votes are four times as likely as 200,000 votes. Finally, he thinks that 250,000 votes are only half as likely as 350,000 votes.

a) What are the probabilities of the five vote counts according to the campaign worker?

b) What has the campaign worker implicitly said concerning the probability of getting greater than 400,000 or fewer than 200,000 votes?

15. The research assistant in the experimental lab of the psychology department has the following data on the functioning of the recording apparatus in the cages of the lab:

Response recorder number	Days functioning	Days out of service
1	244	16
2	252	8
3	237	23
4	208	52
5	254	6

What is the probability of a recorder being out of service on a given day?

16. Classify the following probability estimates as to their type (classical, relative frequency, or subjective):

a) The probability that you will make a B in this course is .75.

b) The probability that a randomly selected family from a particular community has two children is .25.

c) The probability that my candidate will win the election is .60.

d) The probability that a student from this high school will go on to college is .90.

e) The probability of my ticket winning a raffle drawing for which 1,000 tickets were sold is .001.

4 PROBABILITY RULES

Most decision makers who use probabilities are concerned with two conditions:

1. the case where one event *or* another will occur
2. the situation where two or more events will both occur

We are interested in the first case when we ask, "What is the probability that today's demand will exceed our inventory?" To illustrate the second situation, we could ask, "What is the probability that today's demand will exceed our inventory *and* that more than 10 percent of our sales force will not report for work?" In the sections to follow, we shall illustrate methods of determining answers to questions like these under a variety of conditions.

Some commonly used symbols, definitions, and rules

Symbol for a marginal probability. In probability theory, we use symbols to simplify the presentation of ideas. As we discussed earlier in this chapter, the probability of the event *A* would be expressed as:

$$P(A) \quad \text{the } \textbf{probability} \text{ of } \textbf{event } A \text{ happening}$$

A *single* probability means that only one event can take place. It is called a *marginal,* or *unconditional probability.* To illustrate, let us suppose that 50 members of a school class drew tickets to see which student would get a free trip to the National Rock Festival. Any one of the students could calculate his or her chances of winning by the formulation:

$$P(\text{Winning}) = \frac{1}{50}$$
$$= .02$$

In this case, a student's chance is 1 in 50 because we are certain that the possible events are mutually exclusive; that is, only one student can win at a time.

There is a nice diagrammatic way to illustrate this example and other probability concepts. We use a pictorial representation called a *Venn diagram,* after the nineteenth-century English mathematician, John Venn. In these diagrams, the entire sample space is represented by a rectangle, and events are represented by parts of the rectangle. If two events *are* mutually exclusive, their parts of the rectangle will not overlap each other, as shown in Fig. 5·2(*a*). If two events are *not* mutually exclusive, their parts of the rectangle *will* overlap, as in Fig. 5·2(*b*).

Since probabilities behave a lot like areas, we shall let the rectangle have an area of one (because the probability of *something* happening is one). Then the probability of an event is the area of *its* part of the rectangle.

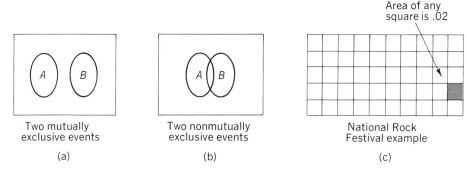

FIG. 5·2
**Some Venn
diagrams**

Area of any
square is .02

Two mutually
exclusive events

(a)

Two nonmutually
exclusive events

(b)

National Rock
Festival example

(c)

Figure 5·2(c) illustrates this for the National Rock Festival example. There the rectangle is divided into 50 equal, nonoverlapping parts.

Addition rule for mutually exclusive events. Often, however, we are interested in the probability that one thing *or* another will occur. If these two events are mutually exclusive, we can express this probability using the addition rule for mutually exclusive events. This rule is expressed symbolically as:

Probability of
one or more
mutually exclusive
events

P(A or B)   the **probability** of either **A** or **B** happening

and is calculated as follows:

$$P(A \text{ or } B) = P(A) + P(B) \qquad [5\cdot2]$$

This addition rule is illustrated by the Venn diagram in Fig. 5·3, where we note that the area in the two circles together (denoting the event *A or B*) is the sum of the areas of the two circles.

FIG. 5·3
**Venn diagram for
the addition rule
for mutually
exclusive events**

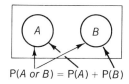

P(A or B) = P(A) + P(B)

Let's use this formula in an example. Five equally capable students are waiting for a summer job interview with a company that has announced that it will hire only one of five by random drawing. The group consists of Bill, Helen, John, Sally, and Walter. If our question is, "What is the probability that John will be the candidate?" we can use Equation 5·1 and give the answer:

$$P(\text{John}) = \frac{1}{5}$$
$$= .2$$

If, however, we ask, "What is the probability that either John *or* Sally will be the candidate?" we would use Equation 5·2:

$$P(\text{John or Sally}) = P(\text{John}) + P(\text{Sally})$$
$$= \frac{1}{5} + \frac{1}{5}$$
$$= \frac{2}{5}$$
$$= .4$$

Let's calculate the probability of two or more events happening once more. Table 5·1 contains data on the size of families in a certain town. We are interested in the question, "What is the probability that a family chosen at random from this town will have 4 or more children (that is, 4, 5, or 6 or more children)?" Using Equation 5·2, we can calculate the answer as:

$$P(4, 5, 6 \text{ or more}) = P(4) + P(5) + P(6 \text{ or more})$$
$$= .15 + .10 + .05$$
$$= .30$$

Number of children	0	1	2	3	4	5	6 or more
Proportion of families having this many children	.05	.10	.30	.25	.15	.10	.05

TABLE 5·1
Family size data

There is an important special case of Equation 5·2. For any event *A*, either *A* happens or it doesn't. So the events *A* and *not A* are exclusive and exhaustive. Applying Equation 5·2 yields the result:

A special case of Equation 5·2

$$P(A) + P(not\,A) = 1$$

or equivalently:

$$P(A) = 1 - P(not\,A)$$

For example, referring back to Table 5·1, the probability of a family having 5 or fewer children is most easily obtained by subtracting from 1 the probability of the family having 6 or more children, and thus is seen to be .95.

Addition rule for events that are not mutually exclusive. If two events are not mutually exclusive, it is possible for both events to occur. In these cases, our addition rule must be modified. For example, what is the probability of drawing either an ace *or* a heart from a deck of cards? Obviously, the events ace and heart can occur together because we could draw the ace of hearts. Thus, ace and heart are not mutually exclusive events. We must adjust our Equation 5·2 to avoid double counting; that is, we have to *reduce* the probability of drawing either an ace or a heart *by the chance* that we could draw both of them together. As a result, the correct equation for the probability of one or more of two events that are not mutually exclusive is:

Probability of one or more events *not* mutually exclusive

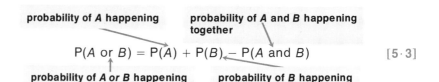

$$P(A \text{ or } B) = P(A) + P(B) - P(A \text{ and } B) \qquad [5\cdot3]$$

A Venn diagram illustrating Equation 5·3 is given in Fig. 5·4. There, the event *A or B* is outlined with a heavy line. The event *A and B* is the cross-hatched wedge in the middle. If we add the areas of circles *A* and *B*, we *double count* the area of the wedge, and so we must subtract it, to make sure it is counted only once.

FIG. 5·4

Venn diagram for the addition rule for two not mutually exclusive events

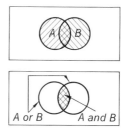

Using Equation 5·3 to determine the probability of drawing either an ace *or* a heart, we can calculate:

$$P(\text{Ace or Heart}) = P(\text{Ace}) + P(\text{Heart}) - P(\text{Ace and Heart})$$

$$= \tfrac{4}{52} + \tfrac{13}{52} - \tfrac{1}{52}$$
$$= \tfrac{16}{52} \text{ or } \tfrac{4}{13}$$

Let's do a second example. The employees of a certain company have elected five of their number to represent them on the employee-management productivity council. Profiles of the five are as follows:

#1	male	age 30	
2	male	32	
3	female	45	
4	female	20	
5	male	40	

This group decides to elect a spokesperson by drawing a name from a hat. Our question is, "What is the probability the spokesperson will be *either* female *or* over 35?" Using Equation 5·3, we can set up the solution to our question like this:

$$P(\text{Female or Over 35}) = P(\text{Female}) + P(\text{Over 35}) - P(\text{Female and Over 35})$$

$$= \tfrac{2}{5} + \tfrac{2}{5} - \tfrac{1}{5}$$
$$= \tfrac{3}{5}$$

We can check our work by inspection and see that, of the five persons in the group, three would fit the requirements of being either female or over 35.

EXERCISES

From the Venn diagrams below, which indicate the number of outcomes of an experiment corresponding to each event and the number of outcomes that do not correspond to either event, give the probabilities indicated:

17. Total outcomes = 60

P(A) =

P(B) =

P(A or B) =

18. Total outcomes = 50

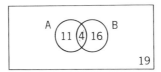

P(A) =

P(B) =

P(A or B) = $P(A) + P(B) - P(A$

19. An urn contains 60 marbles: 40 are blue, and 15 of these blue marbles are swirled. The rest of the marbles are red, and 10 of the red ones are swirled. The marbles that are not swirled are clear. What is the probability of drawing

   a) a red marble from the urn?

   b) a clear marble from the urn?

   c) a red, swirled marble?

   d) a blue, clear marble?

   e) a swirled marble?

20. As the safety officer of an airline, Debbie Best has been asked to give a talk to the press concerning engine safety. As part of her talk, she has decided to include the probability of a two-engine jet having engine failure on a flight. After consulting her records, she finds the following information about last year's operating record for two-engine aircraft:

   Twenty-nine reported failure of the right engine.
   Thirty-three reported failure of the left engine.
   There was one crash attributed to double engine failure.
   There were 345,000 flights during the year.

   What probability should she report to the press?

21. Mary Harper, member of a women's activist group, feels that the Equal Rights Amendment will pass next year in six states if one of two things happens: (1) enough pro-ERA candidates are elected to office (and none of its current supporters voted out of office) or (2) the lobby in each state can change the minds of a certain number of current congressmen who are nonsupporters. The following estimates are available to her:

State	1	2	3	4	5	6
Probability of electing enough pro-ERA congressmen	.025	.10	.15	.15	.05	.025
Probability of changing minds of current nonsupporters	.005	.02	.075	.10	.15	.15

Ms. Harper also believes that there is no chance that both of these events will happen this year in any of the states. In which state is the ERA most likely to be passed in the next year, given these probability estimates?

22. The manager of a chemical plant located on the Mississippi River knows that in an upcoming court case the company may be found guilty of polluting the river. Further, he knows that if found guilty, the company will be required to install a water purification system, pay a fine, or both. Thus far, only 10 percent of the companies involved in similar cases have been both fined and required to install the purification system. In addition, when the court's ruling has not involved both penalties, a company has been three times more likely to be fined than to be required to install the purification system. If 28 percent of the companies have been found guilty thus far, what is the probability that this company will be required to install a purification system?

23. In this section, two expressions were developed for the probability of either of two events, A or B, occurring. Referring to Equations 5·2 and 5·3,

   a) What can you say about the probability of A and B occurring simultaneously when A and B are *mutually exclusive?*

   b) Develop an expression for the probability that at least one of three events A, B, or C could occur; i.e., P(A or B or C). Do *not* assume that A, B, and C are mutually exclusive of each other.

   c) Rewrite your expression for the case in which A and B are mutually exclusive, but A and C and B and C are not mutually exclusive.

   d) Rewrite your expression for the case in which A and B *and* A and C are mutually exclusive, but not B and C.

   e) Rewrite your expression for the case in which A, B, and C are each mutually exclusive of the others.

5 PROBABILITIES UNDER CONDITIONS OF STATISTICAL INDEPENDENCE

Independence defined

When two events happen, the outcome of the first event may or may not have an effect on the outcome of the second event. That is, the events may be either dependent or independent. In this section, we examine events that are *statistically independent:* the occurrence of one event *has no effect* on the probability of the occurrence of any other event. There are three types of probabilities under statistical independence:

1. marginal
2. joint
3. conditional

Marginal probabilities under statistical independence

Marginal probability of independent events

As we explained previously, a marginal or unconditional probability is the simple probability of the occurrence of an event. In a fair coin toss, $P(H)$ = .5 and $P(T)$ = .5; that is, the probability of heads equals .5, and the probability of tails equals .5. This is true for every toss, no matter how many tosses have been made or what their outcomes have been. Every toss stands alone and is in no way connected with any other toss. Thus the outcome of *each* toss of a fair coin is a statistically independent event.

Imagine that we have a biased or unfair coin that has been altered in such a way that heads occurs .90 of the times and tails .10 of the time. On each individual toss, $P(H) = .90$ and $P(T) = .10$. The outcome of any particular toss is completely unrelated to the outcomes of the tosses that may precede or follow it. The outcome of each toss of *this* coin is a statistically independent event, too, even though the coin is biased.

Joint probabilities under statistical independence

The probability of two or more independent events occurring together or in succession is the product of their marginal probabilities. Mathematically, this is stated:

$$P(AB) = P(A) \times P(B) \qquad [5\cdot4]$$

Multiplication rule for joint, independent events

where:
$P(AB)$ = probability of events A and B occurring together or in succession; this is known as a *joint probability*
  $P(A)$ = marginal probability of event A occurring
  $P(B)$ = marginal probability of event B occurring

In terms of the fair coin example, the probability of heads on two successive tosses is the probability of heads on the first toss (which we shall call H_1) times the probability of heads on the second toss (H_2). That is $P(H_1 H_2) = P(H_1) \times P(H_2)$. We have shown that the events are statistically independent because the probability of any outcome is not affected by any preceding outcome. Therefore, the probability of heads on any toss is .5, and $P(H_1 H_2) = .5 \times .5 = .25$. Thus the probability of heads on two successive tosses is .25.

Likewise, the probability of getting 3 heads on three successive tosses is $P(H_1 H_2 H_3) = .5 \times .5 \times .5 = .125$.

Assume next that we are going to toss an unfair coin that has $P(H) = .8$ and $P(T) = .2$. The events (outcomes) are independent because the probabilities of all tosses are exactly the same—the individual tosses are completely separate and in no way affected by any other toss or outcome. Suppose our question is, "What is the probability of getting 3 heads on three successive tosses?" We use Equation 5·4 and discover that:

$$P(H_1 H_2 H_3) = P(H_1) \times P(H_2) \times P(H_3) = .8 \times .8 \times .8 = .512$$

Now let us ask the probability of getting 3 tails on three successive tosses:

$$P(T_1 T_2 T_3) = P(T_1) \times P(T_2) \times P(T_3) = .2 \times .2 \times .2 = .008$$

Note that these two probabilities do not add up to 1 because the events $H_1 H_2 H_3$ and $T_1 T_2 T_3$ do not constitute a collectively exhaustive list. They *are* mutually exclusive, because if one occurs, the other cannot.

We can make the probabilities of events even more explicit using a *probability tree*. *Figure 5·5* is a probability tree showing the possible outcomes and their respective probabilities for one toss of a fair coin.

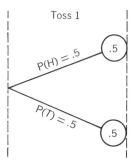

For toss 1 we have two possible outcomes, heads and tails, each with a probability of .5. Assume that the outcome of toss 1 is heads. We toss again. The second toss has two possible outcomes, heads and tails, each with a probability of .5. In Fig. 5·6 we add these two branches of the tree.

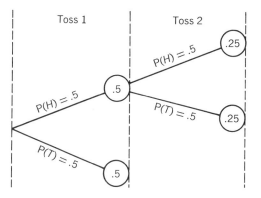

Next we consider the possibility that the outcome of toss 1 is tails. Then the second toss must stem from the second branch representing toss 1. Thus in Fig. 5·7, we add two more branches to the tree. Notice that on two tosses we have four possible outcomes: $H_1 H_2$, $H_1 T_2$, $T_1 H_2$, and $T_1 T_2$ (remember that the subscripts indicate the toss number and that T_2, for example, means tails on toss 2). Thus, after two tosses we may arrive at any one of four possible points. Since we are going to toss three times, we must add more branches to the tree.

Assuming that we have had heads on the first two tosses, we are now ready to begin adding branches for the third toss. As before, the two possible outcomes are heads and tails, each with a probability of .5. The first step

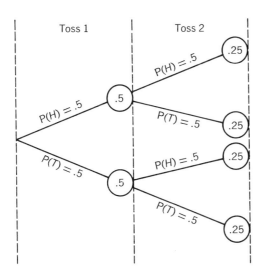

FIG. 5·7
**Probability tree of
two tosses**

is shown in Fig. 5·8. The additional branches are added in exactly the same manner. The completed probability tree is shown in Fig. 5·9. Notice that both heads and tails have a probability of .5 of occurring no matter how far from the origin (first toss) any particular toss may be. **This follows from our definition of independence: no event is affected by the events preceding or following it.**

All tosses
are independent

Suppose we are going to toss a fair coin and want to know the probability that all three tosses will result in heads. Expressing the problem symbolically, we want to know $P(H_1 H_2 H_3)$. From the mathematical definition of the joint probability of independent events, we know that

$$P(H_1 H_2 H_3) = P(H_1) \times P(H_2) \times P(H_3) = .5 \times .5 \times .5 = .125$$

We could have read this answer from the probability tree in Fig. 5·9 by following the branches giving $H_1 H_2 H_3$.

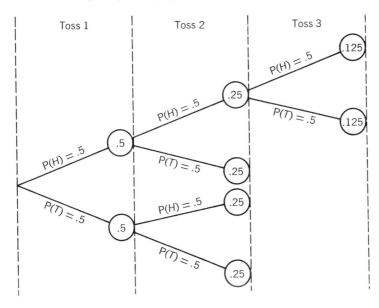

FIG. 5·8
**Probability tree of
partial third toss**

FIG. 5·9
Completed
probability tree

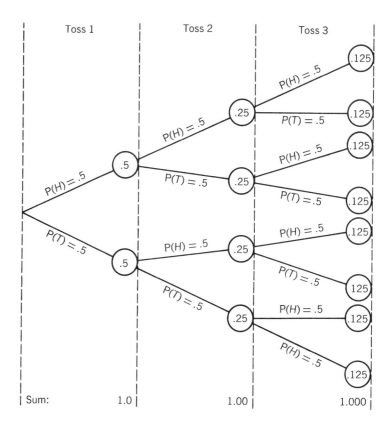

Try solving these problems using the probability tree in Fig. 5·9:

* EXAMPLE 1: What is the probability of getting tails, heads, tails *in that order* on three successive tosses of a fair coin?

* SOLUTION: $P(T_1 H_2 T_3) = P(T_1) \times P(H_2) \times P(T_3) = .125$. Following the prescribed path on the probability tree will give us the same answer.

* EXAMPLE 2: What is the probability of getting tails, tails, heads *in that order* on three successive tosses of a fair coin?

* SOLUTION: If we follow the branches giving tails on the first toss, tails on the second toss, and heads on the third toss, we arrive at the probability of .125. Thus $P(T_1 T_2 H_3) = .125$.

It is important to notice that the probability of arriving at a given point by a given route is *not* the same as the probability of, say, heads on the third toss. $P(H_1 T_2 H_3) = .125$, but $P(H_3) = .5$. The first is a case of *joint probability,* that is, the probability of getting heads on the first toss, tails on the second, and heads on the third. The latter, by contrast, is simply the *marginal probability* of getting heads on a particular toss, in this instance toss 3.

Notice that the sum of the probabilities of all the possible outcomes for each toss is 1. This results from the fact that we have mutually exclusive and collectively exhaustive lists of outcomes. These are given in Table 5·2.

TABLE 5·2

Lists of outcomes

1 Toss		2 Tosses		3 Tosses	
Possible outcomes	Probability	Possible outcomes	Probability	Possible outcomes	Probability
H_1	.5	H_1H_2	.25	$H_1H_2H_3$	.125
T_1	.5	H_1T_2	.25	$H_1H_2T_3$	.125
	1.0	T_1H_2	.25	$H_1T_2H_3$	.125
		T_1T_2	.25	$H_1T_2T_3$	.125
			1.00	$T_1H_2H_3$	.125
				$T_1H_2T_3$	.125
				$T_1T_2H_3$	.125
				$T_1T_2T_3$	.125
					1.000

♦ EXAMPLE 3: What is the probability of *at least* two heads on three tosses?

♦ SOLUTION: Recalling that the probabilities of mutually exclusive independent events are additive, we can note the possible ways that at least two heads on three tosses can occur, and we can sum their individual probabilities. The outcomes satisfying the requirement are $H_1H_2H_3$, $H_1H_2T_3$, $H_1T_2H_3$, and $T_1H_2H_3$. Since each of these has an individual probability of .125, the sum is .5. Thus the probability of at least 2 heads on three tosses is .5.

♦ EXAMPLE 4: What is the probability of *at least* 1 tail on three tosses?

♦ SOLUTION: There is only one case in which no tails occur, namely, $H_1H_2H_3$. Therefore we can simply subtract for the answer:

$$1 - P(H_1H_2H_3) = 1 - .125 = .875$$

The probability of at least 1 tail occurring in three successive tosses is .875.

♦ EXAMPLE 5: What is the probability of *at least* 1 head on two tosses?

♦ SOLUTION: The possible ways a head may occur are H_1H_2, H_1T_2, T_1H_2. Each of these has a probability of .25. Therefore, the probability of at least 1 head on 2 tosses is .75. Alternatively, we could consider the case in which no head occurs—namely, T_1T_2—and subtract its probability from 1, that is,

$$1 - P(T_1T_2) = 1 - .25 = .75$$

Conditional probabilities under statistical independence

Thus far we have considered two types of probabilities, marginal (or unconditional) probability and joint probability. Symbolically, marginal probability is P(A) and joint probability is P(AB). Besides these two, there is

Conditional probability

one other type of probability, known as *conditional* probability. Symbolically, conditional probability is written:

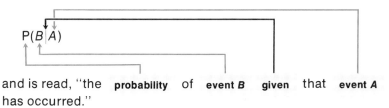

and is read, "the **probability** of **event B** **given** that **event A** has occurred."

Conditional probability is the probability that a second event (B) will occur *if* a first event (A) has already happened.

Conditional
probability of
independent events

For statistically independent events, the conditional probability of event B given that event A has occurred is simply the probability of event B:

$$P(B|A) = P(B) \qquad [5 \cdot 5]$$

At first glance, this may seem contradictory. Remember, however, that by definition independent events are those whose probabilities are in no way affected by the occurrence of each other. In fact, statistical independence is defined symbolically as the condition in which $P(B|A) = P(B)$.

We can understand conditional probability better by solving an illustrative problem. Our question is, "What is the probability that the second toss of a fair coin will result in heads, given that heads resulted on the first toss? Symbolically, this is written as $P(H_2|H_1)$. Remember that for two independent events, the results of the first toss have absolutely no effect on the results of the second toss. Since the probabilities of heads and tails are identical for every toss, the probability of heads on the second toss is .5. Thus, we must say that $P(H_2|H_1) = .5$.

Table 5·3 summarizes the three types of probabilities and their mathematical formulas under conditions of statistical independence.

TABLE 5·3
Probabilities under statistical independence

Type of probability	Symbol	Formula
Marginal	$P(A)$	$P(A)$
Joint	$P(AB)$	$P(A) \times P(B)$
Conditional	$P(B|A)$	$P(B)$

EXERCISES

24. What is the probability that a couple's second child will be
    a) a boy, given that their first child was a girl?
    b) a girl, given that their first child was a girl?

25. In rolling two dice, what is the probability of rolling

   a) a total of 7 on the first roll, followed by a total of 5 on the second roll and another total of 5 on the third roll?

   b) three sets of doubles in three rolls?

   c) two sets of doubles in two rolls?

26. What is the probability that, in selecting two cards, one at a time, from a deck with replacement, the second card is

   a) a spade, given that the first card was a heart?

   b) black, given that the first card was red?

   c) a queen, given that the first card was a queen?

27. Use a probability tree to answer the following questions. Assuming A, B, and C are independent events with marginal probabilities: $P(A) = .2$, $P(B) = .5$, $P(C) = .3$; and that the subscripts represent trial numbers, find

   a) $P(A_1 B_2 C_3)$

   b) $P(C_1 C_2 C_3)$

   c) $P(A_1 C_2 B_3 C_4)$

   d) $P(A_1 B_2)$

   e) $P(B_1 B_2)$

28. Sue Martin, a social psychologist working on crowding behavior, uses human observers and machine recorders as a check to keep track of subject responses in a large project. She knows that 5 percent of the time the human observer is apt to misrecord the number of responses made, and 2 percent of the time the machine will malfunction and misrecord the number of responses.

   a) If Martin finds that the machine recorder was malfunctioning, what is the probability that the human observer will have made a mistake in recording the number of responses?

   b) If she knows that the human observer failed to record the number of responses correctly, what is the probability that she will find that the machine has malfunctioned?

   c) What is the probability that the human observer will misrecord the number of responses and the machine recorder will malfunction during the same experimental session (i.e., at the same time)?

29. A social psychologist plans to use two current topics of interest—abortion and support for nuclear power plants—in a proposed study of attitude changes. He knows from a questionnaire completed at the beginning of the experiment that 35 percent of the subjects favor the construction of nuclear power plants and 50 percent are in favor of federally subsidized abortions. He also knows that individual support for one issue is independent of support for the other issue.

   a) What is the probability that a subject supports both federally funded abortions and the construction of nuclear power plants?

   b) What is the probability that a person supports federally funded abortions or nuclear power plants, but not both?

6 PROBABILITIES UNDER CONDITIONS OF STATISTICAL DEPENDENCE

Statistical dependence exists when the probability of some event is dependent upon or affected by the occurrence of some other event. Just as with independent events, the types of probabilities under statistical dependence are

1. conditional
2. joint
3. marginal

Conditional probabilities under statistical dependence

Conditional and joint probabilities under statistical dependence are more involved than marginal probabilities are. We shall discuss conditional probabilities first, because the concept of joint probabilities is best illustrated by using conditional probabilities as a basis.

Examples of conditional probability of dependent events

Assume that we have one box containing 10 balls distributed as follows:

3 are colored and dotted
1 is colored and striped
2 are gray and dotted
4 are gray and striped

The probability of drawing any one ball from this box is .1, since there are 10 balls, each with equal probability of being drawn. The discussion of the following examples will be facilitated by reference to Table 5·4 and to Fig. 5·10, which shows the contents of the box in diagram form.

◆ QUESTION 1: Suppose someone draws a colored ball from the box. What is the probability that it is dotted? What is the probability it is striped?

◆ SOLUTION: This question can be expressed symbolically as $P(D|C)$, or "What is the conditional probability that this ball is dotted, *given* that it is colored?"

TABLE 5·4
Color and configuration of 10 balls

Event	Probability of event	
1	.1	
2	.1	colored and dotted
3	.1	
4	.1	colored and striped
5	.1	
6	.1	gray and dotted
7	.1	
8	.1	
9	.1	gray and striped
10	.1	

FIG. 5·10
Contents of the box

Gray

2 balls are gray and dotted

Colored

3 balls are colored and dotted

4 balls are gray and striped

1 ball is colored and striped

We have been told that the ball that was drawn is colored. Therefore, to calculate the probability that the ball is dotted, we will ignore *all* the gray balls and concern ourselves with colored only. In diagram form, we consider only what is shown in Fig. 5·11.

From the statement of the problem, we know that there are 4 colored balls, 3 of which are dotted and 1 of which is striped. Our problem is now to find the simple probabilities of dotted and striped. To do so we divide the number of balls in each category by the total number of colored balls:

$$P(D|C) = \frac{3}{4} = .75$$

$$P(S|C) = \frac{1}{4} = \underline{.25}$$

$$\mathbf{1.00}$$

In other words, three-fourths of the colored balls are dotted, and one-fourth of the colored balls are striped. Thus, the probability of dotted, given that the ball is colored, is .75. Likewise, the probability of striped, given that the ball is colored, is .25.

Now we can see how our reasoning will enable us to develop the formula for conditional probability under statistical dependence. We can first assure ourselves that these events *are* statistically dependent by observing that the color of the balls determines the probabilities that they are either striped or dotted. For example, a gray ball is more likely to be striped than a colored ball. Since color affects the probability of striped or dotted, these two events are dependent.

To calculate the probability of dotted given colored, $P(D|C)$, we divided the probability of colored and dotted balls (3 out of 10, or .3) by the probability of colored balls (4 out of 10, or .4):

Formula for conditional probability of dependent events

$$P(D|C) = \frac{P(DC)}{P(C)}$$

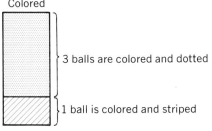

Colored

3 balls are colored and dotted

1 ball is colored and striped

FIG. 5·11
Probability of dotted and striped, given colored

Expressed as a general formula using the letters *A* and *B* to represent the two events, the equation is:

$$P(B|A) = \frac{P(BA)}{P(A)} \qquad [5\cdot6]$$

This *is* the formula for *conditional probability under statistical dependence.*

◆ QUESTION 2: Continuing with our example of the colored and gray balls, let's answer the question, "What is $P(D|G)$?" and, "What is $P(S|G)$?"

◆ SOLUTION:

$$P(D|G) = \frac{P(DG)}{P(G)} = \frac{.2}{.6} = \frac{1}{3}$$

$$P(S|G) = \frac{P(SG)}{P(G)} = \frac{.4}{.6} = \frac{2}{3}$$

$$\mathbf{1.0}$$

The problem is shown diagrammatically in Fig. 5·12.

FIG. 5·12
**Probability of
dotted and striped,
given gray**

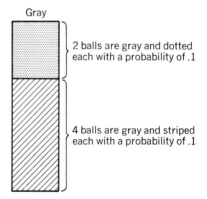

Gray

2 balls are gray and dotted
each with a probability of .1

4 balls are gray and striped
each with a probability of .1

The total probability of gray is .6 (6 out of 10 balls). To determine the probability that the ball (which we know is gray) will be dotted, we divide the probability of gray and dotted (.2) by the probability of gray (.6), or .2/.6 = 1/3. Similarly, to determine the probability that the ball will be striped, we divide the probability of gray and striped (.4) by the probability of gray (.6), or .4/.6 = 2/3.

◆ QUESTION 3: Calculate $P(G|D)$ and $P(C|D)$.

◆ SOLUTION: Figure 5·13 shows the contents of the box arranged according to the striped or dotted markings on the balls. Since we have been told that the ball that was drawn is dotted, we can disregard striped and consider only dotted.

FIG. 5·13
**Contents of the
box arranged by
configuration,
striped and dotted**

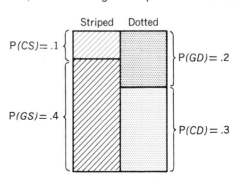

Striped   Dotted

$P(CS) = .1$

$P(GD) = .2$

$P(GS) = .4$

$P(CD) = .3$

Now see Fig. 5·14 showing the probabilities of colored and gray given dotted. Notice that the relative proportions of the two are as .4 is to .6. The calculations used to arrive at these proportions were:

$$P(G|D) = \frac{P(GD)}{P(D)} = \frac{.2}{.5} = .4$$

$$P(C|D) = \frac{P(CD)}{P(D)} = \frac{.3}{.5} = .6$$

$$\overline{1.0}$$

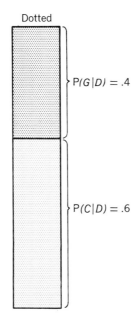

Dotted

$P(G|D) = .4$

$P(C|D) = .6$

FIG. 5·14
**Probability of
colored and gray,
given dotted**

◆ QUESTION 4: Calculate $P(C|S)$ and $P(G|S)$.

◆ SOLUTION:

$$P(C|S) = \frac{P(CS)}{P(S)} = \frac{.1}{.5} = .2$$

$$P(G|S) = \frac{P(GS)}{P(S)} = \frac{.4}{.5} = .8$$

$$\overline{1.0}$$

Joint probabilities under statistical dependence

We have shown that the formula for conditional probability under conditions of statistical dependence is:

$$P(B|A) = \frac{P(BA)}{P(A)} \qquad [5·6]$$

If we solve this for $P(BA)$ by cross multiplication, we have the formula for *joint probability under conditions of statistical dependence:*

joint probability of events _B_ and _A_ happening together or in succession

probability of event _B_ given that event _A_ has happened

$$P(\overset{\downarrow}{B}A) = P(B|A) \times P(A)*$$

[5·7]

probability that event _A_ will happen

Notice that this formula is _not_ $P(BA) = P(B) \times P(A)$, as it would be under conditions of statistical independence.

Converting the general formula $P(BA) = P(B|A) \times P(A)$ to our example and to the terms of colored, gray, dotted, and striped, we have $P(CD) = P(C|D) \times P(D)$, or $P(CD) = .6 \times .5 = .3$. Here .6 is the probability of colored given dotted (computed in example 3 above), and .5 is the probability of dotted (also computed in example 3).

$P(CD) = .3$ can be verified in Table 5·4, where we originally arrived at the probability by inspection: 3 balls out of 10 are colored and dotted.

The following joint probabilities are computed in the same manner and can also be substantiated by reference to Table 5·4.

$$P(CS) = P(C|S) \times P(S) = .2 \times .5 = .1$$
$$P(GD) = P(G|D) \times P(D) = .4 \times .5 = .2$$
$$P(GS) = P(G|S) \times P(S) = .8 \times .5 = .4$$

Marginal probabilities under statistical dependence

Marginal probabilities under statistical dependence are computed by summing up the probabilities of all the joint events in which the simple event occurs. In the above example, we can compute the marginal probability of the event colored by summing the probabilities of the joint events in which colored occurred:

$$P(C) = P(CD) + P(CS) = .3 + .1 = .4$$

Similarly, the marginal probability of the event gray can be computed by summing the probabilities of the joint events in which gray occurred:

$$P(G) = P(GD) + P(GS) = .2 + .4 = .6$$

In like manner, we can compute the marginal probability of the event dotted by summing the probabilities of the joint events in which dotted occurred:

$$P(D) = P(CD) + P(GD) = .3 + .2 = .5$$

*To find the joint probability of events _A_ and _B_, you could also use the formula $P(BA) = P(AB) = P(A|B) \times P(B)$. This is because $BA = AB$.

TABLE 5·5 Probabilities under statistical independence and dependence

Type of probability	Symbol	Formula under statistical independence	Formula under statistical dependence
Marginal	$P(A)$	$P(A)$	Sum of the probabilities of the joint events in which A occurs
Joint	$P(AB)$	$P(A) \times P(B)$	$P(A|B) \times P(B)$
	or $P(BA)$	$P(B) \times P(A)$	$P(B|A) \times P(A)$
Conditional	$P(B|A)$	$P(B)$	$\dfrac{P(BA)}{P(A)}$
	or $P(A|B)$	$P(A)$	$\dfrac{P(AB)}{P(B)}$

And finally, the marginal probability of the event striped can be computed by summing the probabilities of the joint events in which striped occurred:

$$P(S) = P(CS) + P(GS) = .1 + .4 = .5$$

These four marginal probabilities $P(C) = .4$, $P(G) = .6$, $P(D) = .5$, and $P(S) = .5$ can be verified by inspection of Table 5·4.

We have now considered the three types of probability (conditional, joint, and marginal) under conditions of statistical dependence. Table 5·5 provides a resumé of our development of probabilities under both statistical independence and statistical dependence.

EXERCISES

30. In a study of the number of men and women employed at a plant, data show that 65 percent of the employees are males, 40 percent of the employees are production workers, and that the probability that an employee is a male production worker is .30. If a randomly selected employee turns out to be a production worker, what is the probability that the employee is a male?

31. According to a survey, the probability that a family owns 2 cars if their annual income is greater than $15,000 is .70. Of the households surveyed, 50 percent had incomes over $15,000 and 40 percent had 2 cars. What is the probability that a family has 2 cars and an income over $15,000 a year?

32. Two events, A and B, are statistically dependent. $P(A) = .25$, $P(B) = .33$ and $P(A \text{ or } B) = .43$. Find the probability that

    a) neither A nor B will occur

    b) both A and B will occur

    c) B will occur, given that A has occurred

    d) A will occur, given than B has occurred

33. Given that $P(A) = \frac{1}{6}$, $P(B) = \frac{1}{3}$, $P(C) = \frac{4}{9}$, $P(A \text{ and } C) = \frac{1}{12}$, and $P(B|C) = \frac{1}{4}$, find the following probabilities:

    $P(A|C)$; $P(C|A)$; $P(B \text{ and } C)$; $P(C|B)$

34. A university administrator is doing a statistical analysis of what combinations women students choose as their majors and minors. Math and engineering are two fields commonly paired as major and minor. Ten percent of the women at the school major in math, while 5 percent minor in engineering. In addition, 3 percent major in math and minor in engineering.

   a) What is the probability that a woman at the university will be a math major if she is an engineering minor?

   b) What is the probability that a woman will be an engineering minor if she is a math major?

35. The administrator of the food stamp program knows that for families having 6 or more members, one out of 10 will qualify for food stamps, but that only one out of 15 families of this size ever apply for the program. From past records, 90 percent of such families applying for food stamps have been found to qualify. What is the probability that a family of 6 or more members will apply for food stamps and also qualify?

36. A farmer is trying to determine how likely he is to sustain significant damage to his corn crop from the corn earworm. He knows that with crop dusting he will have 2 chances out of 10 of sustaining significant earworm damage, providing that the wind is less than 10 mph when the dusting occurs and providing that it does not rain immediately after the dusting. If the wind is above 10 mph, he will have 5 chances out of 10 of sustaining significant damage. If it rains immediately after the dusting, he will have 7 chances out of 10 of sustaining significant damage. If both the high wind and rain occur, he will have 8 chances out of 10 of sustaining significant damage. Because of the age of his crop, he must dust tomorrow or not at all, and the weatherman has predicted a 60 percent chance of rain for the next day and a 30 percent chance that the wind will be above 10 mph. (Wind and rain may be assumed to be independent of each other.) What is the probability of

   a) wind over 10 mph and rain immediately after dusting?

   b) wind or rain, but not both; wind but not rain; rain but not wind?

   c) neither wind nor rain?

   d) major crop damage, even if the farmer dusts the crop?

37. Referring to problem 36, suppose that the farmer did, in fact, sustain crop damage during the year. What is the probability that

   a) there was high wind during the dusting, but no rain immediately after?

   b) there was rain immediately after the dusting, but no high wind during it?

   c) there were both high wind and rain?

   d) there was neither high wind nor rain?

7 REVISING PRIOR ESTIMATES OF PROBABILITIES: BAYES' THEOREM
At the beginning of the baseball season, the fans of last year's pennant winner thought that their team had a good chance of winning again. As the season progressed, however, injuries sidelined their shortstop and their chief rivals drafted a terrific homerun hitter. The team began to lose. Late in the season, the fans realized that they must alter their prior probabilities of winning.

A similar situation often occurs in business. If a manager of a boutique finds that most of the purple and chartreuse ski jackets that she thought would sell so well are hanging on the rack, she must revise her prior probabilities and order a different color combination.

In both these cases, certain probabilities were altered after the people involved got additional information. The new probabilities are known as revised, or *posterior* probabilities. Because probabilities can be revised as more information is gained, probability theory is of great value in managerial decision making.

Posterior
probabilities
defined

The origin of the concept of obtaining posterior probabilities with limited information is attributable to the Reverend Thomas Bayes (1702–1761), and the basic formula for conditional probability under dependence,

Bayes' theorem

$$P(B|A) = \frac{P(BA)}{P(A)} \qquad [5 \cdot 6]$$

is called *Bayes' theorem.*

Bayes, an Englishman, was a Presbyterian minister and a competent mathematician. He pondered how he might prove the existence of God by examining whatever evidence the world about him provided. Attempting to show "that the Principal End of the Divine Providence . . . is the Happiness of His Creatures," Reverend Bayes used mathematics to study God. Unfortunately, the theological implications of his findings so alarmed the good Reverend Bayes that he refused to permit publication of his work during his lifetime. Nevertheless, his work outlived him, and modern decision theory is often called Bayesian decision theory in his honor.

Bayes' theorem offers a powerful statistical method of evaluating new information and revising our prior estimates (based upon limited information only) of the probability that things are in one state or another. If correctly used, it avoids the necessity of gathering masses of data over long periods of time in order to make decisions based upon probabilities.

Value of
Bayes' theorem

Calculating posterior probabilities

Assume, as a first example of revising prior probabilities, that we have equal numbers of two types of deformed (biased or weighted) dice in a bowl. On half of them, ace (or one dot) comes up 40 percent of the time; therefore, P(ace) = .4. On the other half, ace comes up 70 percent of the time, and P(ace) = .7. Let us call the former type 1 and the latter type 2. One die is drawn, rolled once, and comes up ace. What is the probability that it is a type 1 die? Knowing the bowl contains the same number of both types of dice, we might incorrectly answer that the probability is one-half; but we can do better than this. To answer the question correctly, we set up Table 5·6.

Finding a new
posterior estimate

TABLE 5·6
Finding
the
marginal
probability of
getting an ace

Elementary event	Probability of elementary event	P(ace|elementary event)	P(ace, elementary event)*
Type 1	.5	.4	.4 × .5 = .20
Type 2	.5	.7	.7 × .5 = .35
	1.0		**P(ace) = .55**

*A comma is used to separate joint events. We can join individual letters to indicate joint events without confusion (*AB*, for example), but joining whole words in this way could produce strange looking events (aceelementaryevent) in this table, and they could be confusing.

The sum of the probabilities of the elementary events (drawing either a type 1 or a type 2 die) is 1.0, because there are only two types of dice. The probability of each type is .5. The two types constitute a mutually exclusive and collectively exhaustive list.

The sum of P(ace|elementary event) does *not* equal 1.0. The figures .4 and .7 simply represent the conditional probabilities of getting an ace, given type 1 and type 2, respectively.

The fourth column shows the joint probability of ace and type 1 occurring together (.4 × .5 = .20), and the joint probability of ace and type 2 occurring together (.7 × .5 = .35). The sum of these joint probabilities (.55) is the marginal probability of getting an ace. Notice that in each case the joint probability was obtained by using the formula

$$P(AB) = P(A|B) \times P(B) \tag{5·7}$$

To find the probability that the die we have drawn is type 1, we use the formula for conditional probability under statistical dependence

$$P(B|A) = \frac{P(BA)}{P(A)} \tag{5·6}$$

Converting to our problem, we have

$$P(\text{type 1}|\text{ace}) = \frac{P(\text{type 1, ace})}{P(\text{ace})}$$

or

$$P(\text{type 1}|\text{ace}) = \frac{.20}{.55} = .364$$

Thus the probability that we have drawn a type 1 die is .364.
Let us compute the probability that the die is type 2.

$$P(\text{type 2}|\text{ace}) = \frac{P(\text{type 2, ace})}{P(\text{ace})} = \frac{.35}{.55} = .636$$

What have we accomplished with one additional piece of information made available to us? What inferences have we been able to draw from one roll of the die? Before we rolled this die, the best we could say was that there is a .5 chance it is a type 1 die and a .5 chance it is a type 2 die. However, after rolling the die, we have been able to *alter,* or revise, *our prior probability estimate.* Our new posterior estimate is that there is a higher probability (.636) that the die we have in our hand is a type 2 rather than a type 1 (only .364).

Posterior probabilities with more information

We may feel that one roll of the die is not sufficient to indicate its characteristics (whether it is type 1 or type 2). In this case, we can obtain additional information by rolling the die again. (Obtaining more information in most decision-making situations, of course, is more complicated and time consuming.) Assume that the same die is rolled a second time and again comes up ace. What is the further revised probability that the die is type 1? To determine the answer, see Table 5·7.

Finding a new posterior estimate with more information

We have one new column in this table, P(2 aces|elementary event). This column gives the *joint* probability of 2 aces on two successive rolls if the die is type 1 and if it is type 2: P(2 aces|type 1) = .4 × .4 = .16 and P(2 aces|type 2) = .7 × .7 = .49. In the last column, we see the joint probabilities of 2 aces on two successive rolls and the elementary events (type 1 and type 2). That is, P(2 aces, type 1) equals P(2 aces|type 1) times the probability of type 1, or .16 × .5 = .080, and P(2 aces, type 2) equals P(2 aces|type 2) times the probability of type 2, or .49 × .5 = .245. The sum of these (.325) is the marginal probability of 2 aces on two successive rolls.

We are now ready to compute the probability that the die we have drawn is type 1, given an ace on each of two successive rolls. Using the same general formula as before, we convert to

$$P(\text{type 1}|\text{2 aces}) = \frac{P(\text{type 1, 2 aces})}{P(\text{2 aces})} = \frac{.080}{.325} = .246$$

Similarly,

$$P(\text{type 2}|\text{2 aces}) = \frac{P(\text{type 2, 2 aces})}{P(\text{2 aces})} = \frac{.245}{.325} = .754$$

Elementary event	Probability of elementary event	P(ace|elementary event)	P(2 aces| elementary event)	P(2 aces, elementary event)
Type 1	.5	.4	.16	.16 × .5 = .080
Type 2	.5	.7	.49	.49 × .5 = .245
	1.0			P(2 aces) = .325

TABLE 5·7
**Finding
the marginal
probability of
two aces on two
successive rolls**

What have we accomplished with two rolls? When we first drew the die, all we knew was that there was a probability of .5 that it was type 1 and a probability of .5 that it was type 2. In other words, there was a 50-50 chance that it was either type 1 or type 2. After rolling the die once and getting an ace, we revised these original probabilities to the following:

Probability that it is type 1 = .364
Probability that it is type 2 = .636

After the second roll (another ace), we revised the probabilities again:

Probability that it is type 1 = .246
Probability that it is type 2 = .754

We have thus changed the original probabilities from .5 for each type to .246 for type 1 and .754 for type 2. This means that we can now assign a probability of .754 that if a die turns up ace on two successive rolls, it is type 2.

In both these experiments, we gained new information free of charge. We were able to roll the die twice, observe its behavior, and draw inferences from that behavior without any monetary cost. Obviously, there are few situations in which this is true, and decision makers must not only understand how to utilize new information to revise prior probabilities, but also be able to determine *how much that information is worth* to them before the fact. In many cases, the value of the information obtained may be considerably less than its cost.

A problem with three revisions

Example of posterior probability based on three trials

Now let's consider this problem. A Little League baseball team has been using an automatic pitching machine. If the machine is correctly set up— that is, properly adjusted—it will pitch strikes 85 percent of the time. If it is incorrectly set up, it will pitch strikes only 35 percent of the time. Past experience indicates that 75 percent of the set ups of the machine are correctly done. After the machine has been set up at batting practice one day, it throws three strikes on the first three pitches. What is the revised probability that the setup has been done correctly? Table 5·8 illustrates how we can answer this question.

We can interpret the numbered table headings in Table 5·8 as follows: (1) P(*event*) describes the individual probabilities of correct and incorrect. P(correct) = .75 is given in the problem. Thus we can compute:

P(incorrect) = 1.00 − P(correct) = 1.00 − .75 = .25

(2) P(*1* strike|*event*) represents the probability of a strike given that the setup is correct or incorrect. These probabilities are given in the problem. (3) P(*3* strikes|*event*) is the probability of getting 3 strikes on 3 successive

TABLE 5·8

Posterior probabilities with joint events

Event	P(event) (1)	P(1 strike|event) (2)	P(3 strikes|event) (3)	P(event, 3 strikes) (4)
Correct	.75	.85	.6141	.6141 × .75 = .4606
Incorrect	.25	.35	.0429	.0429 × .25 = .0107
	1.00			**P(3 strikes) = .4713**

pitches, given the event; that is, given correct or incorrect. The probabilities are computed as follows:

$$P(3\text{ strikes}|\text{correct}) = .85 \times .85 \times .85 = .6141$$
$$P(3\text{ strikes}|\text{incorrect}) = .35 \times .35 \times .35 = .0429$$

(4) P(*event, 3* strikes) is the probability of the joint occurrence of the event (correct or incorrect) and 3 strikes. We can compute the probabilities in this problem as follows:

$$P(\text{correct, 3 strikes}) = .6141 \times .75 = .4606$$
$$P(\text{incorrect, 3 strikes}) = .0429 \times .25 = .0107$$

Notice that if A = event and B = strikes, these last two probabilities conform to the general mathematical formula for joint probabilities under conditions of dependence: $P(AB) = P(BA) = P(B|A) \times P(A)$, Equation 5·7.

After finishing the computations in Table 5·8, we are ready to determine the revised probability that the machine is correctly set up. We use the general formula

$$P(A|B) = \frac{P(AB)}{P(B)} \qquad [5·6]$$

and convert it to the terms and numbers in this problem:

$$P(\text{correct}|3\text{ strikes}) = \frac{P(\text{correct, 3 strikes})}{P(3\text{ strikes})}$$
$$= \frac{.4606}{.4713} = .9773$$

The *posterior probability* that the machine is correctly set up is .9773, or 97.73 percent. We have thus revised our original probability of a correct setup from 75 to 97.73 percent, based on 3 strikes being thrown in 3 pitches.

Posterior probabilities with inconsistent outcomes

In each of our problems so far, the behavior of the experiment was consistent—the die came up ace on two successive rolls, and the automatic machine threw strikes on each of the first three pitches. In most situations,

An example with inconsistent outcomes

TABLE 5·9 Posterior probabilities with inconsistent outcomes

Event	P(event)	P(S|event)	P(SBSSS|event)	P(event, SBSSS)
Correct	.75	.85	$.85 \times .15 \times .85 \times .85 \times .85 = .07830$	$.07830 \times .75 = .05873$
Incorrect	.25	.35	$.35 \times .65 \times .35 \times .35 \times .35 = .00975$	$.00975 \times .25 = .00244$
	1.00			P(SBSSS) = .06117

$$P(\text{correct setup}|SBSSS) = \frac{P(\text{correct setup}|SBSSS)}{P(SBSSS)}$$
$$= \frac{.05873}{.06117}$$
$$= .9601$$

we would expect a less consistent distribution of outcomes. In the case of the pitching machine for example, we might find the first five pitches to be: strike, ball, strike, strike, strike. Calculating our posterior probability that the machine is correctly set up in this case is really no more difficult than it was with a set of perfectly consistent outcomes. Using the notation S = strike and B = ball, we have solved this example in Table 5·9.

EXERCISES

38. Given: the probabilities of three events, *A, B,* and *C,* occurring are: $P(A) = .5$, $P(B) = .3$, and $P(C) = .2$. Assuming that *A, B,* or *C* has occurred, the probabilities of another event, *X,* occurring are: $P(X|A) = .6$, $P(X|B) = .8$ and $P(X|C) = .4$. Find $P(A|X)$; $P(B|X)$; $P(C|X)$.

39. Carol Ryan, a social psychologist, knows from past research that 3 common experimental methods of changing attitudes are differentially effective. In a survey of recent attitude change studies published in psychology journals, 60 percent of the studies used speakers to persuade subjects to change their attitudes, 25 percent used counterattitudinal advocacy, and 15 percent had subjects read articles presenting arguments in favor of some issue. The probabilities of changing subjects' attitudes in the desired direction with the three methods are .80, .50, and .40 respectively. After reading an article on the same topic just submitted to her for review, she finds that the subjects' attitudes were significantly changed after the experimental manipulation.

    a) Given this information, what is the probability that the experimenter used a speaker to persuade the subjects?

    b) What is the probability that the experimenter used counterattitudinal advocacy?

    c) What is the probability that the experimenter used an article to try to change subjects' attitudes?

40. A public interest group is planning to make a court challenge to auto insurance rates at one of three cities: Atlanta, Denver, or Indianapolis. The probability that it will choose Atlanta is .40; Denver, .30; Indianapolis, .30. The group also knows that it has a 50 percent chance of a favorable court ruling if it chooses Atlanta, 60 percent if it chooses Denver, and 75 percent if it chooses Indianapolis. If the group did receive a favorable ruling, which city did it most likely choose?

41. In a particular town there are two Sunday newspapers, the *Times* and the *Herald,* each of which has a classified ad section. Twenty percent of the employers in the city place a want

ad only in the *Times,* 10 percent place an ad only in the *Herald,* and 70 percent place an ad in both newspapers. In the past, 75 percent of the ads appearing only in the *Times* have received more than one reply, 65 percent of the ads appearing only in the *Herald* have received more than one reply, and 90 percent of the ads appearing in both newspapers have received more than one reply. If an employer places an ad and receives only one reply, what is the probability that the ad appeared in both papers?

42. An independent research group has been making a study of the chances that an accident at a nuclear power plant will result in radiation leakage. The group considers that the only possible types of accidents at a reactor are fire, material failure, and human error, and that two or more accidents never occur together. It has performed studies that indicate that if there were a fire, a radiation leak would occur 10 percent of the time; if there were a mechanical failure, a radiation leak would occur 40 percent of the time; and if there were human error, a radiation leak would occur 5 percent of the time. Its studies have also shown that the probability of

   - a fire and a radiation leak occurring together is .0005
   - a mechanical failure and a radiation leak occurring together is .0010
   - human error and a radiation leak occurring together is .0007

   a) What are the respective probabilities of having a fire, mechanical failure, and human error upon which the probabilities given above are based?

   b) What are the respective probabilities that a radiation leak would be caused by fire, mechanical failure, or human error?

   c) What is the probability of a radiation leak?

43. Data on readership of a certain magazine indicate that the proportion of male readers over 30 years old is .20. The proportion of male readers under 30 is .40. If the proportion of readers under 30 is .70, what is

   a) the proportion of subscribers that are male?

   b) the probability that a randomly selected male subscriber is under 30?

44. An introductory statistics course is made up of 12 freshmen, 25 sophomores, 18 juniors, and 5 seniors. Three of the freshmen, 5 of the sophomores, 6 of the juniors, and 2 of the seniors received an A in the course. If a student is selected at random from the class and found to have received an A, what is the probability that the student is a (a) junior, (b) senior, (c) sophomore, (d) freshman?

8 TERMS INTRODUCED IN CHAPTER 5

- **A PRIORI PROBABILITY** Probability estimate made prior to receiving new information.

- **BAYES' THEOREM** The formula for conditional probability under statistical dependence.

- **CLASSICAL PROBABILITY** The number of outcomes favorable to the occurrence of an event divided by the total number of possible outcomes.

- **COLLECTIVELY EXHAUSTIVE EVENTS** The list of events that represents all of the possible outcomes of an experiment.

- **CONDITIONAL PROBABILITY** the probability of one event occurring, given that another event has occurred.

- **EVENT** One or more of the possible outcomes of doing something, or one of the possible outcomes of an experiment.

- **EXPERIMENT** The activity that results in, or produces, an event.

- **JOINT PROBABILITY** The probability of two events occurring together or in succession.

- **MARGINAL PROBABILITY** The unconditional probability of one event occurring; the probability of a single event.

- **MUTUALLY EXCLUSIVE EVENTS** Events that cannot happen together.

- **POSTERIOR PROBABILITY** A probability that has been revised after additional information was obtained.

- **PROBABILITY** The chance that something will happen.

- **PROBABILITY TREE** A graphical representation showing the possible outcomes of a series of experiments and their respective probabilities.

- **RELATIVE FREQUENCY OF OCCURRENCE** The proportion of times that an event occurs in the long run when conditions are stable, or the observed relative frequency of an event in a very large number of trials.

- **SAMPLE SPACE** The set of all possible outcomes of an experiment.

- **STATISTICAL DEPENDENCE** The condition when the probability of some event is dependent upon, or affected by, the occurrence of some other event.

- **STATISTICAL INDEPENDENCE** The condition when the occurrence of one event has no effect upon the probability of occurrence of any other event.

- **SUBJECTIVE PROBABILITY** Probabilities based on the personal beliefs of the person making the probability estimate.

- **VENN DIAGRAM** A pictorial representation of probability concepts, in which the sample space is represented as a rectangle and the events in the sample space as portions of that rectangle.

9 EQUATIONS INTRODUCED IN CHAPTER 5

p. 119

$$\text{Probability of an event} = \frac{\text{the number of outcomes where the event occurs}}{\text{the total number of possible outcomes}} \qquad [5 \cdot 1]$$

This is the definition of the *classical* probability that an event will occur.

p. 125

$$P(A) = \text{the probability of event } A \text{ happening}$$

A single probability refers to the probability of one particular event occurring, and it is called *marginal* probability.

p. 126

$$P(A \text{ or } B) = \text{the probability of } either \ A \ or \ B \text{ happening}$$

This notation represents the probability that one event *or* the other will occur.

p. 126

$$P(A \text{ or } B) = P(A) + P(B) \qquad [5 \cdot 2]$$

The probability of either A or B happening when A and B are mutually exclusive equals the sum of the probability of event A happening and of the probability of event B happening. This is the *addition rule for mutually exclusive events*.

p. 128

$$P(A \text{ or } B) = P(A) + P(B) - P(AB) \qquad [5 \cdot 3]$$

The addition rule for events that are not mutually exclusive shows that the probability of *A or B* happening when *A* and *B* are not mutually exclusive is equal to the probability of event *A* happening plus the probability of event *B* happening minus the probability of *A* and *B* happening together, symbolized P(*AB*).

p. 131

$$P(AB) = P(A) \times P(B)$$

[5·4]

where:

P(*AB*) = the joint probability of events *A* and *B* occurring together or in succession
P(*A*) = the marginal probability of event *A* happening
P(*B*) = the marginal probability of event *B* happening

The *joint* probability of two or more *independent* events occurring together or in succession is the product of their marginal probabilities.

p. 136

$$P(B|A) = \text{the probability of event } B, \textit{ given that}$$
$$\text{event } A \text{ has happened}$$

This notation shows *conditional* probability, the probability that a second event (*B*) will occur if a first event (*A*) has already happened.

p. 136

$$P(B|A) = P(B)$$

[5·5]

For *statistically independent* events, the *conditional* probability of event *B*, given that event *A* has occurred, is simply the probability of event *B*. Independent events are those whose probabilities are in no way affected by the occurrence of each other.

p. 140

$$P(B|A) = \frac{P(BA)}{P(A)}$$

[5·6]

and

$$P(A|B) = \frac{P(AB)}{P(B)}$$

For statistically *dependent* events, the *conditional* probability of event *B*, given that event *A* has occurred, is equal to the joint probability of events *A* and *B* divided by the marginal probability of event *A*.

p. 142

$$P(AB) = P(A|B) \times P(B)$$

[5·7]

and

$$P(BA) = P(B|A) \times P(A)$$

Under conditions of statistical *dependence,* the *joint* probability of events *A* and *B* happening together or in succession is equal to the probability of event *A*, given that event *B* has already happened, multiplied by the probability that event *B* will happen.

153

45. Life insurance premiums are higher for older persons, but auto insurance premiums are generally higher for younger individuals. What does this suggest about the risks and probabilities associated with these two areas of the insurance business?

46. "The chance of rain today is 80 percent." Which of the following best explains this statement?

    a) It will rain 80 percent of the day today.

    b) It will rain in 80 percent of the area for which this forecast applies today.

    c) In the past, weather conditions of this sort have produced rain in this area 80 percent of the time.

47. "There is a .25 probability that two people who were married last year will be divorced within 5 years." When social scientists make such statements, how have they arrived at their conclusions?

48. Using probability theory, explain the success of gambling and poker establishments.

49. If we assume a person is equally likely to be born on any day of the week, what are the probabilities of a certain baby being born

    a) on a Tuesday?

    b) on a day beginning with the letter S?

    c) between Wednesday and Friday, inclusive?

    d) What type of probability estimates are these?

50. A real estate agent estimates that your house will go up in market value by 15 percent or more in the next 6 months, with probability .60. He estimates that the probability that my house will increase in market value by 15 percent or more in the next 6 months is .8. He also estimates that the probability of a certain client taking his advice and buying your house is .7. If at the end of 6 months, the client's new home has indeed increased in value by 15 percent or more, what is the probability that the client bought (a) my house, (b) your house?

51. Betty Barnes has worked with the U.S. Postal Service for 12 years. During this time, she has inspected many letters and has made a list of the most common mistakes that people make when addressing letters. Here is her list:

    > No zip code
    > No return address
    > No street address
    > Too much postage
    > Not enough postage

    a) In probability theory, would each item on the list be classified as an "event"?

    b) Are all the items on the list mutually exclusive? Are any of them mutually exclusive?

    c) Is Betty's list collectively exhaustive? (Remember, if you answer no, then you should be able to add at least one more item to the list. Can you?)

52. Which of the following sets of two events are mutually exclusive?

    a) You run an experiment using only male subjects, all under 24 years of age.

b) You find in a study that rewarding children for prosocial behavior increases the number of friendly acts they perform and that the amount of reward influences their expressed liking for their peers.

c) You decide to give a homework assignment to the class to examine their knowledge of multiplication tables, and you decide to give them a test rather than a homework assignment.

d) You give a homework assignment to the class to examine their knowledge of multiplication tables, and you give them a test on the material as well.

e) You intend to run for governor of the state this fall, but not to be a candidate for any elective office this year.

53. The scheduling officer for a local police department is trying to decide whether to schedule additional patrol units in each of two neighborhoods. She knows that on any given day during the past year the probabilities of major crimes and minor crimes being committed in the northern neighborhood were .589 and .342, respectively, and that the corresponding probabilities in the southern neighborhood were .507 and .863.

a) What is the probability that a crime of either type will be committed in the northern neighborhood on a given day?

b) What is the probability that a crime of either type will be committed in the southern neighborhood on a given day?

c) What is the probability that *no* crime will be committed in either neighborhood on a given day?

54. The Environmental Protection Agency is trying to assess the pollution impact of a paper mill that is to be built near Spokane, Washington. In studies of six similar plants built during the last year, the EPA determined the following pollution factors:

Plant	1	2	3	4	5	6
Sulfur dioxide emission in parts per million (ppm)	15	12	18	16	11	19

EPA defines excessive pollution as a sulfur dioxide emission of 18 ppm or greater.

a) Calculate the probability that the new plant will be an excessive sulfur dioxide polluter.

b) Classify this probability according to the three types discussed in the chapter: classical, relative frequency, and subjective.

c) How would you judge the accuracy of your result?

55. The American Cancer Society is planning to mail out questionnaires concerning breast cancer. From past experience with questionnaires, the society knows that only 12 percent of the persons receiving questionnaires will respond. It also knows that 1 percent of the questionnaires mailed out will have a mistake in the address and will never be delivered, that 3 percent will be lost or destroyed by the post office, that 22 percent will be mailed to persons who have moved, and that only 52 percent of the persons who move leave forwarding addresses.

a) Do the percentages given in the problem represent classical, relative frequency, or subjective probability estimates?

b) What is the probability that the society will receive a reply from a given questionnaire?

56. A population researcher has been studying the patterns of population growth in the state. He knows that during the past two years the movement of people into the state has declined by 10 percent and that during this same time period movement into Cobb County has declined by 12 percent. Also, he knows that during this period the movement of people into Drexel County has increased by 2 percent. Is the probability of an increase in the movement of people into the community in the next two-year period greater for Cobb or Drexel County?

57. As the administrator of a hospital, Cindy Turner wants to know what the probability is that a person checking into the hospital will require X-ray treatment and will also have hospital insurance that will cover the X-ray treatment. She knows that during the past 5 years, 12 percent of the people entering the hospital required X-rays and that during the same period, 58 percent of the people checking into the hospital had insurance that covered X-ray treatments. What is the correct probability?

58. The air traffic controller at O'Hare Airport has specific regulations that require him to divert one of two airplanes if the probability of the aircraft meeting at the same point exceeds .225. The controller has two inbound aircraft scheduled to arrive 10 minutes apart. He knows that Flight 100, scheduled to arrive first, has a history of being 5 minutes late 20 percent of the time. He also knows that Flight 200, scheduled to arrive second, has a history of being 5 minutes early 25 percent of the time.

   a) If the controller finds out that Flight 200 will definitely arrive 5 minutes early, should he divert Flight 100?

   b) If the controller finds out that Flight 100 will definitely arrive 5 minutes late, should he divert Flight 200?

59. In a program testing children entering first grade, there is an attempt to screen for gifted and slow children, so that they may be put in special classes aimed at their needs. The results of 1,000 children tested showed the following results:

Total children tested	1,000
Number scoring above 150 (gifted)	4
Number scoring below 80 (slow)	10
Number scoring between 80 and 150	986

   If only children scoring between 80 and 150 are kept in the regular classes, what is the probability that a given child, selected at random, will be put in a gifted or slow class?

60. Which of the following pairs of events are statistically independent?

   a) The number of red-headed children in a class and the number of children in the class who have freckles.

   b) The number of women in the U.S. who earn annual incomes over $20,000 and the number of women who have college degrees.

   c) The number of trials it takes subject A to learn a list of words and the number of trials it takes subject B to learn the list.

   d) The number of people who vote for the presidential candidate of one political party and the number of candidates for office of the same party who are elected in the same year.

   e) The number of students who enroll in introductory chemistry at the state university and the number of students who enroll in developmental psychology.

61. Susan Douglas, an administrator in research and development for the school system, wants to implement a new series of reading materials for fourth, fifth, and sixth grades. She estimates that there is an 80 percent chance that the school board will adopt the new series for sixth grades and a 50 percent chance that if they do adopt it, reading achievement scores will improve. She estimates that there is a 60 percent chance that they'll adopt the new fifth grade program and a 70 percent chance that if they do adopt it, reading scores will improve. Finally, she estimates there is a 40 percent chance that the school board will adopt the proposed series for fourth grades and a 90 percent chance that if they do, reading scores will improve. Suppose that at the end of the year the school system's fourth-, fifth-, and sixth-graders do not show improvement in their reading scores.

   a) For each grade, what is the probability that the new reading program was not adopted?

   b) For each grade, what is the probability that the new program was adopted but that reading scores did not improve?

62. Jill and Robert Taft are considering starting a family in the next few years. Their decision will be influenced by financial considerations such as Robert's being promoted and transferred and whether federal health project with which Jill works is renewed. She knows they will decide to start a family if one of two events occur: (1) Robert is promoted, and they move to another city or (2) the health project with which she works is cancelled. Jill also feels that both of these events won't happen in the same year. Therefore, she has made the following estimates:

   1) The probability of Robert's promotion and transfer within one year is .10, and the probability of the project's cancellation within one year is .05.

   2) The probability of Robert's promotion and transfer sometime during the next two years is .25, and the probability of the project's cancellation during the next two years is .20.

   3) The probability of Robert's promotion and transfer sometime during the next three years is .40, and the probability of the project's cancellation during the next three years is .45.

      (a) What are the probabilities that Jill and Robert will decide to start a family in years 1, 2, or 3, respectively?

      (b) What is the probability that they will decide to start a family at all in the three-year period?

63. Draw Venn diagrams to represent the following situations involving three events, A, B, and C, which are part of a sample space of events but do not include the whole sample space.

   a) Each pair of events (A and B, A and C, and B and C) may occur together, but all three may not occur together.

   b) A and B are mutually exclusive, but not A and C nor B and C.

   c) A, B, and C are all mutually exclusive of one another.

   d) A and B are mutually exclusive, B and C are mutually exclusive, but A and C are not mutually exclusive.

64. As means of double-checking diagnoses, computers are currently being programmed to determine if, based on medical information, a patient has a certain illness. A company that supplies hospitals and clinics with computers claims that they make errors in diagnosing a particular condition only .1 percent of the time.

a) Suppose that in a communitywide screening of 10,000 patients, 6,000 are diagnosed by computer #1 and 4,000 diagnosed by computer #2. If one patient is found to be misdiagnosed, what is the probability that the diagnosis was made by computer #1?

b) Suppose that after three years, records indicate that .2 percent of the diagnoses made by computer #1 were incorrect and .3 percent of the diagnoses made by computer #2 were incorrect. In another screening of 10,000 patients, 6,000 are diagnosed on computer #1 and 4,000 on computer #2. If a patient is found to be misdiagnosed in this group, what is the probability that the diagnosis was done by computer #1?

65. Determine the probability that

a) a person is a heroin addict and smokes marijuana, given that 62 percent of all heroin addicts smoke marijuana and that the probability of a person being a heroin addict is .005.

b) a child in a certain school district comes from an intact family with an income over $20,000, given that 50 percent of the intact families in the district have incomes over $20,000 and 95 percent of the families in the district are intact.

c) a man on campus is a business major and lives in James Dorm, given that 30 percent of the men are business majors and 10 percent of them live in James Dorm.

66. In a recent study of heritability of intelligence, a psychologist, administering I.Q. tests to pairs of siblings less than two years apart in age, found similar patterns of scores. Specifically, 72 percent of the children whose older sibling scored over 120 on the test also scored over 120. If 20 percent of the older siblings scored over 120, what is the probability that a pair of siblings in the group both scored over 120? What information would you need in order to determine the probability of a younger sibling in the group scoring over 120?

Answer true *or* false. *Answers are in the back of the book.*

1. In probability theory, the outcome from some experiment is known as an activity.

2. The probability of two or more statistically independent events occurring together or in succession is equal to the sum of their marginal probabilities.

3. Using Bayes' theorem, we may develop revised probabilities based upon new information; these revised probabilities are also known as posterior probabilities.

4. In classical probability, we can determine a priori probabilities based upon logical reasoning before any experiments take place.

5. The set of all possible outcomes of an experiment is called the sample space for the experiment.

6. Under statistical dependence, a marginal probability may be computed for some simple event by taking the product of the probabilities of all joint events in which the simple event occurs.

7. When a list of events resulting from some experiment includes all possible outcomes, the list is said to be collectively exclusive.

8. An unconditional probability is also known as a marginal probability.

9. A subjective probability may be nothing more than an educated guess.

10. When the occurrence of some event has no effect upon the probability of occurrence of some other event, the two events are said to be statistically independent.

Probability II:
distributions

1. Introduction to probability distributions, 162

2. Random variables, 166

3. Use of expected value in decision making, 172

4. The binomial distribution, 175

5. The Poisson distribution, 185

6. The normal distribution: a distribution of a continuous random variable, 191

7. Expected value decision making with continuous distributions, 205

8. Choosing the correct probability distribution, 213

9. Terms introduced, 214

10. Equations introduced, 215

11. Chapter review exercises, 217

12. Chapter concepts test, 223

Last summer the authors and their children planted a vegetable garden. One of the crops was a hybrid corn, Butter and Honey, which has white and gold kernels on the same ear. As we were putting 6 seeds in each of 15 hills, the children asked, "What is the chance that exactly half of the seeds in any hill will sprout?" The seed packet claimed a germination rate of 80 percent. Although we could not make an exact forecast, the ideas of probability distributions discussed in this chapter enabled us to get a pretty good idea of how many we would see.

Frequency
distributions

Probability
distributions
and frequency
distributions

1 INTRODUCTION TO PROBABILITY

DISTRIBUTIONS In Chapters 2, 3, and 4, we described frequency distributions as a useful way of summarizing variations in observed data. We prepared frequency distributions by listing all the possible outcomes of an experiment and then indicating the observed frequency of each possible outcome. *Probability distributions* are related to frequency distributions. **In fact, we can think of a probability distribution as a theoretical frequency distribution.** Now, what does that mean? A theoretical frequency distribution is a probability distribution that describes how outcomes are *expected* to vary. Since these distributions deal with expectations, they are useful models in making inferences and decisions under conditions of uncertainty. In later chapters, we will discuss the methods we use under these conditions.

Examples of probability distributions

To begin our study of probability distributions, let's go back to the idea of a fair coin, which we introduced in Chapter 5. Suppose we toss a fair coin twice. Table 6·1 illustrates the possible outcomes from this two-toss experiment.

TABLE 6·1
Possible outcomes from two tosses of a fair coin

First toss	Second toss	Number of tails on two tosses	Probability of the four possible outcomes
T	T	2	$.5 \times .5 = .25$
T	H	1	$.5 \times .5 = .25$
H	H	0	$.5 \times .5 = .25$
H	T	1	$.5 \times .5 = \underline{.25}$
			1.00

Now suppose that we are interested in formulating a probability distribution of the number of tails that could possibly result when we toss the coin twice. We would begin by noting any outcome that did *not* contain a tail. With a fair coin, that is only the third outcome in Table 6·1: *H, H.* Then we would note those outcomes containing only one tail (the second and fourth outcomes in Table 6·1), and finally we would note that the first outcome contains two tails. In Table 6·2 we rearrange the outcomes of Table 6·1 to emphasize the number of tails contained in each outcome. We must

TABLE 6·2
Probability distribution of possible number of tails from two tosses of a fair coin

Number of tails T	Tosses	Probability of this outcome P(T)
0	(H, H)	.25
1	(T, H) + (H, T)	.50
2	(T,T)	.25

be careful to note at this point that Table 6·2 is *not* the actual outcome of tossing a fair coin twice. Rather it is a *theoretical* outcome; that is, it represents the way in which we would expect our two-toss experiment to behave over time.

We can illustrate in graphic form the probability distribution in Table 6·2. To do this, we graph the number of tails we might see on two tosses against the probability that this number would happen. We have shown this graph in Fig. 6·1.

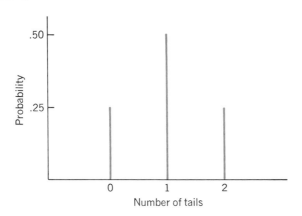

FIG. 6·1
Probability distribution of the number of tails in two tosses of a fair coin

Consider another example. A political candidate for local office is considering the votes she can get in a coming election. Assume that votes can take on only four possible values. If the candidate's assessment is like this:

Voting example

Number of votes	1,000	2,000	3,000	4,000	
Probability this will happen	.1	.3	.4	.2	**Total 1.0**

then the graph of the probability distribution representing her expectations will be like the one shown in Fig. 6·2.

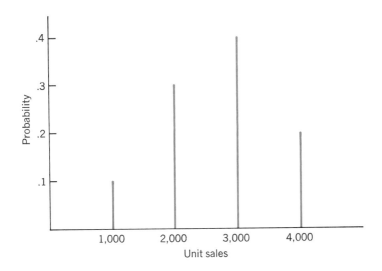

FIG. 6·2
Probability distribution of number of votes

Difference between
frequency
distributions
and probability
distributions

Before we move on to other aspects of probability distributions, we should point out that a *frequency distribution* is a listing of the observed frequencies of all the outcomes of an experiment that actually occurred when the experiment was done, whereas a *probability distribution* is a listing of the probabilities of all of the possible outcomes that *could* result if the experiment is done. Also, as we can see in the two examples we presented in Fig. 6·1 and 6·2, probability distributions can be based on theoretical considerations (the tosses of a coin) or on a subjective assessment of the likelihood of certain outcomes (the candidate's estimate). Probability distributions can also be based on experience. Insurance company actuaries determine insurance premiums, for example, by using long years of experience with death rates to establish probabilities of dying among various age groups.

Types of probability distributions

Discrete
probability
distributions

Probability distributions are classified as either *discrete* or *continuous.* A discrete probability is allowed to take on only a limited number of values. An example of a discrete probability distribution is shown in Fig. 6·2, where we expressed the candidate's ideas about the coming election. There, votes were allowed to take on only four possible values (1,000, 2,000, 3,000, or 4,000). Similarly, the probability that you were born in a given month is also discrete, since there are only twelve possible values (the twelve months of the year).

Continuous
probability
distributions

In a continuous probability distribution, on the other hand, the variable under consideration is allowed to take on any value within a given range. Suppose we were examining the level of effluent in a variety of streams and we measured the level of effluent by parts of effluent per million parts of water. We would expect quite a continuous range of ppm (parts per million), all the way from very low levels in clear mountain streams to extremely high levels in polluted streams. In fact, it would be quite normal for the variable "parts per million" to take on an enormous number of values. We would call the distribution of this variable (ppm) a continuous distribution. Continuous distributions are convenient ways to represent discrete distributions that have many possible outcomes, all very close to each other.

EXERCISES

1. Given the following list of possible outcomes of an experiment and their associated probabilities, draw a graph illustrating this probability distribution.

Score	0	5	10	15	25
Probability of score	.08	.18	.30	.24	.20

2. Based on the following graph of a probability distribution, construct the table that corresponds to the graph.

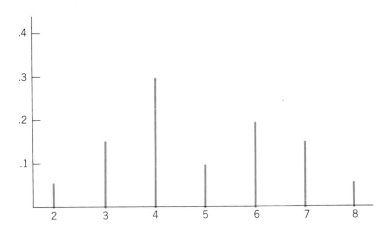

3. In the last chapter, we looked at the possible outcomes of tossing two dice, and we calculated some probabilities associated with various outcomes. Construct a table and a graph of the probability distribution representing the outcomes (in terms of total number of dots showing on both dice) for this experiment.

4. Which of the following statements regarding probability distributions are correct?

    a) A probability distribution provides information about the long run or expected frequency of each outcome of an experiment.

    b) The graph of a probability distribution has the possible outcomes of an experiment marked on the horizontal axis.

    c) A probability distribution lists the probabilities that each outcome is random.

    d) A probability distribution is always constructed from a set of observed frequencies like a frequency distribution.

    e) A probability distribution may be based on subjective estimates of the likelihood of certain outcomes.

5. The Southport Seafood Restaurant offers a number of entrees on its menu. Because of the chef's culinary artistry and also by way of industry custom, Walter Harrison, the manager, has resorted to selling a number of various combination seafood platters. Currently Mr. Harrison, who prides himself on never being out of stock, is worried because of a shortage of clams on the market. The restaurant serves the following platters:

The Angler:	Flounder	Scallops	Clams
The Captain:	Clams	Shrimp	Flounder
The Mate:	Scallops	Shrimp	Oysters
The Mermaid:	Shrimp	Clams	Scallops

Mr. Harrison assigns an equal chance that any of the entrees will be ordered.

    a) What is the probability that any one customer will order an entree that requires a portion of clams?

    b) Assume that two customers come into the restaurant. Construct a table showing the probability distribution of the number of portions of clams they order.

6. Susan Mullin is running for a seat in the state Senate. Her platform is such that she appeals to blocks of voters on the basis of her stance on tax reform. She has hired a sociologist to survey voter opinions and help determine her support base. In her district, she feels she can pick up 16,000 votes with a 60 percent probability. The sociologist feels, however, that

4,000 of these votes are tenuous. He feels there is half as much chance of getting 12,000 votes as there is of getting all 16,000. They also both agree that with a slight change in stance, Mullin could, with some probability, pick up an additional 2,000 votes; but they feel this is the maximum she could obtain. Looking at the votes in blocks as Susan and the consultant are doing, construct a table and draw a graph of the probability distribution of votes.

Random variable defined

2 RANDOM VARIABLES

A random variable is a variable that takes on different values as a result of the outcomes of a random experiment. A random variable can be either discrete or continuous. If a random variable is allowed to take on only a limited number of values, it is a *discrete random variable.* On the other hand, if it is allowed to assume any value within a given range, it is a *continuous random variable.*

Example of discrete random variables

You can think of a random variable as a value or magnitude that changes from occurrence to occurrence in no predictable sequence. A breast cancer screening clinic for example, has no way of knowing exactly how many women will be screened on any one day. So tomorrow's number of patients is a random variable. The values of a random variable are the numerical values corresponding to each possible outcome of the random experiment. If past daily records of the clinic indicate that the values of the random variable range from 100 to 115 patients daily, the random variable is a discrete random variable.

Table 6·3 illustrates the number of times each level has been reached during the last 100 days. Note that Table 6·3 gives a frequency distribution.

TABLE 6·3

Number of women screened daily during 100 days

Number screened	Number of days this level was observed
100	1
101	2
102	3
103	5
104	6
105	7
106	9
107	10
108	12
109	11
110	9
111	8
112	6
113	5
114	4
115	2
	100

Number screened (value of the random variable)	Probability that the random variable will take on this value
100	.01
101	.02
102	.03
103	.05
104	.06
105	.07
106	.09
107	.10
108	.12
109	.11
110	.09
111	.08
112	.06
113	.05
114	.04
115	.02
	1.00

TABLE 6·4
Probability distribution for number of women screened

To the extent that we believe that the experience of the past 100 days has been typical, we can use this historical record to assign a probability to each possible number of patients and find a probability distribution. We have accomplished this in Table 6·4, by normalizing the observed frequency distribution (in this case, dividing each value in the right hand column of Table 6·3 by 100, the total number of days for which the record has been kept). The probability distribution for the random variable "daily number screened" is illustrated graphically in Fig. 6·3. Notice that the probability distribution for a random variable provides a probability for each possible value and that these probabilities must sum to one. Table 6·4 shows that both of these requirements have been met. Furthermore, both Table 6·4 and

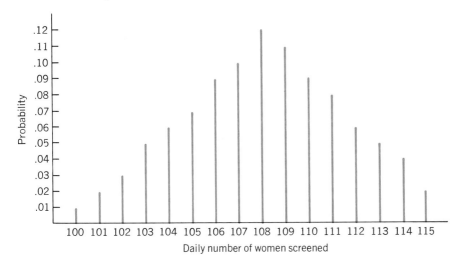

FIG. 6·3
Probability distribution for the discrete random variable "daily number screened"

Fig. 6·3 give us information about the long-run frequency of occurrence of daily patient screenings we would expect to observe if this random "experiment" is repeated.

The expected value of a random variable

Suppose you toss a coin 10 times and get 7 heads like this:

Heads	Tails	Total
7	3	10

Hmm, strange, you say. You then ask a friend to try tossing the coin 20 times; she gets 15 heads and 5 tails. So now you have, in all, 22 heads and 8 tails out of 30 tosses.

What did you expect? Was it something closer to 15 heads and 15 tails (half and half)? Now suppose you turn the tossing over to a machine and get 792 heads and 208 tails out of 1,000 tosses of the same coin. You might now be suspicious of the coin because it didn't live up to what you expected.

Expected value is a fundamental idea in the study of probability distributions. For many years, the concept has been put to considerable practical use in the insurance industry, and in the last twenty years, it has been widely used by many others who must make decisions under conditions of uncertainty.

Expected value defined

To obtain the expected value of a discrete random variable we multiply each value that the random variable can assume by the probability of occurrence of that value and then sum these products. Table 6·5 illustrates this procedure for our clinic problem. The total in Table 6·5 tells us that the expected value of the discrete random variable "number screened" is 108.02 women. What does this mean? It means that over a long period of time, 108.02 is the number of expected daily screenings. Remember that 108.02 does *not* mean that tomorrow exactly 108.02 women will visit the clinic.

The clinic director would base her decisions on the expected value of daily screenings because the expected value is a *weighted average of the outcomes she expects in the future.* Expected value weights each possible outcome by the frequency with which it is expected to occur. Thus, more common occurrences are given more weight than are less common ones. As conditions change over time, the director would recompute the expected value of daily screenings and use this new figure as a basis for decision making.

Deriving expected value

In our clinic example, the director used past patients' records as the basis for calculating the expected value of daily screenings. The expected value can also be derived from the director's subjective assessments of the

TABLE 6·5

Possible values of the random variable (1)	Probability that the random variable will take on these values (2)	(1) × (2)
100	.01	1.00
101	.02	2.02
102	.03	3.06
103	.05	5.15
104	.06	6.24
105	.07	7.35
106	.09	9.54
107	.10	10.70
108	.12	12.96
109	.11	11.99
110	.09	9.90
111	.08	8.88
112	.06	6.72
113	.05	5.65
114	.04	4.56
115	.02	2.30
Expected value of the random variable "daily number screened"→		**108.02**

probability that the random variable will take on certain values. In that case, the expected value represents nothing more than her personal convictions about the possible outcome.

In this section, we have worked with the probability distribution of a random variable in tabular form (Table 6·5) and in graphic form (Fig. 6·3). In many situations, however, we will find it more convenient, in terms of the computations that must be done, to represent the probability distribution of a random variable in *algebraic* form. By doing this, we can make probability calculations by substituting numerical values directly into an algebraic equation. In the following sections, we shall illustrate situations in which this is appropriate and methods for accomplishing it.

EXERCISES

7. Construct a table for a possible probability distribution based on the frequency distribution given below.

Outcome	10	12	14	16	18	20
Frequency	15	20	45	42	18	10

a) Draw a graph of the hypothetical probability distribution.

b) Compute the expected value of the outcome.

8. From the following graph of a probability distribution:

    a) Construct a table of the probability distribution.

    b) Find the expected value of the random variable.

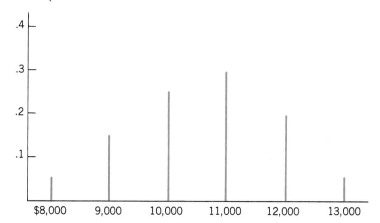

9. A bet involves the following possible outcomes for a player. Find the expected value of the amount of winnings for this bet.

Amount of winnings	$0.00	$5.00	$10.00	$25.00	$50.00
Probability of amount	.25	.40	.20	.10	.05

If someone wishes to bet only when the expected amount of winnings exceeds the cost of the bet, would that person be willing to place $10.00 on this bet?

10. The only information available to you regarding the probability distribution of a set of outcomes is the following list of frequencies:

X	0	1	2	3	4	5
Frequencies	18	48	180	252	72	30

    a) Construct a possible probability distribution for the set of outcomes.

    b) Find the expected value of an outcome.

11. A Las Vegas gambling casino owner has hired a behavior modification expert to help him figure out ways to make larger profits from the slot machines. One specific machine, located in the lobby of the convention hall, is chosen for study. The owner of the casino wants the psychologist to focus on the months of August and September, pivotal months in the gambling casino business. The psychologist hopes to alter the design of the slot machine to take advantage of his knowledge of the clientele and gambling behavior in general. The owner and psychologist agree that a preliminary analysis should be done of past activity on this machine. The psychologist decides to prepare a probability distribution and compute the expected value of gross profits for August and September. The following data are available:

Dollars taken from slot machine

Year	J	F	M	A	M	J	J	A	S	O	N	D
1975	9,500	7,000	7,000	7,500	7,000	8,000	6,500	8,000	10,000	7,500	6,000	8,000
1976	6,500	7,500	6,000	9,000	9,500	7,000	7,500	6,500	6,000	9,000	5,500	6,500
1977	9,000	7,000	7,500	10,000	7,000	8,000						

12. Dr. James Sperling, chief psychiatrist at a mental institution recently became concerned about the institution's treatment of patients having acute anxiety neuroses. Specifically, he felt these patients were being kept in the institution too long. Dr. Sperling specialized in therapy but did not feel that he handled figures very well, so he assigned his assistant, Dr. Alice Thorndike, to compile a summary report to help him investigate his belief. When Dr. Thorndike presented a probability distribution to Dr. Sperling, he was less than appreciative.

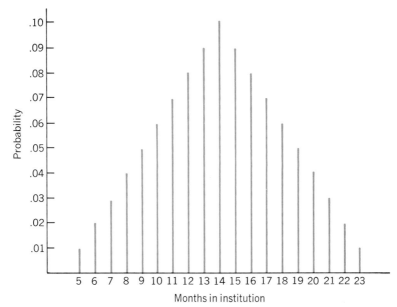

Months in institution

He commented that she had been assigned to compile figures, not to attempt creativity by drawing pictures. She returned to her office with the realization that all Dr. Sperling desired was one figure—the expected length of time one of these patients stayed in the institution. What answer should she give him?

13. A psychologist in California has been running two-hour sensitivity training clinics on weekends. The clinics have been doing quite well. On the basis of five years of data, he computed the following probability distribution for number of participants per weekend:

Participants	3,000	4,000	4,500	5,000
Probability	.2	.4	.2	.2

From this he computed the expected number of participants.

Looking more carefully at his data, he noticed that fewer participants had signed up in the last year than in the years before. Extrapolating, he observed with surprise, "At this rate, I'll be lucky to average 2,500 per weekend this year."

a) What was the expected number of participants, based on the distribution from past data?

b) If the clinic costs each participant $5, and the psychologist spends $4 for material for each person, how much would he gain (or lose) each weekend if he has material printed for the expected number based on past data, but only 2,500 participate?

3 USE OF EXPECTED VALUE IN DECISION MAKING

In the previous section, we calculated the expected value of a random variable and noted that it can have significant value to decision makers. Now we need to take a moment to illustrate how decision makers combine the probabilities that a random variable will take on certain values with the monetary gain or loss that results when it does take on those values. Doing just this enables them to make intelligent decisions under uncertain conditions.

Combining probabilities and monetary values

Laboratory
problem

Let us look at the case of a biological laboratory that uses quite a bit of a certain reagent in its work. This reagent has a very limited useful (full-strength) life. If not used on the day of delivery, it is worthless. This reagent is used in tissue tests, a specialty of the laboratory, which provides significant financial support for other laboratory work. One liter of reagent is required for each test; the reagent costs $20 a liter, and the laboratory receives $50 a test for its work. The laboratory director cannot specify the number of tests that laboratory clients will call for on any one day but her analysis of past records has produced the information in Table 6·6.

TABLE 6·6
Tests done during 100 days

Daily tests	Number of days done	Probability of each number being done
10	15	.15
11	20	.20
12	40	.40
13	25	.25
	100	**1.00**

Types of losses defined

Table of
conditional losses

Two types of losses are incurred by the laboratory: (1) *obsolescence losses* caused by stocking too much reagent on any one day and having to throw it away the next day, and (2) *opportunity losses* caused by being out of the reagent any time that clients of the laboratory want tissue tests requiring the reagent (clients will not wait beyond the day a test is requested).

Table 6·7 is a table of conditional losses for our lab. Each value in the table is conditional on a specific number of liters of reagent being stocked and a specific number of tests being requested. The values in Table 6·7 include not only losses from obsolete reagent, but also those losses resulting from lost revenue when the laboratory is unable to perform requested tests.

Neither of these two types of losses is incurred when the number of liters of reagent stocked on any one day is the same as the number of tests

TABLE 6·7
Conditional
loss table

| Possible requests | Possible stock actions | | | |
for tests	10 liters	11 liters	12 liters	13 liters
10	$ 0	$20	$40	$60
11	30	0	20	40
12	60	30	0	20
13	90	60	30	0

requested. When that happens, the laboratory uses all the reagent, does all the requested tests, and incurs no losses. This situation is indicated by a blue zero in the appropriate column. Figures above any zero represent losses arising from obsolete reagent; in each case here the number of liters of reagent stocked is greater than the number of tests requested. For example, if the lab director stocks 12 liters of reagent, but receives requests for only 10 tests, the lab loses $40, ($20 per liter for obsolete reagent).

Values below the blue zeros represent opportunity losses resulting from requests for tests that cannot be filled. If, for example, only 10 liters of reagent are stocked on a day that 11 requests for tests are received, the laboratory suffers an opportunity loss of $30 for the test it cannot perform, ($50 income per test that would have been received − $20 reagent cost per test = $30 per test).

Opportunity losses

Calculating expected losses

Examining each possible stock action, we can compute the expected loss. We do this by weighting each of the four possible loss figures in each column of Table 6·7 by the probabilities from Table 6·6. For a stock action of 10 cases, the expected loss is computed as in Table 6·8.

TABLE 6·8
Expected loss
from stocking
10 liters

Possible requests for tests	Conditional loss		Probability of this many requests		Expected loss
10	$ 0	×	.15	=	$ .00
11	30	×	.20	=	6.00
12	60	×	.40	=	24.00
13	90	×	.25	=	22.50
			1.00		$52.50

The conditional losses in Table 6·8 are taken from the first column of Table 6·7 for a stock action of 10 liters. The fourth column total in Table 6·8 shows us that if 10 liters are stocked each day, over a long period of time the average or expected loss will be $52.50 a day. There is no guarantee that *tomorrow's* loss will be exactly $52.50.

Meaning of expected loss

Tables 6·9 through 6·11 show the computations of the expected loss resulting from decisions to stock 11, 12, and 13 liters, respectively. The optimum stock action is the one that will minimize expected losses. This action calls for the stocking of 12 liters each day, at which point the expected loss is minimized at $17.50. We could just as easily have solved this problem by taking an alternative approach; that is, *maximizing expected gain* ($50 received per test less $20 for the reagent) instead of minimizing expected loss. The answer, 12 liters, would have been the same.

TABLE 6·9
Expected loss from stocking 11 liters

Possible requests for tests	Conditional loss		Probability of this many requests		Expected loss
10	$20	×	.15	=	$ 3.00
11	0	×	.20	=	.00
12	30	×	.40	=	12.00
13	60	×	.25	=	15.00
			1.00		$30.00

TABLE 6·10
Expected loss from stocking 12 liters

Possible requests for tests	Conditional loss		Probability of this many requests		Expected loss
10	$40	×	.15	=	$ 6.00
11	20	×	.20	=	4.00
12	0	×	.40	=	.00
13	30	×	.25	=	7.50
			1.00		$17.50← optimum stock action

TABLE 6·11
Expected loss from stocking 13 liters

Possible requests for tests	Conditional loss		Probability of this many requests		Expected loss
10	$60	×	.15	=	$ 9.00
11	40	×	.20	=	8.00
12	20	×	.40	=	8.00
13	0	×	.25	=	.00
			1.00		$25.00

EXERCISES

14. The luggage department of Madison Rhodes Department Store featured a special Day-After-Christmas Sale on unsold Christmas merchandise. The luggage brand on sale was Imagemaker. The manager of the luggage department was planning his order. Because the store did not carry Imagemaker during the year, the manager wanted to avoid overstocking; yet because of a special price the manufacturer offered on the line, he also wanted to minimize stockouts. He was currently attempting to decide the number of women's tote

bags to purchase. His estimate of the probable sales, based in part on past performance, is shown below.

Bags	27	28	29	30	31	32	33
Probability	.11	.13	.17	.20	.15	.14	.10

The store is planning to sell the tote bag for $37.30. His cost is $22.00. How many bags should he order for the sale? Use the method of minimizing expected loss to compute your answer.

15. Airport Rent-a-Car is a locally operated business in competition with several major firms. ARC is planning a new deal for prospective customers who want to rent a car for only one day and will return it to the airport. For $8.50 the company will rent a small economy car to a customer, whose only other expense is to fill the car with gas at day's end. ARC is planning to buy a number of small cars from the manufacturer at a reduced price of $2,300. The big question is how many to buy. Company executives have decided on the following probable average demands per day for the service.

Number of cars rented	8	9	10	11	12	13
Probability		.17	.18	.20	.16	.15	.14

The company intends to offer the plan 6 days a week (312 days per year) and anticipates that its variable cost per car per day will be 75¢. After the end of one year, the company expects to sell the cars and recapture 50 percent of the original cost. Disregarding the time value of money and any noncash expenses, use the expected loss method to determine the optimal number of cars for ARC to buy.

4 THE BINOMIAL DISTRIBUTION

One widely used probability distribution of a discrete random variable is the binomial distribution. It describes a variety of processes of interest to managers. The binomial distribution describes discrete, not continuous, data, resulting from an experiment known as a *Bernoulli process* after the seventeenth-century Swiss mathematician Jacob Bernoulli. The tossing of a fair coin a fixed number of times is a Bernoulli process, and the outcomes of such tosses can be represented by the binomial probability distribution. The success or failure of interviewees on an aptitude test may also be described by a Bernoulli process. On the other hand, the frequency distribution of the lives of fluorescent lights in a factory would be measured on a continuous scale of hours and would not qualify as a binomial distribution.

The binomial distribution, a Bernoulli process

Use of the Bernoulli process

We can use the outcomes of a fixed number of tosses of a fair coin as an example of a Bernoulli process. We can describe this process as follows:

Bernoulli process defined

1. Each trial (each toss, in this case) has only *two* possible outcomes: heads or tails, yes or no, success or failure.
2. The probability of the outcome of any trial (toss) remains *fixed* over time. With a fair coin, the probability of heads remains .5 for each toss regardless of the number of times the coin is tossed.

3. The trials are *statistically independent;* that is to say, the outcome of one toss does not affect the outcome of any other toss.

Each Bernoulli process has its own characteristic probability. Take the situation in which historically seven-tenths of all persons who applied for a certain type of job passed the job test. We would say that the characteristic probability here is .7, but we could describe our testing results as Bernoulli only if we felt certain that the proportion of those passing the test (.7) remained constant over time. The other characteristics of the Bernoulli process would also have to be met, of course. Each test would have to have only two outcomes (success or failure), and the results of each test would have to be statistically independent.

In more formal language the symbol p represents the probability of a success (in our example .7), and the symbol q, $(q = 1 - p)$, the probability of a failure (.3). To represent a certain number of successes, we will use the symbol r, and to symbolize the total number of trials, we use the symbol n. In the situations we will be discussing, the number of trials is fixed before the experiment is begun.

Using this language in a simple problem, we can calculate the chances of getting exactly two heads (in any order) on three tosses of a fair coin. Symbolically, we express the values as follows:

Characteristic probability defined

p = characteristic probability or probability of success = .5
$q = 1 - p$ = probability of failure = .5
r = number of successes desired = 2
n = number of trials undertaken = 3

We can solve the problem by using the *binomial formula:*

Binomial formula

$$\text{Probability of } r \text{ successes in } n \text{ trials} = \frac{n!}{r!(n-r)!} p^r q^{n-r} \qquad [6 \cdot 1]$$

Although this formula may look somewhat complicated, it can be used quite easily. The symbol *!* means *factorial,* which is computed as follows: 3! means $3 \times 2 \times 1$, or 6. To calculate 5!, we multiply $5 \times 4 \times 3 \times 2 \times 1 = 120$. Mathematicians define 0! as equal to 1. Using the binomial formula to solve our problem, we discover:

$$\text{Probability of 2 successes in 3 trials} = \frac{3!}{2!(3-2)!} (.5^2)(.5^1)$$

$$= \frac{3 \times 2 \times 1}{(2 \times 1)(1 \times 1)} (.5^2)(.5^1)$$

$$= \frac{6}{2} (.25)(.5)$$

$$= .375$$

Thus there is a .375 probability of getting two heads on three tosses of a fair coin.

By now you've probably recognized that we can use the binomial distribution to determine the probabilities for the number of sprouts that resulted from the hybrid corn we planted at the beginning of this chapter. Recall that historically eight-tenths of the seeds produced did germinate (successes). If we want to compute the probability of getting exactly 3 of 6 seeds to germinate in a given hill, we would first define our symbols:

$$p = .8$$
$$q = .2$$
$$r = 3$$
$$n = 6$$

and then use the binomial formula as follows:

$$\text{Probability of } r \text{ successes in } n \text{ trails} = \frac{n!}{r!(n-r)!} p^r q^{n-r} \qquad [6\cdot1]$$

$$\text{Probability of 3 sprouts out of 6} = \frac{6 \times 5 \times 4 \times 3 \times 2 \times 1}{(3 \times 2 \times 1)(3 \times 2 \times 1)} (.8^3)(.2^3)$$

$$= \frac{720}{(6 \times 6)} (.512)(.008)$$

$$= (20)(.512)(.008)$$

$$= .08192$$

Of course, we *could* have solved these two problems using the probability trees we developed in Chapter 5; but for larger problems, trees become quite cumbersome. In fact, using the binomial formula (Equation 6·1) is no easy task when we have to compute the value of something like 46 factorial. For this reason, binomial probability tables have been developed, and we shall use them shortly.

Some graphic illustrations of the binomial distribution

To this point, we have dealt with the binomial distribution only in terms of the binomial formula, but the binomial, like any other distribution, can be expressed graphically as well.

To illustrate several of these distributions, consider a situation at Kerr Elementary School where students are often late. Five students are in kindergarten. The principal has studied the situation over a period of time and has determined that there is a .4 chance of any one student being late and that students arrive independently of one another. How would we draw a binomial probability distribution illustrating the probabilities of 0, 1, 2, 3, 4, or 5 students being late simultaneously? To do this we would need to use the binomial formula where:

$$p = .4$$
$$q = .6$$
$$n = 5^*$$

and to make a separate computation for each r, from 0 through 5. Remember that mathematically any number to the zero power is defined as being equal to one. Beginning with our binomial formula,

Using the formula
to derive
the binomial
probability
distribution

Probability of r late arrivals out of n students

$$= \frac{n!}{r!(n-r)!} p^r q^{n-r} \qquad [6\cdot1]$$

For $r = 0$, we get:

$$P(0) = \frac{5!}{0!(5-0)!}(.4^0)(.6^5)$$

$$= \frac{5 \times 4 \times 3 \times 2 \times 1}{(1)(5 \times 4 \times 3 \times 2 \times 1)}(1)(.6^5)$$

$$= \frac{120}{120}(1)(.07776)$$

$$= (1)(1)(.07776)$$

$$= .07776$$

For $r = 1$, we get:

$$P(1) = \frac{5!}{1!(5-1)!}(.4^1)(.6^4)$$

$$= \frac{5 \times 4 \times 3 \times 2 \times 1}{(1)(4 \times 3 \times 2 \times 1)}(.4)(.6^4)$$

$$= \frac{120}{24}(.4)(.1296)$$

$$= (5)(.4)(.1296)$$

$$= .2592$$

For $r = 2$, we get:

$$P(2) = \frac{5!}{2!(5-2)!}(.4^2)(.6^3)$$

$$= \frac{5 \times 4 \times 3 \times 2 \times 1}{(2 \times 1)(3 \times 2 \times 1)}(.4^2)(.6^3)$$

$$= \frac{120}{12}(.16)(.216)$$

$$= (10)(.03456)$$

$$= .3456$$

*When we define n, we look at the number of students. The fact that there is a possibility that none will be late does not alter our choice of $n = 5$.

For $r = 3$, we get:

$$P(3) = \frac{5!}{3!(5-3)!} (.4^3)(.6^2)$$
$$= \frac{5 \times 4 \times 3 \times 2 \times 1}{(3 \times 2 \times 1)(2 \times 1)} (.4^3)(.6^2)$$
$$= (10)(.064)(.36)$$
$$= .2304$$

For $r = 4$, we get:

$$P(4) = \frac{5!}{4!(5-4)!} (.4^4)(.6^1)$$
$$= \frac{5 \times 4 \times 3 \times 2 \times 1}{(4 \times 3 \times 2 \times 1)(1)} (.4^4)(.6)$$
$$= (5)(.0256)(.6)$$
$$= .0768$$

Finally, for $r = 5$, we get:

$$P(5) = \frac{5!}{5!(5-5)!} (.4^5)(.6^0)$$
$$= \frac{5 \times 4 \times 3 \times 2 \times 1}{(5 \times 4 \times 3 \times 2 \times 1)(1)} (.4^5)(1)$$
$$= (1)(.01024)(1)$$
$$= .01024$$

The binomial distribution for this example is shown graphically in Figure 6·4.

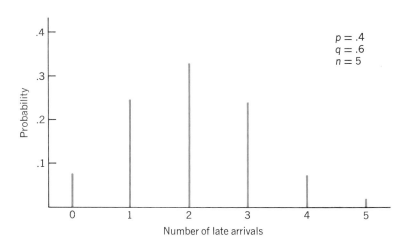

$p = .4$
$q = .6$
$n = 5$

FIG. 6·4
**Binomial
probability
distribution of late
arrivals**

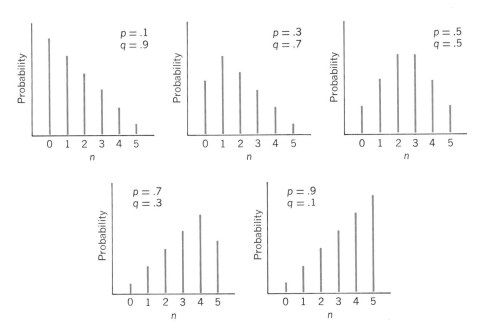

Without doing all the calculations involved, we can illustrate the general appearance of a family of binomial probability distributions. In Fig. 6·5, for example, each distribution represents $n = 5$. In each case, the p and q have been changed and are noted beside each distribution. From Fig. 6·5, we can make the following generalizations:

1. When p is small (.1), the binomial distribution is skewed to the right.
2. As p increases (to .3, for example), the skewness is less noticeable.
3. When $p = .5$, the binomial distribution is symmetrical.
4. When p is larger than .5, the distribution is skewed to the left.
5. The probabilities for .3, for example, are the same as those for .7 except that the values of p and q are *reversed*. This is true for any pair of complementary p and q values (.3 and .7), (.4 and .6), and (.2 and .8).

Let us examine graphically what happens to the binomial distribution when p stays constant but n is increased. Figure 6·6 illustrates the general shape of a family of binomial distributions with a constant p of .4 and n's from 5 to 30. As n increases, the vertical lines not only become more numerous but also tend to bunch up together to form a *bell shape*. We shall have more to say about this bell shape shortly.

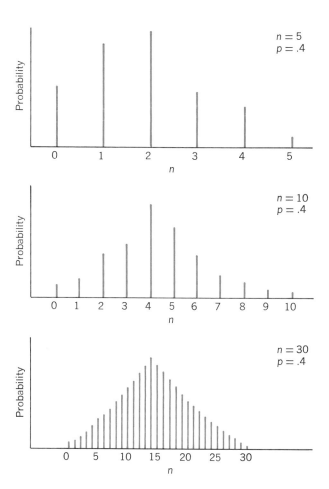

Using the binomial tables

Earlier we recognized that it is tedious to calculate probabilities using the binomial formula when *n* is a large number. Fortunately, we can use Appendix Table 3 to determine binomial probabilities quickly.

To illustrate the use of the binomial tables, consider this problem. What is the probability that 8 or more of the 15 registered Democrats on Prince Street will fail to vote in the coming primary if the probability of any individual's not voting is .30, and if people decide independently of each other whether or not to vote? First we represent the elements in this problem in binomial distribution notation:

Solving problems using the binomial tables

$n = 15$ number of registered Democrats
$p = .30$ probability that any one individual won't vote
$r = 8$ number of individuals who will fail to vote

Then, since the problem involves 15 trials, we must find the table corresponding to $n = 15$. Since the probability of an individual's not voting is .30, we must look through the $n = 15$ table until we find the column where $p = .30$. (This is denoted as 30.) We then move down that column until we are opposite the $r = 8$ row. The answer there is 0500, which can be interpreted as being a probability value of .0500. This represents the probability of 8 or more nonvoters, since the tables are so constructed.

Our problem asked for the probability of 8 or more nonvoters. If it had asked for the probability of more than 8 nonvoters we would have looked up the probability of 9 or more nonvoters. Had the problem asked for the probability of exactly 8 nonvoters we would have subtracted .0152 (the probability of 9 or more nonvoters) from .0500 (the probability of 8 or more nonvoters). The answer would be .0348 = the probability of exactly 8 nonvoters. Finally, if the problem had asked for the probability of fewer than 8 nonvoters we would have subtracted .0500 (the probability of 8 or more nonvoters) from 1.0 for an answer of .9500. (Note that Appendix Table 3 only goes up to $p = .50$. Instructions for using the table when p is larger than .50 are found on the first page of Appendix Table 3.)

Measures of central tendency and dispersion for the binomial distribution

Computing the mean and the standard deviation

Earlier in this chapter we encountered the concept of the expected value or mean of a probability distribution. The binomial distribution has an expected value or mean (μ) and a standard deviation (σ), and we should be able to compute both of these statistical measures. Intuitively, we can reason that if a certain machine produces good parts with a $p = .5$, then, over time, the mean of the distribution of the good parts in the output would be .5 times the total output. If there is a .5 chance of tossing a head with a fair coin, over a large number of tosses the mean of the binomial distribution of the number of heads would be .5 times the total number of tosses.

Symbolically, we can represent the mean of a binomial distribution as:

$$\mu = np \qquad [6 \cdot 2]$$

where:
n = number of trials
p = probability of success

And we can calculate the standard deviation of a binomial distribution by using the formula:

$$\sigma = \sqrt{npq} \qquad [6 \cdot 3]$$

where:
n = number of trials
p = probability of success
q = probability of failure = $1 - p$

To see how to use Equations 6·2 and 6·3, take the case of a packaging machine that produces 20 percent defective packages. If we take a random sample of 10 packages, we can compute the mean and the standard deviation of the binomial distribution of that process like this:

$$\mu = np \qquad\qquad [6\cdot2]$$
$$= (10)(.2)$$
$$= 2$$

$$\sigma = \sqrt{npq} \qquad\qquad [6\cdot3]$$
$$= \sqrt{(10)(.2)(.8)}$$
$$= \sqrt{1.6}$$
$$= 1.265$$

Meeting the conditions for using the Bernoulli process

We need to be careful in the use of the binomial probability distribution to make certain that the three conditions necessary for a Bernoulli process introduced on pp. 175-76 are met, particularly conditions 2 and 3. Condition 2 requires the probability of the outcome of any trial to remain fixed over time. In many industrial processes, however, it is extremely difficult to guarantee that this is indeed the case. Each time an industrial machine produces a part, for instance, there is some infinitesimal wear on the machine. If this wear accumulates beyond a reasonable point, the proportion of acceptable parts produced by the machine will be altered, and condition 2 for the use of the binomial distribution may be violated. This problem is not present in a coin toss experiment, but it is an integral consideration of all real applications of the binomial probability distribution.

Condition 3 requires that the trials of a Bernoulli process be statistically independent; that is, the outcome of one trial cannot affect in any way the outcome of any other trial. Here, too, we can encounter some problems in real applications. Consider an interviewing process in which high-potential candidates are being screened for top political positions. If the interviewer has talked with five unacceptable candidates in a row, he may not view the sixth with complete impartiality. The trials, therefore, would not be statistically independent.

Applying the binomial distribution to real-life situations

EXERCISES

16. For a binomial distribution with $n = 6$ and $p = .3$, find
    a) $P(r = 5)$ b) $P(r > 4)$ c) $P(r < 2)$ d) $P(r \geqslant 3)$

17. For a binomial distribution with $n = 15$ and $p = .2$, use Appendix Table 3 to find
    $P(r = 6)$ $P(r > 9)$ $P(r \leqslant 12)$

18. Find the mean and standard deviation of the following binomial distributions:

    a) $n = 12$, $p = .25$

    b) $n = 25$, $q = .4$

    c) $n = 500$, $p = .10$

    d) $n = 40$, $p = .05$

    e) $n = 2{,}250$, $p = .95$

19. For $n = 10$ trials, compute the probability that $r \geq 1$ for each of the following values of p:

    a) $p = .2$

    b) $p = .4$

    c) $p = .5$

    d) $p = .7$

20. Jane Kendall, a sociologist, wants a random sample of four people taken from a population consisting of all inhabitants of a certain block in New York City. She knows from previous research that 60 percent of the inhabitants are male.

    a) What is the probability that her sample will contain exactly two males? Do not use the tables.

    b) What is the probability that there will be four males in the sample? Solve this without use of tables, using the binomial formula.

21. Historical data supplied by the North Carolina Department of Motor Vehicles show that 80 percent of beginning students in driver education classes were unable to engage the clutch of a training automobile satisfactorily on the first attempt. (Students are judged by in-the-car instructors who rate performances acceptable or unacceptable.) Find the following probabilities (to 4 decimal places) without the use of tables.

    a) Of the eight students in a car per day, what is the probability that exactly 6 will unacceptably engage the clutch of the automobile on the first try?

    b) Exactly 7?

22. An educational psychologist was arguing heatedly with the dean of his college about the effect of changing the entrance criterion for admittance to their university. The psychologist argued that raising the entrance criterion so that 90 percent of those admitted had a combined SAT score of 1,000 or above (rather than the current 80 percent) would reduce the chance of getting exactly 2 scores below 1,000 in a sample of 5. The dean disagreed, saying that the probability would be the same, that it had to be the same, since all the probabilities had to add up to 1.00. The psychologist presented the dean with a graph and chart that reflected the university's current acceptance criterion. His attempt to explain how the increased admittance level would change the probabilities was, however, unsuccessful.

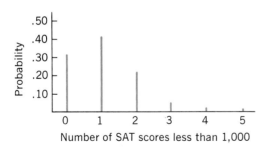

Probability distribution of SAT scores

Number of scores less than 1,000	Probability
0	.3277
1	.4096
2	.2048
3	.0512
4	.0064
5	.0003
	1.0000

    a) Draw a similar graph for an entrance policy that requires 90 percent of scores above 1,000, and construct a similar chart of the probabilities.

    b) What will be the difference, if any, in the probability of obtaining exactly 2 scores below 1,000 from a sample of 5, once the admittance process is changed?

5 THE POISSON DISTRIBUTION

There are many discrete probability distributions, but our discussion will focus on only two: the *binomial,* which we have just concluded, and the *Poisson,* which is the subject of this section. The Poisson distribution is named for Siméon Denis Poisson (1781–1840), a Frenchman who developed the distribution from studies during the latter part of his lifetime.

Examples of
Poisson
distributions

The Poisson distribution is used to describe a number of processes, including the distribution of telephone calls going through a switchboard system, the demand (needs) of patients for service at a health institution, the arrivals of trucks and cars at a toll booth, and the number of accidents at an intersection. These examples all have a common element: they can be described by a discrete random variable that takes on integer (whole) values (0, 1, 2, 3, 4, 5, and so on). The number of patients who arrive at a physician's office in a given interval of time will be 0, 1, 2, 3, 4, 5, or some other whole number. Similarly, if you count the number of cars arriving at a toll booth on the New Jersey Turnpike during some 10-minute period, the number will be 0, 1, 2, 3, 4, 5, and so on.

Characteristics of processes that produce a Poisson probability distribution

Conditions leading
to a Poisson
probability
distribution

The number of vehicles passing through a single turnpike toll booth at rush hour serves as an illustration of Poisson probability distribution characteristics:

1. The average (mean) arrivals of vehicles per rush hour can be estimated from past traffic data.
2. If we divide the rush hour into periods (intervals) of one second each, we will find these statements to be true:
   a) The probability that exactly one vehicle will arrive at the single booth per second is a very small number and is constant for every one-second interval.
   b) The probability that two or more vehicles will arrive within a one-second interval is so small that we can assign it a zero value.
   c) The number of vehicles that arrive in a given one-second interval is independent of the time at which that one-second interval occurs during the rush hour.
   d) The number of arrivals in any one-second interval is not dependent on the number of arrivals in any other one-second interval.

Now, we can generalize from these four conditions described for our toll booth example and apply them to other processes. If these new processes meet the same four conditions, then we can use a Poisson probability distribution to describe them.

Calculating probabilities, using the Poisson distribution

The Poisson probability distribution, as we have shown, is concerned with certain processes that can be described by a discrete random variable. The letter X usually represents that discrete random variable, and X can take on integer values (0, 1, 2, 3, 4, 5, and so on). We use capital X to represent the random variable and lower-case x to represent a specific value that capital X can take. The probability of *exactly* x occurrences in a Poisson distribution is calculated with the formula.

Poisson
distribution
formula

$$P(x) = \frac{\lambda^x \times e^{-\lambda}}{x!}$$

[6·4]

Look more closely at each part of this formula:

Lambda (the mean number of occurrences per interval of time) raised to the x power

e, or 2.71828 (the base of the Naperian or natural logarithm system), raised to the negative lambda power

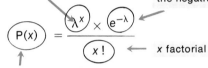

x factorial

$P(x)$

Probability of exactly
x occurrences

Suppose that we are investigating the safety of a dangerous intersection. Past police records indicate a mean of 5 accidents per month at this intersection. The number of accidents is distributed according to a Poisson distribution and the Highway Safety Division wants us to calculate the probability in any month of exactly 0, 1, 2, 3, and 4 accidents. We can use Appendix Table 4 to avoid having to calculate e's to negative powers. Applying the formula:

An example using
the Poisson formula

$$P(x) = \frac{\lambda^x \times e^{-\lambda}}{x!}$$

[6·4]

we can calculate the probability of exactly 0 accidents:

$$P(0) = \frac{(5^0)(e^{-5})}{0!}$$

$$= \frac{(1)(.00674)}{1}$$

$$= .00674$$

For exactly 1 accident:

$$P(1) = \frac{(5^1)(e^{-5})}{1!}$$

$$= \frac{(5)(.00674)}{1}$$

$$= .03370$$

For exactly 2 accidents:

$$P(2) = \frac{(5^2)(e^{-5})}{2!}$$

$$= \frac{(25)(.00674)}{2 \times 1}$$

$$= .08425$$

For exactly 3 accidents:

$$P(3) = \frac{(5^3)(e^{-5})}{3!}$$

$$= \frac{(125)(.00674)}{3 \times 2 \times 1}$$

$$= \frac{.8425}{6}$$

$$= .14042$$

Finally, for exactly 4 accidents:

$$P(4) = \frac{(5^4)(e^{-5})}{4!}$$

$$= \frac{(625)(.00674)}{4 \times 3 \times 2 \times 1}$$

$$= \frac{4.2125}{24}$$

$$= .17552$$

Our calculations will answer several questions. Perhaps we want to know the probability of there being 0, 1, or 2 accidents in any month. We find this by adding together the probabilities of exactly 0, 1, and 2 accidents like this:

What the answers to the formula mean

$$P(0) = .00674$$
$$P(1) = .03370$$
$$P(2) = .08425$$
$$P(0,1,2) = \mathbf{.12469}$$

We will take action to improve the intersection if the probability of more than 3 accidents per month exceeds .65. Should we act? To solve this problem, we need to calculate the probability of having 0, 1, 2, or 3 accidents and then subtract the sum from 1.0 to get the probability for more than 3 accidents. We begin like this:

$$P(0) = .00674$$
$$P(1) = .03370$$
$$P(2) = .08425$$
$$P(3) = \underline{.14042}$$
$$P(3 \text{ or fewer}) = \mathbf{.26511}$$

Because the Poisson probability of 3 or fewer accidents is .26511, the probability of more than 3 must be .73489, (1.00000 − .26511). Since .73489 exceeds .65, steps should be taken to improve the intersection.

Constructing
a Poisson
probability
distribution

We could continue calculating the probabilities for more than 4 accidents and eventually produce a Poisson probability distribution of the number of accidents per month at this intersection. Table 6·12 illustrates such a distribution. To produce this table, we have used Formula 6·4. Try doing the calculations yourself for the probabilities beyond exactly 4 accidents. Figure 6·7 illustrates graphically the Poisson probability distribution of the number of accidents.

TABLE 6·12
Poisson probability distribution of accidents per month

x = number of accidents	P(x) = Probability of exactly that number
0	.00674
1	.03370
2	.08425
3	.14042
4	.17552
5	.17552
6	.14627
7	.10448
8	.06530
9	.03628
10	.01814
11	.00824
	.99486 ←Probability for 0 through 11 accidents
12 or more	.00514 ← Probability for 12 or more (1.0 − .99486)
	1.00000

FIG. 6·7
Poisson probability distribution of the number of accidents

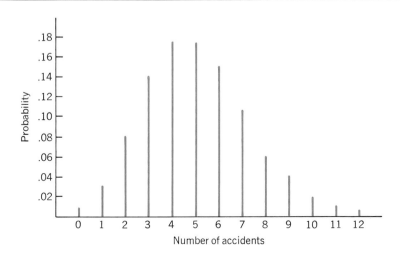

Poisson distribution as an approximation
of the binomial distribution

189

Section 5
POISSON
DISTRIBUTION

Sometimes, if we wish to avoid the tedious job of calculating binomial probability distributions, we can use the Poisson instead. The Poisson distribution can be a reasonable approximation of the binomial, but only under certain conditions. These conditions are when *n* is large and *p* is small; that is, when the number of trials is large and the binomial probability of success is small. The rule most often used by statisticians is that the Poisson is a good approximation of the binomial when *n* is equal to, or greater than, 20 and *p* is equal to or less than .05. In cases that meet these conditions, we can substitute the mean of the binomial distribution (*np*) in place of the mean of the Poisson distribution (λ), so that the formula becomes:

A modification of
the Poisson formula

$$P(x) = \frac{(np)^x \times e^{-np}}{x!} \qquad [6 \cdot 5]$$

Let us use both the binomial probability formula (6·1) and the Poisson approximation formula (6·5) on the same problem to determine the extent to which the Poisson is a good approximation of the binomial. Say that we have a hospital with 20 kidney dialysis machines and that the chance of any one of them malfunctioning during any day is .02. What is the probability that exactly 3 machines will be out of service on the same day? Table 6·13 shows the answers to this question. As we can see, the difference between the two probability distributions is slight (only about a 10 percent error in this example).

Comparing
the Poisson and
binomial formulas

Poisson approach	*Binomial approach*
$P(x) = \dfrac{(np)^x \times e^{-np}}{x!}$ [6·5]	$P(r) = \dfrac{n!}{r!(n-r)!} p^r q^{n-r}$ [6·1]
$P(3) = \dfrac{(20 \times .02)^3 e^{-(20 \times .02)}}{3!}$	$P(3) = \dfrac{20!}{3!(20-3)!}(.02^3)(.98^{17})$
$= \dfrac{(.4^3)(e)^{-.4*}}{(3 \times 2 \times 1)}$	$= .0065$ (from Appendix Table 3)
$= \dfrac{(.064)(.67032)^*}{6}$	
$= .00715$	

*Use Appendix Table 4 to find the value of $(e)^{-.4}$

TABLE 6·13
Comparison of Poisson and binomial probability approaches to the kidney dialysis situation

EXERCISES

23. If the receptionist's phone rings an average of 4 times an hour, find the probability of

a) no calls in a randomly selected hour
b) 2 calls

c) 4 calls
d) 5 or more calls

24. Given $\lambda = 3.5$, for a Poisson distribution, find

    a) $P(X \leqslant 2)$ b) $P(X \geqslant 4)$ c) $P(X = 6)$

25. Given a binomial distribution with $n = 25$ trials and $p = .02$, use the Poisson approximation to the binomial to find

    a) $P(r = 20)$ b) $P(r = 5)$ c) $P(r = 2)$

26. Given a binomial distribution with $n = 20$ trials and $p = .04$, use the Poisson approximation to the binomial to find

    a) $P(r \geqslant 2)$ b) $P(r < 5)$ c) $P(r = 0)$

27. Sally Shuping had been the sole undergraduate sociology adviser on the faculty of a small university. She had complained to the departmental head that she was overworked in her advisory capacity, and needed help. Sally kept a record showing that the mean number of students coming to see her each day was 10. The department head agreed (without knowing of her record) that if there were 7 or more students on a particular day on which he kept track, he would appoint another advisor. If the probability that there will be more than 7 students is .7798, what is the probability that Sally will get an assistant?

28. Psychologist James Olsen is running a reinforcement experiment with pigeons. He has 8 pigeons, each of which has been trained to peck a key continuously in return for intermittent reinforcement with food pellets. James has some question about using one of the pigeons, however, which is an older bird. Unlike the others, this pigeon stops for short rest periods an average of 4.1 times per hour. The rest period is a fairly consistent 3 minutes each time. James thinks this pigeon perhaps should be replaced with another. He decides that if the probability of the pigeon resting for 12 minutes or more per hour is greater than .50, he will replace it with a younger bird. Should he do so?

29. In a student bookstore, the shelf holding books for a course called "Computing in the Social Sciences" is quite small. Owing to a change in graduation requirement, enrollment in the course for this semester is 20 times as great as ever before. Because of this, during book rush the shelf must be restocked as students come in to buy the book. On the average, the shelf empties 5.4 times a day.

    a)   What is the probability that the shelf will be emptied exactly 5 times?

    b)   If the probability that the shelf will be emptied 4 or fewer times is .3733, what is the probability that the shelf will be emptied more than 5 times?

30. A psycholinguist who developed a new approach to learning foreign language has been quite pleased with the success of his program. Over several years of testing only 4 percent of students have had significant trouble reading short sentences after a year of training. He wants to demonstrate his program to a team of experts by randomly selecting students who have taken the program, and testing them before the experts. He is somewhat worried, though, that he might, by chance, get one or more of the students who had trouble. He believes that whether or not a student has trouble with the test is a Bernoulli process, and he is convinced the probability of failure really is about .04.

    a)   Assuming the psycholinguist picks exactly 50 students to test before the experts, and using the Poisson distribution as an approximation to the binomial, what chance will he have of getting at least one student who had trouble reading the language after a year?

    b)   No students who had trouble?

31. Southcentral Telephone Company employs the Boynton Delivery Service to deliver its telephone books. Southcentral has been pleased with the service because over the years Boynton has delivered telephone books to 97 percent of the names that were supplied to it

by the phone company. Nevertheless, Southcentral continues to make spot checks, randomly calling numbers that should be supplied with new telephone books.

a) What is the probability that out of 100 calls made, exactly 3 people will not have received telephone books?

b) Exactly one person?

6 THE NORMAL DISTRIBUTION: A DISTRIBUTION OF A CONTINUOUS RANDOM VARIABLE

So far in this chapter, we have been concerned with discrete probability distributions. In this section, we shall turn to cases in which the variable can take on *any* value within a given range and in which the probability distribution is continuous.

Continuous distribution defined

A very important continuous probability distribution is the *normal* distribution. Several mathematicians were instrumental in its development, among them the eighteenth century mathematician-astronomer Karl Gauss. In honor of his work, the normal probability distribution is often called the Gaussian distribution.

Normal distribution

There are two basic reasons why the normal distribution occupies such a prominent place in statistics. First, it has some properties that make it applicable to a great many situations in which it is necessary to make inferences by taking samples. In Chapter 7, we will find that the normal distribution is a useful sampling distribution. Second, the normal distribution comes close to fitting the actual observed frequency distributions of many phenomena, including human characteristics (weights, heights, and I.Q.'s), outputs from physical processes (dimensions and yields), and other measures of interest to decision makers in both the social and natural sciences.

Characteristics of the normal probability distribution

Look for a moment at Fig. 6·8. This diagram suggests several important features of a normal probability distribution:

The normal curve described

1. The curve has a single peak; thus, it is unimodal. It has the bell shape that we described earlier.
2. The mean of a normally distributed population lies at the center of its normal curve.

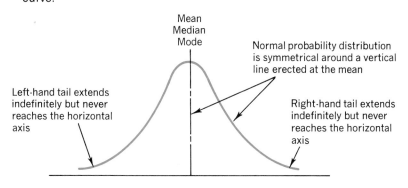

Mean
Median
Mode

Normal probability distribution is symmetrical around a vertical line erected at the mean

Left-hand tail extends indefinitely but never reaches the horizontal axis

Right-hand tail extends indefinitely but never reaches the horizontal axis

FIG. 6·8

Frequency curve for the normal probability distribution

3. Because of the symmetry of the normal probability distribution, the median and the mode of the distribution are also at the center; thus for a normal curve, the mean, median, and mode are the same value.

4. The two tails of the normal probability distribution extend indefinitely and never touch the horizontal axis (graphically, of course, this is impossible to show).

Significance of
the two parameters

Most real-life populations do not extend forever in both directions; but for such populations, the normal distribution is a convenient approximation. There is not a single normal curve but rather a family of normal curves. To define a particular normal probability distribution, we need only two parameters: the mean (μ) and the standard deviation (σ). In Table 6·14 each of the populations is described only by the mean and the standard deviation, and each has a particular normal curve.

TABLE 6·14
Different normal probability distributions

Nature of the population	*Its mean*	*Its standard deviation*
Annual earnings of employees at one plant	$10,000/year	$1,000
Length of standard 8′ building lumber	8′	.5″
Air pollution in one community	2,500 particles per million	750 particles per million
Per capita income in a single developing country	$1,400	$300
Violent crimes per year in a given city	8,000	900

Figure 6·9 shows three normal probability distributions, each of which has the same mean but a different standard deviation. Although these curves differ in appearance, all three are "normal curves."

FIG. 6·9
Normal probability distributions with identical means but different standard deviations

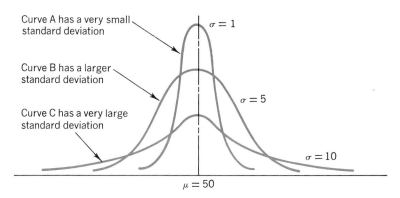

Figure 6·10 illustrates a "family" of normal curves, all with the same standard deviation but each with a different mean.

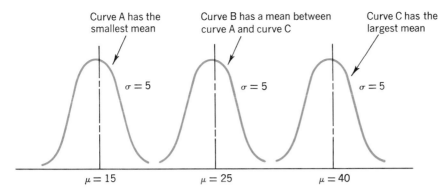

Curve A has the smallest mean

Curve B has a mean between curve A and curve C

Curve C has the largest mean

$\sigma = 5$ $\sigma = 5$ $\sigma = 5$

$\mu = 15$ $\mu = 25$ $\mu = 40$

FIG. 6·10

Normal probability distributions with different means but the same standard deviations

Finally, Fig. 6·11 shows three different normal probability distributions, each with a different mean *and* a different standard deviation. The normal probability distributions illustrated in Figs. 6·9, 6·10, and 6·11 demonstrate that the normal curve can describe a large number of populations, differentiated only by the mean and/or the standard deviation.

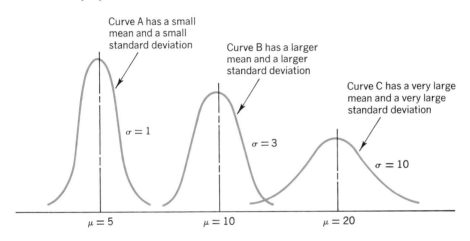

Curve A has a small mean and a small standard deviation

Curve B has a larger mean and a larger standard deviation

Curve C has a very large mean and a very large standard deviation

$\sigma = 1$ $\sigma = 3$ $\sigma = 10$

$\mu = 5$ $\mu = 10$ $\mu = 20$

FIG. 6·11

Three normal probability distributions, each with a different mean and a different standard deviation

Areas under the normal curve

No matter what the values of μ and σ are for a normal probability distribution, the total area under the normal curve is 1.00, so that we may think of areas under the curve as probabilities. Mathematically, it is true that:

Measuring the area under a normal curve

1. Approximately 68 percent of all the values in a normally distributed population lie within 1 standard deviation (plus and minus) from the mean.
2. Approximately 95.5 percent of all the values in a normally distributed population lie within 2 standard deviations (plus and minus) from the mean.
3. Approximately 99.7 percent of all the values in a normally distributed population lie within 3 standard deviations (plus and minus) from the mean.

These three statements are shown graphically in Fig. 6·12.

FIG. 6·12

Relationship between the area under the curve and the distance from the mean measured in standard deviations for a normal probability distribution

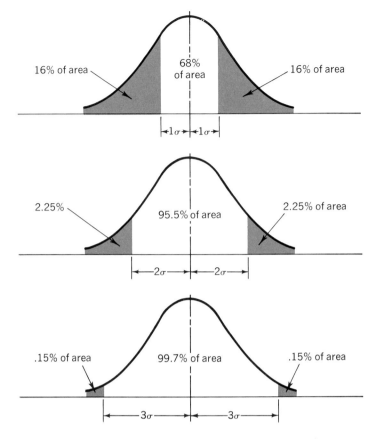

16% of area

68% of area

16% of area

$\leftarrow 1\sigma \rightarrow \leftarrow 1\sigma \rightarrow$

2.25%

95.5% of area

2.25% of area

$\leftarrow 2\sigma \rightarrow \leftarrow 2\sigma \rightarrow$

.15% of area

99.7% of area

.15% of area

$\leftarrow 3\sigma \rightarrow \leftarrow 3\sigma \rightarrow$

Figure 6·12 shows three different ways of measuring the area under the normal curve. However, very few of the applications we shall make of the normal probability distribution involve intervals of *exactly* 1, 2, or 3 standard deviations (plus and minus) from the mean. What should we do about all these other cases? Fortunately, we can refer to statistical tables constructed for precisely these situations. They indicate portions of the area under the normal curve that are contained within any number of standard deviations (plus and minus) from the mean.

Standard normal probability distribution

It is not possible or necessary to have a different table for every possible normal curve. Instead, we can use a table of areas under the curve to find a **standard normal probability distribution.** With this table, we can determine the area, or probability, that the normally distributed random variable will lie within certain distances from the mean. These distances are defined in terms of standard deviations.

We can better understand the concept of the standard normal probability distribution by examining the special relationship of the standard deviation to the normal curve. Look at Fig. 6·13. Here we have illustrated two normal probability distributions, each with a different mean and a different standard deviation. Both area *a* and area *b,* the shaded areas under

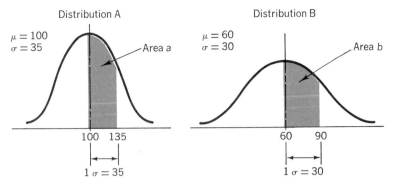

FIG. 6·13
Two intervals,
each one standard
deviation to the
right of the mean

the curves, contain the *same* proportion of the total area under the normal curve. Why? Because both of these areas are defined as being the area between the mean and one standard deviation to the right of the mean. *All* intervals containing the same number of standard deviations from the mean will contain the same proportion of the total area under the curve for *any* normal probability distribution. This makes possible the use of only one standard normal probability distribution table.

Let's find out what proportion of the total area under the curve is represented by both colored areas in Fig. 6·13. In Fig. 6·12, we saw that an interval of one standard deviation (plus *and* minus) from the mean contained about 68 percent of the total area under the curve. In Fig. 6·13, however, we are interested only in the area between the mean and one standard deviation to the *right* of the mean (plus, *not* plus and minus). This area must be half of 68 percent, or 34 percent, for both distributions.

Deriving the percentage of the total area under the curve

One more example will reinforce our point. Look at the two normal probability distributions in Fig. 6·14. Each of these has a different mean and a different standard deviation. The colored area under *both* curves, however, contains the same proportion of the total area under the curve. Why? Because the problem states that both colored areas fall within 2 standard deviations plus and minus from the mean. Two standard deviations plus and minus from the mean include the same proportion of the total area under any normal probability distribution. In this case, we can refer to Fig. 6·12 again and see that the colored area in both distributions in Fig. 6·14 contains about 95.5 percent of the total area under the curve.

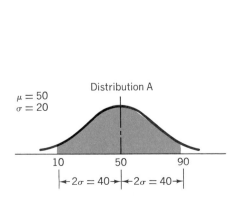
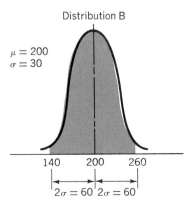

FIG. 6·14
Two intervals,
each two standard
deviations plus
and minus from
the mean

Using the standard normal probability distribution table

Formula for
measuring distances
under the
normal curve

Appendix Table 1 shows the area under the normal curve between the mean and any value of the normally distributed random variable. Notice in this table the location of the column labeled z. The value for z is derived from the formula:

$$z = \frac{x - \mu}{\sigma}$$ [6·6]

where:

x = value of the random variable with which we are concerned
μ = mean of the distribution of this random variable
σ = standard deviation of this distribution
z = number of standard deviations from x to the mean of this distribution

Why do we use z rather than "the number of standard deviations"? Normally distributed random variables take on many *different units* of measure: dollars, inches, parts per million, pounds, time. Since we shall use one table, Table 1 in the Appendix, we talk in terms of *standard units* (which really means standard deviations) and we give them a symbol of z.

Using z values

We can illustrate this graphically. In Fig. 6·15 we see that the use of z is just a change of the scale of measurement on the horizontal axis.

FIG. 6·15
Normal distribution illustrating comparability of z values and standard deviations

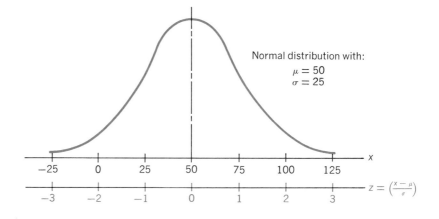

Standard Normal
Probability
Distribution Table

The Standard Normal Probability Distribution Table, Appendix Table 1, is organized in terms of standard units, or z values. It gives the values for only *half* the area under the normal curve, beginning with 0.0 at the mean. Since the normal probability distribution is symmetrical (return to Fig. 6·8 to review this point), the values true for one half of the curve are true for the other. We can use this one table for problems involving both sides of the normal curve. Working a few examples will help us to feel comfortable with the table.

We have a training program designed to upgrade the teaching skills of special education teachers. Because the program is self-administered, teachers require different numbers of hours to complete the program. A study of past participants indicates that the mean length of time spent on the program is 500 hours and that this normally distributed random variable has a standard deviation of 100 hours.

Using the table to find probabilities (an example)

◆ QUESTION 1: What is the probability that a participant selected at random will require more than 500 hours to complete the program?

◆ SOLUTION: In Fig. 6·16 we see that half of the area under the curve is located on either side of the mean of 500 hours. Thus, we can deduce that the probability that the random variable will take on a value higher than 500 is the colored half, or .5.

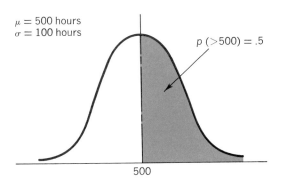

$\mu = 500$ hours
$\sigma = 100$ hours

$p\,(>500) = .5$

500

FIG. 6·16

Distribution of time required to complete the training program, with interval more than 500 hours in the colored area

◆ QUESTION 2: What is the probability that a candidate selected at random will take between 500 and 650 hours to complete the training program?

◆ SOLUTION: We have shown this situation graphically in Fig. 6·17. The probability that will answer this question is represented by the colored area between the mean (500 hours) and the x value in which we are interested (650 hours). Using Equation 6·6, we get a z value of:

$$z = \frac{x - \mu}{\sigma} \qquad\qquad [6 \cdot 6]$$

$$= \frac{650 - 500}{100}$$

$$= \frac{150}{100}$$

$$= 1.5 \text{ standard deviations}$$

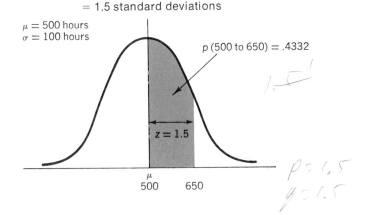

$\mu = 500$ hours
$\sigma = 100$ hours

$p\,(500 \text{ to } 650) = .4332$

$z = 1.5$

μ
500   650

FIG. 6·17

Distribution of time required to complete the training program, with interval 500 to 650 hours in color

197

If we look up $z = 1.5$ in Appendix Table 1, we find a probability of .4332. Thus, the chance that a candidate selected at random would require between 500 and 650 hours to complete the training program is slightly higher than .4.

◆ QUESTION 3: What is the probability that a candidate selected at random will take more than 700 hours to complete the program?

◆ SOLUTION: This situation is different from our previous examples. Look at Fig. 6·18. We are interested in the colored area to the right of the value "700 hours." How can we solve this problem? We can begin by using Equation 6·6:

$$z = \frac{x - \mu}{\sigma} \qquad [6·6]$$

$$= \frac{700 - 500}{100}$$

$$= \frac{200}{100}$$

$$= 2 \text{ standard deviations}$$

Looking in Appendix Table 1 for a z value of 2.0, we find a probability of .4772. That represents the probability the program will require *between* 500 and 700 hours. However, we want the probability it will take *more than* 700 hours (the colored area in Fig. 6·18). Since the right half of the curve (between the mean and the right-hand tail), represents a probability of .5, we can get our answer (the area to the right of the 700-hour point), if we subtract .4772 from .5 (.5000 − .4772 = .0228). Therefore, there are just over 2 chances in 100 that a participant chosen at random would take more than 700 hours to complete the course.

FIG. 6·18
Distribution of time required to complete the training program, with interval above 700 hours in color

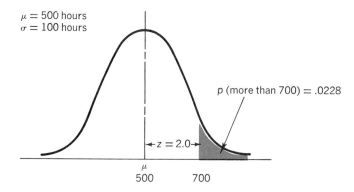

$\mu = 500$ hours
$\sigma = 100$ hours

p (more than 700) $= .0228$

$z = 2.0$

μ
500   700

◆ QUESTION 4: Suppose the training program director wants to know the probability that a participant chosen at random would require between 550 and 650 hours to complete the required work.

◆ SOLUTION: This probability is represented by the colored area in Fig. 6·19. This time, our answer will require two steps. First, we calculate a z value for the 650-hour point as follows:

$$z = \frac{x - \mu}{\sigma} \qquad [6·6]$$

$$= \frac{650 - 500}{100}$$

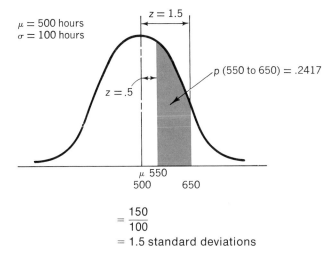

$$z = 1.5$$

$\mu = 500$ hours
$\sigma = 100$ hours

$z = .5$

p (550 to 650) = .2417

μ 550
500   650

FIG. 6·19
Distribution of time
required to
complete the
training program,
with interval
between 550 and
650 hours in color

$$= \frac{150}{100}$$
$$= 1.5 \text{ standard deviations}$$

When we look up a z of 1.5 in Appendix Table 1, we see a probability value of .4332 (the probability that the random variable will fall between the mean and 650 hours). Now for step two. We calculate a z value for our 550-hour point like this:

$$z = \frac{x - \mu}{\sigma} \qquad [6·6]$$
$$= \frac{550 - 500}{100}$$
$$= \frac{50}{100}$$
$$= .5 \text{ standard deviations}$$

In Appendix Table 1, the z value of .5 has a probability of .1915 (the chance that the random variable will fall between the mean and 550 hours). To answer our question we must subtract as follows:

.4332   Probability that the random variable
         will lie between the mean and 650 hours
−.1915   Probability that the random variable
         will lie between the mean and 550 hours
─────
.2417 ← Probability that the random variable
         will lie between 550 and 650 hours

Thus, the chance of a candidate selected at random taking between 550 and 650 hours to complete the program is a bit less than one in four.

◆ QUESTION 5: What is the probability that a candidate selected at random will require less than 580 hours to complete the program?

◆ SOLUTION: This situation is illustrated in Fig. 6·20. Using Equation 6·6 to get the appropriate z value for 580 hours we have:

$$z = \frac{x - \mu}{\sigma}$$
$$= \frac{580 - 500}{100}$$
$$= \frac{80}{100}$$
$$= .8 \text{ standard deviations}$$

[6·6]

199

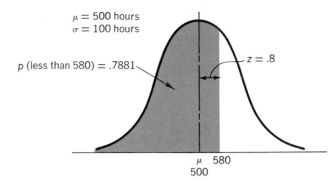

FIG. 6·20

**Distribution of time
required to
complete the
training program,
with interval less
than 580 hours in
color**

$\mu = 500$ hours
$\sigma = 100$ hours

p (less than 580) = .7881

$z = .8$

$\mu \quad 580$
500

Looking in Appendix Table 1 for a z value of .8, we find a probability of .2881—the
probability that the random variable will lie between the mean and 580 hours. We
must add to this the probability that the random variable will be between the left-
hand tail and the mean. Since the distribution is symmetrical with half the area on
each side of the mean, we know this value must be .5. As a final step, then, we add
the two probabilities:

.2881	Probability that the random variable will lie between the mean and 580 hours
+.5000	Probability that the random variable will lie between the left-hand tail and the mean
.7881 ←	Probability that the random variable will lie between the left-hand tail and 580 hours

Thus, the chances of a candidate requiring less than 580 hours to complete the
program are slightly higher than 75 percent.

  ◆ QUESTION 6: What is the probability that a candidate chosen at random will
take between 420 and 570 hours to complete the program?

  ◆ SOLUTION: Figure 6·21 illustrates the interval in question, from 420 to 570
hours. Again the solution requires two steps. First, we calculate a z value for the 570-
hour point:

$$z = \frac{x - \mu}{\sigma} \qquad [6 \cdot 6]$$

$$= \frac{570 - 500}{100}$$

$$= \frac{70}{100}$$

$$= .7 \text{ standard deviations}$$

We look up the z value of .7 in Appendix Table 1 and find a probability value of .2580.
Second, we calculate the z value for the 420-hour point:

FIG. 6·21

**Distribution of time
required to
complete the
training program,
with interval
between 420 and
570 hours in color**

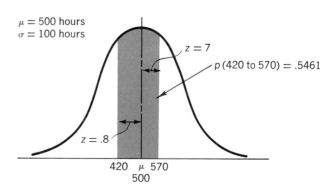

$\mu = 500$ hours
$\sigma = 100$ hours

$z = 7$

p (420 to 570) = .5461

$z = .8$

420 μ 570
500

$$z = \frac{x - \mu}{\sigma}$$ [6·6]

$$= \frac{420 - 500}{100}$$

$$= \frac{-80}{100}$$

$$= -.8 \text{ standard deviations}$$

Since the distribution is symmetrical, we can disregard the sign and look for a *z* value of .8. The probability associated with this *z* value is .2881. We find our answer by adding these two values as follows:

.2580	Probability that the random variable will lie between the mean and 570 hours
+.2881	Probability that the random variable will lie between the mean and 420 hours
.5461 ←	Probability that the random variable will lie between 420 and 570 hours

Thus, there is slightly better than a 50 percent chance that a participant chosen at random will take between 420 and 570 hours to complete the training program.

Shortcomings of the normal probability distribution

Earlier in this section we noted that the tails of the normal distribution approach but never touch the horizontal axis. This implies that there is *some* probability (although it may be very small) that the random variable can take on enormous values. It is possible for the right-hand tail of a normal curve to assign a minute probability of a person weighing 2,000 pounds. Of course, no one would believe that such a person exists. (A weight of one ton or more would lie about 50 standard deviations to the right of the mean and would have a probability that began with 250 zeros to the right of the decimal point!) **We do not lose much accuracy by ignoring values far out in the tails. But in exchange for the convenience of using this theoretical model, we must accept the fact that it can assign impossible empirical values.**

Theory and practice

The normal distribution as an approximation of the binomial distribution

Although the normal distribution is continuous, it is interesting to note that it sometimes can be used to approximate discrete distributions. To see how we can use it to approximate the binomial distribution, suppose we would like to know the probability of getting 5, 6, 7, or 8 heads in 10 tosses of a fair coin. We could use Appendix Table 3 to find this probability as follows:

Sometimes the normal is used to approximate the binomial

$$\begin{array}{ccc} \text{probability of 5, 6,} & \text{probability of} & \text{probability of} \\ \text{7, or 8 heads} = & \text{5 or more heads} - & \text{9 or more heads} \\ = & .6230 & - & .0107 \\ = & .6123 & & \end{array}$$

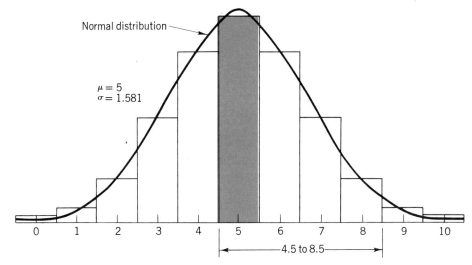

Normal distribution

$\mu = 5$
$\sigma = 1.581$

4.5 to 8.5

Two distributions with the same means and standard deviations

Continuity correction factors

Figure 6·22 shows the binomial distribution for $n = 10$ and $p = \frac{1}{2}$ with a normal distribution superimposed on it with the *same* mean ($\mu = np = 10(\frac{1}{2}) = 5$) and the *same* standard deviation ($\sigma = \sqrt{npq} = \sqrt{10(\frac{1}{2})(\frac{1}{2})} = \sqrt{2.5} = 1.581$.

Look at the area under the normal curve between $5 + \frac{1}{2}$ and $5 - \frac{1}{2}$. We see that this area is *approximately* the same size as the area of the colored bar representing the binomial probability of getting 5 heads. The two $\frac{1}{2}$'s that we add to, and subtract from, 5 are called *continuity correction factors* and are used to improve the accuracy of the approximation.

Using the continuity correction factors, we see that the binomial probability of 5, 6, 7, or 8 heads can be approximated by the area under the normal curve between 4.5 and 8.5. Compute that probability by finding the z values corresponding to 4.5 and 8.5.

$$\text{at } x = 4.5, z = \frac{x - \mu}{\sigma} \qquad [6·6]$$

$$= \frac{4.5 - 5}{1.581}$$

$$= -0.32 \text{ standard deviations}$$

$$\text{at } x = 8.5, z = \frac{x - \mu}{\sigma} \qquad [6·6]$$

$$= \frac{8.5 - 5}{1.581}$$

$$= 2.21 \text{ standard deviations}$$

Now, from Appendix Table 1, we find

.1255   the probability that z will be between -0.32 and 0 (and correspondingly, that x will be between 4.5 and 5)

+ .4864   the probability that z will be
between 0 and 2.21 (and correspondingly, that x will be between
_____   5 and 8.5)
= .6119   the probability that x will be between
4.5 and 8.5

Comparing the binomial probability of .6123 (which we got on page 201
from Appendix Table 3) with this normal approximation of .6119, we can see
that the error in the approximation is less than $\frac{1}{10}$ of 1 percent.

The normal approximation to the binomial distribution is very convenient, since it enables us to solve the problem without extensive tables of
the binomial distribution. (You might note that Appendix Table 3, which
gives binomial probabilities for values of n up to 15, already is 13 pages
long.) We should note that some care needs to be taken in using this
approximation, but it is quite good whenever both np and nq are at
least 5.

EXERCISES

32.   Given that a random variable, X, has a normal distribution with mean 5.6 and standard
deviation 1.4, find

a) $P(5.0 < x < 6.0)$ c) $P(x < 4.4)$

b) $P(x > 7.0)$ d) $P((x < 3.4)$ or $(x > 6.4))$

33.   Given that a random variable has a binomial distribution with $n = 80$ trials and $p = .40$, use
the normal approximation to the binomial to find

a) $P(x > 25)$ c) $P(x < 35)$

b) $P(x > 40)$ d) $P(30 < x < 36)$

34.   In a normal distribution with a standard deviation of 4.0, the probability that an observation
selected at random exceeds 30 is .06.

a)   Find the mean of the distribution.

b)   Find the value below which 10 percent of the values in the distribution lie.

35.   The mean of a normal distribution is 72. If 85 percent of the values in the distribution lie
below 82, what is the standard deviation of the distribution? What is the value above which
5 percent of the values in the distribution lie?

36.   Use the normal approximation to compute the following binomial probabilities

a) $n = 50$ $p = .36$, between 10 and 20 successes

b) $n = 64$ $p = .81$, 48 or more successes

c) $n = 19$ $p = .44$, at most 7 successes

d) $n = 26$ $p = .52$, between 9 and 12 successes

37. A company manufacturing psychological equipment is coming out with a new color wheel for visual perception experiments. The wheels are being made to fit on a spinner machine that is also marketed by this same company. The spinner holds disks of diameters 4.2 ± .05 inches. In the first batch of color wheels, the mean diameter is 4.18″ with a standard deviation of .06″. What is the probability of a color wheel from this batch fitting on the spinner?

38. James Westall, a schoolteacher, wanted to compare his students' scores with those of students in other schools who were taking the same standardized test. He took the scores of his class to a testing expert and explained, "I know the mean on this test is 200, but I don't know the standard deviation. All I can remember is that one student last year made a 68 on this test, and they told him the probability of scoring that or lower was .1210." The test specialist very quickly computed the standard deviation. What was it?

39. Susan James, a sociologist studying crowding, has set up a large number of "mice communities." She believes the size of the mice affects crowding. Furthermore, she thinks brown mice have different crowding reactions than do white mice. Because of this, she doesn't want any of the mice communities in her study to have more than one-sixth white mice by weight. The average total weight of the communities is 3 lbs. On the average, there are .4 lbs of white mice in each community. One of Susan's research assistants tells her that in 68 percent of the communities, the total weight of the white mice added to the brown mice community is within a range of ±.05 lbs of the average. What is the probability that Susan will exclude a mouse community from her study because of weight composition?

40. At a health food camp for athletes who want to avoid processed foods for a time to purify their bodies, the cook plans menus which average 2,400 calories per day, with a standard deviation of 450 calories. The cook is concerned about the rising cost of health foods and also about the fact that some of the residents are gaining weight during their stay. He is considering cutting back on food to design menus with an average of 2,350 calories per day.

   a) What is the probability that a day's menu will still be over 2,400 calories if he does cut back, given that the standard deviation remains the same?

   b) Since the brochure advertises a 2,400-calories-a-day plan, what is the probability that a day's menu will still contain at least 98 percent of the advertised amount?

41. James Young, the professor of a psychological statistics class, has designed his class so that it is entirely self-paced. Students may hand in homework and take tests at any time during the semester, so long as all the work is eventually done. Professor Young's class has been especially popular in the summer, since students in summer school feel they can work hard and finish early. In the last several years, average completion of the class has been 8 weeks, with a standard deviation of 2 weeks.

   a) What is the probability a student will finish the class between 40 and 51 days?

   b) What is the probability of finishing the class in fewer than 35 days?

   c) Fewer than 34 or more than 84 days?

42. On the basis of past experience, automobile inspectors in New Jersey have noticed that 7 percent of all cars coming in for their annual inspection fail to pass. Using the normal approximation to the binomial distribution, find the probability that between 10 and 20 of the next 200 cars to enter the Eatontown, N.J., inspection station will fail the inspection.

7 EXPECTED VALUE DECISION MAKING WITH CONTINUOUS DISTRIBUTIONS

Limitations of the discrete method

In making decisions under conditions of uncertainty, it is often cumbersome to use conditional loss and expected loss tables (such as Tables 6·7 through 6·11) because of the number of computations required. In section 3 of this chapter, the laboratory had only 4 possible levels of requests for tests and 4 possible stock actions. These resulted in a conditional loss table containing only 16 possibilities. If we had 300 possible values for requests and 300 possible stock actions, the number of calculations would be almost prohibitive. Fortunately, another approach avoids this problem.

Excessive computations

Marginal analysis

Marginal analysis is based on the fact that when an additional liter of reagent (in our laboratory problem) is stocked, two fates are possible: the reagent *will* be used in a test, or it *will not* be used. The probabilities of these two events must sum to 1. For example, if the probability of using the additional liter is .6, then the probability of not using it must be .4.

If we let *p* represent the probability of using one additional liter of reagent, then $1 - p$ must be the probability of not using it. If the laboratory uses the additional liter of reagent in a test, it realizes an increase in its financial benefits of $30, (the $50 it receives from the test less the $20 it must pay for the liter of reagent that it uses).

Probabilities of use and non-use

We refer to the $30 as the *marginal gain* or, symbolically, *MG*. Table 6·15 illustrates this point. If the laboratory stocks 10 liters each day and the daily demand is for 10 or more tests, our conditional gain is $300 per day. Now we decide to stock 11 liters each day. If the eleventh liter is used (and this is the case when demand is for 11, 12, or 13 tests), our conditional gain is increased to $330 per day. Notice that the increase in conditional gain does *not* follow merely from *stocking* the eleventh liter. Under the conditions assumed in the problem, this increase in gain will result only when demand is for 11 or more tests. This will be the case 85 percent of the time.

Possible requests for tests	Probability of this many requests	Possible stock actions			
		10 liters	11 liters	12 liters	13 liters
10	.15	$300	$280	$260	$240
11	.20	300	330	310	290
12	.40	300	330	360	340
13	.25	300	330	360	390

TABLE 6·15
Conditional gain table

We must also consider the effect on gain of stocking an additional liter and *not* using it. This reduces our conditional gain. The amount of the reduction is referred to as the *marginal loss* (*ML*) resulting from the stocking of an item that is not used. In our previous example, the marginal loss was $20 per liter, the cost of that liter.

Table 6·15 also illustrates marginal loss. Once more we decide to stock 11 liters. If the eleventh liter (the marginal liter) is not used the conditional gain is $280. The $300 conditional gain when 10 liters were stocked and 10 were used is reduced by $20, the cost of the unused liter of reagent.

Additional liters should be stocked so long as the expected marginal gain from stocking each of them is greater than the expected marginal loss from stocking each. The size of each day's order should be increased up to that point where the expected marginal gain from stocking one more liter of reagent is just equal to the expected marginal loss from stocking that liter if it remains unused and must be discarded.

In our illustration, the probability distribution of requests for tests is:

Requests for tests	Probability of that many requests
10	.15
11	.20
12	.40
13	.25
	1.00

This distribution tells us that as we increase our stock, the probability of using 1 additional liter (this is *p*) decreases. If we increase our stock from 10 to 11 liters the probability of using all 11 is .85. This is the probability that requests will be for 11 tests or more. Here is the computation:

Probability that requests will be for 11	.20
Probability that requests will be for 12	.40
Probability that requests will be for 13	.25
Probability that requests will be for 11 or more	**.85**

If we add a twelfth liter, the probability of using all 12 liters is reduced to .65 (the sum of the probabilities of requests for 12 or 13 tests). Finally, the addition of a thirteenth liter carries with it only a .25 probability of our using all 13 liters, because requests will be for 13 tests only 25 percent of the time.

Deriving the minimum probability equation

The expected marginal gain from stocking and using an additional liter is the marginal gain of the test multiplied by the probability that the test will be done. This is $p(MG)$. The expected marginal loss from stocking and not using an additional liter is the marginal loss incurred if the test is not done multiplied by the probability that the test will not be done; this is

$(1 - p)(ML)$. We can generalize that the laboratory director should stock reagent each day up to the point at which:

$$p(MG) = (1 - p)(ML) \qquad [6 \cdot 7]$$

This equation describes the point at which the expected gain from stocking an additional liter $p(MG)$, is equal to the expected loss from stocking the liter $(1 - p)(ML)$. As long as $p(MG)$ is larger than $(1 - p)(ML)$, additional liters should be stocked because the expected gain from such a decision is greater than the expected loss.

In any given decision problem, there will be only *one* value of p for which the maximizing equation will be true. We must determine that value in order to know the optimum stock action to take. We can do this by taking our maximizing equation and solving it for p in the following manner:

$$p(MG) = (1 - p)(ML) \qquad [6 \cdot 7]$$

Multiplying the two terms on the right side of the equation, we get

$$p(MG) = ML - p(ML)$$

Collecting terms containing p, we have

$$p(MG) + p(ML) = ML$$

or

$$p(MG + ML) = ML$$

Dividing both sides of the equation by $MG + ML$ gives

$$p = \frac{ML}{MG + ML} \qquad [6 \cdot 8]$$

Minimum probability equation

The letter p represents the minimum required probability of using at least an additional liter to justify the stocking of that additional liter. The laboratory director should stock additional liters so long as the probability of using at least an additional liter is greater than p.

We can now compute p for our problem. The marginal gain per unit is $30 (the test price minus the cost); the marginal loss per unit is $20 (the cost of each liter); thus

$$p = \frac{ML}{MG + ML} = \frac{\$20}{\$30 + \$20} = \frac{\$20}{\$50} = .4$$

This value of .4 for p means that in order to make the stocking of an additional liter justifiable, we must have at least a .4 *cumulative* probability of using that liter. In order to determine the probability of using each additional liter we consider stocking, we must compute a series of cumulative probabilities as we have done in Table 6·16.

TABLE 6·16
Cumulative probabilities of reagent use

Requests for tests	Probability of this many requests	Cumulative probability that requests will be at this level or greater
10	.15	1.00
11	.20	.85
12	.40	.65
13	.25	.25

Derivation of cumulative probabilities

The cumulative probabilities in the right-hand column of Table 6·16 represent the probabilities that requests will reach or exceed each of the four levels. For example, the 1.00 that appears beside the 10 request level means that we are 100 percent certain of doing 10 or more tests. This must be true because our problem assumes that one of the four levels will *always* occur.

The .85 probability value beside the 11 request figure means that we are only 85 percent sure of doing 11 or more tests. This can be calculated in two ways. First, we could add the chances of doing 11, 12, and 13 tests:

$$
\begin{array}{ll}
11 \text{ requests} & .20 \\
12 \text{ requests} & .40 \\
13 \text{ requests} & +.25 \\
\hline
& \mathbf{.85} = \text{probability of doing 11 or more tests}
\end{array}
$$

Or we could reason that requests of 11 or more tests include all possible outcomes except requests of 10 tests which has a probability of .15.

$$
\begin{array}{ll}
\text{All possible outcomes} & 1.00 \\
\text{Probability of doing 10 tests} & -.15 \\
\hline
& \mathbf{.85} = \text{probability of doing 11 or more}
\end{array}
$$

The cumulative probability value of .65 assigned to 12 or more tests can be established in similar fashion. Twelve or more tests must mean 12 tests or 13 tests, so:

$$
\begin{array}{ll}
\text{Probability of 12 tests} & .40 \\
\text{Probability of 13 tests} & +.25 \\
\hline
& \mathbf{.65} = \text{probability of 12 or more tests}
\end{array}
$$

And, of course, the cumulative probability of doing 13 tests is still .25 because we have assumed that requests will never exceed 13.

As we mentioned previously, the value of p decreases as the levels of requests increase. This causes the expected marginal gain to decrease and the expected marginal loss to increase until, at some point, our stocking of an additional liter would not be warranted.

We have said that additional liters should be stocked so long as the probability of using at least an additonal liter is greater than p. We can now apply this rule to our probability distribution of requests and determine how many liters should be stocked.

In this case, the probability of using 11 or more liters is .85, a figure clearly greater than our p of .40; thus, we should stock an eleventh liter. The expected marginal gain from stocking this liter is greater than the expected marginal loss from stocking it. We can verify this as follows:

$$p(MG) = .85(\$30) = \$25.50 \text{ expected marginal gain}$$
$$(1 - p)(ML) = .15(\$20) = \$\ 3.00 \text{ expected marginal loss}$$

A twelfth liter should be stocked because the probability of using 12 or more liters (.65) is greater than the required p of .40. Such action will result in the following expected marginal gain and expected marginal loss:

Optimum stocking level for this problem

$$p(MG) = .65(\$30) = \$19.50 \text{ expected marginal gain}$$
$$(1 - p)(ML) = .35(\$20) = \$\ 7.00 \text{ expected marginal loss}$$

Twelve is the *optimum* number of liters to stock because the addition of a thirteenth liter carries with it only a .25 probability that it will be used and that is less than our required p of .40. The following figures reveal why the thirteenth liter should not be stocked:

$$p(MG) = .25(\$30) = \$\ 7.50 \text{ expected marginal gain}$$
$$(1 - p)(ML) = .75(\$20) = \$15.00 \text{ expected marginal loss}$$

If we stock a thirteenth liter, we add more to expected loss than we add to expected gain.

Notice that the use of marginal analysis leads us to the same conclusion that we reached with the use of conditional gain and expected gain tables. Both methods of analysis suggest that the laboratory should stock 12 liters each period.

Our strategy, to stock 12 liters every day, assumes that daily requests is a random variable. In actual practice, however, daily requests often take on recognizable patterns, depending upon the particular day of the week. In situations with recognizable patterns in daily usage we can apply the techniques we have learned by computing an optimal stocking level for each day of the week. For Saturday, we would use as our input data past experience for Saturdays only. Each of the other six days could be treated in the same fashion. Essentially this approach represents nothing more than recognition of, and reaction to, discernible patterns in what may first appear to be a completely random environment.

Adjusting the optimal stocking level

Using the standard normal probability distribution

We first learned the concept of the standard normal probability distribution in Section 6. We can now use this idea to help us solve a decision problem employing a continuous distribution.

Solving a problem
using
marginal analysis

A clinic uses a nonemergency discretionary vaccine with an approximately normally distributed past usage with a mean of 50 units daily and a standard deviation in daily doses of 15 units. The clinic purchases the vaccine for $4 per dose and bills it for $9 per dose. If the vaccine is not used on the purchase day it is worth nothing. Using the marginal method of calculating optimum purchase levels, we can calculate our required p:

$$p = \frac{ML}{MG + ML} \qquad [6\cdot8]$$

$$= \frac{\$4}{\$5 + \$4} = .44$$

This means that the clinic must be .44 sure of using at least an additional dose before they stock that dose. Let us reproduce the curve of past usage and determine how to incorporate the marginal method with continuous distributions of past daily usage.

Standard normal
probability
distribution
combined with
marginal analysis

Now refer to Fig. 6·23. If we erect a vertical line *b* at 50 units, the area under the curve to the right of this line is one-half the total area. This tells us that the probability of selling 50 or more units is .5. *The area to the right of any such vertical line represents the probability of using that quantity or more.* As the area to the right of any vertical line decreases, so does the probability that we will use that quantity or more.

Suppose the clinic considers stocking 25 doses, line *a.* Most of the entire area under the curve lies to the right of the vertical line drawn at 25; thus the probability is great that the clinic will use 25 doses or more. If the clinic considers stocking 50 doses (the mean), one-half the entire area

FIG. 6·23
Normal distribution of past daily use of doses of vaccine

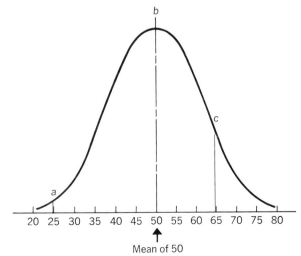

20 25 30 35 40 45 50 55 60 65 70 75 80

↑
Mean of 50

under the curve lies to the right of vertical line *b;* they are .5 sure of using the 50 doses or more. Now if they consider stocking 65 units, only a small portion of the entire area under the curve lies to the right of line *c;* thus the probability of using 65 or more doses is quite small.

Figure 6·24 illustrates the .44 probability that must exist before it no longer pays to stock another dose. They will stock additional doses until they reach point *Q.* If they stock a larger quantity, the colored area under the curve drops below .44 and the probability of using another dose or more falls below the required .44. How can we locate point *Q?* As we saw in Section 6, we can use Appendix Table 1 to determine how many standard deviations it takes to include any portion of the area under the curve measuring from the mean to any points such as *Q.* In this particular case, since we know that the colored area must be .44 of the total area, then the area from the mean to point *Q* must be .06 (the total area from the mean to the right tail is .50). Looking in the body of the table, we find that .06 of the area under the curve is located between the mean and a point .15 standard deviations to the right of the mean. Thus we know that point *Q* is .15 standard deviations to the right of the mean (50).

We have been given the information that 1 standard deviation for this distribution is 15 doses, so .15 times this would be 2.25 doses. Since point *Q* is 2.25 doses to the right of the mean (50), it must be at about 52 doses. This is the optimum order for the clinic to place: 52 doses per day.

Optimum solution for this problem

Another problem utilizes the continuous probability distribution. This time, assume the following situation for a normally distributed daily record:

Mean of past daily usage	60 doses
Standard deviation of past daily use distribution	10 doses
Cost per dose	$20
Billing price per dose	$32
Value if not used on selling day	$ 2 (a value derived from sales to an animal research lab, which can tolerate lower average dosage quality)

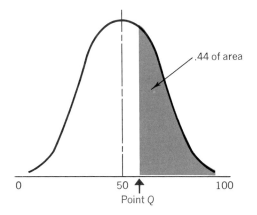

FIG. 6·24

Normal probability distribution, with .44 of the area under the curve colored

.44 of area

0   50 ↑   100
Point *Q*

As we did in the previous problem, we first calculate the p that is required to justify the stocking of an additional dose. In this instance,

$$= \frac{ML}{MG + ML} \qquad\qquad [6 \cdot 8]$$

$$= \frac{\$20 - \$2}{\$12 + (\$20 - \$2)}$$

← Notice that a salvage value of $2 is deducted from the cost of $20 to obtain the *ML*.

$$= \frac{\$18}{\$12 + \$18}$$

$$= \frac{\$18}{\$30} = .6$$

We can now illustrate the probability on a normal curve by marking off .6 of the area under the curve, starting from the right-hand end of the curve, as in Fig. 6·25.

The clinic will want to increase the size of the order this time, until it reaches point Q. Notice that point Q lies to the *left* of the mean, whereas in the preceding problem it lay to the *right* of the mean. How can we locate point Q? Since .5 of the area under the curve is located between the mean and the right-hand tail, .1 of the colored area must be to the left of the mean. We look for .1 in the body of Appendix Table 1. The nearest value to .1 is .0987, so we want to find a point Q with .0987 of the area under the curve contained between the mean of the curve and that point Q. The table indicates point Q to be .25 standard deviations from the mean. We can now solve for point Q as follows:

$$.25 \times \text{standard deviation} = .25 \times 10 \text{ doses} = 2.50 \text{ doses}$$
$$\text{Point } Q = \text{mean } \textbf{less } 2.50 \text{ doses}$$
$$= 60 - 2.50 \text{ doses} = 57.50, \text{ or } 57 \text{ doses}$$

Optimum solution

The optimum stock for the clinic to order each day is 57 doses.

FIG. 6·25
Normal probability distribution with .6 of the area under the curve colored

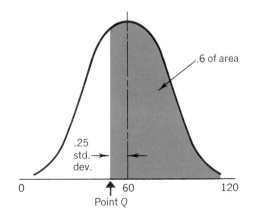

.6 of area

.25
std.→
dev.

0   ↑ 60   120
    Point Q

43. Highway road construction in North Carolina is concentrated in the months from May through September. To provide some measure of protection to the crews at work on the highways, the Department of Transportation (DOT) requires that large, orange MEN WORKING signs be placed well in advance of any construction. Because of vandalism, wear and tear, and theft, each year DOT purchases a number of new signs to be put into service. The signs are actually made under the auspices of the Department of Corrections, but because of interdepartmental budgeting and accounting, DOT must pay as much as if it were to buy the signs from an outside source. The interdepartmental charge is $11 per sign if more than 30 of the same kind are ordered. Otherwise the cost per sign is $15. Because of budget pressures, DOT attempts to minimize its costs by not buying too many signs but ordering more than 30, to take advantage of the $11 price. In recent years the department has averaged purchases of 84 signs per year with a standard deviation of 11. Use marginal analysis to determine the number of signs DOT should purchase.

44. Bike Wholesale Parts was established in the early 1970s in response to demands of several small and newly established bicycle shops that needed access to a wide variety of inventory but were not able to finance it themselves. The company carries a wide variety of replacement parts and accessories but does not maintain any stock of completed bicycles. Management is preparing to order $27'' \times 1\frac{1}{4}''$ rims from the Flexspin Company in anticipation of a business upturn expected in about two months. Flexspin makes a superior product, but the lead time required necessitates that wholesalers place only one order, which must last through the critical summer months. In the past, Bike Wholesale Parts has sold an average of 120 rims per summer, with a standard deviation of 28. The company expects that its stock of rims will be depleted by the time the new order arrives. Bike Wholesale Parts has been quite successful and plans to move its operations to a larger plant during the winter. Management feels that the combined cost of moving some items such as rims and the existing cost of financing them is at least equal to the firm's purchase cost of $7.30. Accepting management's hypothesis that any unsold rims at the end of the summer season are permanently unsold, use marginal analysis to determine the number of rims the company should order if the selling price is $8.10.

8 CHOOSING THE CORRECT PROBABILITY DISTRIBUTION
If we plan to use a probability distribution to describe a situation, we must be careful to choose the right one. We need to be certain that we are not using the *Poisson* probability distribution when it is the *binomial* that more nearly describes the situation we are studying. Remember that the binomial distribution is applied when the number of trials is fixed before the experiment begins, and each trial is independent and can result in only two mutually exclusive outcomes (success/failure, either/or, yes/no). Like the binomial, the Poisson distribution applies when each trial is independent. But although the probabilities in a Poisson distribution approach zero after the first few values, the possible values are infinite. The results are not limited to two mutually exclusive outcomes.

Under some conditions, the Poisson distribution can be used as an approximation of the binomial, but not always. All the assumptions that form the basis of a distribution must be met if our use of that distribution is to produce usable results.

Even though the normal probability distribution is the only continuous distribution we have discussed in this chapter, we should realize that there are other useful continuous distributions. In the chapters to come, we shall study three additional continuous distributions, each of interest to decision makers who solve problems using statistics.

EXERCISES

45. Which probability distribution is most likely the appropriate one to use for the following variables: binomial, Poisson, or normal?

   a) distribution of customers arriving at a complaint office

   b) distribution of scores on an intelligence test

   c) number of sales made in 10 house calls by a sales representative

   d) amount of daily rainfall

46. What characteristics of a situation help to determine which is the appropriate distribution to use?

47. Explain in your own words the difference between discrete and continuous random variables. What difference does such classification make in determining the probabilities of future events?

48. In practice, social scientists see many different types of distributions. Often, the nature of these distributions is not as apparent as are some of the examples provided in this book. What alternatives are open to students, teachers, and researchers who want to use probability distributions in their work but who are not sure exactly which distributions are appropriate for given situations?

9   TERMS INTRODUCED IN CHAPTER 6

• BERNOULLI PROCESS   A process in which each trial has only two possible outcomes, the probability of the outcome of any trial remains fixed over time, and the trials are statistically independent.

• BINOMIAL DISTRIBUTION   A discrete distribution describing the results of an experiment known as a Bernoulli process.

• CONTINUITY CORRECTION FACTOR   Corrections used to improve the accuracy of the approximation of a binomial distribution by a normal distribution.

• CONTINUOUS PROBABILITY DISTRIBUTION   A probability distribution in which the variable is allowed to take on any value within a given range.

• CONTINUOUS RANDOM VARIABLE   A random variable allowed to take on any value within a given range.

- **DISCRETE PROBABILITY DISTRIBUTION** A probability distribution in which the variable is allowed to take on only a limited number of values.
- **DISCRETE RANDOM VARIABLE** A random variable that is allowed to take on only a limited number of values.
- **EXPECTED VALUE** A weighted average of the outcomes of an experiment.
- **EXPECTED VALUE OF A RANDOM VARIABLE** The sum of the products of each value of the random variable with that value's probability of occurrence.
- **NORMAL DISTRIBUTION** A distribution of a continuous random variable with a single-peaked, bell-shaped curve. The mean lies at the center of the distribution, and the curve is symmetrical around a vertical line erected at the mean. The two tails extend indefinitely, never touching the horizontal axis.

- **POISSON DISTRIBUTION** A discrete distribution in which the probability of the occurrence of an event within a very small time period is a very small number, the probability that two or more such events will occur within the same small time interval is effectively 0, and the probability of the occurrence of the event within one time period is independent of where that time period is.
- **PROBABILITY DISTRIBUTION** A list of the outcomes of an experiment with the probabilities we would expect to see associated with these outcomes.
- **RANDOM VARIABLE** A variable that takes on different values as a result of the outcomes of a random experiment.
- **STANDARD NORMAL PROBABILITY DISTRIBUTION** A normal probability distribution, the standard deviations being expressed in standard units.

10   EQUATIONS INTRODUCED IN CHAPTER 6

p. 176:
$$\text{Probability of } r \text{ successes in } n \text{ Bernoulli or binomial trials} = \frac{n!}{r!(n-r)!}p^r q^{n-r}$$ [6·1]

where:

r = number of successes desired
n = number of trials undertaken
p = probability of success (characteristic probability)
q = probability of failure ($q = 1 - p$)

This *binomial formula* enables us to calculate algebraically the probability of success. We can apply it to any Bernoulli process, where (1) each trial has only two possible outcomes—a success or a failure; (2) the probability of success remains the same trial after trial; and (3) the trials are statistically independent.

p. 182:
$$\mu = np$$ [6·2]

The *mean* of a *binomial distribution* is equal to the number of trials multiplied by the probability of success.

p. 182:

$$\sigma = \sqrt{npq}$$

[6·3]

The *standard deviation* of a *binomial distribution* is equal to the square root of the product of (1) the number of trials, (2) the probability of a success, and (3) the probability of a failure (found by taking $q = 1 - p$).

p. 186:

$$P(x) = \frac{\lambda^x \times e^{-\lambda}}{x!}$$

[6·4]

This formula enables us to calculate the probability of a discrete random variable occurring in a *Poisson distribution*. The formula states that the probability of *exactly x* occurrences is equal to λ, or lambda (the mean number of occurrences per interval of time in a Poisson distribution), raised to the x power and multiplied by e, or 2.71828 (the base of the natural logarithm system), raised to the negative lambda power, and the product divided by x factorial. The table of values for $e^{-\lambda}$ is Appendix Table 4.

p. 189:

$$P(x) = \frac{(np)^x \times e^{-np}}{x!}$$

[6·5]

If we substitute in Equation 6·4 the mean of the binomial distribution (np) in place of the mean of the Poisson distribution (λ), we can use the Poisson probability distribution as a reasonable approximation of the binomial. The approximation is good when n is equal to or greater than 20 and p is equal to or less than .05.

p. 196:

$$z = \frac{x - \mu}{\sigma}$$

[6·6]

where:

x = value of the random variable with which we are concerned
μ = mean of the distribution of this random variable
σ = standard deviation of this distribution
z = number of standard deviations from x to the mean of
this distribution

Once we have derived z using this formula, we can use the Standard Normal Probability Distribution Table (which gives the values for half the area under the normal curve, beginning with 0.0 at the mean) and determine the probability that the random variable with which we are concerned is within that distance from the mean of this distribution.

p. 207:

$$p(MG) = (1 - p)(ML)$$

[6·7]

This equation describes the point at which the *expected gain* from stocking an additional unit, $p(MG)$, is equal to the *expected loss* from stocking the unit, $(1 - p)(ML)$. As long as $p(MG)$ is larger than $(1 - p)(ML)$, additional units should be stocked, because the expected gain from such a decision is greater than the expected loss.

p. 207:

$$p = \frac{ML}{MG + ML}$$

[6·8]

This is the *minimum probability equation.* The letter p represents the minimum required probability of using at least an additional unit to justify the stocking of that additional unit. As long as the probability of using one additional unit is greater than p, the unit should be stocked.

11   CHAPTER REVIEW EXERCISES

49. Capital City Coach maintains a fleet of buses and operates as a commercial carrier with scheduled buses and charters. Over the last several years, Capital City Coach has maintained an excellent safety record and has averaged only one accident for every 250,000 bus miles (including fender-benders and bus station mishaps). For the week of June 17, Capital City Coach has buses scheduled (including charters) for 50,000 bus miles.

    a) What is the probability that in the seven-day period, Capital City Coach will experience only one accident?

    b) No accidents?

50. Leslie Jarrett, an experimental psychologist, recently patented a device that she has been using to measure reaction times. It is a small, portable device, and she believes it to be as reliable as their larger, nonportable machine. She had 100 of the devices produced for testing. For each of these, she took a reading for a reaction time that had also been measured by the larger machine. As long as the reading from her reaction timer was within $\pm.0004$ seconds of the larger machine's, she considered it acceptable. Otherwise, her portable machine was rejected. She had heard of the Bernoulli process and thought it might be applicable if she could establish the probability of a defect. As described, do you think the production of the portable machines is a Bernoulli process? Why or why not?

51. The regional office of the Environmental Protection Agency annually hires second-year law students as summer interns to help the agency prepare court cases. The agency is under a budget and wishes to keep its costs at a minimum. However, hiring student interns is less costly than hiring full-time employees. Accordingly, the agency wishes to hire the maximum number of students without overstaffing. On the average, it takes two interns all summer to research a case. The interns turn their work over to staff attorneys, who prosecute the case in the fall when the Circuit Court convenes. The legal staff coordinator has to place his budget request in June of the preceding summer for the number of positions he wishes to maintain. It is therefore impossible for him to know with certainty how many cases will be researched in the following summer. The data from preceding summers are as follows:

Year	1968	1969	1970	1971	1972	1973	1974	1975	1976	1977
Number of cases	6	4	8	7	5	6	4	5	4	5

Using these data as his probability distribution for the number of cases, the legal staff coordinator wishes to hire enough interns to research the expected number of cases that will arise. How many intern positions should be requested in the budget?

52. Label the following probability distributions as discrete or continuous.

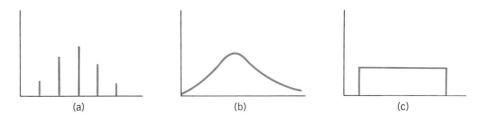

(a)   (b)   (c)

53. Which probability distribution would you use in the following situations: binomial, Poisson, or normal?

    a)   10 trials, probability of success .5

    b)   200 trials, probability of success .01

    c)   500 trials, probability of success .04

    d)   30 trials, probability of success .10

54. The town of Green Lake, Wisconsin, is preparing for the celebration of the 79th Annual Milk and Dairy Day. As a fund-raising device, the city council once again plans to sell souvenir T-shirts. The T-shirts printed in 6 colors, will have a picture of a cow and the words "79th Annual Milk and Dairy Day" on the front. The city council purchases heat transfer patches for $.75 each and plain white cotton T-shirts for $1.50 each. A local merchant supplies the appropriate heating device and also purchases all unsold white cotton T-shirts. The council plans to set up a booth on Main Street and sell the shirts for $3.25. The transfer of the color to the shirt will be completed when the sale is made. In the past year, similar shirt sales have averaged 200 with a standard deviation of 34. The council knows that there will be no market for the patches after the celebration. How many patches should the city council buy?

55. A traffic engineer assigned his assistant, James Wadley, to count the number of cars passing a certain intersection between 8 in the morning and 5 in the afternoon. James's job was so tedious that he tried to figure out ways to make it more interesting. Since he really liked red cars, he counted these separately from the other cars. He found that out of every 30 cars, an average of 4.6 of them were red.

    a)   What is the probability that James will see exactly 6 red cars in the next 30?

    b)   Exactly 9 red cars?

56. British Recording Laboratories produces long-playing records from master tapes supplied to it by recording studios. British Recording Laboratories takes the magnetic tape, cuts a master copy of the record, and makes a mold to press other records. Blanks are purchased from a plastics manufacturer. Periodically, blanks are inserted improperly into the press, causing imperfections in the record grooves and distortion when the record is played. Ian Cambridge, production manager for British Recording Laboratories, reports that on the average, 3 records are mispressed per day.

    a)   What is the probability that exactly 4 records will be mispressed in one day?

b) After the records are pressed, appropriate paper labels are glued to each side. Cambridge noted that on the average, 4.2 records are mislabeled every day. What is the probability that exactly 3 records will be mislabeled?

57. Jean Barton likes to give a competency test to her class at the start of each semester. If a student has trouble on the test, she makes a note of it in the grade book and gives the student extra handout material during the term. Her past results show that 30 percent of all students fail the competency test and need extra attention. Given that test taking is a Bernoulli process, out of 15 tests what is the probability that

a) Exactly 3 will not pass?

b) Exactly 10?

c) Fewer than 4?

d) More than 6?

58. United States Customs agents check the documents of incoming foreigners to see if each person entering the country has been vaccinated for smallpox. Departmental records show that 50 percent of all foreigners entering the United States have been vaccinated.

a) From a sample of 15, what is the probability that 6 or more will not have been vaccinated?

b) 8 or more?

c) Fewer than 5?

59. A farmer on the Delmarva Peninsula recently planted 15 hills of Half-Runner green beans. Both the seed producer and the local seed store guarantee an 80 percent fertility rate based on years of past experience with that particular brand of seed.

a) What is the probability that more than 12 hills will come up?

b) 12 or more fertile hills?

c) Fewer than 12 fertile hills?

d) 8 or fewer fertile hills?

60. The Virginia Department of Health and Welfare publishes a pamphlet, *A Guide to Selecting Your Doctor.* Free copies are available to individuals, institutions, and organizations who are willing to pay the postage. Most of the copies have gone to a small number of groups who, in turn, have disseminated the literature. Mailings for 5 years have been as follows:

Year	1974	1975	1976	1977	1978
Virginia Medical Association	6,000	4,000	1,000	—	3,000
Octogenarian clubs	2,000	1,000	1,000	2,000	1,000
Virginia Federation of Women's Clubs	3,000	2,000	1,000	5,000	2,000
Medical College of Virginia	500	300	600	800	1,000
U.S. Department of Health, Education and Welfare	1,000	—	—	—	1,000

Additionally, an average of 1,500 copies per year were mailed or given to walk-in customers. Assistant Secretary Susan Fleming has to estimate the number of pamphlets to print for 1979. She knows that the pamphlet will be revised in 1979 and a new edition published. She feels that the demand in 1979 will most likely resemble that of 1977; however, she has constructed this assessment of the probabilities:

Year	**1974**	**1975**	**1976**	**1977**	**1978**
Probability that 1979 will resemble this year	.15	.20	.10	.40	.25

a) Construct a table of a probability distribution of demand for the pamphlet, and draw a graph representing that distribution.

b) Assuming Fleming's assessment of the probabilities was correct, how many pamphlets should she order to be certain there will be enough for 1979?

61. Mitch Johnson, a big-league outfielder, has been eating chocolate candy bars for energy ever since he began playing baseball. On August 1, he heard that the company making his favorite candy bar, Choc-O-Late, had run out of almonds and was closing down for a few weeks. Mitch feels he plays best only when he can have a pregame Choc-O-Late bar, and on checking his supply, he finds he has 2 dozen left. Mitch knows how many Choc-O-Late bars he has eaten in the last 2 years, and he consults his records for these figures.

Monthly consumption of Choc-O-Late bars

	J	*F*	*M*	*A*	*M*	*J*	*J*	*A*	*S*	*O*	*N*	*D*
1976	31	30	28	29	30	29	30	32	30	27	33	32
1977	31	26	31	28	27	29	30	34	28	33	30	32

a) From Mitch's data, compute the probability of each value indicated for the number of Choc-O-Late Bars he has eaten in a given month.

b) Compute the expected number of Choc-O-Late Bars he eats per month.

c) What is Mitch's expected energy loss for August in Choc-O-Late Bar units?

62. Dr. Levy has kept track over many semesters of how many students have enrolled in his honors anthropology class. Here are his figures:

Number enrolled	22	23	24	25	26	27	28	29	30	31	32	33	34	35
Frequency	1	2	3	4	4	5	7	6	5	4	3	3	2	1

a) Using this past data, compute the probability distribution for the random variable "number enrolled per class."

b) How many textbooks should Dr. Levy order if he uses expected value to make his decisions?

63. Maria Caruso is a driving safety expert doing research on the effects of long distance driving on human performance. She believes that distance, more than time, fatigues drivers, and that fast drivers tire much more quickly than slow drivers. To help test her theory, Maria had a group of volunteers drive at a slow highway speed for 12 hours a day over a 12-day period. She did not define "slow" but found that in this time drivers averaged 6,250 miles with a standard deviation of 178 miles. She does not really want to use the data from drivers who drove less than 6,000 miles. What percentage of the drivers drove 6,000 miles or more?

64. A child psychologist specializes in treating juvenile delinquents below the age of 12. He knows that in an average week, he will see 30 patients. In the summer, however, when children are out of school, he knows there are more problems than at any other time of the year. Because of this, he expects to see 45 patients every week in June.

a) Does the psychologist's action provide any insight into expected value computation?

b) Assume the number of patients the psychologist sees is normally distributed with a given mean (30 patients) and some standard deviation. Does the fact that the probability of seeing 45 patients in a week is very low cause him not to expect them? Is this situation similar to the one previously described?

65. The purchasing agent in charge of procuring automobiles for the state of Minnesota's interagency motor pool was considering two different models. Both were 4-door, 6-cylinder cars with comparable service warranties. The decision was to choose the automobile that achieved the best mileage per gallon. The state had done some tests of its own, which produced the following results for the two automobiles in question.

	Average MPG	Standard deviation
Automobile A	23	7
Automobile B	25	1

The purchasing agent was uncomfortable with the standard deviations, so she set her own decision criterion for the car that would be most likely to get more than 26 miles per gallon.

a) Using the data provided in combination with the purchasing agent's decision criterion, which car should she choose?

b) If the purchasing agent's decision criterion was to reject the automobile that most often obtained fewer than 24 mpg, which car should she buy?

66. Alice Adams, an educational psychologist, has been studying the way students write term papers. She has interviewed a number of psychology and sociology majors, in particular, and has found that it takes, on the average, 40 hours of research and writing to complete a paper that will receive a grade of B or above, with a standard deviation of 10 hours.

a) What is the probability that a student will spend between 35 and 42 hours writing a term paper graded B or above?

b) What is the probability that it will take between 38 and 50 hours to write a term paper that good?

67. A team of experimental psychologists, running a massive study on effective reinforcement, plans to run rats through mazes for 25,000 trials and to record the trial time for each success. Every trial the rat reaches a goal box, it receives a food pellet. On the basis of past experiments, the team estimates the following probability distribution of success:

Number of successes	14,000	15,000	16,000	17,000	18,000	19,000	20,000
Probability	.10	.15	.20	.30	.15	.05	.05

Since pellets must be ordered well in advance, the team has been trying to decide how many to order. The head of the project feels they should be sure that all successes can be rewarded. His assistant feels that a 95 percent assurance level (5 percent chance of more successes than pellets) is sufficient.

a) How many food pellets does the head of the project want to order (based on the validity of the distribution given above)?

b) How many does the assistant want to order?

c) If the team decided on a 90 percent assurance level, how many pellets should be ordered?

d) If the team decided on an 80 percent assurance level, how many should be ordered?

68. Surveys by the Federal Deposit Insurance Corporation have shown that the life of a regular savings account maintained in one of its member banks averages 18 months, with a standard deviation of 6.45 months.

   a) If a depositor opened an account at a bank that was a member of the FDIC, what is the probability that there will still be money in that account in 22 months?

   b) What is the probability that the account will have been closed before 2 years?

69. Sensurex Productions, Incorporated, has recently patented and developed an ultrasensitive smoke detector for use in both residential and commercial buildings. Whenever a detectable amount of smoke is in the air, a wailing siren is set off. In recent tests conducted in a 20′ × 15′ × 8′ room, the smoke levels that activate the smoke detector averaged 372 parts per million (ppm) of smoke in the room, with a standard deviation of 13 ppm.

   a) If a cigarette introduces 75 ppm into the atmosphere of a 20′ × 15′ × 8′ room, what is the probability that 5 people simultaneously smoking cigarettes will set off the alarm?

   b) Three people?

70. Rework problem 58, using the normal approximation. Compare the approximate and the exact answer.

71. Try to use the normal approximation for problem 59. Notice that nq is only 3. Comment on the accuracy of the approximation.

Answer true *or* false. *Answers are in the back of the book.*

1. The expected value of an experiment is obtained by computing the arithmetic average value over all possible outcomes of the experiment.

2. The value of z for some point x lying in a normal distribution is the area between x and the mean of the distribution.

3. The right and left tails of the normal distribution extend indefinitely, never touching the horizontal axis.

4. For a normal distribution, the mean always lies in between the mode and the median.

5. All but about three-tenths of one percent of the area in a normal distribution lies within plus-and-minus three standard deviations from the mean.

6. Developing a conditional loss table is cumbersome when there are many possible actions and outcomes, because the loss resulting from every action/outcome pair must be included in the table.

7. The area under the curve of a normal distribution between the mean and a point 1.8 standard deviations above the mean is greater for a distribution having a mean of 100 than it is for a distribution having a mean of 0.

8. The normal distribution may be used to approximate the binomial distribution when the number of trials, n, is equal to, or greater than, 60.

9. When using marginal analysis in a stocking problem, the optimal stocking action (stock X units) is the one in which the expected marginal gain of stocking the Xth unit equals the expected marginal loss of stocking it.

10. When the probability of success in a Bernoulli process is 50 percent ($p = .5$), its binomial distribution is symmetrical.

7 Sampling and sampling distributions

1. Introduction to sampling, 226

2. Random sampling, 228

3. Introduction to sampling distributions, 235

4. Sampling distributions in more detail, 238

5. An operational consideration in sampling: the relationship between sample size and standard error, 248

6. Terms introduced, 253

7. Equations introduced, 254

8. Chapter review exercises, 255

9. Chapter concepts test, 257

The Gallup Poll may interview only 2,500 registered voters selected at random. Suppose 72 percent indicate they will vote for the same candidate in the next election. Gallup may then generalize that when the entire voting population goes to the polls, that candidate will win. Why select only 2,500 people out of perhaps 85 million voters? Because time and the average cost of an interview (about $25) prohibit pollsters from attempting to reach millions of people. And since polls are reasonably accurate, interviewing everyone is not necessary. This chapter will answer questions such as: "How many people should be queried? How should they be selected? How do we know when our sample will yield an accurate reflection of the whole population?"

1 INTRODUCTION TO SAMPLING   Shoppers often sample a small piece of cheese before purchasing any. They decide from one piece what the larger chunk will taste like. A chemist does the same thing when he takes a sample of whiskey from a vat, determines that it is 90 proof, and infers that all whiskey in the vat is 90 proof. If the chemist tests all the whiskey or the shoppers taste all the cheese, there will be none to sell. Testing all of the product destroys it and is unnecessary. To determine the characteristics of the whole, we have to sample only a portion.

Reasons for sampling

Suppose, as the personnel director of a state agency, you need to write a report describing all the employees who have voluntarily left the company in the last ten years. You would have a difficult task locating all these thousands of people. They are not easily accessible as a group—many have died, moved from the community, left the country, or acquired a new name by marriage. How do you write the report? The best idea is to locate a representative sample and interview them, in order to generalize about the entire group.

Time is also a factor when decision makers need information quickly in order to adjust an operation or change a policy. Take an automatic machine that sorts thousands of pieces of mail daily. Why wait for an entire day's output to check whether the machine is working accurately (whether the *population characteristics* are those required by the postal service)? Instead, samples can be taken at specific intervals, and if necessary, the machine can be adjusted right away.

Census or sample

Sometimes it is possible and practical to examine every person or item in the population we wish to describe. We call this a *complete enumeration,* or *census.* We use sampling when it is not possible to count or measure every item in the population.

Statisticians use the word *population* **to refer not only to people but to all items that have been chosen for study.** In the cases we have just mentioned, the populations are all the cheese in the chunk, all the whiskey in the vat, all the employees of the state agency who voluntarily left in the last 10 years, all registered voters, and all mail sorted by the automatic machine since the previous sample check. **Statisticians use the word** *sample* **to describe a portion chosen from the population.**

Statistics and parameters

Function of statistics and parameters

Mathematically, we can describe samples and populations by using measures such as the mean, median, mode, and standard deviation, which we introduced in Chapters 3 and 4. When these terms describe the characteristics of a sample, they are called *statistics.* When they describe the characteristics of a population, they are called *parameters.* **A statistic is a characteristic of a sample, and a parameter is a characteristic of a population.**

Suppose that the mean height in inches of all tenth graders in the

United States is 60 inches. In this case, 60 inches is a characteristic of the population "all tenth graders" and can be called a *population parameter.* On the other hand, if we say that the mean height in Ms. Jones's tenth-grade class in Bennetsville is 60 inches, we are using 60 inches to describe a characteristic of the sample "Ms. Jones's tenth graders." In that case, 60 inches would be a *sample statistic.* If we are convinced that the mean height of Ms. Jones's tenth graders is an accurate estimate of the mean height of all tenth graders in the United States, we could use the sample statistic "mean height of Ms. Jones's tenth graders" to estimate the population parameter "mean height of all U.S. tenth graders" without having to count all the millions of tenth graders in the United States.

Using statistics to estimate parameters

To be consistent, statisticians use lower-case Roman letters to denote sample statistics and Greek or capital letters for population parameters. Table 7·1 lists these symbols and summarizes the definitions we have studied so far in this chapter.

N, μ, σ, and n, x, s: standard symbols

	Population	*Sample*
Definition	Collection of items being considered	Part or portion of the population chosen for study
Characteristics	"parameters"	"statistics"

Symbols	population size = N
	population mean = μ
	  (pronounced *myoo*)
	population standard deviation
	  = σ (pronounced **sig**-*ma*)
		sample size = n
		sample mean = $\overline{x}$ (called "x bar")
		sample standard deviation
		  = s

TABLE 7·1
Differences between populations and samples

Types of sampling

There are two methods of selecting samples from populations: *non-random,* or *judgment,* sampling and *random,* or *probability,* sampling. In probability sampling, all the items in the population have a chance of being chosen in the sample. In judgment sampling, personal knowledge and opinion are used to identify those items from the population that are to be included in the sample. A sample selected by judgment sampling is based on someone's expertise about the population. A forest ranger, for example, would have a judgment sample if he decided ahead of time which parts of a large forested area he would walk through to estimate the total board feet of lumber that could be cut. Sometimes a judgment sample is used as a pilot or trial sample to decide how to take a random sample later. Judgment samples avoid the statistical analysis that is necessary to make probability samples. They are more convenient and can be used successfully even though we are unable to measure their validity. But if a study uses judgment sampling and loses a significant degree of "representativeness," it will have purchased convenience at too high a price.

Judgment and probability sampling

1. What is the major drawback of judgment sampling?

2. Depending on the extent of conclusions drawn from a statistical analysis, a given set of observations may be thought of as a sample or as a population. Explain.

3. List the advantages of sampling over complete enumeration, or census.

4. What are some of the disadvantages of probability sampling versus judgment sampling?

5. County supervisor of education Jeff Cochran was in a rather heated discussion with state board member Bob Milgram about some statistical tests made by the Department of Education. Milgram was in charge of evaluation in the county where, as in all counties in the state, children are given standard achievement tests each year. Cochran's county tested 100 percent of the children in the county at the end of each year. For this reason, Cochran argued that he was dealing with a population, as far as statistics was concerned, and that the county's average score and standard deviation were parameters. Milgram argued that since the county was only one of several in which children were given the standard achievement tests, the information from the county was simply a sample of a larger set of scores; and accordingly, any statistics generated were sample statistics, not parameters. Who was right?

6. Jean McKinney, who was hired to explain contraceptive methods to women in an urban area, met with some difficulty after reporting on the effectiveness of the program. In her first six weeks she collected pilot data through interviews with women in some neighborhoods. McKinney's plan was based on statistical sampling, and from the beginning data it was clear (or so Jean thought) that the publicity department was not getting information to enough women. Jean's report was shrugged off with the comment, "This is no good. Nobody can make statements about the success of our publicity campaigns when she talks to only a little over 15 percent of those women we are trying to reach. Everyone knows you have to check at least 50 percent to have any idea what's going on. We didn't hire you to make guesses about the success of our program." Is there any defense for Jean's position?

7. The university ethics committee wants to conduct a follow-up investigation of possible aftereffects on subjects involved in a large scale study of induced stress. The committee currently has two trained interviewers to work on the project and is considering hiring two more. The only drawback is that it is very expensive and time-consuming to train interviewers and to conduct the interviews. The committee considered interviewing just a sample of the total subject population to determine possible detrimental aftereffects of participation in the study but was undecided about what course of action to take. It had heard from a fairly reliable source that a census was always better than a sample, since it provided more accurate information. The committee thought that since money in the budget could be used to pay more interviewers, it might be wise to hire them, rather than spend the money on other research. Evaluate the information from the reliable source and advise the committee what action to take.

2 RANDOM SAMPLING

In a random or probability sample, we know what the chances are that an element of the population will or will not be included in the sample. As a result, we can assess objectively the estimates of the population characteristics that result from our sample; that is, we can describe mathematically how objective our estimates are. Let us begin our explanation of this process by introducing 4 methods of random sampling:

simple random sampling
2. systematic sampling
3. stratified sampling
4. cluster sampling

Simple random sampling

Simple random sampling selects samples by methods that allow *each possible sample to have an equal probability of being picked* and *each item in the entire population to have an equal chance of being included in the sample.* We can illustrate these requirements with an example. Suppose we have a population of 4 students in a seminar and we want samples of 2 students at a time for interviewing purposes. Table 7·2 illustrates the possible combinations of samples of 2 students in a population size of 4, the probability of each sample being picked, and the probability that each student will be in the sample.

An example of simple random sampling

Our example illustrated in Table 7·2 uses a *finite* population of 4 students. By *finite,* we mean that the population has a stated or limited size; that is to say, there is a whole number (*N*) that tells us how many items there are in the population. Certainly, if we sample without "replacing" the student we shall soon exhaust our small population group. Notice, too, that if

Defining finite *and* replacement

Students A, B, C, and D

Possible samples of two persons: AB, AC, AD, BC, CD, BD
Probability of drawing this sample of two persons must be:

$$AB = \tfrac{1}{6}$$
$$AC = \tfrac{1}{6}$$
$$AD = \tfrac{1}{6} \quad \text{(There are only six possible}$$
$$BC = \tfrac{1}{6} \quad \text{samples of two persons)}$$
$$CD = \tfrac{1}{6}$$
$$BD = \tfrac{1}{6}$$

Probability of this student being in the sample must be:

$$A = \tfrac{1}{2} \quad \text{(In Chapter 5 we saw that the}$$
$$B = \tfrac{1}{2} \quad \text{marginal probability is equal}$$
$$C = \tfrac{1}{2} \quad \text{to the } sum \text{ of the joint}$$
$$D = \tfrac{1}{2} \quad \text{probabilities of the events}$$
$$\text{within which the simple}$$
$$\text{event is contained:}$$
$$P(A) = P(AB + AC + AD) = \tfrac{1}{2})$$

TABLE 7·2
Chances of selecting samples of 2 students from a population of 4 students

we sample *with replacement* (that is, if we replace the sampled student immediately after he or she is picked and before the second student is chosen), the same person could appear twice in the same sample.

An infinite
population

We have used this example only to help us think about sampling from an infinite population. An *infinite* population is a population in which it is theoretically impossible to observe all the elements. Although many populations appear to be exceedingly large, no truly infinite population of physical objects actually exists. After all, given unlimited resources and time, we could enumerate any finite population, even the grains of sand on the beaches of North America. As a practical matter, then, we will use the term *infinite population* when we are talking about a population that could not be enumerated in a reasonable period of time. In this way, we will use the theoretical concept of infinite population as an approximation of a large finite population, just as we earlier used the theoretical concept of continuous random variable as an approximation of a discrete random variable that could take on many closely-spaced values.

How to do random sampling

The easiest way to select a sample randomly is to use random numbers. These numbers can be generated either by a computer programmed to scramble numbers or by a table of random numbers, which should properly be called a *table of random digits*.

Table 7·3 illustrates a portion of such a table. Here we have 1,250 random digits in sets of 10 digits. These numbers have been generated by a completely random process. The probability that any one digit from 0 through 9 will appear is the same as that for any other digit, and the probability of one sequence of digits occurring is the same as that for any other sequence of the same length.

Using a table of
random digits

To see how to use this table, suppose that we have 100 employees in a laboratory and wish to interview a randomly chosen sample of 20. We could get such a random sample by assigning every employee a number from 00 to 99, consulting Table 7·3, and picking a systematic method of selecting two-digit numbers. In this case, let's do the following:

1. Go from the top to the bottom of the columns beginning with the left-hand column, and read only the first two digits in each row. Notice that our first number using this method would be 15, the second 09, the third 41, and so on.

2. If we reach the bottom of the last column on the right and have come short of our desired 20 two-digit numbers of 99 and under, we can go back to the beginning (the top of the left-hand column) and start reading the third and fourth digits of each number. These would begin 81, 28, and 12.

Using
slips of paper

Another way to select our employees would be to write the name of each one on a slip of paper and deposit the slips in a box. After mixing them thoroughly, we could draw 20 slips at random. This method works well with

TABLE 7·3
1,250 random digits*

1581922396	2068577984	8262130892	8374856049	4637567488
0928105582	7295088579	9586111652	7055508767	6472382934
4112077556	3440672486	1882412963	0684012006	0933147914
7457477468	5435810788	9670852913	1291265730	4890031305
0099520858	3090908872	2039593181	5973470495	9776135501
7245174840	2275698645	8416549348	4676463101	2229367983
6749420382	4832630032	5670984959	5432114610	2966095680
5503161011	7413686599	1198757695	0414294470	0140121598
7164238934	7666127259	5263097712	5133648980	4011966963
3593969525	0272759769	0385998136	9999089966	7544056852
4192054466	0700014629	5169439659	8408705169	1074373131
9697426117	6488888550	4031652526	8123543276	0927534537
2007950579	9564268448	3457416988	1531027886	7016633739
4584768758	2389278610	3859431781	3643768456	4141314518
3840145867	9120831830	7228567652	1267173884	4020651657
0190453442	4800088084	1165628559	5407921254	3768932478
6766554338	5585265145	5089052204	9780623691	2195448096
6315116284	9172824179	5544814339	0016943666	3828538786
3908771938	4035554324	0840126299	4942059208	1475623997
5570024586	9324732596	1186563397	4425143189	3216653251
2999997185	0135968938	7678931194	1351031403	6002561840
7864375912	8383232768	1892857070	2323673751	3188881718
7065492027	6349104233	3382569662	4579426926	1513082455
0654683246	4765104877	8149224168	5468631609	6474393896
7830555058	5255147182	3519287786	2481675649	8907598697

*Source: Dudley J. Cowden and Mercedes S. Cowden, *Practical Problems in Business Statistics*, 2d ed., Englewood Cliffs, N.J., Prentice-Hall, Inc., 1960.

a small group of people but presents problems if the people in the population number in the thousands. There is the added problem, too, of not being certain that the slips of paper are mixed well. In the draft lottery of 1970, for example, when capsules were drawn from a bowl to determine by birthdays the order of selecting draftees for the armed services, December birthdays appeared more often than the probabilities would have suggested. As it turned out, the December capsules had been placed in the bowl last, and the capsules had not been mixed properly. Thus, December capsules had the highest probability of being drawn.

Systematic sampling

In systematic sampling, elements are selected from the population at a uniform interval that is measured in time, order, or space. If we wanted to interview every twentieth student on a college campus, we would choose a random starting point in the first twenty names in the student directory and then pick every twentieth name thereafter.

Systematic sampling differs from simple random sampling in that each *element* has an equal chance of being selected but each *sample* does *not* have an equal chance of being selected. This would have been the case if,

Characteristics of systematic sampling

in our earlier example, we had assigned numbers between 001 and 100 to our employees and then had begun to choose a sample of ten by picking every tenth number beginning 1, 11, 21, 31, and so forth. Employees numbered 2, 3, 4, and 5 would have had no chance of being selected.

Shortcomings of the systematic approach

In systematic sampling, there is the problem of introducing an error into the sampling process. Suppose we were sampling paper waste produced by households, and we decided to sample a hundred households every Monday. Chances are high that our sample would not be representative because Monday's trash would very likely include the Sunday newspaper. Thus, the amount of waste would be biased upward by our choice of this sampling procedure.

Systematic sampling has advantages too, however. Even though systematic sampling may be inappropriate when the elements lie in a sequential pattern, this method may require less time and sometimes results in lower costs than the simple random sampling method.

Stratified sampling

Two ways to take stratified samples

To use stratified sampling, we divide the population into relatively homogeneous groups, called *strata.* Then we use one of two approaches. Either we select at random from each stratum a specified number of elements corresponding to the proportion of that stratum in the population as a whole, or we draw an equal number of elements from each stratum and give weight to the results according to the stratum's proportion of total population. With either approach, stratified sampling guarantees that every element in the population has a chance of being selected.

Stratified sampling is appropriate when the population is already divided into groups of different sizes and we wish to acknowledge this fact. Suppose that the clients of a government social service agency are divided into four groups according to age, as shown in Table 7·4. The agency must measure the use of prescribed medicines by persons in this population. To obtain an estimate of this characteristic of the population, the agency could take a random sample from each of the four age groups and give weight to the samples according to the percentage of its clients in that group. This would be an example of a stratified sample.

The advantage of stratified samples is that when they are properly designed, they more accurately reflect characteristics of the population from which they were chosen than do other kinds of sampling.

TABLE 7·4
**Composition of
clients by age**

Age group	Percentage of total clients
Birth–19 years	30%
20–39 years	40
40–59 years	20
60 years and older	10

In cluster sampling, we divide the population into groups, or *clusters,* and then select a random sample of these clusters. We assume that these individual clusters are representative of the population as a whole. If a survey team is attempting to determine by sampling the average number of television sets per household in a large city, they could use a city map to divide the territory into blocks and then choose a certain number of households (a cluster) in each block for interviewing. Every one of these households would be interviewed. A well-designed cluster sampling procedure can produce a more precise sample at considerably less cost than that of simple random sampling.

With both stratified and cluster sampling, the population is divided into well-defined groups. We use *stratified* sampling when each group has small variation wtihin itself, but there is wide variation between the groups. We use *cluster* sampling in the opposite case—when there is considerable variation within each group, but the groups are essentially similar to each other.

Comparison of stratified and cluster sampling

Basis of statistical inference: simple random sampling

Systematic sampling, stratified sampling, and cluster sampling attempt to approximate simple random sampling. All are methods that have been developed for their precision, economy, or physical ease. Even so, assume for the rest of the examples and problems in this book that we obtain our material by simple random sampling. This is necessary because the principles of simple random sampling are the foundation for *statistical inference,* the process of making inferences about populations from information contained in samples. Once these principles have been developed for simple random sampling, their extension to the other sampling methods is conceptually quite simple but somewhat involved mathematically. If you understand the basic ideas involved in simple random sampling, you will have a good grasp of what is going on in the other cases, even if you must leave the technical details to the professional statistician.

EXERCISES

8. In the example below, probability distributions for 3 natural subgroups of a larger population are shown. For which situation would you recommend stratified sampling?

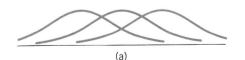

(a)

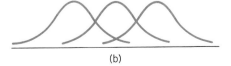

(b)

9. If we have a population of 1,000 individuals and we wish to sample 25 randomly, use the random number table (Table 7·3) to select 25 individuals from the 1,000. List the numbers of those elements selected, based on the random number table.

10. Using a calendar, systematically sample every 14th day of one year, beginning with January 3rd.

11. A population is made up of groups that have wide variation within each group but little variation from group to group. The appropriate type of sampling for this population is:

   a) stratified

   b) systematic

   c) cluster

   d) judgment

12. The Council of Graduate Schools has appointed a committee to find out how physical sciences teachers rate graduate programs in their fields, which include biology, chemistry, and physics. The study will include all universities that award doctorates in these areas. It will sample a large number of faculty in various university departments in the United States. The committee is using a list of department chairpersons from which to draw its sample. Is it a random sample?

13. Romney J. Peterson from the borough of the Bronx in New York City is running for a seat in the House of Representatives. Peterson, anxious to know how his campaign is progressing, plans to conduct a telephone poll of residents of the area. He feels that such a poll will give him a good idea of who will vote for him. He plans to look in the telephone book and select random numbers with addresses that correspond to his congressional district. Will Peterson's sample be a random one?

14. Consult Table 7·3. What is the probability that a 3 will appear as the rightmost digit in each set of 10 digits? That a 6 will appear? 8? How many times would you expect to see each of those digits in the rightmost position? How many times is each found in that position? Can you explain any differences in the number found and the number expected?

15. In an extensive study of language, the research assistant is required to listen to, and transcribe, tapes of subjects recalling nonsense syllables after a distracting task. She usually does 10 tape transcriptions per day before changing to another type of work, since there are certain fatigue effects involved in listening to the tapes. The chief investigator on the project wishes to do a reliability check on her transcriptions and is considering a statistical method of sampling to check their accuracy. Would systematic sampling be appropriate?

16. The state highway safety department has decided to do a study of automobile accidents in a four-county area to examine some of the variables involved in accidents, e.g., the kind of violation, ages of the drivers, type of location (interstate highway, crowded intersection), and time of day. It has been decided that 250 of the 2,500 accidents reported last year in the area will be sampled. The accident reports are filed by date in a large filing cabinet. Bill Mullins, one of the employees of the department, has proposed that the survey use a systematic sampling technique and select every tenth report in the file for the sample. Would Mullins's plan of systematic sampling be appropriate here? Explain.

17. Bob Curtis, a criminologist, is interested in looking at the types of crime problems that exist in the country. His assistant, Mary Williamson, has recommended a stratified random sampling process where the cities and communities studied are broken into substrata, depending on the size and nature of the community. Mary Williamson proposes the following classification:

Category	Type of community
Urban	Inner city (population 100,000+)
Suburban	Outlying areas of city or smaller communities (pop. 20,000 to 100,000)
Rural	Small communities (fewer than 20,000 residents)

Is stratified random sampling appropriate here?

3 INTRODUCTION TO SAMPLING

DISTRIBUTIONS In Chapters 3 and 4, we introduced methods by which we can use sample data to calculate statistics such as the mean and the standard deviation. So far in this chapter, we have examined how samples can be taken from populations. If we apply what we have learned and take several samples from a population, the statistics we would compute for each sample need not be the same and most probably would vary from sample to sample.

Statistics differ among samples from the same population

Suppose our samples each consist of ten 25-year-old women from a city with a population of 100,000 (an infinite population, according to our usage). By computing the mean height and standard deviation of that height for each of these samples, we would quickly see that the mean of each sample and the standard deviation of each sample would be different. A probability distribution of all the possible means of the samples is a distribution of the sample means. Statisticians call this a sampling distribution of the mean.

Sampling distribution defined

We could also have a sampling distribution of a proportion. Assume that we have determined the proportion of beetle infested pine trees in samples of 100 trees taken from a very large forest. We have taken a large number of those 100-item samples. If we plot a probability distribution of the possible proportions of infested trees in all these samples, we would see a distribution of the sample proportions. In statistics, this is called a *sampling distribution of the proportion.* (Notice that the term *proportion* refers to the proportion that is infested.)

Describing sampling distributions

Any probability distribution (and, therefore, any sampling distribution) can be partially described by its mean and standard deviation. Table 7·5 illustrates several populations. Beside each, we have indicated the sample taken from that population, the sample statistic we have measured, and the sampling distribution that would be associated with that statistic.

Now, how would we describe each of the sampling distributions in Table 7·5? In the first example, the sampling distribution of the mean can be partially described by its mean and standard deviation. The sampling

	Population	Sample	Sample statistic	Sampling distribution
	Water in a river	10-gallon containers of water	Mean number of parts of mercury per million parts of water	Sampling distribution of the mean
	All professional basketball teams	Groups of 5 players	Median height	Sampling distribution of the median
	All parts produced by a manufacturing process	50 parts	Proportion defective	Sampling distribution of the proportion

distribution of the median in the second example can be partially described by the mean and standard deviation of the distribution of the medians. And in the third, the sampling distribution of the proportion can be partially described by the mean and standard deviation of the distribution of the proportions.

Concept of standard error

Simplification of terms

Derivation of the term standard error

Rather than say "standard deviation of the distribution of sample means" to describe a distribution of sample means, statisticians refer to the *standard error of the mean.* Similarly, the "standard deviation of the distribution of sample proportion" is shortened to the *standard error of the proportion.* The term *standard error* is used because it conveys a specific meaning. An example will help explain the reason for the name. Suppose we wish to learn something about the height of freshmen at a large state university. We could take a series of samples and calculate the mean height for each sample. It is highly unlikely that all of these sample means would be the same; we expect to see some variability in our observed means. This variability in the sample statistic results from *sampling error* due to chance; that is, there are differences between each sample and the population, and among the several samples, owing solely to the elements we happened to choose for the samples.

The standard deviation of the distribution of sample means measures the extent to which we expect the means from the different samples to vary because of this chance error in the sampling process. Thus **the standard deviation of the distribution of a sample statistic is known as the standard error of the statistic.**

Size of the standard error

The standard error indicates not only the size of the chance error that has been made but also the accuracy we are likely to get if we use a sample statistic to estimate a population parameter. A distribution of sample means that is less spread out (that has a small standard error) is a better estimator of the population mean than a distribution of sample means that is widely dispersed and has a larger standard error.

Table 7·6 indicates the proper use of the term *standard error.* In Chapter 8, we shall discuss how to estimate population parameters using sample statistics.

TABLE 7·6
Conventional terminology used to refer to sample statistics

When we wish to refer to the:	We use the conventional term:
Standard deviation of the distribution of sample means	Standard error of the mean
Standard deviation of the distribution of sample proportions	Standard error of the proportion
Standard deviation of the distribution of sample medians	Standard error of the median
Standard deviation of the distribution of sample ranges	Standard error of the range

EXERCISES

18. The systolic blood pressure of 18-year-old women is known to have a mean of 120 and a standard deviation of 12. Two nurses who work in the student infirmary of a particular university were discussing the probability that in the screening of freshmen women who had not turned in medical forms, the 30 women examined would have an average blood pressure of 120. One nurse who had been practicing nursing for years said that it was entirely chance that the average blood pressure of the sample would equal the mean of the population. The younger nurse thought there was very great probability that the sample would average exactly 120, since that was the expected blood pressure. Which of the nurses is right? Explain.

19. The term *error,* in standard error of the mean, refers to what type of error?

20. A research assistant took a random sample from a population known to have a distribution with a mean of 100 and a standard deviation of 15. The sample had a sample mean of 105, leading him to assume that his sample must not have been representative. Is his conclusion correct?

21. A sampling distribution of the mean for a certain population has been determined to have a standard error of 5 and a mean of 24.2. Two different samples are selected at random, and the means are 20 and 27, respectively. The assistant in charge of data collection concludes that the second sample is the better one because it is better to overestimate than underestimate the true mean. Comment. Is one of the sample means "better" in some way, given the true population mean?

22. A survey researcher has mailed out questionnaires to a large sample of people in cities across the nation. Different numbers of questionnaires are sent to different geographic regions, based on the population density of the area. Not all of the questionnaires sent out are returned. Upon receiving the ones that were completed, he sorts them according to region and computes the mean response for each region. He then plots the frequency distribution of these means. Does this represent a sampling distribution of the mean for the population? Why or why not?

23. Marilyn Katz is a biologist studying the blooming time for a certain hybrid rye plant. From a sample of plants, she recorded the length of time it took them to produce their blooms. This hybrid has been studied by others, and Marilyn is attempting to present the results of these studies to a conference. She has with her two graphs that, she explained, represent

frequency distributions of the length of time it took the hybrid plants to bloom. Her audience, however, does not get the point. Can you explain to the conference group the significance of the two graphs?

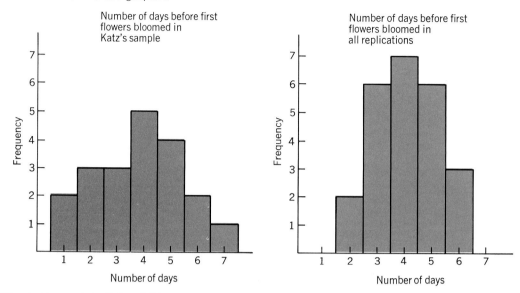

24. In times of declining SAT scores and problems of functional illiteracy, the admissions committee of a prestigious university is concerned with keeping high standards of admission. Each year, after decisions on acceptance are made, the committee publishes and distributes statistics on students admitted, giving, for example, the average SAT score. On the report containing the statistics are the words, "Standard Error of the Mean." The secretary who types the report has known that for several years the average SAT score was about 1,200 and has assumed that the standard error of the mean was how much the committee allowed an admitted student's score to deviate from the mean. Is the assumption correct? Explain.

4 SAMPLING DISTRIBUTIONS IN MORE DETAIL In the last section of this chapter, we introduced the idea of a sampling distribution. We examined the reasons why sampling from a population and developing a distribution of these sample statistics would produce a sampling distribution, and we introduced the concept of standard error. Now we will study these concepts further, so that we will not only be able to understand them conceptually but also be able to handle them computationally.

Conceptual basis for sampling distributions

Deriving the sampling distribution of the mean

Figure 7·1 will help us examine sampling distributions without delving too deeply into statistical theory. We have divided this illustration into three parts. Part a of Fig. 7·1 illustrates a *population distribution*. Let's assume that this population is all the filter screens in a large pollution monitoring

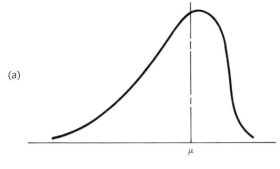

(a)

The population distribution:
This distribution is the distribution of the operating hours of *all* the filter screens. It has:

μ = the mean of this distribution

σ = the standard deviation of this distribution

If somehow we were able to take *all* the possible samples of a given size from this *population distribution,* they would be represented graphically by these four samples below. Although we have shown only four such samples, there would actually be an enormous number of them.

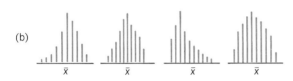

(b)

The sample frequency distributions:
These only *represent* the enormous number of sample distributions possible. *Each* sample distribution is a discrete distribution and has:

$\bar{x}$ = its own mean called "x bar"

s = its own standard deviation

Now, if we were able to take the means from all the *sample distributions* and produce a distribution of these sample means, it would look like this:

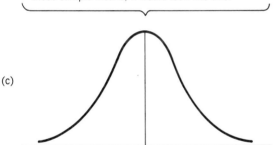

(c)

The sampling distribution of the mean:
This distribution is the distribution of all the sample means and has:

$\mu_{\bar{x}}$ = mean of the sampling distribution of the means called "mu sub x bar"

$\sigma_{\bar{x}}$ = standard error of the mean (standard deviation of the sampling distribution of the mean) called "*sigma* sub x bar"

FIG. 7·1   Conceptual population distribution, sample distributions, and sampling distribution

system and that this distribution is the operating hours before a screen becomes clogged. The distribution of operating hours has a mean μ *(mu)* and a standard deviation σ *(sigma)*.

Suppose that somehow we are able to take all the possible samples of 10 screens from the population distribution (actually, there would be far too many for us to consider). Next we would calculate the mean and the standard deviation for each one of these *samples* as represented in part b of Fig. 7·1. As a result, *each* sample would have its own mean $\bar{x}$ (x bar) and its own standard deviation s. All the individual sample means would *not* be the same as the population mean. They would tend to be near the population mean, but only rarely would they be exactly that value.

As a last step, we would produce a distribution of all the means from every sample that could be taken. This distribution, called the *sampling distribution of the mean,* is illustrated in part c of Fig. 7·1. This distribution of the sample means, or sampling distribution, would have its own mean $\mu_{\bar{x}}$ (mu sub x bar) and its own standard deviation, or standard error, $\sigma_{\bar{x}}$ (*sigma sub x bar*).

In statistical terminology, the sampling distribution we would obtain by taking all the samples of a given size is a *theoretical sampling distribution.* Part c of Fig. 7·1 describes such an example. In practice, the size and character of most populations prohibit decision makers from taking all the possible samples from a population distribution. Fortunately, statisticians have developed formulas for estimating the characteristics of these theoretical sampling distributions, making it unnecessary for us to collect large numbers of samples. In most cases, decision makers take only one sample from the population, calculate statistics for that sample, and from those statistics infer something about the parameters of the entire population. We shall illustrate this shortly.

Why we use the
sampling
distribution
of the mean

In each example of sampling distributions in the remainder of this chapter, we shall use the sampling distribution of the mean. We could study the sampling distribution of the median, range, or proportion, but we will stay with the mean for the continuity it will add to the explanation. Once you develop an understanding of how to deal computationally with the sampling distribution of the mean, you will be able to apply it to the distribution of any other sample statistic.

Sampling from normal populations

Sampling
distribution of the
mean from normally
distributed
populations

Suppose we draw samples from a normally distributed population with a mean of 100 and a standard deviation of 25, and that we start by drawing samples of 5 items each and by calculating their means. The first mean might be 95, the second 106, the third 101, and so on. Obviously, there is just as much chance for the sample means to be above the population mean of 100 as there is for it to be below 100. Since we are *averaging* 5 items to get each sample mean, very large values in the sample would be averaged down and very small values up. We would reason that we would get less spread among the sample means than we would among the individual items in the original population. That is the same as saying that the standard error of the mean, or standard deviation of the sampling distribution of the mean, would be less than the standard deviation of the *individual* items in the population. Figure 7·2 illustrates this point graphically.

Now suppose we increase our sample size from 5 to 20. This would not change the standard deviation of the items in the original population. But with samples of 20, we have increased the effect of averaging in each sample and would expect even less dispersion among the sample means. Figure 7·3 illustrates this point.

The sampling distribution of a mean of a normally distributed popula-

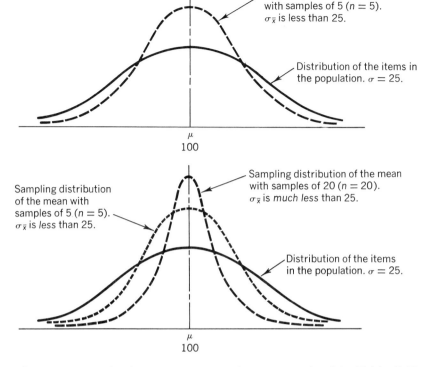

Sampling distribution of the mean with samples of 5 ($n = 5$). $\sigma_{\bar{x}}$ is less than 25.

Distribution of the items in the population. $\sigma = 25$.

μ
100

FIG. 7·2

Relationship between the population distribution and the sampling distribution of the mean for a normal population

Sampling distribution of the mean with samples of 5 ($n = 5$). $\sigma_{\bar{x}}$ is less than 25.

Sampling distribution of the mean with samples of 20 ($n = 20$). $\sigma_{\bar{x}}$ is *much less* than 25.

Distribution of the items in the population. $\sigma = 25$.

μ
100

FIG. 7·3

Relationship between the population distribution and sampling distribution of the mean with increasing n's

tion demonstrates the important properties summarized in Table 7·7. An example will further illustrate these properties. A bank calculates that its individual savings accounts are normally distributed with a mean of $2,000 and a standard deviation of $600. If the bank takes a random sample of 100 accounts, what is the probability that the sample mean will lie between $1,900 and $2,050? This is a question about the sampling distribution of the mean; therefore, we must first calculate the standard error of the mean. In this case, we shall use the equation for the standard error of the mean designed for situations in which the population is infinite (later, we shall introduce an equation for finite populations):

Properties of the sampling distribution of the mean

$$\sigma_{\bar{x}} = \frac{\sigma}{\sqrt{n}} \qquad [7·1]$$

where:

$\sigma_{\bar{x}}$ = standard error of the mean

σ = population standard deviation

n = sample size

Finding the standard error of the mean for infinite populations

Applying this to our example, we get:

$$\sigma_{\bar{x}} = \frac{\$600}{\sqrt{100}}$$
$$= \frac{\$600}{10}$$
$$= \$60 \leftarrow \text{standard error of the mean}$$

241

TABLE 7·7

**Properties
of the sampling
distribution of
the mean when
the population
is normally
distributed**

	Property	*Illustrated symbolically*
	The sampling distribution has a mean equal to the population mean	$\mu_{\bar{x}} = \mu$
	The sampling distribution has a standard deviation (a standard error) equal to the population standard deviation divided by the square root of the sample size	$\sigma_{\bar{x}} = \dfrac{\sigma}{\sqrt{n}}$
	The sampling distribution is normally distributed	

Next, we need to use the table of z values (Appendix Table 1) and Equation 6·6, which enables us to use the Standard Normal Probability Distribution Table. With these we can determine the probability that the sample mean will lie between $1,900 and $2,050.

$$z = \frac{x - \mu}{\sigma} \qquad [6 \cdot 6]$$

Equation 6·6 tells us that to convert any normal random variable to a standard normal random variable, we must subtract the mean of the variable being standardized and divide by the standard error (the standard deviation of that variable). Thus, in this particular case, Equation 6·6 becomes:

Converting the
sample mean
to a z value

Sample mean ⟶ **Population mean**

$$z = \frac{\bar{x} - \mu}{\sigma_{\bar{x}}} \qquad [7 \cdot 2]$$

Standard error of the mean $= \dfrac{\sigma}{\sqrt{n}}$

Now we are ready to compute the two z values as follows:

For $\bar{x} = $1,900$

$$z = \frac{\bar{x} - \mu}{\sigma_{\bar{x}}} \qquad [7 \cdot 2]$$

$$= \frac{\$1,900 - \$2,000}{\$60}$$

$$= -\frac{100}{60}$$

$$= -1.67 \leftarrow \text{standard deviations from the mean of a standard normal probability distribution}$$

For $\bar{x} = $2,050$

$$z = \frac{\bar{x} - \mu}{\sigma_{\bar{x}}} \qquad [7 \cdot 2]$$

$$= \frac{\$2,050 - \$2,000}{\$60}$$

$$= \frac{50}{60}$$

$$= .83 \leftarrow \text{standard deviations from the mean of a standard normal probability distribution}$$

Appendix Table 1 gives us an area of .4525 corresponding to a z-value of −1.67, and it gives an area of .2967 for a z-value of .83. If we add these two together, we get .7492 as the total probability that the sample mean will lie between $1,900 and $2,050. We have shown this problem graphically in Figure 7·4.

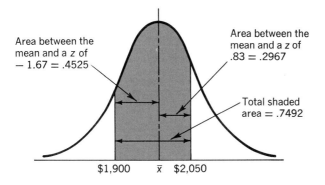

Area between the
mean and a z of
− 1.67 = .4525

Area between the
mean and a z of
.83 = .2967

Total shaded
area = .7492

$1,900   x̄ $2,050

FIG. 7·4
Probability of sample mean lying between $1,900 and $2,050

Sampling from non-normal populations

In the previous section, we concluded that when the population is normally distributed, the sampling distribution of the mean is also normal. Yet decision makers must deal with many populations that are not normally distributed. How does the sampling distribution of the mean react when the population from which the samples are drawn is *not* normal? An illustration will help us answer this question.

Consider the data in Table 7·8, concerning five motorcycle owners and the lives of their tires. Since only five people are involved, the population is too small to be approximated by a normal distribution. We'll take all of the possible samples of the owners in groups of three, compute the sample means ($\bar{x}$), list them, and compute the mean of the sampling distribution ($\mu_{\bar{x}}$). We have done this in Table 7·9. These calculations show that even in a case in which the population is not normally distributed, $\mu_{\bar{x}}$, the mean of the sampling distribution, is still equal to the population mean μ.

The mean of the sampling distribution of the mean equals population mean

Owner	Carl	Debbie	Elizabeth	Frank	George	
Tire life (months)	3	3	7	9	14	**Total: 36 months**

$$\text{Mean} = \frac{36}{5} = 7.2 \text{ months}$$

TABLE 7·8
Experience of five motorcycle owners with life of tires

TABLE 7·9
Calculation
of sample
mean tire life
with $n = 3$

Samples of three	Sample data (tire lives)	Sample mean
EFG*	7 + 9 + 14	10
DFG	3 + 9 + 14	8⅔
DEG	3 + 7 + 14	8
DEF	3 + 7 + 9	6⅓
CFG	3 + 9 + 14	8⅔
CEG	3 + 7 + 14	8
CEF	3 + 7 + 9	6⅓
CDF	3 + 3 + 9	5
CDE	3 + 3 + 7	4⅓
CDG	3 + 3 + 14	6⅔
		72 months

$$\mu_{\bar{x}} = \frac{72}{10}$$
$$= 7.2 \text{ months}$$

*Names abbreviated by first initial

Increase in number
of samples leads to
a more normal
sampling
distribution

Now look at Fig. 7·5. Part a is the population distribution of tire lives for the five motorcycle owners, a distribution that is anything but normal in shape. In part b of Fig. 7·5, we have shown the sampling distribution of the mean for a sample size of three, taking the information from Table 7·9. Notice the difference between the probability distributions in a and b. In part b, the distribution looks a little more like the bell shape of the normal distribution.

FIG. 7·5
Population
distribution and
sampling
distribution of the
mean for tire life

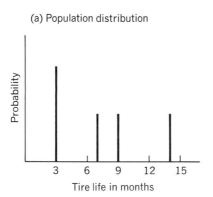

(a) Population distribution

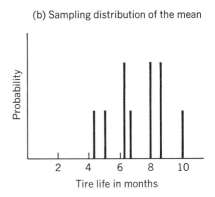

(b) Sampling distribution of the mean

If we had a long time and much space, we could repeat this example and enlarge the population size to twenty. Then we could take samples of *every* size. Next we would plot the sampling distribution of the mean that would occur in *each* case. Doing this would show quite dramatically how quickly the sampling distribution of the mean approaches normality, regardless of the shape of the population distribution. Figure 7·6 simulates this process graphically without all the calculations.

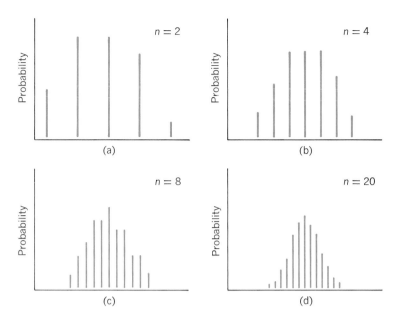

The central limit theorem

The example in Table 7·9 and the two probability distributions in Fig. 7·5 should suggest several things to you. First, the mean of the sampling distribution of the mean will equal the population mean regardless of the sample size, even if the population is not normal. Second, as the sample size increases, the sampling distribution of the mean will approach normality, regardless of the shape of the population distribution.

Results of increasing sample size

This relationship between the shape of the population distribution and the shape of the sampling distribution of the mean is called the *central limit theorem*. The central limit theorem is perhaps the most important theorem in all of statistical inference. It assures us that the sampling distribution of the mean approaches normal as the sample size increases. There are theoretical situations in which the central limit theorem fails to hold, but they are almost never encountered in practical decision making. Actually, a sample does not have to be very large for the sampling distribution of the mean to approach normal. Statisticians use the normal distribution as an approximation to the sampling distribution whenever the sample size is at least 30, but the sampling distribution of the mean can be nearly normal with samples of even half that size. The significance of the central limit theorem is that it permits us to use sample statistics to make inferences about population parameters without knowing anything about the shape of the frequency distribution of that population other than what we can get from the sample. Putting this ability to work is the subject of much of the material in the subsequent chapters of this book.

Significance of the central limit theorem

Let's illustrate the use of the central limit theorem. The distribution of annual earnings of all dental hygienists with 5 years' experience is skewed negatively, as shown in part a of Fig. 7·7. This distribution has a mean of

$15,000 and a standard deviation of $2,000. If we draw a random sample of 30 hygienists, what is the probability that their earnings will average more than $15,750 annually? In part b of Fig. 7·7, we show the sampling distribution of the mean that would result, and we have colored the area representing "earnings over $15,750."

FIG. 7·7

Population distribution and sampling distribution for hygienists' earnings

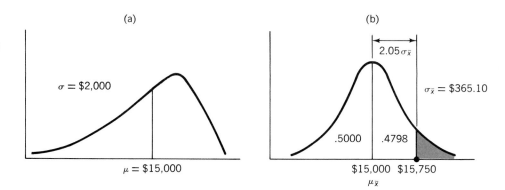

Our first task is to calculate the standard error of the mean from the population standard deviation as follows:

$$\sigma_{\bar{x}} = \frac{\sigma}{\sqrt{n}} \qquad [7·1]$$

$$= \frac{\$2,000}{\sqrt{30}}$$

$$= \frac{\$2,000}{5.477}$$

$$= \$365.16 \leftarrow \text{standard error of the mean}$$

Since we are dealing with a sampling distribution, we must now use Equation 7·2 and the table of z values (Appendix Table 1):

For $\bar{x} = \$15,750$

$$z = \frac{\bar{x} - \mu}{\sigma_{\bar{x}}} \qquad [7·2]$$

$$= \frac{\$15,750 - \$15,000}{\$365.16}$$

$$= \frac{\$750.00}{\$365.16}$$

$$= 2.05 \quad \leftarrow \begin{array}{l}\text{standard deviations from the mean of a standard} \\ \text{normal probability distribution}\end{array}$$

This gives us an area of .4798 for a z value of 2.05. We show this area in Fig. 7·7 as the area between the mean and $15,750. Since half, or .5000, of

the area under the curve lies between the mean and the right-hand tail, the
colored area must be:

.5000   area between the mean and the right-hand tail
−.4798   area between the mean and $15,750
_____
.0202 ← area between the right-hand tail and $15,750

Thus, we have determined that there is slightly more than a 2 percent
chance of average earnings being more than $15,750 annually in a group of
30 dental hygienists.

EXERCISES

25. In a sample of 36 observations from a normal distribution with a mean of 125 and a variance
of 225, what is

   a) $P(\bar{x} < 127)$ b) $P(\bar{x} > 130)$
   If, instead of 36 observations, 81 observations are taken, find
   c) $P(\bar{x} < 127)$ d) $P(\bar{x} > 130)$

26. In a sample of 9 observations from a normal distribution with mean 76.8 and standard
deviation 4.8, what is

   a) $P(75 < \bar{x} < 80)$
   b) Find the corresponding probability, given a sample of 25.

27. In a normal distribution with mean 72 and standard deviation of 10, how large a sample
must be taken so that there will be a 90 percent chance that its mean is greater than 70?

28. In a normal distribution with mean of 250 and standard deviation of 20, how large a sample
must be taken so that the probability will be .95 that the sample mean falls between 240
and 260?

29. The obstetrics ward of Filmore Memorial Hospital averages delivering 15 babies per day
with a variance of 16 babies squared. What is the probability that

   a) in a sample of 1 day, between 14 and 16 babies will be delivered?

   b) in a sample of 2 days, the average number of babies delivered will be between 14 and
   16?

   c) in a sample of 3 days, the average number of babies delivered will be between 14 and
   16?

   d) Explain the difference in your answers.

30. A sociologist studying income levels of residents of a Chicago suburb finds that among the
residents of the suburb on the northeast side, the average household income is $25,000
with a standard deviation of $1700.

   a) What is the probability that a family will have an income of at least $26,000?

   b) Is the probability that the average income of a sample of two families will be at least
   $26,000 greater or less than the probability of one family having an income that large?
   Explain.

31. Robertson Employment Service customarily gives standard intelligence and aptitude tests
to all persons who seek employment through the firm. The firm has collected data for
several years and has found that the distribution of scores is not normal, but is skewed to

the left with a mean of 83 and a standard deviation of 18. What is the probability that in a sample of 75 applicants taking the test, the mean score will be less than 82.5 or greater than 84?

32. In the experimental lab of the biology department, the university has backup recording apparatus to keep track of animal responses continually and to prevent losing important data. One particular piece of equipment has an average life of 4,300 hours with a standard deviation of 730 hours. In addition to the primary machine, the lab has set up two standby recorders, which are duplicates of the primary one. In the case of malfunction of one of the recorders, another will automatically take over in its place to keep records continuously. The operating life of each recorder is independent of the others.

a) What is the probability that a given set of recorders will last at least 13,000 hours?

b) At most, 12,630 hours?

33. In a study of cooperation in play among children, rewards given to each group are based on the number of cooperating responses they exhibit. Prior research shows that the average number of tokens received by a group is 10, with a variance of 35.5 tokens squared. If a sample of 7 groups is taken, what is the probability that there will be

a) more than 16 or fewer than 12 tokens received?

b) fewer than 16 and more than 5 tokens received?

34. Bats use a process known as echolocation to hunt insects for food. This process involves sending out high frequency sounds and then listening for their echos. As soon as an insect is detected, the signal intervals suddenly decrease, so that the bat can focus on its prey's location. Scientists have been able to measure this "detection" distance, the distance between a bat and its insect prey when the bat first is aware of the insect and the signals become more rapid. The average detection distance has been found to be 50.2 centimeters, with a standard deviation of 1.2 centimeters. If a sample of 20 bats is taken and this detection distance measured, can you say that with a 98 percent probability the average distance between a bat and its insect prey at the time of location will be between 49.5 and 50.7 centimeters?

5 AN OPERATIONAL CONSIDERATION IN SAMPLING: THE RELATIONSHIP BETWEEN SAMPLE SIZE AND STANDARD ERROR

Precision of the sample mean

We saw earlier in this chapter that the standard error, $\sigma_{\bar{x}}$, is a measure of dispersion of the sample means around the population mean. If the dispersion decreases (if $\sigma_{\bar{x}}$ becomes smaller), then the values taken by the sample mean tend to cluster *more* closely around μ. Conversely, if the dispersion increases (if $\sigma_{\bar{x}}$ becomes larger), the values taken by the sample mean tend to cluster *less* closely around μ. We can think of this relationship this way: as the standard error decreases, the value of any sample mean will probably be closer to the value of the population mean. Statisticians describe this phenomenon in another way: as the standard error decreases, the *precision* with which the sample mean can be used to estimate the population mean increases.

248

If we refer to Equation 7·1, we can see that as n increases, $\sigma_{\bar{x}}$ decreases. This happens because in Equation 7·1 the larger denominator on the right side produces a smaller $\sigma_{\bar{x}}$ on the left side. Two examples will show this relationship, both of which assume the same population standard deviation σ of 100.

249

Section 5
RELATIONSHIP
BETWEEN
SAMPLE SIZE AND
STANDARD ERROR

$$\sigma_{\bar{x}} = \frac{\sigma}{\sqrt{n}} \qquad [7\cdot1]$$

When $n = 10$,

$$\sigma_{\bar{x}} = \frac{100}{\sqrt{10}}$$

$$= \frac{100}{3.162}$$

$$= 31.63 \leftarrow \text{standard error of the mean}$$

Increasing
the sample size:
diminishing returns

And when $n = 100$

$$\sigma_{\bar{x}} = \frac{100}{\sqrt{100}}$$

$$= \frac{100}{10}$$

$$= 10 \leftarrow \text{standard error of the mean}$$

What have we shown? As we increased our sample size from 10 to 100 (a tenfold increase), the standard error dropped from 31.63 to 10, which is only about one-third of its former value. **Our examples suggest that, due to the fact that $\sigma_{\bar{x}}$ varies inversely with the square root of n, there is a diminishing return in sampling.**

It is true that sampling more items will decrease the standard error, but this benefit may not be worth the cost. A statistician would say, "The increased precision is not worth the additional sampling cost." In a statistical sense, it seldom pays to take excessively large samples. Decision makers should always assess *both* the worth and the cost of the additional precision they will obtain from a larger sample before they commit resources to take it.

The finite population multiplier

To this point in our discussions of sampling distributions, we have used Equation 7·1 to calculate the standard error of the mean:

Modifying
Equation 7·1

$$\sigma_{\bar{x}} = \frac{\sigma}{\sqrt{n}} \qquad [7\cdot1]$$

This equation is designed for situations in which the population is infinite, or in which we sample from a finite population with replacement (that is to say, after each item is sampled it is put back into the population before the next item is chosen, so that the same item can possibly be chosen more than once). If you will refer back to page 241 where we introduced Equation 7·1, you will recall our parenthesized note, which said, "Later we shall introduce an equation for finite populations." Introducing this new equation is the purpose of this section.

Many of the populations that decision makers examine are finite; that is, of stated or limited size. Example of these include the employees in a given company, the clients of a city social services agency, the students in a specific class, and a day's production in a given manufacturing plant. Not one of these populations is infinite, so we need to modify Equation 7·1 to deal with them. The formula designed to find the standard error of the mean when the population is *finite* is:

$$\sigma_{\bar{x}} = \frac{\sigma}{\sqrt{n}} \times \sqrt{\frac{N - n}{N - 1}} \qquad [7\cdot3]$$

Finding the
standard error
of the mean for
finite populations

where: N = size of the population
              n = size of the sample

This new term on the right-hand side, which we multiply by our original standard error, is called the *finite population multiplier:*

$$\text{finite population multiplier} = \sqrt{\frac{N - n}{N - 1}} \qquad [7\cdot4]$$

A few examples will help us become familiar with interpreting and using Equation 7·3. Suppose we are interested in a population of 20 municipal governments of the same size, all of which are experiencing excessive labor turnover problems. Our study indicates that the standard deviation of the distribution of annual turnover is 75 employees. If we sample 5 of these municipal governments and wish to compute the standard error of the mean, we would use Equation 7·3 as follows:

$$\sigma_{\bar{x}} = \frac{\sigma}{\sqrt{n}} \times \sqrt{\frac{N - n}{N - 1}} \qquad [7\cdot3]$$

$$= \frac{75}{\sqrt{5}} \times \sqrt{\frac{20 - 5}{20 - 1}}$$

$$= \frac{75}{2.236} \times \sqrt{\frac{15}{19}}$$

$$= 33.54 \times \sqrt{.789}$$

$$= (33.54)(.888)$$

$$= 29.8 \leftarrow \text{standard error of the mean of a finite population}$$

In this example, a finite population multiplier of .888 reduced the standard error from 33.54 to 29.8.

In cases in which the population is very large in relation to the size of the sample, this finite population multiplier is close to 1 and has little effect on the calculation of the standard error. Say that we have a population of 1,000 items and that we have taken a sample of 20 items. If we use Equation 7·4 to calculate the finite population multiplier, the result would be:

Sometimes the finite population multiplier is close to 1

$$\text{finite population multiplier} = \sqrt{\frac{N - n}{N - 1}} \qquad [7 \cdot 4]$$

$$= \sqrt{\frac{1,000 - 20}{1,000 - 1}}$$

$$= \sqrt{\frac{980}{999}}$$

$$= \sqrt{.981}$$

$$= .99$$

Using this multiplier of .99 would produce little effect on the calculation of the standard error of the mean.

This last example shows that when we sample a small fraction of the entire population (that is, when the population size N is very large relative to the sample size n), the finite population multiplier takes on a value close to 1.0. Statisticians refer to the fraction n/N as the *sampling fraction* because it *is* the fraction of the population N that is contained in the sample.

Sampling fraction defined

When the sampling fraction is small, the standard error of the mean for finite populations is so close to the standard error of the mean for infinite populations that we might as well use the same formula for both, namely Equation 7·1: $\sigma_{\bar{x}} = \dfrac{\sigma}{\sqrt{n}}$. The generally accepted rule is: when the sampling fraction is less than .05, the finite population multiplier need not be used.

When we use Equation 7·1, σ is constant, and so the measure of sampling precision $\sigma_{\bar{x}}$ depends only on the sample size n and not on the proportion of the population sampled. That is, to make $\sigma_{\bar{x}}$ smaller, it is necessary to make only n larger. Thus it turns out that it is the absolute size of the sample that determines sampling precision, not the fraction of the population sampled.

Sample size determines sampling precision

35. Every unit in the sample for a certain study costs $2. The information value of various sample sizes may be figured according to the formula $6,400/$\sigma_{\bar{x}}$. If a researcher wants to increase the sample until cost equals information value, how many individuals should she sample if the population standard deviation is 200?

36. Given a population of size $N = 65$ with a mean of 12 and a standard deviation of 2.1, what is the probability that a sample of size 16 will have a mean between 11.5 and 12.5?

37. From a population of 145 items with a mean of 120 and a standard deviation of 15, 64 items were chosen.

    a) What is the estimated standard error of the mean?

    b) What is the $P(122 < \bar{x} < 124)$?

38. For a population of size $N = 120$ with a mean of 7.5 and a standard deviation of 1.5, find the estimated standard error of the mean for the following sample sizes:

    a) $n = 9$ b) $n = 25$ c) $n = 49$

39. George Bransford has 25 five-acre plots of corn. He has been considering retiring because his health has worsened and also because he feels the crop is not as good as it once was. Over the last several years years the mean yield for each of the 25 plots has been 21,000 bushels with a standard deviation of 3,400 bushels. George has said that if the first 5 plots harvested do not yield 100,000 bushels, he will sell the farm this year. What is the probability that George will sell the farm?

40. Cynthia Hyde, a statistician in the Australian Ministry of the Interior, is worried about the proliferation of koala bears since the species has been protected. She fears that there will soon be more koalas than the continent's eucalyptus trees can support, so she wants to determine the leaf consumption of the adult koala. Once she has found this figure, she can recommend policies for managing the size of the koala population. She believes that the daily consumption is normally distributed with unknown mean μ and a standard deviation of about 2 ounces.

    a) If Hyde takes a sample of 16 adult koala bears and records their consumption of eucalyptus leaves, what is the probability that the sample mean is within one ounce of the population mean?

    b) How large a sample must she take in order to be 98 percent certain that the sample mean is within one ounce of the population mean?

41. Sara Gordon is heading a fund-raising drive for Milford College. She wishes to concentrate on the current tenth reunion class, and hopes to get contributions from 40 percent of the 160 members of that class. Past data indicate that those who contribute to the tenth year reunion gift will donate 3 percent of their annual salaries. Sara believes that the reunion class members have an average annual salary of $18,000 with a standard deviation of $8,000. If her expectations are met (40 percent of the class donate 3 percent of their salaries), what is the probability that the tenth reunion gift will be between $32,000 and $38,400?

42. As part of a nationwide study of the growth of cities, a population researcher has named 100 cities that he predicts will grow an average of 12 percent during the next three years, with a standard deviation of 4 percent. He is challenging his colleagues to choose 10 cities at random, from the list, to see if his prediction comes true. If his expectations are correct, what is the probability that a colleague following his instructions will see an average population increase of 10 to 15 percent for the three-year period?

6 TERMS INTRODUCED IN CHAPTER 7

- **CENSUS** The measurement or examination of every element in the population.

- **CENTRAL LIMIT THEOREM** A rule assuring that the sampling distribution of the mean approaches normal as the sample size increases, regardless of the shape of the population distribution from which the sample is selected.

- **CLUSTERS** Within a population, groups that are essentially similar to each other, although the groups themselves have wide internal variation.

- **CLUSTER SAMPLING** A method of random sampling in which the population is divided into groups, or clusters of elements, and then a random sample of these clusters is selected.

- **FINITE POPULATION** A population having a stated or limited size.

- **FINITE POPULATION MULTIPLIER** A factor used to correct an estimate of the standard error of the mean for studying a population of finite size that is small in relation to the size of the sample.

- **INFINITE POPULATION** A population in which it is theoretically impossible to observe all the elements.

- **JUDGMENT SAMPLING** A method of selecting a sample from a population in which personal knowledge or expertise are used to identify those items from the population that are to be included in the sample.

- **PARAMETERS** Values that describe the characteristics of a population.

- **PRECISION** The degree of accuracy with which the sample mean can estimate the population mean, as revealed by the standard error of the mean.

- **RANDOM OR PROBABILITY SAMPLING** A method of selecting a sample from a population in which all the items in the population have an equal chance of being chosen in the sample.

- **SAMPLE** A portion of the elements in a population chosen for direct examination or measurement.

- **SAMPLING DISTRIBUTION OF A STATISTIC** For a given population, a probability distribution of all the possible values a statistic may take on for a given sample size.

- **SAMPLING DISTRIBUTION OF THE MEAN** A probability distribution of all the possible means of samples of a given size, n, from a population.

- **SAMPLING ERROR** Error or variation among sample statistics due to chance: i.e., differences between each sample and the population, and among several samples, which are due solely to the elements we happened to choose for the sample.

- **SAMPLING FRACTION** The fraction or proportion of population contained in a sample.

- **SIMPLE RANDOM SAMPLING** Methods of selecting samples that allow each possible sample an equal probability of being picked *and* each item in the entire population an equal chance to be included in the sample.

- **STANDARD ERROR** The standard deviation of the sampling distribution of a statistic.

- **STANDARD ERROR OF THE MEAN** The standard deviation of the sampling distribution of the mean; a measure of the extent to which we expect the means from different samples to vary from the population mean, owing to the chance error in the sampling process.

- **STATISTICAL INFERENCE** The process of making inferences about populations from information contained in samples.

- **STATISTICS** Measures describing the characteristics of a sample.

- **STRATA** Groups within a population formed in such a way that each group is relatively homogeneous, but wider variability exists among the separate groups.

- **STRATIFIED SAMPLING** A method of random sampling in which the population is divided into homogeneous groups, or strata, and elements within each stratum are selected at random according to one of two rules: (1) a specified number of elements is drawn from each stratum corresponding to the proportion of that stratum in the population, or (2) an equal number of elements is drawn from each stratum, and the results are weighted according to the stratum's proportion of the total population.

- **SYSTEMATIC SAMPLING** A method of random sampling in which elements are selected from the population at a uniform interval that is measured in time, order, or space.

p. 241:
$$\sigma_{\bar{x}} = \frac{\sigma}{\sqrt{n}}$$
[7·1]

Use this formula to derive the *standard error of the mean* when the population is *infinite,* that is, when the elements of the population cannot be enumerated in a reasonable period of time or when we sample with replacement. This equation explains that the sampling distribution has a standard deviation, which we also call a standard error, equal to the population standard deviation divided by the square root of the sample size.

p. 242:
$$z = \frac{\bar{x} - \mu}{\sigma_{\bar{x}}}$$
[7·2]

A modified version of Equation 6·6, this formula allows us to determine the distance of the *sample mean* $\bar{x}$ from the population mean μ when we divide the difference by the standard error of the mean $\sigma_{\bar{x}}$. Once we have derived a z value, we can use the Standard Normal Probability Distribution Table and compute the probability that the sample mean will be that distance from the population mean. Because of the Central Limit Theorem, we can use this formula for non-normal distributions if the sample size is at least 30.

p. 250:
$$\sigma_{\bar{x}} = \frac{\sigma}{\sqrt{n}} \times \sqrt{\frac{N - n}{N - 1}}$$
[7·3]

where: N = size of the population
n = size of the sample

This is the formula for finding the *standard error of the mean* when the population is *finite,* that is, of stated or limited size.

p. 251:
$$\text{finite population multiplier} = \sqrt{\frac{N - n}{N - 1}}$$
[7·4]

In Equation 7·3, the term $\sqrt{(N - n)/(N - 1)}$, which we multiply times the standard error from Equation (7·1), is called the *finite population multiplier.* When the population is small in relation to the size of the sample, the finite population multiplier reduces the size of the standard error. Any decrease in the standard error increases the precision with which the sample mean can be used to estimate the population mean.

43. Dr. Donna Sutherland, a clinical psychology professor, is in charge of training Ph.D. students in psychotherapy and counseling techniques. Dr. Sutherland arbitrarily observes the students in their sessions with patients. Since she has other responsibilities in the department, she is not able to observe all of the patient sessions. She does, however, observe from adjoining rooms sometimes, without forewarning students that they will be observed. She feels that this makes her testing and evaluations objective. Is Dr. Sutherland using random sampling or judgment sampling?

44. Jim Neybors, owner of a large farm, gradually built it up so that his son could help him run it when he got out of school. Over the years, Jim has inspected his soil to determine what fertilizers to use on different areas. He feels that he has been farming long enough to recognize what type of enrichment the soil needs. His son, however, having learned all the modern techniques in agriculture school, wants to do soil analyses on random samples of soil from the farm, to determine the best fertilizer for different areas. The soil analysis will be quite costly and must be done every year. Jim is sure that his experience with crops and the soil enable him to decide what is needed, and so there has been some disagreement between the two. Can you defend either position?

45. A candidate for the state senate needs to find out the attitudes of his potential constituents: which are major issues, what are their feelings on these issues. The information is necessary in planning ways of gaining voters' support. The district includes voters of various age and income levels, but a large proportion are middle-class residents between the ages of 30 and 50. As yet, he is unsure of whether there are differences among age groups or income levels in their support of issues. Would stratified random sampling be appropriate here?

46. Bronson University is attempting to find out which issues the faculty sees as important and in need of attention on the campus. To assess the opinions of its 37 departments, the administration is considering a sampling plan. It has been recommended to the vice-chancellor of the university that the administration adopt a cluster sampling plan. The administration would choose 6 departments and interview all the faculty members to sound out each one on critical issues. Upon collecting and assessing the data, Bronson University could then develop new projects and plan for areas of work. Is a cluster sampling plan appropriate in this situation?

47. A sociologist interested in housing patterns and educational levels did a survey of a number of upper-middle class areas in several cities. She found that the average level of education for the heads of households in these areas was 16 years. The standard deviation of the distribution of number of years of education for head of household was 4 years. How large would a sample have to be for the probability to be 95.44 percent that the mean of the sample would fall between 15.2 and 16.8 years of education?

48. The Centerville Fire Department recently conducted a campaign in the city school system to alert children to potential fire hazards in the home and teach them proper fire safety. As part of the plan, the students were to complete a questionnaire at home with their parents, in order to include the entire family in the learning process. The fire department now wants to see how many parents participated in the program, and it is considering telephoning households that should have participated. The fire department has lists of all the school children and their telephone numbers, and key personnel in the department feel that the task could be accomplished by on-duty firemen in their spare time. Since the fire department apparently has both the time and the manpower, are there any reasons for it to poll a sample of school children's households rather than the entire population?

49. The average circumference of the skulls of ancient Romans from a particular period is known to be 20 inches, based on finds from a number of archeological sites. Martin Billings, a member of a current expedition to Italy, announced as he came upon another location with artifacts and remains from the same period that the sample mean should be lower than 20. He said, "For any sample, the mean should be lower because the sampling mean always understates the population mean because of sample variation." Billings was ignored by the other expedition members. Is there any truth to what Billings said?

50. Several weeks later at an archeological conference, Martin Billings again demonstrated his expertise in statistics. He had drawn a graph and presented it to the group, saying, "This is a sampling distribution of means. It is a normal curve and represents a distribution of all observations in each possible sample combination." Is Billings right? Explain.

51. In designing heart pacemakers and other medical devices for heart patients, doctors like to have estimates of the level of activity of the patients for whom different sizes and types are designed. In one study of elderly women, Dr. Nell Smith used a battery-operated heart rate monitor worn by the subject to measure average daily energy expenditure. Recent results have shown that for the women sampled, the average energy expenditure is 1424 kcals per day with a standard deviation of 240 kcals. Dr. Smith estimates that the benefits she obtains with a sample size of 25 are worth $1440. She expects to reduce the standard error by half of its current value and double the benefit. If it costs $20 for every woman in the sample, should Smith reduce her standard error?

52. The United States Customs Agency routinely checks all passengers arriving from foreign countries as they enter the United States. The department reports that on the average 35 people per day, with a standard deviation of 6, are found to be carrying contraband material as they enter the U.S. through the John F. Kennedy Airport in New York. What is the probability that in 4 days at that airport, the average number of passengers found carrying contraband will exceed 40?

53. HAL Corporation manufactures large computer systems and has always prided itself on the reliability of its System 666 central processing units. In fact, past experience has shown that the monthly downtime of System 666 CPU's averages 40 minutes, with a standard deviation of 8 minutes. The computer center at a large state university maintains an installation built around five System 666 CPU's. James Batter, the director of the computer center, feels that a satisfactory level of service is provided to the university community if the average downtime on the five CPU's is less than three-quarters of an hour per month. In any given month, what is the probability that Batter will be satisfied with the level of service?

54. Members of the Organization for Consumer Action send more than 100 volunteers a day all over the state to increase support for a consumer protection bill that is currently before the state legislature. Usually, each volunteer will visit a household and talk briefly with the resident, in the hope that the resident will sign a petition to be given to the state legislature. On the average, a volunteer will obtain 4.2 signatures for the petition each day, with a standard deviation of .6. What is the probability that a sample of 70 volunteers will result in an average between 4.05 and 4.1 signatures per day?

55. In an evaluation of the effectiveness of a federally funded remedial reading class, the project chairwoman, Jill Sommers, is attempting to obtain evidence that will enable the project to continue receiving funds. For this evaluation, it costs $3 for every unit included in the sample. Since precision is a desirable characteristic in presenting persuasive statistical evidence, Sommers figures the benefits she will receive for various sample sizes as being according to a formula, benefits = $4,608/$\sigma_{\bar{x}}$. If Sommers wants to increase her sample until the cost equals the benefit, how many units should she sample? The population standard deviation is 240.

256

56. A second-grade teacher is interested in the size and composition of the families of the 29 children in her class. She wonders how these factors relate to the children's academic performance. Each child fills out a questionnaire asking for background information about the number of brothers and sisters, birth order, whether both parents live at home, etc. She finds that the average size family among the 29 children is 4 members with a standard deviation of 2. What is the probability that, in a sample of 3 children, the mean will be more than 6 family members?

57. The 56 men who reported for the U.S. Olympic Team volleyball trials at a certain university had a mean weight of 195 lbs with a standard deviation of 18 lbs, a mean height of 6'2" with a standard deviation of 4.3", and a mean vertical jumping ability of 27" with a standard deviation of 5.6". The coach plans to have the men scrimmage on 6-man teams to observe the talent. What is the probability that the

   a) average weight of a scrimmaging team is below 190?

   b) mean jumping height of a team will be above 28"?

   c) mean height of a scrimmaging team will be between 6' and 6'4"?

9 CHAPTER CONCEPTS TEST

Answer true *or* false. *Answers are in the back of the book.*

1. When the items included in a sample are based upon the judgment of the individual conducting the sample, the sample is said to be nonrandom.

2. A statistic is a characteristic of a population.

3. A sampling plan that selects members from a population at uniform intervals in time, order, or space is called stratified sampling.

4. As a general rule, it is not necessary to include a finite population multiplier in a computation for standard error of the mean when the size of the sample is greater than 50.

5. The probability distribution of all the possible means of samples is known as the sampling distribution of the mean.

6. The theoretical foundation for statistical inference is based upon the principles of simple random sampling.

7. The standard error of the mean is the standard deviation of the distribution of sample means.

8. A sampling plan that divides the population into well-defined groups from which random samples are drawn is known as cluster sampling.

9. With increasing sample size, the sampling distribution of the mean approaches normality, regardless of the distribution of the population.

10. The standard error of the mean decreases in direct proportion to sample size.

8 Estimation

1. Introduction, 260

2. Point estimates, 264

3. Interval estimates;
   basic concepts, 267

4. Interval estimates and
   confidence intervals, 270

5. Calculating interval
   estimates of the mean from
   large samples, 273

6. Calculating interval
   estimates of the proportion
   —large samples, 277

7. Interval estimates using
   the *t* distribution, 280

8. Determining the sample
   size in estimation, 286

9. Terms introduced, 291

10. Equations introduced, 291

11. Chapter review exercises, 292

12. Chapter concepts test, 295

As part of its budget request for next year, the Public Works Department of Detroit must know how much salt it will need for de-icing roads. Last year, the department was short of salt, so it is reluctant to request only that amount again. However, the department does feel that past usage data will enable it to *estimate* the amount of salt to order. A random sample of 10 working days chosen from the last 5 years gave a mean usage of 11,400 pounds per day, with a sample standard deviation of 700 pounds a day. With this data and the procedures to be discussed in this chapter, the department can make a sensible estimate of the size of its order this year.

Reasons for
estimates

Making statistical
inferences

Estimating
population
parameters

1 INTRODUCTION

Everyone makes estimates. When you get ready to cross a street, you estimate the speed of any car that is approaching, the distance between you and that car, and your own speed. Having made these quick estimates, you decide whether to wait or run.

All decision makers must make quick estimates, too. The outcome of these estimates can affect their organizations as seriously as the outcome of your decision as to whether to cross the street. Department heads make estimates of next fall's enrollment in statistics. Credit managers estimate whether a purchaser will eventually pay his bills. Prospective home buyers make estimates concerning the behavior of interest rates in the mortgage market. All these people make estimates without worry about whether they are scientific but with the hope that the estimates bear a reasonable resemblance to the outcome.

Decision makers use estimates because in all but the most trivial decisions, they must make rational decisions without complete information and with a great deal of uncertainty about what the future will bring. As educated citizens and professionals you will be able to make more useful estimates by applying the techniques described in this and subsequent chapters.

The material on probability theory covered in Chapters 5, 6, and 7 forms the foundation for *statistical inference,* the branch of statistics concerned with using probability concepts to deal with uncertainty in decision making. Statistical inference is based on *estimation,* which we shall introduce in this chapter, and *hypothesis testing,* which is the subject of Chapters 9 and 10. In both estimation and hypothesis testing, we shall be making inferences about characteristics of populations from information contained in samples.

How do decision makers use sample statistics to estimate population parameters? The department head attempts to estimate enrollments next fall from current enrollments in prerequisite courses and preregistration data. The credit manager attempts to estimate the creditworthiness of prospective customers from a sample of their past payment habits. The home buyer attempts to estimate the future course of interest rates by observing the current behavior of those rates. In each case, somebody is trying to infer something about a population from information taken from a sample.

This chapter introduces methods that enable us to estimate with reasonable accuracy the *population proportion* (the proportion of the population that possesses a given characteristic) and the *population mean.* To calculate the exact proportion or the exact mean would be an impossible goal. Even so, we will be able to make an estimate, make a statement about the error that will probably accompany this estimate, and implement some controls to avoid as much of the error as possible. As decision makers, we will be forced at times to rely on blind hunches. Yet in other situations in which information is available and we apply statistical concepts, we can do better than that.

Types of estimates

We can make two types of estimates about a population: a *point* estimate and an *interval* estimate. A point estimate is a single number that is used to estimate an unknown population parameter. If, while watching the first members of a football team come onto the field, you say, "Why, I bet their line must weigh 250 pounds," you have made a point estimate. A department chairwoman would make a point estimate if she said, "Our current data indicate that this course will have 350 students in the fall."

Point estimate defined

A point estimate is often insufficient because it is either right or wrong. If you are told only that the chairwoman's point estimate of enrollment is wrong, you do not know how wrong it is, and you cannot be certain of the estimate's reliability. If you learn that it is off by only 10 students you would accept 350 students as a good estimate of future enrollment. But if the estimate is off by 90 students, you would reject it as an estimate of future enrollment. Therefore a point estimate is much more useful if it is accompanied by an estimate of the error that might be involved.

Shortcomings of point estimates

An interval estimate is a range of values used to estimate a population parameter. It indicates the error in two ways: by the extent of its range and by the probability of the true population parameter lying within that range. In this case, the department chairwoman would say something like, "I estimate that the true enrollment in this course in the fall will be between 330 and 380 and that it is very likely that the exact enrollment will fall within this interval." The chairwoman has a better idea of the reliability of her estimate. If the course is taught in sections of about 100 students each, and if the chairwoman had tentatively scheduled 5 sections, on the basis of her estimate, she can now cancel one of those sections and offer an elective instead.

Interval estimate defined

Estimator and estimates

Any sample statistic that is used to estimate a population parameter is called an estimator; that is, an estimator is a sample statistic used to estimate a population parameter. The sample mean $\bar{x}$ can be an estimator of the population mean μ, and the sample proportion can be used as an estimator of the population proportion. We can also use the sample range as an estimator of the population range.

Estimator defined

When we observe a specific numerical value of our estimator, we call that value an estimate. In other words, an estimate is a specific observed value of a statistic. We form an estimate by taking a sample and computing the value taken by our estimator in that sample. Suppose that we calculate the mean odometer reading (mileage) from a sample of used taxis and find

Estimate defined

it to be 98,000 miles. If we use this specific value to estimate the mileage for a whole fleet of used taxis, the value 98,000 miles would be an estimate. Table 8·1 illustrates several populations, estimators, and estimates.

TABLE 8·1 **Populations, population parameters, estimators, and estimates**

Population in which we are interested	Population parameter we wish to estimate	Sample statistic we will use as an estimator	Estimate we make
Employees in a large hospital	Mean turnover per year	Mean turnover for a period of 1 month	8.9% turnover per year
Applicants for town manager of Chapel Hill	Mean formal education (years)	Mean formal education of every 5th applicant	17.9 years of formal education
Teenagers in a given community	Proportion who have a criminal record	Proportion of a sample of 50 teenagers who have a criminal record	.02, or 2%, have a criminal record

Criteria of a good estimator

Qualities of a
good estimator

Some statistics are better estimators than are others. Fortunately, we can evaluate the quality of a statistic as an estimator by using four criteria:

1. **Unbiasedness.** This is a desirable property for a good estimator to have. The term *unbiasedness* refers to the fact that a sample mean is an unbiased estimator of a population mean because the mean of the sampling distribution of sample means taken from the same population is equal to the population mean itself. We can say that a statistic is an unbiased estimator if, on the average, it tends to assume values that are above the population parameter being estimated as frequently and to the same extent as it tends to assume values that are below the population parameter being estimated.

2. **Efficiency.** Another desirable property of a good estimator is that it be efficient. Efficiency refers to the size of the standard error of the statistic. If we compare two statistics from a sample of the same size and try to decide which one is the more efficient estimator, we would pick the statistic that has the smaller standard error, or standard deviation of the sampling distribution. Suppose we choose a sample of a given size and must decide whether to use the sample mean or the sample median to estimate the population mean. If we calculate the standard error of the sample mean and find it to be 1.05 and then calculate the standard error of the sample median and find it to be 1.6, we would say that the sample mean is a *more efficient estimator* of the population mean *because its standard error is smaller*. It makes sense that an estimator with a smaller standard error or with less

variation will have more chance of producing an estimate nearer to the population parameter under consideration.

3. **Consistency.** A statistic is a consistent estimator of a population parameter if *as the sample size increases, it becomes almost certain that the value of the statistic comes very close to the value of the population parameter.* If an estimator is consistent, it becomes more reliable with large samples. Thus, if you are wondering whether to increase the sample size to get more information about a population parameter, find out first whether your statistic is a consistent estimator. If it is not, you will waste time and money by taking larger samples.

4. **Sufficiency.** An estimator is sufficient if it makes so much use of the information in the sample that no other estimator could extract from the sample additional information about the population parameter being estimated. Although we will not mention all four of these criteria again, we present them here to make you aware of the care that statisticians must use in picking an estimator.

A given sample statistic is not always the best estimator of its analogous population parameter. Consider a symmetrically distributed population in which the values of the median and the mean coincide. In this instance, the sample mean would be an *unbiased* estimator of the population median because it would assume values that on the average would equal the population median. Also, the sample mean would be a *consistent* estimator of the population median because, as the sample size increases, the value of the sample mean would tend to come very close to the population median. And the sample mean would be a more *efficient* estimator of the population median than the sample median itself because in large samples the sample mean has a smaller standard error than the sample median. At the same time, the sample median in a symmetrically distributed population would be an unbiased and consistent estimator of the population mean but *not the most efficient* estimator because in large samples its standard error is larger than that of the sample mean.

Finding the best estimator

EXERCISES

1. What two basic tools are used in making statistical inferences?

2. Why do decision makers often measure samples rather than entire populations? What is the disadvantage?

3. Explain a shortcoming that occurs in a point estimate but not in an interval estimate. What measure is included with a point estimate to compensate for this problem?

4. What is an estimator? How does an estimate differ from an estimator?

5. List and describe briefly the criteria of a good estimator.

6. What role does consistency play in determining sample size?

Using the
sample mean
to estimate the
population mean

2 POINT ESTIMATES

The sample mean $\bar{x}$ is the best estimator of the population mean μ. It is unbiased, consistent, the most efficient estimator, and, as long as the sample is sufficiently large, its sampling distribution can be approximated by the normal distribution.

If we know the sampling distribution of $\bar{x}$, we can make statements about any estimate we may make from sampling information. Let's look at a medical supplies company that produces disposable hypodermic syringes. Each syringe is wrapped in a sterile package and then jumble-packed in a large corrugated carton. Jumble packing causes the cartons to contain differing numbers of syringes. Since the syringes are sold on a per unit basis, the company needs an estimate of the number of syringes per carton for billing purposes. We have taken a sample of 35 cartons at random and recorded the number of syringes in each carton. Table 8·2 illustrates our results. Using the results of Chapter 3, we can obtain the sample mean $\bar{x}$ by finding the sum of all our results, Σx, and dividing this total by n, the number of samples we have taken:

Finding the
sample mean

$$\bar{x} = \frac{\Sigma x}{n} \qquad [3 \cdot 2]$$

TABLE 8·2
**Results of
sample of
35 cartons of
hypodermic
syringes
(syringes per
carton)**

101	103	112	102	98	97	93
105	100	97	107	93	94	97
97	100	110	106	110	103	99
93	98	106	100	112	105	100
114	97	110	102	98	112	99

Using this equation to solve our problem, we get:

$$\bar{x} = \frac{3570}{35}$$
$$= 102 \text{ syringes}$$

Thus, using the sample mean $\bar{x}$ as our estimator, the point estimate of the population mean μ is 102 syringes per carton. Since the manufactured price of a disposable hypodermic syringe is quite small (about 25¢), both the buyer and seller would accept the use of this point estimate as the basis for billing, and the manufacturer can save the time and expense of counting each syringe that goes into a carton.

Point estimate of the population variance and standard deviation

Using the sample
standard deviation
to estimate
the population
standard deviation

Suppose the management of the medical supplies company wants to estimate the variance and/or standard deviation of the distribution of the number of packaged syringes per carton. The most frequently used estimator of the population standard deviation σ is the sample standard deviation s. We can calculate the sample standard deviation as in Table 8·3 and discover that the sample standard deviation is 6.01 syringes.

Values of x (needles per carton) (1)	x^2 (2)	Sample mean $\bar{x}$ (3)	$(x - \bar{x})$ (4) = (1) − (3)	$(x - \bar{x})^2$ (5) = $(4)^2$	TABLE 8·3 Calculation of sample variance and standard deviation for syringes per carton
101	10,201	102	−1	1	
105	11,025	102	3	9	
97	9,409	102	−5	25	
93	8,649	102	−9	81	
114	12,996	102	12	144	
103	10,609	102	1	1	
100	10,000	102	−2	4	
100	10,000	102	−2	4	
98	9,604	102	−4	16	
97	9,409	102	−5	25	
112	12,544	102	10	100	
97	9,409	102	−5	25	
110	12,100	102	8	64	
106	11,236	102	4	16	
110	12,100	102	8	64	
102	10,404	102	0	0	
107	11,449	102	5	25	
106	11,236	102	4	16	
100	10,000	102	−2	4	
102	10,404	102	0	0	
98	9,604	102	−4	16	
93	8,649	102	−9	81	
110	12,100	102	8	64	
112	12,544	102	10	100	
98	9,604	102	−4	16	
97	9,409	102	−5	25	
94	8,836	102	−8	64	
103	10,609	102	1	1	
105	11,025	102	3	9	
112	12,544	102	10	100	
93	8,649	102	−9	81	
97	9,409	102	−5	25	
99	9,801	102	−3	9	
100	10,000	102	2	4	
99	9,801	102	−3	9	
3,570	**365,368**		Sum of all the squared differences	$\Sigma(x - \bar{x})^2 \to$ **1,228**	

[4.11] $s^2 = \dfrac{\Sigma x^2}{n - 1} - \dfrac{n\bar{x}^2}{n - 1}$

←or→ Sum of the squared differences divided by 34, the number of items in the sample − 1 (sample variance) $\dfrac{\Sigma(x - \bar{x})^2}{n - 1} \to$ **36.12**

$= \dfrac{365,368}{34} - \dfrac{35(102)^2}{34}$

$= \dfrac{1228}{34}$

$= 36.12$

[4.12] $s = \sqrt{s^2}$

$= \sqrt{36.12}$

$= 6.01$ syringes

Sample standard deviation s $\sqrt{\dfrac{\Sigma(x - \bar{x})^2}{n - 1}} \to$ 6.01 syringes

If instead of considering

$$s^2 = \frac{\Sigma(x - \bar{x})^2}{n - 1}$$ [4·11]

as our sample variance, we had considered

$$s^2 = \frac{\Sigma(x - \bar{x})^2}{n}$$

the result would have some *bias* as an estimator of the population variance; specifically, it would tend to be too low. Using a divisor of $n - 1$ gives us an unbiased estimator of σ^2. Thus, we will use s^2 (as defined in Equation 4·11) and s (as defined in Equation 4·12) to estimate σ^2 and σ.

Point estimate of the population proportion

Using the
sample proportion
to estimate
the population
proportion

The proportion of units that have a particular characteristic in a given population is symbolized p. If we know the proportion of units in a sample that has that same characteristic (symbolized $\bar{p}$), we can use this $\bar{p}$ as an estimator of p. It can be shown that $\bar{p}$ has all the desirable properties we discussed earlier; it is unbiased, consistent, efficient, and sufficient.

Let's continue our example of the manufacturer of medical supplies to try to estimate the population proportion from the sample proportion. Suppose the management wishes to estimate the number of cartons that will arrive damaged, owing to poor handling in shipment after the cartons leave the factory. We can check a sample of 50 cartons from their shipping point to the arrival at their destination and then record the presence or absence of damage. If, in this case, we find that the proportion of damaged cartons in the sample is .08, we would say that:

$$\bar{p} = .08 \leftarrow \text{sample proportion damaged}$$

And since the sample proportion $\bar{p}$ is a convenient estimator of the population proportion p, we can estimate that the proportion of damaged cartons in the population will also be .08.

EXERCISES

7. Calculate the sample mean for the following set of data.

.61 .70 .63 .76 .67 .72 .64 .82 .88 .82
.78 .84 .83 .82 .74 .85 .73 .85 .87 .75

8. Find point estimates of the mean and the variance of the population from which the following sample came.

17.0 25.0 13.0 8.5 27.5 20.0 18.5 17.0 16.0 12.0

9. Below are a set of performance scores for 35 individuals. Treating these as (a) a sample and (b) a population, compute the standard deviation.

4.7	5.0	8.0	3.5	5.0	4.3	7.0
5.6	8.1	8.0	4.0	7.8	6.0	10.0
6.8	2.0	5.9	7.3	5.8	4.7	6.1
3.9	8.0	5.0	8.0	6.4	7.0	8.0
6.8	4.4	7.0	5.5	6.4	5.0	4.2

10. In a sample of 500 voters, 284 expressed support for a certain candidate, indicating an intention to vote for him. Give a point estimate of the proportion of voters supporting the candidate.

3 INTERVAL ESTIMATES: BASIC CONCEPTS

The purpose of gathering samples is to learn more about a population. We can compute this information from the samples as either *point* estimates, which we have just discussed, or as *interval* estimates, the subject of the rest of this chapter. *An interval estimate describes a range of values within which a population parameter is likely to lie.*

Suppose the director of a Consumer Protection Agency needs an estimate in months of the average life of car batteries purchased at a large retail chain. We select a random sample of 200 batteries, record the car owners' names and addresses as listed in store records, and interview these owners about the battery life they have experienced. Our sample of 200 users had a mean battery life of 36 months. If we use the point estimate of the sample mean $\overline{x}$ as the best estimator of the population mean μ, we would report that the mean life of the store's batteries is 36 months.

Finding the point estimate

But the director also asks for a statement about the uncertainty that will be likely to accompany this estimate; that is, a statement about the range within which the unknown population mean is likely to lie. To provide such a statement, we need to find *the standard error of the mean.*

Finding the likely error of this estimate

We learned from Chapter 7 that if we select and plot a large number of sample means from a population, the distribution of these means will approximate a normal curve. Furthermore, the mean of the sample means will be the same as the population mean. Our sample size of 200 is large enough so that we can apply the central limit theorem, as we have done graphically in Fig. 8·1. To measure the spread, or dispersion, in our distribution of sample means, we can use the following formula and calculate the standard error of the mean:

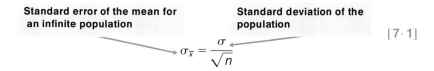

Standard error of the mean for an infinite population	**Standard deviation of the population**	
		[7·1]

$$\sigma_{\overline{x}} = \frac{\sigma}{\sqrt{n}}$$

*We have not used the finite population multiplier to calculate the standard error of the mean because the population of batteries is large enough to be considered infinite.

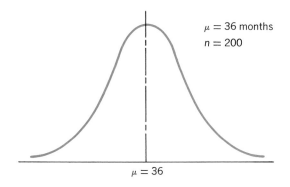

FIG. 8·1
Sampling
distribution of the
mean for samples
of 200 batteries

$\mu = 36$ months

$n = 200$

$\mu = 36$

In this case, the director has already estimated the standard deviation of the population of the batteries and reported that it is 10 months. Using this standard deviation and the first equation from Chapter 7, we can calculate the standard error of the mean:

$$\sigma_{\bar{x}} = \frac{\sigma}{\sqrt{n}}$$ [7·1]

$$= \frac{10}{\sqrt{200}}$$

$$= \frac{10}{14.14}$$

$$= .707 \text{ months} \leftarrow \text{one standard error of the mean}$$

We could now report to the director our estimate of the life of the store's batteries is 36 months, and the standard error that accompanies this estimate is .707. In other words, the actual mean life for all the batteries *may* lie somewhere in the interval estimate of from 35.293 to 36.707 months. This is helpful but insufficient information for the director. Next we need to calculate the chance that the actual life will lie in this interval *or* in other intervals of different widths that we might choose, $\pm 2\sigma$, $(2 \times .707)$, $\pm 3\sigma$, $(3 \times .707)$, etc.

Probability of the true population parameter falling within the interval estimate

To begin to solve this problem, we should review relevant parts of Chapter 6. There we worked with the normal probability distribution and learned that specific portions of the area under the normal curve are located between plus and minus any given number of standard deviations from the mean. In Fig. 6·12 (page 194), we saw how to relate these portions to specific probabilities.

Finding the chance the mean will fall in this interval estimate; constructing intervals

Fortunately, we can apply these properties to the standard error of the mean and make the following statement about the range of values in an interval estimate for our battery problem.

The probability is .955 that the mean of a sample size of 200 will be within plus and minus 2 standard errors of the population mean. Stated differently, 95.5 percent of all the sample means are within plus and minus 2 standard errors from μ, and hence μ is within plus and minus 2 standard errors of 95.5 percent of all the sample means. Theoretically, if we select 1,000 samples at random from a given population and then construct an interval of plus and minus 2 standard errors around the mean of each of these samples, about 955 of these intervals will include the population mean. Similarly, the probability is .683 that the mean of the sample will be within plus or minus one standard error of the population mean, and so forth. This theoretical concept is basic to our study of interval construction and of statistical inference. In Fig. 8·2, we have illustrated the concept graphically, showing five such intervals. Only the interval constructed around the sample mean $\bar{x}_4$ does not contain the population mean. In words, statisticians would describe the interval estimate represented in Fig. 8·2 by saying, "The population mean μ will be located within plus or minus 2 standard errors from the sample mean 95.5 percent of the time."

As far as any particular interval in Fig. 8·2 is concerned, it either contains the population mean or it does not, because the population mean is a fixed parameter and does not vary. Since we know that in 95.5 percent of all samples the interval will contain the population mean, we say that we are 95.5 percent confident that the interval contains the population mean.

Applying this to the battery example, we can now report to the director. Our best estimate of the life of the store's batteries is 36 months, *and* we are 68.3 percent confident that the life lies in the interval from 35.293 to 36.707 months ($36 \pm 1\sigma_{\bar{x}}$). Similarly, we are 95.5 percent confident that the life falls within the interval of 34.586 to 37.414 months ($36 \pm 2\sigma_{\bar{x}}$).

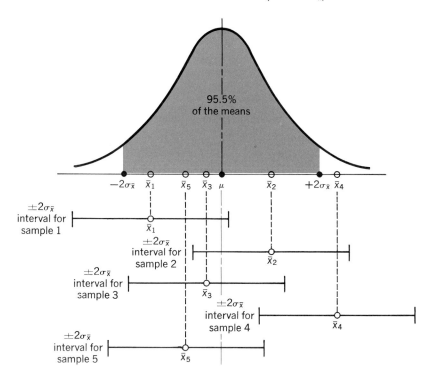

FIG. 8·2
A number of intervals constructed around sample means, all intervals except one including the population mean

11. From a population known to have a standard deviation of .9, a sample of 36 individuals is taken. The mean for this sample is found to be 9.6.

    a) Find the standard error of the mean.

    b) Establish an interval estimate around the mean, using one standard error of the mean.

12. A sample of 81 individuals is taken from a population whose standard deviation is known to be 3.6. The mean for this sample turns out to be 24.5.

    a) Find the standard error of the mean.

    b) What is the interval around the sample mean that will include the population mean 95.5 percent of the time?

13. For a population with a known variance of 196, a sample of 49 leads to 210 as an estimate of the mean.

    a) Find the standard error of the mean.

    b) Establish an interval estimate that should include the population mean 68.3 percent of the time.

14. A sample of 25 individuals produces a sample mean of 68. The population standard deviation has been established as 4.25.

    a) Give an interval estimate that has a 68.3 percent probability of including the population mean.

    b) Give an interval estimate that has a 99.7 percent chance of including the population mean.

15. Sue Willingham, an anthropologist studying children in Indonesia, has randomly sampled 100 children for her study. One of the variables she records is the hand width, measured in inches from the tip of the thumb to the tip of the little finger. She finds an average width of 6 inches. If the population standard deviation is known to be .8 inches, estimate an interval that has a 95.5 percent chance of containing the true mean.

16. An ethologist, studying the swarming patterns of mosquitos, randomly samples 36 different locations and finds an average of 36,000 mosquitos in each location. He knows the standard deviation of this type of data to be 12,000.

    a) Establish an interval estimate for the average number of mosquitos, so that we are 68.3 percent certain that the population mean lies within this interval.

    b) Establish an interval estimate for the average number of mosquitos, so that we are 95.5 percent certain that the population mean lies within this interval.

4 INTERVAL ESTIMATES AND CONFIDENCE INTERVALS

In using interval estimates, we are not confined to plus and minus 1, 2, and 3 standard errors. According to Appendix Table 1, for example, plus and minus 1.64 standard errors includes about 90 percent of the area under the curve; it includes .4495 of the area on either side of the mean in a normal distribution. Similarly, plus and minus 2.58

standard errors includes about 99 percent of the area, or 49.51 percent on each side of the mean.

In statistics, the probability that we associate with an interval esti- Confidence level defined
mate is called the confidence level. This probability, then, indicates how confident we are that the interval estimate will include the population parameter. A higher probability means more confidence. In estimation, the most commonly used confidence levels are 90 percent, 95 percent, and 99 percent, but we are free to apply *any* confidence level. In Fig. 8·2, for example, we used a 95.5 percent confidence level.

The confidence interval is the range of the estimate we are making. Confidence intervals and confidence limits
If we report that we are 90 percent confident that the mean of the population of incomes of persons in a certain community will lie between $8,000 and $14,000, then the range $8,000–$14,000 is our confidence interval. More often, however, we will express the confidence interval in standard errors rather than in numerical values. Thus, we would normally express confidence intervals like this: $\overline{x} \pm 2.3\sigma_{\overline{x}}$, where:

$$\overline{x} + 2.3\sigma_{\overline{x}} = \text{upper limit of the confidence interval}$$
$$\overline{x} - 2.3\sigma_{\overline{x}} = \text{lower limit of the confidence interval}$$

Thus, confidence limits are the upper and lower limits of the confidence interval. In this case, $\overline{x} + 2.3\sigma_{\overline{x}}$ is called the *upper confidence limit,* and $\overline{x} - 2.3\sigma_{\overline{x}}$ is the *lower confidence limit.*

Relationship between confidence level and confidence interval

You may think that we should use a high confidence level, such as 99 The shortcoming of high confidence levels
percent, in all estimation problems. After all, a high confidence level seems to signify a high degree of accuracy in the estimate. In practice, however, high confidence levels will produce large confidence intervals, and such large intervals are not precise; they give very fuzzy estimates.

Consider an appliance store customer who inquires about the delivery of a new washing machine. In Table 8·4 are several of the questions the customer might ask and the likely responses. This table indicates the direct relationship that exists between the confidence level and the confidence interval for any estimate. As the customer sets a tighter and tighter confidence interval, the store manager admits a lower and lower confidence level. Notice, too, that when the confidence interval is too wide, as is the case with a one-year delivery, the estimate may have little real value, even though the store manager attaches a 99 percent confidence level to that estimate. Similarly, if the confidence interval is too narrow ("Will my washing machine get home before I do?"), the estimate is associated with such a low confidence level (1 percent) that we question its value.

TABLE 8·4

Illustration of the relationship between confidence level and confidence interval

Customer's question	Store manager's response	Implied confidence level	Implied confidence interval
Will I get my washing machine within 1 year?	I am absolutely certain of that.	Better than 99%	1 yr
Will you deliver the washing machine within 1 month?	I am almost positive it will be delivered this month.	At least 95%	1 mo
Will you deliver the washing machine within a week?	I am pretty certain it will go out within this week.	About 80%	1 wk
Will I get my washing machine tomorrow?	I am not certain we can get it to you then.	About 40%	1 day
Will my washing machine get home before I do?	There is little chance it will beat you home.	Near 1%	1 hr

Using sampling and confidence interval estimation

Estimating from only one sample

In our discussion of the basic concepts of interval estimation, particularly in Fig. 8·2, we described samples being drawn repeatedly from a given population in order to estimate a population parameter. We also mentioned selecting a large number of sample means from a population. In practice, however, it is often difficult to take more than one sample from a population. Based on just one sample, we estimate the population parameter. We must be careful, then, about interpreting the results of such a process.

If we calculate from one sample in our battery example the following confidence interval and confidence level: "We are 95 percent confident that the mean battery life of the population lies within 30 and 42 months," **this statement does not mean that the chance is .95 that the mean life of all our batteries falls within the interval established from this one sample. Instead, it means that if we select many random samples of this sample size and if we calculate a confidence interval for each of these samples, then in about 95 percent of these cases the population mean will lie within that interval.**

EXERCISES

17. Define the confidence level for an interval estimate.

18. Define the confidence interval.

19. Suppose you wish to use a confidence level of 80 percent. Give the upper limit of the confidence interval in terms of the sample mean, $\bar{x}$, and the standard error, $\sigma_{\bar{x}}$.

20. In what way may an estimate be less meaningful because of:

   a) a high confidence level?

   b) a narrow confidence interval?

21. Suppose that using one random sample of 50 elements of a population, you have established a confidence interval of 70–90 around the sample mean, with a confidence level of 95 percent that the population mean falls within this interval. What statement can you make concerning the population mean if 99 other 50-element random samples are also taken from this population?

22. Is the confidence level for an estimate based on the interval constructed from one sample?

23. Given the following confidence levels, express the lower and upper limits of the confidence interval for these levels in terms of $\overline{x}$ and $\sigma_{\overline{x}}$.

   a) 50%   b) 75%   c) 85%   d) 98%

5 CALCULATING INTERVAL ESTIMATES OF THE MEAN FROM LARGE SAMPLES

Finding a 95 percent confidence interval

A federal government interagency motor pool needs an estimate of the mean life it can expect from windshield wiper blades under typical driving conditions. Already, management has determined that the standard deviation of the population life is 6 months. When we select a simple random sample of 100 wiper blades and collect data on their useful lives we obtain these results:

$n = 100 \leftarrow$ sample size
$\overline{x} = 21$ months $\leftarrow$ sample mean
$\sigma = 6$ months $\leftarrow$ population standard deviation

Since the interagency motor pool uses tens of thousands of these wiper blades annually, it requests that we find an interval estimate with a confidence level of 95 percent. Since the sample size is greater than 30, we can use the normal distribution as our sampling distribution and calculate the standard error of the mean by using Equation 7·1:

$$\sigma_{\overline{x}} = \frac{\sigma}{\sqrt{n}} \qquad [7 \cdot 1]$$

$$= \frac{6 \text{ months}}{\sqrt{100}}$$

$$= \frac{6}{10}$$

$$= .6 \text{ months} \leftarrow \begin{array}{l} \text{standard error of the mean} \\ \text{for an infinite population} \end{array}$$

Next, we consider the confidence level with which we are working. Since a 95 percent confidence level will include 47.5 percent of the area on either side of the mean of the sampling distribution, we can search in the body of Appendix Table 1 for the .475 value. We discover that .475 of the area under the normal curve is contained between the mean and a point 1.96 standard errors to the right of the mean. Therefore, we know that (2)(.475) = .95 of the area is located between plus and minus 1.96 standard errors from the mean and that our confidence limits are:

$$\overline{x} + 1.96\sigma_{\overline{x}} \leftarrow \text{upper confidence limit}$$
$$\overline{x} - 1.96\sigma_{\overline{x}} \leftarrow \text{lower confidence limit}$$

Then we substitute numerical values into these two expressions:

$$\overline{x} + 1.96\sigma_{\overline{x}} = 21 \text{ months} + 1.96(.6 \text{ months})$$
$$= 21 + 1.18 \text{ months}$$
$$= 22.18 \text{ months} \leftarrow \text{upper confidence limit}$$

$$\overline{x} - 1.96\sigma_{\overline{x}} = 21 \text{ months} - 1.96(.6 \text{ months})$$
$$= 21 - 1.18 \text{ months}$$
$$= 19.82 \text{ months} \leftarrow \text{lower confidence limit}$$

We can now report that we estimate the mean life of the population of wiper blades to be between 19.82 and 22.18 months with 95 percent confidence.

When the population standard deviation is unknown

Finding a 90 percent confidence interval

A more complex interval estimate problem comes from a social service agency in a local government. It is interested in estimating the mean annual income of 700 families living in a four square block section of a community. We take a simple random sample and gain these results:

$$n = 50 \leftarrow \text{sample size}$$
$$\overline{x} = \$4,800 \leftarrow \text{sample mean}$$
$$s = \$950 \leftarrow \text{sample standard deviation}$$

The agency asks us to calculate an interval estimate of the mean annual income of all 700 families so that it can be 90 percent confident that the population mean falls within that interval. Since the sample size is over 30, we can use the normal distribution as the sampling distribution.

Estimating the population standard deviation

Notice that one part of this problem differs from our previous examples: we do not know the population standard deviation, and so we will use the sample standard deviation to estimate the *population standard deviation:*

$$\hat{\sigma} = s = \sqrt{\frac{\Sigma(x - \overline{x})^2}{n - 1}} \qquad [8 \cdot 1]$$

The value $950.00 is our estimate of the standard deviation of the popula-
tion. We can also symbolize this estimated value by $\hat{\sigma}$, which is called *sigma hat*.

Now we can estimate the standard error of the mean. Since we have a
finite population size of 700, we will use the formula for deriving the standard error of the mean of finite populations:

$$\sigma_{\bar{x}} = \frac{\sigma}{\sqrt{n}} \times \sqrt{\frac{N - n}{N - 1}}$$

But since we are calculating the standard error of the mean using an *estimate* of the standard deviation of the population, we rewrite this equation
so that it is correct symbolically:

Estimating
the standard error
of the mean

Symbol that indicates an
estimated value

Estimate of the population
standard deviation

$$\hat{\sigma}_{\bar{x}} = \frac{\hat{\sigma}}{\sqrt{n}} \times \sqrt{\frac{N - n}{N - 1}} \qquad [8.2]$$

$$= \frac{\$950.00}{\sqrt{50}} \times \sqrt{\frac{700 - 50}{700 - 1}}$$

$$= \frac{\$950.00}{7.07} \sqrt{\frac{650}{699}}$$

$$= \$134.37 \sqrt{.9299}$$

$$= (\$134.37)(.9643)$$

$$= \$129.57 \leftarrow \text{estimate of the standard error of the mean}$$
of a finite population (derived from an
estimate of the population standard deviation)

Next we consider the 90 percent confidence level, which would include
45 percent of the area on either side of the mean of the sampling distribution. Looking in the body of Appendix Table 1 for the .45 value, we find that
about .45 of the area under the normal curve is located between the mean
and a point 1.64 standard errors from the mean. Therefore, 90 percent of
the area is located between plus *and* minus 1.64 standard errors from the
mean, and our confidence limits are:

$$\bar{x} + 1.64\sigma_{\bar{x}} = \$4,800 + 1.64(\$129.57)$$
$$= \$4,800 + \$212.50$$
$$= \$5,012.50 \leftarrow \text{upper confidence limit}$$

$$\bar{x} - 1.64\sigma_{\bar{x}} = \$4,800 - 1.64(\$129.57)$$
$$= \$4,800 - \$212.50$$
$$= \$4,587.50 \leftarrow \text{lower confidence limit}$$

Our report to the social service agency would be: with 90 percent confidence, we estimate that the average annual income of all 700 families living in this four square block section falls between $4,587.50 and $5,012.50.

EXERCISES

24. For a population with a standard deviation of 12, a sample of 64 yields a mean of 120.

   a) Compute the standard error of the mean.

   b) Construct a 90 percent confidence interval for the true population mean.

25. Upon collecting a sample of size 100 from a population with known standard deviation of 4.96, the mean is found to be 68.4.

   a) Find a 95 percent confidence interval for the mean.

   b) Find a 99 percent confidence interval for the mean.

26. From a population of size 1,800, a sample of 145 elements is taken. The mean of the sample was 25, and the standard deviation was 7.2.

   a) Calculate the estimated standard error of the mean.

   b) Construct a 90 percent confidence interval for the mean.

27. From a population of size 240, a sample of 49 individuals is taken. From this sample, the mean is found to be 15.8 and the standard deviation 4.2.

   a) Find the estimated standard error of the mean.

   b) Construct a 98 percent confidence interval for the mean.

28. In an automotive safety test conducted by the North Carolina Highway Safety Research Center, the average tire pressure in a sample of 81 tires was found to be 26 pounds per square inch, and the standard deviation was 1.8 pounds per square inch.

   a) Calculate the estimated population standard deviation for this population. (There are about a million cars registered in North Carolina.)

   b) Calculate the estimated standard error of the mean.

   c) Construct a 90 percent confidence interval for the population mean.

29. From a random sample of 36 New York City civil service personnel, the average age and the sample standard deviation were found to be 38 years and 4.5 years, respectively. Construct a 95 percent confidence interval for the mean age of civil servants in New York City.

30. James Seymore, a specialist on crowding behavior, has been using butterflies in some of his experiments, but many butterflies die before he can collect the data he wants. His friend Bill Johnson, a biologist, agreed to try to develop a special breed of butterflies that would live longer. From the 1,500 butterflies of this new breed, Seymore sampled 36 to test for longevity. The butterflies in his sample had a mean life expectancy of 1800 hours and a sample standard deviation of 150 hours.

   a) Estimate the population standard deviation from the sample standard deviation.

   b) Estimate the standard error of the mean for this finite population.

   c) Construct a 98 percent confidence interval for the mean life expectancy of the butterflies.

6
CALCULATING INTERVAL ESTIMATES OF THE PROPORTION—LARGE SAMPLES

277

Section 6
INTERVAL
ESTIMATES OF
THE PROPORTION

Statisticians often use a sample to estimate a *proportion* of occurrences in a population. For example, the government estimates by a sampling procedure the unemployment rate, or the proportion of unemployed persons, in the United States work force.

In Chapter 6, we introduced the binomial distribution, a distribution of discrete, not continuous, data. Also, we presented the two formulas for deriving the mean and the standard deviation of the binomial distribution:

Review of the binomial distribution

$$\mu = np \qquad\qquad [6\cdot2]$$

$$\sigma = \sqrt{npq} \qquad\qquad [6\cdot3]$$

where:

n = number of trials
p = probability of a success
q = probability of a failure found by taking $1 - p$

Theoretically, the binomial distribution is the correct distribution to use in constructing confidence intervals to estimate a population proportion.

Because the computation of binomial probabilities is so tedious (recall that the probability of r successes in n trials is $[n!/r!(n - r)!][p^r q^{n-r}]$, using the binomial distribution to estimate a population proportion is a complex proposition. Fortunately, as the sample size increases, the binomial can be approximated by an appropriate normal distribution, which we can use to approximate the sampling distribution. Statisticians recommend that in estimation, n be more than 30 and np and nq *each* be at least 5 when you use the normal distribution as a substitute for the binomial.

Shortcomings of the binomial distribution

Symbolically, let's express the proportion of successes in a sample by $\overline{p}$ (pronounced p-bar). Then let's modify Equation 6·2 so that we can use it to derive the *mean of the sampling distribution of the proportion of successes*. In words, $\mu = np$ shows that the mean of the binomial distribution is equal to the product of the number of trials, n, and the probability of success p; that is, np equals the mean number of successes. To change this *number* of successes to the *proportion* of successes, we divide np by n and get p alone. The mean in the left-hand side of the equation becomes $\mu_{\overline{p}}$, or the mean of the sampling distribution of the proportion of successes:

Finding the mean and variance of the sample

$$\mu_{\overline{p}} = p \qquad\qquad [8\cdot3]$$

Similarly, we can modify the formula for the standard deviation of the binomial distribution $\sqrt{npq}$, which measures the standard deviation in the number of successes. To change number of successes to proportion of successes, we divide $\sqrt{npq}$ by n and get $\sqrt{pq/n}$. In statistical terms, the standard deviation for the proportion of successes in a sample is symbolized:

$$\sigma_{\bar{p}} = \sqrt{\frac{pq}{n}} \qquad [8\cdot4]$$

and is called the *standard error of the proportion.*

When the population proportion is unknown

Finding
the standard error
of the proportion

We can illustrate how to use these formulas if we estimate for a very large organization what proportion of the employees prefer to provide their own retirement benefits in lieu of a company sponsored plan. First, we conduct a simple random sample of 75 employees and find that .4 of them are interested in providing their own retirement plan. Our results are:

$n = 75 \leftarrow$ sample size
$\bar{p} = .4 \leftarrow$ sample proportion in favor
$\bar{q} = .6 \leftarrow$ sample proportion not in favor

Next, management requests that we use this sample to find an interval about which they can be 99 percent confident that it contains the true population proportion.

Estimating
a population
proportion

But what are p and q for the population? We can estimate the population parameters by substituting the corresponding sample statistics $\bar{p}$ and $\bar{q}$ (*p-bar* and *q-bar*) in the formula for the standard error of the proportion.* Doing this, we get:

Symbol indicates that the standard error of the proportion is estimated

Sample statistics

$$\hat{\sigma}_{\bar{p}} = \sqrt{\frac{\bar{p}\,\bar{q}}{n}} \qquad [8\cdot5]$$

$$= \sqrt{\frac{(.4)(.6)}{75}}$$

$$= \sqrt{\frac{.24}{75}}$$

$$= \sqrt{.0032}$$

$$= .057 \leftarrow \text{ estimated standard error of the proportion}$$

*Notice that we do not use the finite population multiplier, because our population size is so large compared with the sample size.

Now we can provide the estimate management needs by using the same procedure we have used previously. A 99 percent confidence level would include 49.5 percent of the area on either side of the mean in the sampling distribution. The body of Appendix Table 1 tells us that .495 of the area under the normal curve is located between the mean and a point 2.58 standard errors from the mean. Thus, 99 percent of the area is contained between plus *and* minus 2.58 standard errors from the mean. Our confidence limits then become:

Computing the confidence limits

$$\overline{p} + 2.58\sigma_{\overline{p}} = .4 + 2.58(.057)$$
$$= .4 + .147$$
$$= .547 \leftarrow \text{upper confidence limit}$$
$$\overline{p} - 2.58\sigma_{\overline{p}} = .4 - 2.58(.057)$$
$$= .4 - .147$$
$$= .253 \leftarrow \text{lower confidence limit}$$

Thus, we estimate from our sample of 75 employees that with 99 percent confidence we believe that the proportion of the total population of employees who wish to establish their own retirement plans lies between .253 and .547.

EXERCISES

31. In a gubernatorial campaign, a preelection survey of 100 voters found the incumbent, Mr. Ervin, to be favored by 60 percent of the electorate. His one opponent carried the remaining 40 percent of the sample.

a) Estimate the standard error of the proportion of voters in favor of Mr. Ervin.

b) Construct a 98 percent confidence interval for the proportion of the electorate that favors Mr. Ervin.

32. When 64 patients of a group of clinical psychologists were surveyed, it was found that 72 percent of them had been referred by other psychologists.

a) Estimate the standard error of the proportion of patients referred to the group by other psychologists.

b) Find the upper and lower confidence limits for this proportion, given a confidence level equal to .90.

33. A statewide sample of 1,200 shows that 780 people support candidate A in an upcoming election.

a) Estimate the standard error of the proportion of people supporting candidate A.

b) Construct a 95 percent confidence interval for the true population proportion.

34. In a door-to-door survey, a researcher finds that 36 percent of the 500 people questioned favor increased sex education in the schools.

a) Estimate the standard error of the proportion.

b) Construct a 90 percent confidence interval for the true population proportion.

35. In the examination of 400 patients at a screening clinic, 64 were found to exhibit symptoms of anemia.

   a) Estimate the standard error of the proportion.

   b) Construct a 99 percent confidence interval for the true proportion.

36. In a study of 120 items from a population of size 2,500, 72 percent are found to meet certain standards for approval.

   a) Find a 95 percent confidence interval for the proportion meeting standards.

   b) Based on part a, what kind of interval estimate might you give for the absolute number of items in the population that meet the standards, keeping the same 95 percent confidence level?

37. A psychology department has a pool of 1,200 rats. A graduate student, Joe Dunn, has determined that 45 percent in a sample of 64 rats have become "maze educated"; that is, they have run mazes often enough that data taken from them is contaminated. Construct a 95 percent confidence interval for this proportion.

38. By randomly surveying 49 of the 500 families in a certain neighborhood, Ron Keeton, a sociologist, found that 80 percent of those families were happy with the neighborhood in which they lived. Give the upper and lower limits, with a confidence level of .96, for the proportion of residents who are happy with this neighborhood.

7 INTERVAL ESTIMATES USING THE *t* DISTRIBUTION

In our three examples so far, the sample sizes were all larger than 30. We sampled 100 windshield wiper blades, 50 families living in a four square block section of a community, and 75 employees of a very large organization. Each time, the normal distribution was the appropriate sampling distribution to use to determine confidence intervals.

However, this is not always the case. How can we handle estimates where the normal distribution is *not* the appropriate sampling distribution; that is, when we are estimating the population standard deviation and the sample size is 30 or less? For example, in our chapter-opening problem of salt usage in Detroit, we had data from only 10 days. Fortunately, another distribution exists that is appropriate in these cases. It is called the *t distribution*.

Background of the *t* distribution

Early theoretical work on *t* distributions was done by a man named W.S. Gossett in the early 1900s. Gossett was employed by the Guinness Brewery in Dublin, Ireland, which did not permit employees to publish research findings under their own names. So Gossett adopted the pen name "Student" and published under that name. Consequently, the *t* distribution is commonly called *Student's t distribution,* or simply *Student's distribution.*

Conditions for using the *t* distribution

Since it is used when the sample size is 30 or less, statisticians often associate the *t* distribution with small sample statistics. This is misleading because the size of the sample is only *one* of the conditions that leads us to use the *t* distribution. The second condition is that the population standard deviation must be unknown. Use of the t distribution for estimating is required whenever the sample is 30 or less and the population standard deviation is not known. Furthermore, in using the t distribution, we assume that the population is normal or approximately normal.

Characteristics of the t distribution

Without deriving the *t* distribution mathematically, we can gain an intuitive understanding of the relationship between the *t* distribution and the *normal* distribution. Both are symmetrical. In general, the *t* distribution is flatter than the normal distribution, and there is a different *t* distribution for every possible sample size. Even so, as the sample size gets larger, the shape of the *t* distribution loses its flatness and becomes approximately equal to the normal distribution. In fact, for sample sizes of more than 30, the *t* distribution is so close to the normal distribution that we will use the normal to approximate the *t*.

t distribution compared to normal distribution

Figure 8·3 compares one normal distribution with two *t* distributions of different sample sizes. This figure shows two characteristics of *t* distributions: **a *t* distribution is lower at the mean and higher at the tails than a normal distribution.** The figure also demonstrates how the *t* distribution has proportionally more area in its tails than the normal does. This is the reason why it will be necessary to go farther out from the mean of a *t* distribution to include the same area under the curve. Interval widths from *t* distributions are, therefore, wider than those based on the normal distribution.

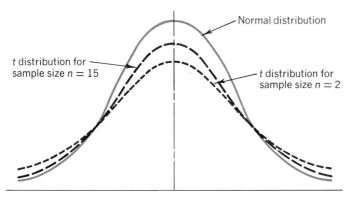

t distribution for sample size n = 15

Normal distribution

t distribution for sample size n = 2

**FIG. 8·3
Normal distribution, *t* distribution for sample size *n* = 15, and *t* distribution for sample size *n* = 2**

Degrees of freedom

We said earlier that there is a separate *t* distribution for each sample size. In proper statistical language, we would say, "There is a different *t* distribution for each of the possible *degrees of freedom*." **What are degrees of freedom? We can define them as the number of values we can choose freely.**

Degrees of freedom defined

Assume that we are dealing with two sample values, *a* and *b*, and we know that they have a mean of 18. Symbolically, the situation is:

$$\frac{a + b}{2} = 18$$

How can we find what values *a* and *b* can take on in this situation? The answer is that *a* and *b* can be any two values whose sum is 36, because 36 ÷ 2 = 18.

Suppose we learn that *a* has a value of 10. Now *b* is no longer free to take on any value but must become the value of 26, because:

$$
\begin{aligned}
\text{if} \quad a &= 10 \\
\text{then} \quad \frac{10 + b}{2} &= 18 \\
\text{so} \quad 10 + b &= 36 \\
\text{and} \quad b &= 26
\end{aligned}
$$

This example shows that when there are two elements in a sample and we know the sample mean of these two elements, we are free to specify only one of the elements because the other element will be determined by the fact that the two elements sum to twice the sample mean.

Let's look at another example, there are 7 elements in our sample, and we learn that the mean of these elements is 16. Symbolically, we have this situation:

$$
\frac{a + b + c + d + e + f + g}{7} = 16
$$

In this case, the degrees of freedom, or the number of variables we can specify freely, are 7 − 1 = 6. We are free to give numbers to 6 values, and then we are no longer free to specify the seventh value. It is determined automatically.

With 2 sample values we had 1 degree of freedom (2 − 1 = 1), and with 7 sample values we had 6 degrees of freedom. In each of these two examples, then, we had *n* − 1 degrees of freedom, assuming *n* is the sample size. Similarly, a sample of 23 would give us 22 degrees of freedom.

Function of
degrees of freedom

We will use degrees of freedom when we select a *t* distribution to estimate a population mean, and we will use *n* − 1 degrees of freedom, letting *n* equal the sample size. If, for example, we use a sample of 20 to estimate a population mean, we will use 19 degrees of freedom in order to select the appropriate *t* distribution.

Using the t distribution table

t table
compared to *z* table:
3 differences

The table of *t* distribution values (Appendix Table 2) differs in construction from the *z* table we have used previously. The *t* table is more compact and shows areas and *t* values for only a few percentages (10, 5, 2, and 1 percent). Since there is a different *t* distribution for each number of degrees of freedom, a more complete table would be quite lengthy. Although we can conceive of the need for a more complete table, in fact Appendix Table 2 contains all the commonly used values of the *t* distribution.

A second difference in the *t* table is that it does *not* focus on the chance that the population parameter being estimated will fall within our confidence interval. Instead, it measures the chance that the population parameter we are estimating will *not* be within our confidence interval (that is, that it will lie *outside* it). If we are making an estimate at the 90 percent confidence level, we would look in the *t* table under the .10 column (100% − 90% = 10%). This .10 chance of error is symbolized by α, which is the Greek letter *alpha*. We would find the appropriate *t* values for confidence intervals of 95%, 98%, and 99% under the α columns headed .05, .02, and .01, respectively.

A third difference in using the *t* table is that we must specify the degrees of freedom with which we are dealing. Suppose we make an estimate at the 90 percent confidence level with a sample size of 14, which is 13 degrees of freedom. Look in Appendix Table 2 under the .10 column until you encounter the row labeled 13 *df* (degrees of freedom). Like a *z* value, the *t* value there of 1.771 shows that if we mark off plus and minus 1.771 $\hat{\sigma}_{\bar{x}}$'s (estimated standard errors of $\bar{x}$) on either side of the mean, the area under the curve between these two limits will be 90 percent, and the area outside these limits (the chance of error) will be 10 percent (see Fig. 8·4).

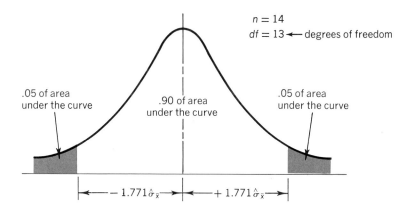

.05 of area under the curve

.90 of area under the curve

.05 of area under the curve

$n = 14$
$df = 13$ ← degrees of freedom

$-1.771\hat{\sigma}_{\bar{x}}$ $+1.771\hat{\sigma}_{\bar{x}}$

FIG. 8·4

A *t* distribution for 13 degrees of freedom, showing confidence interval of 90 percent

Recall that in our chapter-opening problem, Detroit wanted to estimate its average daily consumption of road salt, and it took a sample by measuring salt usage on 10 working days. The sample data are summarized below:

$n = 10$ days ← sample size
$df = 9$ ← degrees of freedom
$\bar{x} = 11,400$ pounds ← sample mean
$s = 700$ pounds ← sample standard deviation

The public works department wants an interval estimate of the mean daily consumption, and they want to be 95 percent confident that the mean consumption falls within that interval. This problem requires the use of a t distribution because the sample size is less than 30 and the population standard deviation is unknown.

Using the *t* table to compute confidence limits

As a first step in solving this problem, recall that we *estimate* the population standard deviation with the sample standard deviation; thus:

$$\hat{\sigma} = s$$
$$= 700$$

[8·1]

Using this estimate of the population standard deviation, we can estimate the standard error of the mean by modifying Equation 8·2 to omit the finite population multiplier (because the population of winter days is infinite):

$$\hat{\sigma}_{\bar{x}} = \frac{\hat{\sigma}}{\sqrt{n}}$$

[8·6]

$$= \frac{700}{\sqrt{10}}$$

$$= \frac{700}{3.162}$$

$$= 221.38 \text{ pounds} \leftarrow \text{estimated standard error of the mean of an infinite population}$$

Now we look in Appendix Table 2 down the .05 column (100% − 95% = 5%) until we encounter the row of 9 degrees of freedom (10 − 1 = 9). There we see the *t* value 2.262 and can set our confidence limits accordingly:

$$\bar{x} + 2.262\hat{\sigma}_{\bar{x}} = 11{,}400 \text{ pounds} + 2.262(221.38 \text{ pounds})$$
$$= 11{,}400 + 500.76$$
$$= 11{,}901 \text{ pounds} \leftarrow \text{upper confidence limit}$$
$$\bar{x} - 2.262\hat{\sigma}_{\bar{x}} = 11{,}400 \text{ pounds} - 2.262(221.38 \text{ pounds})$$
$$= 11{,}400 - 500.76$$
$$= 10{,}899 \text{ pounds} \leftarrow \text{lower confidence limit}$$

FIG. 8·5
Salt problem: a *t* distribution with 9 degrees of freedom and a confidence interval of 95 percent

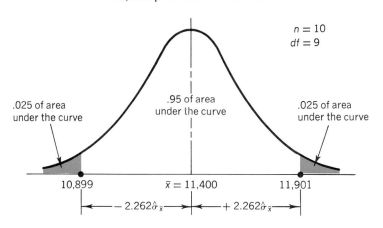

$n = 10$
$df = 9$

.025 of area under the curve

.95 of area under the curve

.025 of area under the curve

10,899

$\bar{x} = 11{,}400$

11,901

$-2.262\hat{\sigma}_{\bar{x}}$

$+2.262\hat{\sigma}_{\bar{x}}$

We can report to the public works director with 95 percent confidence that the mean daily usage of salt lies between 10,899 and 11,901 pounds, and we can use the 11,901 pound figure to estimate how much salt to order.

The only difference between the process we used to make this salt usage estimate and the previous estimating problems is the use of the *t* distribution as the appropriate distribution. Remember that in any estimation problem in which the sample size is 30 or less *and* the standard deviation of the population is unknown, we use the *t* distribution.

Summary of confidence limits under various conditions

Table 8·5 summarizes the various approaches to estimation introduced in this chapter and the confidence limits appropriate for each.

TABLE 8·5 **Summary of formulas for confidence limits estimating mean and proportion**

	When the population is finite	When the population is infinite
Estimating μ (the population mean) **When σ (the population standard deviation) is known**	upper limit: $\bar{x} + z\,\dfrac{\sigma}{\sqrt{n}} \times \sqrt{\dfrac{N-n}{N-1}}$ lower limit: $\bar{x} - z\,\dfrac{\sigma}{\sqrt{n}} \times \sqrt{\dfrac{N-n}{N-1}}$	$\bar{x} + z\,\dfrac{\sigma}{\sqrt{n}}$ $\bar{x} - z\,\dfrac{\sigma}{\sqrt{n}}$
When σ (the population standard deviation) is not known [$\hat{\sigma} = s$] **When n (the sample size) is larger than 30**	upper limit: $\bar{x} + z\,\dfrac{\hat{\sigma}}{\sqrt{n}} \times \sqrt{\dfrac{N-n}{N-1}}$ lower limit: $\bar{x} - z\,\dfrac{\hat{\sigma}}{\sqrt{n}} \times \sqrt{\dfrac{N-n}{N-1}}$	$\bar{x} + z\,\dfrac{\hat{\sigma}}{\sqrt{n}}$ $\bar{x} - z\,\dfrac{\hat{\sigma}}{\sqrt{n}}$
When n (the sample size) is 30 or less	upper limit: $\bar{x} + t\,\dfrac{\hat{\sigma}}{\sqrt{n}} \times \sqrt{\dfrac{N-n}{N-1}}$ lower limit: $\bar{x} - t\,\dfrac{\hat{\sigma}}{\sqrt{n}} \times \sqrt{\dfrac{N-n}{N-1}}$	$\bar{x} + t\,\dfrac{\hat{\sigma}}{\sqrt{n}}$* $\bar{x} - t\,\dfrac{\hat{\sigma}}{\sqrt{n}}$
Estimating p (the population proportion) **When n (the sample size) is larger than 30**	upper limit: $\bar{p} + z\hat{\sigma}_{\bar{p}} \times \sqrt{\dfrac{N-n}{N-1}}$ lower limit: $\bar{p} - z\hat{\sigma}_{\bar{p}} \times \sqrt{\dfrac{N-n}{N-1}}$	$\bar{p} + z\hat{\sigma}_{\bar{p}}$ $\bar{p} - z\hat{\sigma}_{\bar{p}}$

$$\left[\hat{\sigma}_{\bar{p}} = \sqrt{\dfrac{\bar{p}\,\bar{q}}{n}}\right]$$

*Remember that the appropriate *t* distribution to use is the one with $n - 1$ degrees of freedom.

EXERCISES

39. For the following sample sizes and confidence levels, find the appropriate *t* values for constructing confidence intervals.
    a) $n = 5$, 99%
    b) $n = 18$, 99%
    c) $n = 27$, 95%
    d) $n = 16$, 95%
    e) $n = 18$, 95%
    f) $n = 14$, 90%

40. Given the following sample sizes and *t* values used to construct confidence intervals, find the corresponding confidence levels:
    a) $n = 20$; $t = \pm 1.729$
    b) $n = 12$; $t = \pm 2.201$
    c) $n = 7$; $t = \pm 3.707$

41. In a sample of 24 observations, the mean was 27.3 and the standard deviation 1.9. Construct a 98 percent confidence interval for the mean of the population.

42. A sample of size 15 had a mean of 56 and a standard deviation of 12. Construct a 95 percent confidence interval for the mean.

43. The following sample of 8 observations is from an infinite population:

$$12.1 \quad 11.9 \quad 12.4 \quad 12.3 \quad 11.9 \quad 12.1 \quad 12.4 \quad 12.1$$

a) Find the mean.

b) Estimate the population standard deviation.

c) Construct a 90 percent confidence for the mean.

44. Six housewives were randomly sampled, and it was determined that they walked an average of 34.6 miles per week in their housework, with a sample standard deviation of 2.8 miles per week. Construct a 95 percent confidence interval around this estimate.

8 DETERMINING THE SAMPLE SIZE

IN ESTIMATION In all our discussions so far, we have used for sample size the symbol n instead of a specific number. Now we need to know how to determine what number to use. How large should the sample be? If it is too small, we may fail to achieve the objectives of our analysis. But if it is too large, we waste resources when we gather the sample.

What sample size is adequate?

Some sampling error will arise because we have not studied the whole population. Whenever we sample, we always miss *some* helpful information about the population. If we want a high level of precision (that is, if we want to be quite sure of our estimate), we have to sample enough of the population to provide the required information. Sampling error is controlled by selecting a sample that is adequate in size. In general, the more precision you want, the larger the sample you will need to take. Let us examine some methods that are useful in determining what sample size is necessary for any specified level of precision.

Sample size for estimating a mean

Suppose a university is performing a survey of the annual earnings of last year's graduates from its law school. It knows from past experience that the standard deviation of the annual earnings of the entire population (1,000) of these graduates is about $1,500. How large a sample size should the university take in order to estimate the mean annual earnings of last year's class within plus and minus $500 and at a 95 percent confidence level?

Two ways to express a confidence limit

Exactly what is this problem asking? The university is going to take a sample of some size, determine the mean of the sample $\overline{x}$, and use it as a point estimate of the population mean. It wants to be 95 percent certain that the true mean annual earnings of last year's class are not more than $500 above or below the point estimate. Row *a* in Table 8·6 summarizes in symbolic terms how the university is defining its confidence limits for us. Row *b* shows symbolically how we normally express confidence limits for an infinite population. When we compare these two sets of confidence limits, we can see that:

$$z\sigma_{\bar{x}} = \$500$$

Lower confidence limit	Upper confidence limit
a. $\bar{x} - \$500$	a. $\bar{x} + \$500$
b. $\bar{x} - z\sigma_{\bar{x}}$	b. $\bar{x} + z\sigma_{\bar{x}}$

TABLE 8·6
Comparison of two ways of expressing the same confidence limits

Thus, the university is actually saying that it wants $z\sigma_{\bar{x}}$ to be equal to $500. If we look in Appendix Table 1, we find that the necessary z value for a 95 percent confidence level is 1.96. Step by step:

$$\text{If } z\sigma_{\bar{x}} = \$500$$
$$\text{and } z = 1.96$$
$$\text{Then } 1.96\,\sigma_{\bar{x}} = \$500$$
$$\text{and } \sigma_{\bar{x}} = \frac{\$500}{1.96}$$
$$= \$255 \leftarrow \text{standard error of the mean}$$

Remember that the formula for the standard error is Equation 7·1:

Population standard deviation

$$\sigma_{\bar{x}} = \frac{\sigma}{\sqrt{n}}$$

[7·1]

Using Equation 7·1, we can substitute our known population standard deviation value of $1,500 and our calculated standard error value of $255 and solve for *n*:

Finding an adequate sample size

$$\sigma_{\bar{x}} = \frac{\sigma}{\sqrt{n}}$$

[7·1]

$$\$255 = \frac{\$1,500}{\sqrt{n}}$$

$$(\sqrt{n})(\$255) = \$1,500$$

$$\sqrt{n} = \frac{\$1,500}{\$255}$$

$$\sqrt{n} = 5.882 \leftarrow \text{then square both sides}$$

$$n = 34.6 \leftarrow \text{sample size for precision specified}$$

Therefore, since *n* must be greater than, or equal to, 34.6, the university should take a sample of 35 law school graduates to get the precision it wants in estimating the class's mean annual earnings.

FIG. 8·6
Approximate
relationship
between the range
and population
standard deviation

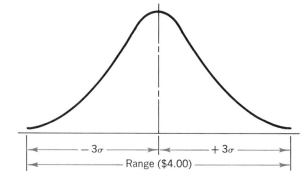

Range ($4.00)

Estimating the
standard deviation
from the range

In this example, we knew the standard deviation of the population, but in many cases the standard deviation of the population is not available. Remember, too, that we have not yet taken the sample, and we are trying to decide how large to make it. We cannot estimate the population standard deviation using methods from the first part of this chapter. If we have a notion about the range of the population, we can use that to get a crude, but workable, estimate.

Suppose we are estimating hourly wage rates in a city and are fairly confident that there is a $4.00 difference between the highest and lowest wage rates. We know that plus and minus 3 standard deviations include 99.7 percent of all the area under the normal curve; that is, plus 3 standard deviations and minus 3 standard deviations include almost all of the distribution. To symbolize this relationship, we have constructed Fig. 8·6, in which $4.00 (the range) equals 6 standard deviations (plus 3 and minus 3). Thus, a rough estimate of the population standard deviation would be:

$$6\sigma = \$4.00$$
$$\sigma = \frac{\$4.00}{6}$$
$$\sigma = \$0.667$$

Our estimate of the population standard deviation using this rough method is not precise, but it may mean the difference between getting a working idea of the required sample size and knowing nothing about that sample size.

Sample size for estimating a proportion

The procedure for determining sample sizes for estimating a population proportion are similar to those for estimating a population mean. Suppose we wish to poll voters in a certain city prior to election day. We want to determine what proportion of them intend to vote in favor of a school bond referendum. We would like a sample size that will enable us to be 90 percent certain of estimating the true proportion that will vote in favor of the referendum within plus or minus .02.

We begin to solve this problem by looking in Appendix Table 1 to find the z value for a 90 percent confidence level. That value is plus and minus

1.64 standard errors from the mean. Since we want our estimate to be within .02, we can symbolize the step-by-step process like this:

$$\text{If } z\sigma_{\bar{p}} = .02$$
$$\text{and } z = 1.64$$
$$\text{Then } 1.64\,\sigma_{\bar{p}} = .02$$

If we now substitute the right side of Equation 8·4 for σ_p, we get:

$$1.64\left(\sqrt{\frac{pq}{n}}\right) = .02$$

$$\sqrt{\frac{pq}{n}} = .0122 \leftarrow \text{now square both sides}$$

$$\frac{pq}{n} = .0001488 \leftarrow \text{multiply both sides by } n$$

$$pq = .0001488n$$

$$\frac{pq}{.0001488} = n$$

To find n, we still need an estimate of the population parameters p and q. If we have strong feelings about the actual proportion in favor of the referendum, we can use that as our best guess to calculate n. But if we have no idea what p is, then our best strategy is to guess at p in such a way that we choose n in a conservative manner (that is, so that the sample size *is* large enough to supply at least the precision we require no matter what p actually is). At this point in our problem, n is equal to the product of p and q divided by .0001488. The way to get the largest n is to generate the largest possible numerator of that expression, or to pick $p = .5$ and $q = .5$. Then n becomes:

$$n = \frac{pq}{.0001488}$$
$$= \frac{(.5)(.5)}{.0001488}$$
$$= \frac{.25}{.0001488}$$
$$= 1680 \leftarrow \text{sample size for precision specified}$$

As a result, to be 90 percent certain of estimating the true proportion within .02, we should pick a simple random sample of 1,680 people to interview.

In the problem we have just solved, we picked a value for p that represented the most conservative strategy. The value .5 generated the largest possible sample. We would have used another value of p *if* we had known its actual value in the population *or* if we had been able to estimate one *or* if we had a strong feeling about one. Whenever all of these solutions are absent, assume the most conservative possible value for p, or .5.

To illustrate that .5 yields the largest possible sample, Table 8·7 solves the referendum problem using several different values of p. You can see from the sample sizes associated with these different values that for the range of p's from .3 to .7, the change in the appropriate sample size is relatively small. Therefore, even if you knew that the true population proportion was .3 and you used a value of .5 for p anyway, you would have sampled only 269 more people ($1,680 - 1,411$) than was actually necessary for the desired degree of precision. Obviously, guessing values of p in cases like this is not as critical as it seems at first glance.

TABLE 8·7
Sample size n associated with different values of p and q

Choose this value for p	Value of q or $1 - p$	Indicated sample size n	$\left(\dfrac{pq}{.0001488}\right)$	
.2	.8	$\dfrac{(.2)(.8)}{.0001488}$	=	1,075
.3	.7	$\dfrac{(.3)(.7)}{.0001488}$	=	1,411
.4	.6	$\dfrac{(.4)(.6)}{.0001488}$	=	1,612
.5	.5	$\dfrac{(.5)(.5)}{.0001488}$	=	1,680 ← most conservative
.6	.4	$\dfrac{(.6)(.4)}{.0001488}$	=	1,612
.7	.3	$\dfrac{(.7)(.3)}{.0001488}$	=	1,411
.8	.2	$\dfrac{(.8)(.2)}{.0001488}$	=	1,075

EXERCISES

45. If the population standard deviation is 200, find the sample size necessary to estimate the true mean within plus-or-minus 100 points for a confidence level of 90 percent.

46. In an election, find the sample size needed to estimate the true proportion of voters favoring a certain candidate within plus-or-minus .03 at the 95 percent confidence level. Assume you have no strong feeling about what the proportion is.

47. Given a population with a standard deviation of .8, what size sample is needed to estimate the mean of the population within plus-or-minus .25 with 98 percent confidence?

48. We have strong indications that the proportion is around .75. Find the sample size needed to estimate the proportion within $\pm.04$ with confidence level 90 percent.

49. Anthropologists studying inhabitants of a South Sea island have noticed that the natives have remarkable consistency in their life span. They believe that the upper and lower limits of life span differ by no more than 600 weeks (about 11½ years) for those who die by natural causes. For a confidence level of 90 percent, how large a sample should they look at to find the average life span of these natives within ±30 weeks?

50. Helen Johnson, an experimental psychologist, needs to know how long rats can live with only water for nourishment. She wants to use a very hardy breed for her study. From previous studies, she knows the population standard deviation is 9 days. How large should the

sample be for 95 percent confidence that the sample average is within ±2 days of the true average?

51. A speed reading course guarantees a certain reading rate increase within 2 days. The teacher knows a few people will not be able to achieve this increase; so before stating the guaranteed reading rate increase, he wants to be 95 percent confident that the percentage has been estimated to within ±3 percent of the true value. What is the most conservative sample size needed for this problem?

9 TERMS INTRODUCED IN CHAPTER 8

- **CONFIDENCE INTERVAL** A range of values that has some designated probability of including the true population parameter value.
- **CONFIDENCE LEVEL** The probability that we associate with an interval estimate of a population parameter indicating how confident we are that the interval estimate will include the population parameter.
- **CONFIDENCE LIMITS** The upper and lower boundaries of a confidence interval.
- **CONSISTENT ESTIMATOR** An estimator that yields values more closely approaching the population parameter as the sample size increases.
- **DEGREES OF FREEDOM** The number of values in a sample we can specify freely, once we know something about that sample.
- **EFFICIENT ESTIMATOR** An estimator with a smaller standard error than some other estimator of the population parameter; i.e., the smaller the standard error of an estimator, the more efficient that estimator is.
- **ESTIMATE** A specific observed value of an estimator.

- **ESTIMATOR** A sample statistic used to estimate a population parameter.
- **INTERVAL ESTIMATE** A range of values used to estimate an unknown population parameter.
- **POINT ESTIMATE** A single number that is used to estimate an unknown population parameter.
- **STUDENT'S t-DISTRIBUTION** A family of probability distributions distinguished by their individual degrees of freedom, similar in form to the normal distribution, and used when the population standard deviation is unknown and the sample size is relatively small ($n < 30$).
- **SUFFICIENT ESTIMATOR** An estimator that contains all the information available in the data concerning a parameter.
- **UNBIASED ESTIMATOR** An estimator of a population parameter that, on the average, assumes values above the population parameter as often, and to the same extent, as it tends to assume values below the population parameter.

10 *EQUATIONS INTRODUCED IN CHAPTER 8*

p. 274 Estimator of the population standard deviation $$\hat{\sigma} = s = \sqrt{\frac{\Sigma(x - \bar{x})^2}{n - 1}}$$ [8·1]

This formula indicates that the sample standard deviation can be used as an estimator of the population standard deviation.

p. 275

$$\hat{\sigma}_{\bar{x}} = \frac{\hat{\sigma}}{\sqrt{n}} \times \sqrt{\frac{N - n}{N - 1}}$$

[8·2]

This formula enables us to derive an *estimated* standard error of the mean of a *finite* population from an *estimate* of the population standard deviation. The symbol ˆ called a hat, indicates that the value is estimated. Equation 8·6 is the same formula for an infinite population.

p. 277

$$\mu_{\bar{p}} = p$$

[8·3]

Use this formula to derive the *mean* of the sampling distribution *of the proportion* of successes. The right-hand side, p, is equal to $(n \times p)/n$, where the numerator is the product of the number of trials and the probability of successes and the denominator is the number of trials. Symbolically, the proportion of successes *in a sample* is written $\bar{p}$ and is pronounced *p-bar*.

p. 278

$$\sigma_{\bar{p}} = \sqrt{\frac{pq}{n}}$$

[8·4]

To get the *standard error of the proportion,* take the square root of the product of the probabilities of success and failure divided by the number of trials.

p. 278

$$\hat{\sigma}_{\bar{p}} = \sqrt{\frac{\bar{p}\,\bar{q}}{n}}$$

[8·5]

This is the formula to use to derive an *estimated* standard error of the proportion when the population proportion is unknown and you are forced to use $\bar{p}$ and $\bar{q}$, the sample proportions of successes and failures.

p. 284

$$\hat{\sigma}_{\bar{x}} = \frac{\hat{\sigma}}{\sqrt{n}}$$

[8·6]

This formula enables us to derive an *estimated* standard error of the mean of an *infinite* population from an *estimate* of the population standard deviation. It is exactly like Equation 8·2 except that it lacks the multiplier.

11 CHAPTER REVIEW EXERCISES

52. The mean grade point average for a sample of 49 students at a university law school is 2.45. Previous studies have determined that the population standard deviation is .7. If we wanted to construct intervals around sample means that would include the true mean grade point average 99.7 percent of the time, what interval would we construct for this particular sample?

53. What are the advantages of using an interval estimate over a point estimate?

54. Why is the size of a statistic's standard error important in its use as an estimator? To which characteristic of estimators does this relate?

55. Liz Griffin is a psychologist studying school children. She wants to determine at a confidence level of 95 percent what proportion (within plus or minus .04) of third graders have

IQ's below 85. Conservatively, how many third graders should she test to find this proportion?

56. A 95 percent confidence interval for the population mean is given by (84,116) while a 75 percent confidence interval is given by (90.96, 109.04). What are the advantages and disadvantages of each of these interval estimates?

57. In a random selection of 81 of the 2,200 intersections in a small midwestern city, the mean number of automobile accidents per year was determined to be 3.2 and the sample standard deviation .9.

    a) Estimate the standard deviation of the population from the sample standard deviation.

    b) Estimate the standard error of the mean for this finite population.

    c) If the desired confidence level is .95, what will be the upper and lower limits of the confidence interval for the mean number of accidents per intersection per year?

58. Given a sample mean of 96, a population standard deviation of 4.8, and a sample of size 36, find the confidence level associated with each of the following intervals:

    a) (94.4, 97.6)   b) (94, 98)   c) (95.328, 96.672)

59. Based on knowledge about the desirable qualities of estimators, for what reasons might $\overline{x}$ be considered the "best" estimator of the true population mean?

60. A school psychologist is studying the relation between the lunch menus and absenteeism, to determine whether there are fewer students absent on days with good meals. She looks at results from 35 random days on which there was cake for dessert. For this sample, the average proportion absent each day is .045, and the associated sample standard deviation is .0120.

    a) Give a point estimate for the proportion of students absent on any given day with cake on the menu.

    b) Estimate the population standard deviation associated with this absentee rate.

61. Given the following expressions for the limits of a confidence interval, find the confidence level associated with the interval:

    a) $\overline{x} - 1.5\sigma_{\overline{x}}$ to $\overline{x} + 1.5\sigma_{\overline{x}}$ b) $\overline{x} - 1.7\sigma_{\overline{x}}$ to $\overline{x} + 1.7\sigma_{\overline{x}}$ c) $\overline{x} - 2.3\sigma_{\overline{x}}$ to $\overline{x} + 2.3\sigma_{\overline{x}}$

62. From previous studies, the population standard deviation for a placement test has been determined to be 12.4; the test is scored on a scale of 0–100. Peggy Wall, a psychologist, wants to be 98 percent certain that the average test score of a sample falls within plus or minus 3 points of the population's average score. How large a sample should she select?

63. A medical research team feels confident that a serum they have developed will cure about 75 percent of the patients suffering from a certain disease. How large should the sample size be for the team to be 98 percent certain that the sample proportion of cures is within plus and minus .04 of the proportion of all cases that the serum will cure?

64. A ski resort manager in Vermont wants to know the resort's average daily registration. The following table presents the number of guests registered each of 30 randomly selected days. Calculate the sample mean.

$$
\begin{array}{cccccccccc}
60 & 58 & 52 & 61 & 63 & 56 & 55 & 57 & 62 & 63 \\
58 & 51 & 57 & 61 & 56 & 59 & 63 & 62 & 61 & 58 \\
62 & 53 & 53 & 55 & 60 & 52 & 54 & 61 & 58 & 59
\end{array}
$$

65. Using the information in problem 64 as a:

    a) sample, find the sample standard deviation

    b) population, find the population standard deviation

66. In evaluating the effectiveness of a federal rehabilitation program, a survey of 49 of a prison's 800 inmates found that 48 percent were repeat offenders.

   a) Estimate the standard error of the proportion of repeat offenders.

   b) Construct a 99 percent confidence interval for the proportion of repeat offenders among the inmates of this prison.

67. In a Utah sample of 64 automotive repair jobs covered by warranty, the average cost was found to be $43. Previous studies in that state had determined a population standard deviation of $24.

   a) Calculate the standard error of the mean.

   b) Establish an interval estimate around the mean using one standard error of the mean.

68. From a random sample of 64 buses, Montreal's mass transit office has calculated the mean number of passengers per kilometer to be 3.5. From previous studies, the population standard deviation is known to be 1.6 passengers per kilometer.

   a) Find the standard error of the mean. (Assume the bus fleet is very large.)

   b) Construct a 95 percent confidence interval for the mean number of passengers per kilometer for the population.

69. A social worker in a city welfare office has calculated that the average monthly welfare payment is $360, and the corresponding standard deviation $30 for a random sample of 40 families. Using this information, estimate the average monthly payment and the associated population standard deviation for welfare recipients in that city last month.

70. Nancy Murray, a specialist on aging, has sampled 100 individuals who lived in a county renowned for the longevity of its inhabitants. The population from which she sampled was composed of all people who had lived in the county at least 20 years, had been deceased for at least 10 years, and who were older than 50 when they died. She found the mean lifetime to be 28,640 days. The population standard deviation was known to be 850 days. For the mean lifetime of this county's residents, construct a confidence interval of (a) 95 percent, (b) 98 percent.

71. Bill Wenslaff, an engineer on the staff of a water purification plant, measures the chlorine content in 100 different samples daily. Over a period of years, he has established the population standard deviation to be 1.2 milligrams of chlorine per liter. The latest samples averaged 4.8 milligrams of chlorine per liter.

   a) Find the standard error of the mean.

   b) Establish the interval around 5.0, the population mean, which will include the sample mean with a probability of 68.7 percent.

72. Jack Laughery, a psychologist, sampled 5 truck drivers from the 200 in a particular union. He measured the time in seconds it took each to react to a warning blinker. He got the following data: 1.8, 2.4, 2.2, 2.6, and 1.6 seconds.

   a) Calculate the mean reaction time and the corresponding standard deviation for the sample.

   b) Estimate the population standard deviation.

   c) Construct a 98 percent confidence interval for the mean reaction time.

73. Larry Culler, the federal grain inspector at a seaport, found the proportion of waste material contained in a random selection of 100 samples of wheat shipments to be 35 percent. Construct a 95 percent confidence interval for him for the actual proportion of waste in shipments from that port.

74. A U.S. Coast Guard survey of 300 small boats in the Cape Cod area found 120 in violation of one or more major safety regulations. Given a confidence level equal to .98, construct a confidence interval for the proportion of unsafe small boats.

75. For a hunger experiment, cats were deprived of food for one week. Then the cats were permitted to drink as much milk as they desired, and the experimenters measured the total number of hours it took them to drink all the milk from a gallon container. For the study, 144 containers of milk were randomly sampled, and the experimenters assumed each container held the same amount. After the experiment, however, they learned from the milk company that the mean filling was 128.4 fluid ounces, with a standard deviation of .6 ounces.

    a) Find the standard error of the mean.

    b) What is the interval around the sample mean that would contain the population mean 95.5 percent of the time?

76. Rishie Baroff is running for the city council. She has calculated that for a random sample of 8 donations, the average donation to her campaign was $10.50, and the standard deviation was $2.50. Construct a 95 percent confidence interval for the size of the average donation.

12 CHAPTER CONCEPTS TEST

Answer true *or* false. *Answers are in the back of the book.*

1. A statistic is said to be an efficient estimator of a population parameter if, with increasing sample size, it becomes almost certain that the value of the statistic comes very close to that of the population parameter.

2. An interval estimate is a range of values used to estimate the shape of a population's distribution.

3. If a statistic tends to assume values higher than the population parameter as frequently as it tends to assume values that are lower, we say that the statistic is an unbiased estimate of the parameter.

4. The probability that a population parameter will lie within a given interval estimate is known as the confidence level.

5. With increasing sample size, the *t* distribution tends to become flatter in shape.

6. We must always use the *t* distribution, rather than the normal, whenever the standard deviation of the population is not known.

7. We may obtain a crude estimate of the standard deviation of some population if we have some information about its range.

8. When using the *t* distribution in estimation, we must assume that the population is approximately normal.

9. Using high confidence levels is not always desirable, because high confidence levels produce large confidence intervals.

10. There is a different *t* distribution for each possible sample size.

9

Testing
hypotheses

1.   Introduction, 298

2.   Concepts basic to the hypothesis
     testing procedure, 299

3.   Testing hypotheses, 301

4.   Hypothesis testing of means—
     samples with population standard
     deviations known, 309

5.   Measuring the power of a
     hypothesis test, 314

6.   Hypothesis testing of
     proportions— large
     samples, 317

7.   Hypothesis testing of
     means under different
     conditions, 322

8.   Hypothesis testing for
     differences between means
     and proportions, 325

9.   Terms introduced, 346

10. Equations introduced, 347

11. Chapter review exercises,
     348

12. Chapter concepts test, 353

The roofing contract for a new sports complex in San Francisco has been awarded to Parkhill Associates, a large architectural firm. Building specifications call for a moveable roof covered by approximately 10,000 sheets of .04 inch thick aluminum. The aluminum sheets cannot be appreciably thicker than .04 inches because the structure could not support the additional weight. Nor can the sheets be appreciably thinner than .04 inches because the strength of the roof would be inadequate. Because of this restriction on thickness, Parkhill carefully checks the aluminum sheets from its supplier. Of course, Parkhill does not want to measure each sheet, so it randomly samples 100. The sheets in the sample have a mean thickness of .0408 inches. From past experience with this supplier, Parkhill believes that these sheets come from a thickness population with a standard deviation of .004 inches. On the basis of this data, Parkhill must decide whether the 10,000 sheets meet specifications. In Chapter 8, we used sample statistics to estimate population parameters. Now, to solve problems like Parkhill's, we shall learn how to use characteristics of samples to test an assumption we have about the population from which that sample came. Our test for Parkhill, later in the chapter, may lead Parkhill to accept the shipment, or it may indicate that Parkhill should reject the aluminum sheets sent by the supplier.

Function of
hypothesis testing

1 INTRODUCTION Hypothesis testing begins with an assumption, called a *hypothesis,* that we make about a population parameter. Then we collect sample data, produce sample statistics, and use this information to decide how likely it is that our hypothesized population parameter is correct. Say that we assume a certain value for a population mean. To test the validity of our assumption, we gather sample data and determine the difference between the hypothesized value and the actual value of the sample mean. Then we judge whether the difference is significant. The smaller the difference, the greater the likelihood that our hypothesized value for the mean is correct. The larger the difference, the smaller the likelihood.

Unfortunately, the difference between the hypothesized population parameter and the actual sample statistic is more often neither so large that we automatically reject our hypothesis nor so small that we just as quickly accept it. So in hypothesis testing as in most significant real life decisions, clear-cut solutions are the exception, not the rule.

When to accept
or reject
the hypothesis

Suppose a manager of a large government organization tells us that the average work efficiency of her employees is 90 percent. How can we test the validity of her hypothesis? Using the sampling methods we learned in Chapter 7, we could calculate the efficiency of a *sample* of her employees. If we did this and the sample statistic came out to be 93 percent, we would readily accept the manager's statement. However, if the sample statistic were 46 percent, we would reject her assumption as untrue. We can interpret both of these outcomes, 93 percent and 46 percent, using our common sense.

The basic problem
will be dealing
with uncertainty

Now suppose that our sample statistic reveals an efficiency of 81 percent. This value is relatively close to 90 percent. But is it close enough for us to accept the manager's hypothesis? Whether we accept or reject the manager's hypothesis, we cannot be absolutely certain that our decision is correct; therefore we will have to learn to deal with uncertainty in our decision making. We cannot accept or reject a hypothesis about a population parameter simply by intuition. Instead, we need to learn how to decide objectively, on the basis of sample information, whether to accept or reject a hunch.

EXERCISES

1. Why must we be required to deal with uncertainty in our decisions, even when using statistical techniques?

2. Theoretically speaking, how might one go about testing the hypothesis that a coin is fair? That a die is fair?

3. Is it possible that a false hypothesis will be accepted? How would you explain this?

4. Describe the hypothesis testing process.

5. How would you explain a large difference between a hypothesized population parameter and a sample statistic if, in fact, the hypothesis is true?

2 CONCEPTS BASIC TO THE HYPOTHESIS TESTING PROCEDURE

Before we introduce the formal statistical terms and procedures, we'll work our chapter opening sports complex problem all the way through. Recall that the aluminum roofing sheets have a claimed average thickness of .04 inches and that they will be unsatisfactory if they are too thick *or* too thin. The contractor takes a sample of 100 sheets and determines that the sample mean thickness is .0408 inches. On the basis of past experience, he knows that the population standard deviation is .004 inches. Does this sample evidence indicate that the batch of 10,000 sheets of aluminum is suitable for constructing the roof of the new sports complex?

Sports complex problem

If we assume that the true mean thickness is .04 inches, and we know that the population standard deviation is .004 inches, how likely is it that we would get a sample mean of .0408 or more from that population? In other words, if the true mean is .04 inches, and the standard deviation is .004 inches, what are the chances of getting a sample mean that differs from .04 inches by .0008 inches or more?

Formulating the hypothesis

These questions show that to determine whether the population mean is actually .04 inches, we must calculate the probability that a random sample with a mean of .0408 inches will be selected from a population with a μ of .04 inches and a σ of .004 inches. This probability will indicate whether it is *reasonable* to observe a sample like this if the population mean is actually .04 inches. If this probability is far too low, we must conclude that the aluminum company's statement is false and that the mean thickness of the aluminum sheets is not .04 inches.

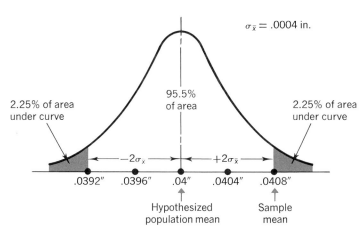

$\sigma_{\bar{x}} = .0004$ in.

95.5% of area

2.25% of area under curve

2.25% of area under curve

$-2\sigma_{\bar{x}}$ $+2\sigma_{\bar{x}}$

.0392" .0396" .04" .0404" .0408"

Hypothesized population mean

Sample mean

FIG. 9·1
Probability that $\bar{x}$ will differ from hypothesized μ by 2 standard errors or more

Let's answer the question illustrated in Fig. 9·1: if the hypothesized population mean is .04 inches and the population standard deviation is .004 inches, what are the chances of getting a sample mean (.0408 inches) that differs from .04 inches by .0008 inches? First we calculate the standard error of the mean from the population standard deviation:

Calculating
the standard error
of the mean

$$\sigma_{\bar{x}} = \frac{\sigma}{\sqrt{n}} \qquad [7 \cdot 1]$$

$$= \frac{.004 \text{ in.}}{\sqrt{100}}$$

$$= \frac{.004 \text{ in.}}{10}$$

$$= .0004 \text{ in.}$$

Next we use Equation 7·2 to discover that the mean of our sample (.0408 inches) lies 2 standard errors to the right of the hypothesized population mean:

$$z = \frac{\bar{x} - \mu}{\sigma_{\bar{x}}} \qquad [7 \cdot 2]$$

$$= \frac{.0408 - .04}{.0004}$$

$$= 2 \leftarrow \text{standard errors of the mean}$$

Interpreting
the probability
associated with
this difference

Using Appendix Table 1, we learn that 4.5 percent is the *total chance* of our sample mean differing from the population mean by 2 or more standard errors; that is, the chances that the sample mean would be .0408 inches or larger or .0392 inches or smaller are only 4.5 percent. With this low a chance, Parkhill could conclude that a population with a true mean of .04 inches would not be likely to produce a sample like this. The project supervisor would reject the aluminum company's statement about the mean thickness of the sheets.

The decision
maker's role
in formulating
hypotheses

In this case, the difference between the sample mean and the hypothesized population mean is too large, and the chance that the population would produce such a random sample is far too low. Why this probability of 4.5 percent is too low, or wrong, is a judgment for decision makers to make. Certain situations demand that decision makers be very sure about the characteristics of the items being tested, and then 4.5 percent is too high to be attributable to chance. Other processes allow for a wider latitude or variation, and a decision maker might accept a hypothesis with a 20 percent probability of chance variation. In each situation, we must try to determine the costs resulting from an incorrect decision and the precise level of risk we are willing to assume.

Risk of rejection

In our example, we rejected the aluminum company's contention that the population mean is .04 inches. But suppose for a moment that the population mean is *actually* .04 inches. If we then stuck to our rejection rule of 2 standard errors or more (the 4.5 percent probability or less in the tails of Fig. 9·1), we would reject a perfectly good lot of aluminum sheets 4.5 percent of the time. Therefore, **our minimum standard for an acceptable probability, 4.5 percent, is** *also* **the** *risk* **we take of** *rejecting a hypothesis that is true.* In this or any decision making, there can be no risk-free tradeoff.

6. What do we mean when we reject a hypothesis on the basis of a sample?

7. Explain why there is no single level of probability used to reject or accept in hypothesis testing.

8. If we reject a hypothesized value because if differs from a sample statistic by more than one standard error, what is the probability that we have rejected a hypothesis that is in fact true?

9. How many standard deviations around the hypothesized value should we use to be 95.5 percent certain that we accept the hypothesis when it is correct?

10. Given a sample mean of 17.0 based on 25 elements from a population with a standard deviation of 2.4, determine whether we can accept the hypothesis that the population mean is 18.3 (using 2 standard errors as the criterion).

11. For a certain population with a standard deviation of 12, the mean is hypothesized to be 84. If a sample of 64 observations is taken and yields a mean of 87.2, determine whether such a sample estimate is reasonable (within 2 standard errors) if, in fact, the hypothesis is true.

12. An automobile manufacturer claims that a particular model gets 24 miles to the gallon. The Environmental Protection Agency, using a sample of 36 automobiles of this model, finds the sample mean to be 23.1 miles per gallon. From previous studies, the population standard deviation is known to be 3 miles per gallon. Could we reasonably expect (within 2 standard deviations) that we could select such a sample if indeed the population mean is actually 24 miles per gallon?

3 TESTING HYPOTHESES

In hypothesis testing, we must state the assumed or hypothesized value of the population parameter *before* we begin sampling. The assumption we wish to test is called the *null hypothesis* and is symbolized H_0, or "H sub-zero."

Making a formal statement of the null hypothesis

Suppose we want to test the hypothesis that the population mean is equal to 500. We would symbolize it as follows and read it, "the null hypothesis is that the population mean is equal to 500":

$$H_0: \mu = 500$$

The term *null* hypothesis arises from earlier agricultural and medical applications of statistics. In order to test the effectiveness of a new fertilizer or drug, the tested hypothesis (the null hypothesis) was that it had *no effect;* that is, there was no difference between treated and untreated samples.

If we use a hypothesized value of a population mean in a problem, we would represent it symbolically as:

$$\mu_{H_0}$$

This is read "the hypothesized value of the population mean."

If our sample results fail to support the null hypothesis, we must conclude that something else is true. **Whenever we reject the null hypothesis, the conclusion we do accept is called the** *alternative hypothesis* **and is symbolized H₁ ("H sub-one").** For the null hypothesis:

$$H_0: \mu = 200 \text{ (Read: "the null hypothesis is that}$$
$$\text{the population mean is equal to 200.")}$$

we will consider three possible alternative hypotheses:

$H_1: \mu \neq 200 \leftarrow$ "the alternative hypothesis is that the population mean is *not equal* to 200"

$H_1: \mu > 200 \leftarrow$ "the alternative hypothesis is that the population mean is *greater than* 200"

$H_1: \mu < 200 \leftarrow$ "the alternative hypothesis is that the population mean is *less than* 200"

Interpreting the significance level

Prior to the test,
selecting the level
of probability
and risk

The purpose of hypothesis testing is not to question the computed value of the sample statistic but to make a judgment about the *difference* between that sample statistic and a hypothesized population parameter. The next step after stating the null and alternative hypotheses, then, is to decide what criterion to use for deciding whether to accept or reject the null hypothesis.

In our sports complex example, we decided that a difference observed between the sample mean $\bar{x}$ and the hypothesized population mean μ_{H_0} had only a 4.5 percent, or .045, chance of occurring. Therefore, we *rejected* the null hypothesis that the population mean was .04 inches (H₀: μ = .04 inches). In statistical terms, the value .045 is called the *significance level.*

What if we test a hypothesis at the 5 percent level of significance? This means that we will reject the null hypothesis if the difference between the sample statistic and the hypothesized population parameter is so large that it or a larger difference would occur, on the average, only 5 times or fewer in every 100 samples when the hypothesized population parameter is correct. **Assuming the hypothesis is correct, then, the significance level indicates the percentage of sample means that is outside of certain limits.** (In estimation, you remember, the confidence level indicated the percentage of sample means that fell *within* the defined confidence limits.)

Figure 9·2 illustrates how to interpret a 5 percent level of significance. Notice that 2.5 percent of the area under the curve is located in each tail. From Appendix Table 1, we can determine that 95 percent of all the area under the curve is included in an interval extending $1.96\sigma_{\bar{x}}$ on either side of the hypothesized mean. In 95 percent of the area, then, there is no significant difference between the sample statistic and the hypothesized popula-

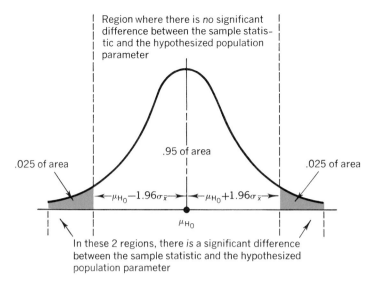

Region where there is *no* significant difference between the sample statistic and the hypothesized population parameter

.95 of area

.025 of area

.025 of area

$\leftarrow \mu_{H_0} - 1.96\sigma_{\bar{x}} \rightarrow$ $\leftarrow \mu_{H_0} + 1.96\sigma_{\bar{x}} \rightarrow$

μ_{H_0}

In these 2 regions, there *is* a significant difference between the sample statistic and the hypothesized population parameter

FIG. 9·2
Regions of significant difference and of no significant difference at a 5 percent level of significance

tion parameter. In the remaining 5 percent (the colored regions in Fig. 9·2) a significant difference does exist.

Figure 9·3 examines this same example in a different way. Here, the .95 of the area under the curve is where we would accept the null hypothesis. The two colored parts under the curve, representing a total of 5 percent of the area, are where we would reject the null hypothesis.

Also called the area where we accept the null hypothesis

A word of caution is appropriate here. Even if our sample statistic in Fig. 9·3 does fall in the nonshaded region (that region comprising 95 percent of the area under the curve), this *does not prove* that our null hypothesis (H_0) is true; it simply does not provide statistical evidence to reject it. Why? Because the only way in which the hypothesis can be accepted with certainty is for us to know the population parameter, and unfortunately this is not possible. Therefore, whenever we say that we accept the null hypothesis, we actually mean that there is not sufficient statistical evidence to reject it. Use of the term *accept,* instead of *do not reject,* has become standard. It means simply that when sample data do not cause us to reject a null hypothesis, we behave as though that hypothesis is true.

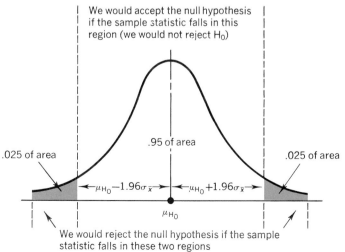

We would accept the null hypothesis if the sample statistic falls in this region (we would not reject H_0)

.95 of area

.025 of area

.025 of area

$\leftarrow \mu_{H_0} - 1.96\sigma_{\bar{x}} \rightarrow$ $\leftarrow \mu_{H_0} + 1.96\sigma_{\bar{x}} \rightarrow$

μ_{H_0}

We would reject the null hypothesis if the sample statistic falls in these two regions

FIG. 9·3
A 5 percent level of significance, with acceptance and rejection regions designated

Selecting a significance level

There is no single standard or universal level of significance for testing hypotheses. In some instances, a 5 percent level of significance is used. Published research results often test hypotheses at the 1 percent level of significance. It is possible to test a hypothesis at *any* level of significance. But remember that our choice of the minimum standard for an acceptable probability, or the significance level, is also the risk we assume of rejecting a null hypothesis when it is true. **The higher the significance level we use for testing a hypothesis, the higher the probability of rejecting a null hypothesis when it is true.**

Examining this concept, we refer to Fig. 9·4. Here we have illustrated a hypothesis test at three different significance levels: .01, .10, and .50. Also, we have indicated the location of the same sample mean $\bar{x}$ on each distribution. In parts *a* and *b,* we would accept the null hypothesis that the population mean is equal to the hypothesized value. But notice that in part *c* we would reject this same null hypothesis. Why? Our significance level there of .50 is so high that we would rarely accept a null hypothesis when it is *not* true but, at the same time, frequently reject one when it *is* true.

**FIG. 9·4
Three different
levels of
significance**

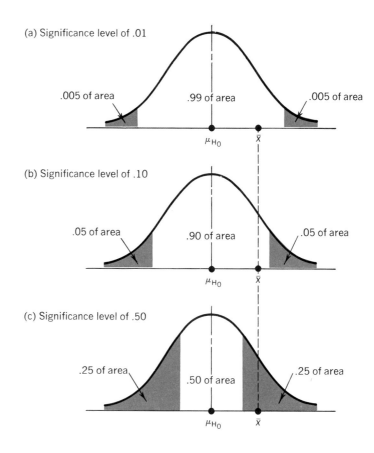

(a) Significance level of .01

.005 of area   .99 of area   .005 of area

μ_{H_0} $\bar{x}$

(b) Significance level of .10

.05 of area   .90 of area   .05 of area

μ_{H_0} $\bar{x}$

(c) Significance level of .50

.25 of area   .50 of area   .25 of area

μ_{H_0} $\bar{x}$

Statisticians give specific definitions and symbols to the concept illustrated in Fig. 9·4. Rejecting a null hypothesis when it is true is called a Type I error, and its probability (which, as we have seen, is also the significance level of the test) is symbolized α (alpha). Alternately, accepting a null hypothesis when it is false is called a Type II error, and its probability is symbolized β (beta). There is a trade-off between these two types of errors: the probability of making one type of error can only be reduced if we are willing to increase the probability of making the other type of error. Notice in part *c,* Fig. 9·4, that our acceptance region is quite small (.50 of the area under the curve). With an acceptance region this small, we will rarely accept a null hypothesis when it is not true, but as a cost of being this sure, we will frequently reject a null hypothesis when it is true. Put another way, in order to get a low β, we will have to put up with a high α. To deal with this trade-off in personal and professional situations, decision makers decide the appropriate level of significance by examining the costs or penalties attached to both types of errors.

Type I and
Type II errors
defined

Suppose that making a Type I error (rejecting a null hypothesis when it is true) involves the time and trouble of reworking a batch of chemicals that should have been accepted. At the same time, making a Type II error (accepting a null hypothesis when it is false) means taking a chance that an entire group of users of this chemical compound will be poisoned. Obviously, the management of this company will prefer a Type I error to a Type II error and, as a result, will set very high levels of significance in its testing to get low β's.

Preference for
a Type I error

Suppose, on the other hand, that making a Type I error involves disassembling an entire engine at the factory, but making a Type II error involves relatively inexpensive warranty repairs by the dealers. Then the manufacturer is more likely to prefer a Type II error and will set low significance levels in its testing.

Preference for
a Type II error

Deciding which distribution to use in hypothesis testing

After deciding what level of significance to use, our next task in hypothesis testing is to determine the appropriate probability distribution. We have a choice between the normal distribution, Appendix Table 1, and the *t* distribution, Appendix Table 2. The rules for choosing the appropriate distribution are similar to those we encountered in Chapter 8 on estimation. Table 9·1 summarizes when to use the normal and *t* distributions in making tests of means. Later in this chapter, we shall examine the distributions appropriate for testing hypotheses about proportions.

Remember one more rule when testing the hypothesized value of a mean. As in estimation, use the *finite population multiplier* whenever the

Prior to the test,
selecting
the correct
distribution

TABLE 9·1
Conditions for using the normal and *t* distributions in testing hypothesis about means

	When the population standard deviation is known	When the population standard deviation is not known
Sample size *n* is larger than 30	Normal distribution, *z* table	Normal distribution, *z* table
Sample size *n* is 30 or less and we assume the population is normal or approximately so	Normal distribution, *z* table	*t* distribution, *t* table

population is finite in size, sampling is done without replacement, and the sample is more than 5 percent of the population.

Two-tailed and one-tailed tests of hypotheses

Description of a two-tailed hypothesis test

In the tests of hypothesized population means that follow, we shall illustrate two-tailed tests and one-tailed tests. These new terms need a word of explanation. A *two-tailed test* of a hypothesis will reject the null hypothesis if the sample mean is significantly higher than *or* lower than the hypothesized population mean. Thus, in a two-tailed test, there are *two* rejection regions. This is illustrated in Fig. 9·5.

A two-tailed test is appropriate when the null hypothesis is $\mu = \mu_{H_0}$ (μ_{H_0} being some specified value) and the alternative hypothesis is $\mu \neq \mu_{H_0}$. Assume that a manufacturer of light bulbs wants to produce bulbs with a mean life of $\mu = \mu_{H_0} = 1{,}000$ hours. If the lifetime is shorter, he will lose customers to his competition; if the lifetime is longer, he will have a very high production cost because the filaments will be excessively thick. In order to see if his production process is working properly, he takes a sample of the output to test the hypothesis $H_0: \mu = 1{,}000$. Since he does not want to deviate significantly from 1,000 hours *in either direction,* the appropriate alternative hypothesis is $H_1: \mu \neq 1{,}000$, and he uses a two-tailed test. That is, he rejects the null hypothesis if the mean life of bulbs in the sample is *either too far above* 1,000 hours *or too far below* 1,000 hours.

FIG. 9·5
Two-tailed test of a hypothesis, showing the two rejection regions

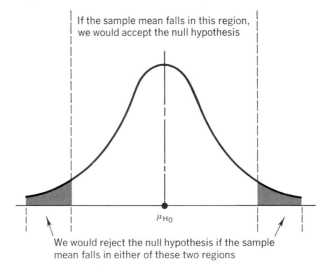

If the sample mean falls in this region, we would accept the null hypothesis

μ_{H_0}

We would reject the null hypothesis if the sample mean falls in either of these two regions

However, there are situations in which a two-tailed test is not appropriate, and we must use a one-tailed test. Consider the case of a government agency that buys light bulbs from the manufacturer discussed above. The agency buys bulbs in large lots and does not want to accept a lot of bulbs unless their mean life is 1,000 hours. As each shipment arrives, the agency tests a sample to decide whether it should accept the shipment. The government agency is going to reject the shipment only if it feels that the mean life is below 1,000 hours. If it feels that the bulbs are better than expected (with a mean life above 1,000 hours), it certainly will not reject the shipment, because the longer life comes at no extra cost to the agency. So the agency's hypotheses are: H_0: $\mu = 1,000$ hours and H_1: $\mu < 1,000$ hours. It rejects H_0 only if the mean life of the sampled bulbs is significantly *below* 1,000 hours. This situation is illustrated in Fig. 9·6. From this figure we can see why this test is called a *left-tailed test* (or a *lower-tailed test*).

In general, a left-tailed (lower-tailed) test is used if the hypotheses are H_0: $\mu = \mu_{H_0}$ and H_1: $\mu < \mu_{H_0}$. In such a situation, it is sample evidence with the sample mean significantly below the hypothesized population mean that leads us to reject the null hypothesis in favor of the alternative hypothesis. Stated differently, the rejection region is in the lower tail (left tail) of the distribution of the sample mean, and that is why we call this a lower-tailed test.

Conditions when a two-tailed test may not be appropriate and we must use a one-tailed test

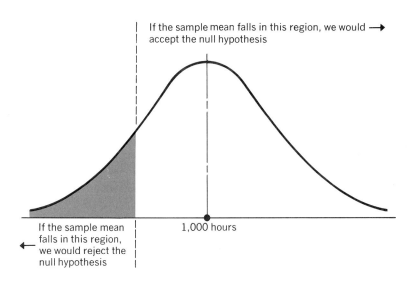

If the sample mean falls in this region, we would → accept the null hypothesis

If the sample mean falls in this region, we would reject the null hypothesis ←

1,000 hours

FIG. 9·6
Left-tailed test (a lower-tailed test) with the rejection region on the left side (lower side)

A left-tailed test is one of the two kinds of one-tailed tests. As you have probably guessed by now, the other kind of one-tailed test is a *right-tailed test* (or an *upper-tailed test*). An upper-tailed test is used when the hypotheses are H_0: $\mu = \mu_{H_0}$ and H_1: $\mu > \mu_{H_0}$. Only values of the sample mean that are *significantly above* the hypothesized population mean will cause us to reject the null hypothesis in favor of the alternative hypothesis. This is called an upper-tailed test because the rejection region is in the upper tail of the distribution of the sample mean.

Left-tailed tests and right-tailed tests

The following situation is illustrated in Fig. 9·7; it calls for the use of an upper-tailed test. A political campaign manager has asked her field workers to observe a limit on traveling expenses. The manager hopes to keep expenses to an average of $100 per field worker per day. One month after the limit is imposed, a sample of submitted daily expenses is taken to see if the limit is being observed. The null hypothesis is H_0: $\mu = \$100.00$, but the campaign manager is only concerned with excessively high expenses. Thus, the appropriate alternative hypothesis is H_1: $\mu > \$100.00$, and an upper-tailed test is used. The null hypothesis is rejected (and corrective measures taken) only if the sample mean is significantly higher than $100.00.

Finally, we should remind you again that in each example of hypothesis testing, when we accept a null hypothesis on the basis of sample information, we are really saying that there is no statistical evidence to reject it. We are not saying that the null hypothesis is true. The only way to prove a null hypothesis is to know what the population parameter is, and that is not possible with sampling. Thus, we accept the null hypothesis and behave as though it is true simply because we can find no evidence to reject it.

FIG. 9·7
Right-tailed
(upper-tailed) test

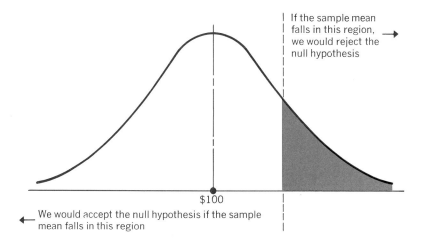

If the sample mean falls in this region, we would reject the null hypothesis →

$100

← We would accept the null hypothesis if the sample mean falls in this region

EXERCISES

13. Formulate the null and alternative hypotheses to test whether the mean lifetime for men is 68 years.

14. Describe what the null and alternative hypotheses typically represent in the hypothesis testing process.

15. Define the term *significance level*.

16. Define Type I and Type II errors.

17. In a trial, the null hypothesis is that an individual is innocent of a certain crime. Would the legal system prefer to commit a Type I or a Type II error with this hypothesis?

18. What is the relationship between the significance level of a test and Type I error?

19. If our goal is to accept a null hypothesis with 99 percent certainty when it's true, and our sample size is more than 30, diagram the acceptance and rejection regions for the following alternative hypothesis:

   a) $\mu \neq 0$ b) $\mu < 0$ c) $\mu > 0$
   Specify the percentage of the area under each region of the curves.

20. For the following cases, specify which probability distribution to use in a hypothesis test:

   a) $H_0: \mu = 25$ $H_1: \mu > 25, \bar{x} = 28.2, \sigma = 4, n = 12$

   b) $H_0: \mu = 1024$ $H_1: \mu \neq 1024, \bar{x} = 976, \sigma = 60, n = 30$

   c) $H_0: \mu = 100$ $H_1: \mu > 100, \bar{x} = 107, s = 3.2, n = 16$

   d) $H_0: \mu = 500$ $H_1: \mu > 500, \bar{x} = 508, s = 4, n = 40$

   e) $H_0: \mu = 6$ $H_1: \mu \neq 6, \bar{x} = 5.4, s = .5, n = 25$

21. Our hypothesis is that a bridge will safely withstand 50 tons of traffic.

   a) Would we rather commit a Type I or a Type II error?

   b) Based on your answer to part a, should we use a high or a low significance level?

22. Under what conditions is it appropriate to use a one-tailed test? a two-tailed test?

23. If you have decided that a one-tailed test is the appropriate test to use, how do you decide whether it should be a lower-tailed test or an upper-tailed test?

24. Martha Inman, a highway safety engineer, decides to test the load-bearing capacity of a bridge that is 20 years old. Considerable data are available from similar tests on the same type of bridge. Which is appropriate, a one-tailed or a two-tailed test? If the minimum load-bearing capacity of this bridge must be 10 tons, what are the null and alternative hypotheses?

4 HYPOTHESIS TESTING OF MEANS— SAMPLES WITH POPULATION STANDARD DEVIATIONS KNOWN

Two-tailed tests of means

A manufacturer supplies the rear axles for U.S. Postal Service mail trucks. These axles must be able to withstand 80,000 pounds per square inch in stress tests, but an excessively strong axle raises production costs significantly. Long experience indicates that the standard deviation of the strength of its axles is 4,000 pounds per square inch. The manufacturer selects a sample of 100 axles from the latest production run, tests them, and

Setting up the problem symbolically

finds that the mean stress capacity of the sample is 79,600 pounds per square inch. Written symbolically, the data in this case are:

$\mu_{H_0} = 80,000 \leftarrow$ hypothesized value of the population mean
$\sigma = 4,000 \leftarrow$ population standard deviation
$n = 100 \leftarrow$ sample size
$\overline{x} = 79,600 \leftarrow$ sample mean

If the axle manufacturer uses a significance level (α) of .05 in testing, will the axles meet his stress requirements? Symbolically, we can state the problem:

$H_0: \mu = 80,000 \leftarrow$ null hypothesis: the true mean is 80,000 pounds per square inch
$H_1: \mu \ne 80,000 \leftarrow$ alternative hypothesis: the true mean is not 80,000 pounds per square inch
$\alpha = .05 \leftarrow$ level of significance for testing this hypothesis

Calculating
the standard error
of the mean

Since we know the population standard deviation, and since the size of the population is large enough to be treated as infinite, we can use the normal distribution in our testing. First, we calculate the standard error of the mean using Equation 7·1:

$$\sigma_{\overline{x}} = \frac{\sigma}{\sqrt{n}} \qquad [7\cdot1]$$

$$= \frac{4,000}{\sqrt{100}}$$

$$= \frac{4,000}{10}$$

$= 400$ pounds per square inch $\leftarrow$ standard error of the mean

Illustrating
the problem

Figure 9·8 illustrates this problem, showing the significance level of .05 as the two shaded regions that each contain .025 of the area. The .95 acceptance region contains two equal areas of .475 each. From the normal distri-

FIG. 9·8
**Two-tailed
hypothesis test at
.05 significance
level**

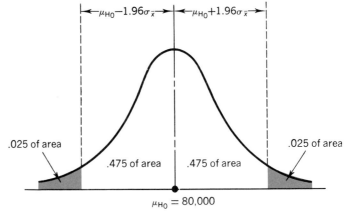

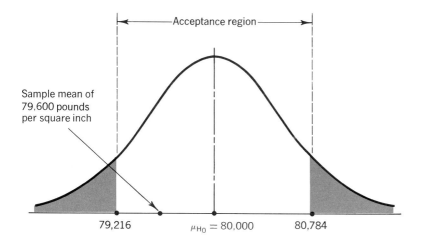

FIG. 9·9
**Two-tailed
hypothesis test at
.05 significance
level, showing
acceptance region
and sample mean**

Sample mean of
79,600 pounds
per square inch

Acceptance region

79,216 $\mu_{H_0} = 80,000$ 80,784

bution table (Appendix Table 1), we can see that the appropriate z value for .475 of the area under the curve is 1.96. Now we can determine the limits of the acceptance region:

$$\mu_{H_0} + 1.96\sigma_{\bar{x}} = 80,000 + 1.96(400)$$
$$= 80,000 + 784$$
$$80,784 \text{ pounds per square inch} \leftarrow \text{upper limit}$$

and: $\mu_{H_0} - 1.96\sigma_{\bar{x}} = 80,000 - 1.96(400)$
$$= 80,000 - 784$$
$$= 79,216 \text{ pounds per square inch} \leftarrow \text{lower limit}$$

These two limits of the acceptance region (80,784 and 79,216) are shown in Fig. 9·9. Also, we have indicated the sample mean (79,600 pounds per square inch). Obviously, the sample mean lies within the acceptance region; the manufacturer should accept the null hypothesis, because there is no significant difference between the hypothesized mean of 80,000 and the observed mean of the sample of axles. On the basis of this sample, the manufacturer should accept the production run as meeting the stress requirements.

One-tailed tests of means

For a one-tailed test of a mean, suppose a hospital uses large quantities of packaged doses of a particular drug. The individual dose of this drug is 100 cubic centimeters (100 cc). The action of the drug is such that the body will harmlessly pass off excessive doses. On the other hand, insufficient doses do not produce the desired medical effect, and they interfere with patient treatment. The hospital has purchased its requirements of this drug from the same manufacturer for a number of years and knows that the population standard deviation is 2 cc. The hospital inspects 50 doses of this drug at random from a very large shipment and finds the mean of these doses to be 99.75 cc.

$\mu_{H_0} = 100 \leftarrow$ hypothesized value of the population mean
$\sigma = 2 \leftarrow$ population standard deviation
$n = 50 \leftarrow$ sample size
$\overline{x} = 99.75 \leftarrow$ sample mean

If the hospital sets a .10 significance level and asks us whether the dosages in this shipment are too small, how can we find the answer?

To begin, we can state the problem symbolically:

$H_0: \mu = 100 \leftarrow$ null hypothesis: the mean of the shipments' dosages is 100 cc.

$H_1: \mu < 100 \leftarrow$ alternative hypothesis: the mean is less than 100 cc.
$\alpha = .10 \leftarrow$ level of significance for testing this hypothesis

Calculating
the standard error
of the mean

Then we can calculate the standard error of the mean, using the known population standard deviation and Equation 7·1 (because the population size is large enough to be considered infinite):

$$\sigma_{\overline{x}} = \frac{\sigma}{\sqrt{n}}$$ [7·1]

$$= \frac{2}{\sqrt{50}}$$

$$= \frac{2}{7.07}$$

$$= .2829 \text{ cc.} \leftarrow \text{standard error of the mean}$$

Illustrating
the problem

Determining
the limit of the
acceptance region

The hospital wishes to know whether the actual dosages are 100 cc or whether, in fact, the dosages are too small. The hospital must determine that the dosages are *more* than a certain amount, or it must reject the shipment. This is a *left-tailed* test, which we have shown graphically in Fig. 9·10. Notice that the colored region corresponds to the .10 significance level. Also notice that the acceptance region consists of 40 percent on the left side of the distribution *plus* the entire right side (50 percent), for a total area of 90 percent. Since we know the population standard deviation, and *n* is larger than 30, we can use the normal distribution. From Appendix Table 1, we can determine that the appropriate *z* value for 40 percent of the area under the curve is 1.28. Using this information, we can calculate the acceptance region's *lower* limit:

$$\mu_{H_0} - 1.28\sigma_{\overline{x}} = 100 - 1.28 \,(.2829)$$

$$= 100 - .36$$

$$= 99.64 \text{ cc} \leftarrow \text{lower limit}$$

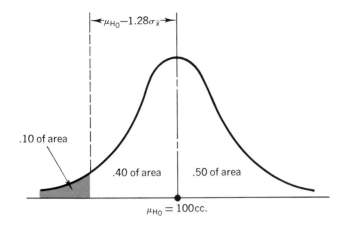

FIG. 9·10
**Left-tailed
hypothesis test at
.10 significance
level**

.10 of area

.40 of area

.50 of area

$\mu_{H_0} = 100$cc.

$\mu_{H_0} - 1.28\sigma_{\bar{x}}$

This lower limit of the acceptance region, 99.64 cc, and the sample mean, 99.75, are both shown in Fig. 9·11. In this figure, we can see that the sample mean lies within the acceptance region. Therefore, the hospital should accept the null hypothesis because there is no significant difference between our hypothesized mean of 100 cc and the observed mean of the sample. On the basis of this sample of 50 doses, the hospital should accept the doses in the shipment as being sufficient.

Interpreting
the results

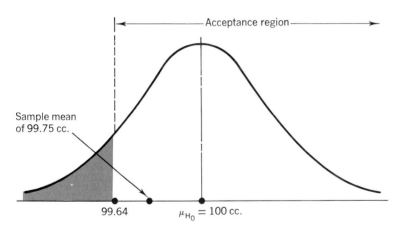

FIG. 9·11
**Left-tailed
hypothesis test at
.10 significance
level, showing
acceptance region
and the sample
mean**

Acceptance region

Sample mean
of 99.75 cc.

99.64

$\mu_{H_0} = 100$ cc.

EXERCISES

25. From a sample of size 25, a mean of 32.4 was found. Given that the population standard deviation is 4.2, test the hypothesis that the population mean is 34, using a two-tailed test, at the .05 level of significance.

26. For a certain population, the mean is hypothesized to be 115.2. A sample of 49 yielded a mean of 117.6. If the population standard deviation is known to be 8.4, test the hypothesis that the true mean is 115.2 against the alternative that it is greater than 115.2, using the .04 significance level.

27. A sample of size 16 from a certain population shows a sample mean of 12,000. The population standard deviation is known to be 2,000. Test whether a hypothesized population mean of 13,000 is acceptable at the .01 level of significance, or whether we should conclude that the population mean is less than 13,000.

28. The average score on a nationally administered aptitude test was 76, and the corresponding standard deviation was 8. In order to evaluate a state's education system, the scores of 100 of the state's students were randomly selected. These students had an average score of 72.

   a) The state wishes to know if there is a significant difference between the local scores and the national scores. What are the null and alternative hypotheses?

   b) At a significance level of .05, test your hypothesis in part a.

29. Boston's welfare department has stated that the average annual per capita income for a particularly depressed section of the city is $4,800, and the population standard deviation is $900. A citizen's group has sampled 40 families from that section and found the average per capita income to be $4,500. At a 5 percent level of significance, would you accept or reject the welfare department's figures?

30. It is an established phenomenon that one member of the rodent family hoards particular foods when the temperature drops to a certain point. In a fixed period of time, and given a fixed food supply, the animals hoard an average of 9.5 grams of foodstuff with a standard deviation of 4 grams. An experimenter interested in the effects of early food deprivation on such hoarding behavior takes a sample of 36 of the rodents, keeps them in a deprived situation for a period while they are young, and then observes their hoarding behavior under controlled conditions. He finds the average amount of foodstuffs stored for these animals is 10.5 grams. Using a significance level of .02, should the experimenter reject the hypothesis that the deprived rodents hoard the same amount of food on the average as animals who have not been deprived?

31. In parapsychology studies, experimenters examine the possibility of extrasensory perception (ESP), mental telepathy, and other extraordinary mental processes. In one of the standard ESP tasks, a subject attempts to identify cards with special symbols on them. From a deck of 60 cards, the average number of cards a subject can identify is 15, with a standard deviation of 5 cards. It is proposed that hypnosis might improve concentration and therefore improve extrasensory abilities, so a study is run with 100 subjects who are hypnotized before attempting the ESP task. The average number of cards identified for these subjects is 16 of the 60. At the .10 level of significance, do the researchers have reason to believe that hypnotism aids extrasensory perception?

5 MEASURING THE POWER OF A HYPOTHESIS TEST

What should a good hypothesis test do?

Now that we have considered two examples of hypothesis testing, a step back is appropriate, to discuss what a good hypothesis test *should* do. Ideally, α and β (the probabilities of Type I and Type II errors) should both be small. Recall that a Type I error occurs when we reject a null hypothesis that is true, and α (the significance level of the test) *is* the probability of making a Type I error. In other words, once we decide upon the significance level, there is nothing else we can do about α. A Type II error occurs when we accept a null hypothesis that is false; the probability of a Type II error is β. What can we say about β?

Meaning of β and $1 - \beta$

Suppose the null hypothesis *is* false. Then decision makers would like the hypothesis test to reject it all the time. Unfortunately, hypothesis tests cannot be foolproof; sometimes when the null hypothesis is false, a test does not reject it, and thus a Type II error is made. When the null hypothesis is false, μ (the *true* population mean) does not equal μ_{H_0} (the *hypothesized* population mean); instead, μ equals some other value. For each possible value of μ for which the alternative hypothesis is true, there is a different probability (β) of incorrectly accepting the null hypothesis. Of course, we would like this β (the probability of accepting a null hypothesis when it is false) to be as small as possible, or equivalently, we would like $1 - \beta$ (the probability of rejecting a null hypothesis when it is false) to be as large as possible.

Interpreting
the values of $1 - \beta$

Since rejecting a null hypothesis when it is false is exactly what a good test ought to do, a high value of $1 - \beta$ (something near 1.0) means the test is working quite well (it is rejecting the null hypothesis when it is false); a low value of $1 - \beta$ (something near 0.0) means that the test is working very poorly (it's not rejecting the null hypothesis when it is false). Since the value $1 - \beta$ is the measure of how well the test is working, it is known as the *power of the test.* If we plot the values of $1 - \beta$ for each value of μ for which the alternative hypothesis is true, the resulting curve is known as a *power curve.*

Computing
the values
of $1 - \beta$

In part *a* of Fig. 9·12, we have reproduced the left-tailed test first introduced in Fig. 9·10. In part *b* of Fig. 9·12, we show the power curve that is associated with this test. Computing the values of $1 - \beta$ to plot the power curve is not difficult; three such points are shown in *b*, Fig. 9·12. Recall that with this test we were deciding whether or not to accept a drug shipment. Our test dictated that we should reject the null hypothesis if the sample mean dosage is less than 99.64 cc.

Consider point *C* on the power curve in *b*, Fig. 9·12. The population mean dosage is 99.42 cc. Given that the population mean is 99.42 cc, we must compute the probability that the mean of a random sample of 50 doses from this population will be less than 99.64 cc (the point below which we decided to reject the null hypothesis). Now look at *c*, Fig. 9·12. On page 312 we computed the standard error of the mean to be .2829 cc, so 99.64 cc is $(99.64 - 99.42)/.2829$, or .78 standard errors above 99.42 cc. Using Appendix Table 1, we can see that the probability of observing a sample mean less than 99.64 cc and thus rejecting the null hypothesis is .7823, the colored area in *c*, Fig. 9·12. Thus the power of the test ($1 - \beta$) at $\mu = 99.42$ is .7823. This simply means that at $\mu = 99.42$, the probability that this test will reject the null hypothesis when it is false is .7823.

Now look at point *D* in *b*, Fig. 9·12. For this population mean dosage of 99.61 cc, what is the probability that the mean of a random sample of 50 doses from this population will be less than 99.64 cc and thus cause the test to reject the null hypothesis? Look at *d*, Fig. 9·12. Here we see that 99.64 is $(99.64 - 99.61)/.2829$, or .11 standard errors above 99.61 cc. Using Appendix Table 1 again, we can see that the probability of observing a sample mean less than 99.64 cc and thus rejecting the null hypothesis is .5438, the colored area in *d*, Fig. 9·12. Thus the power of the test ($1 - \beta$) at $\mu = 99.61$ cc is .5438.

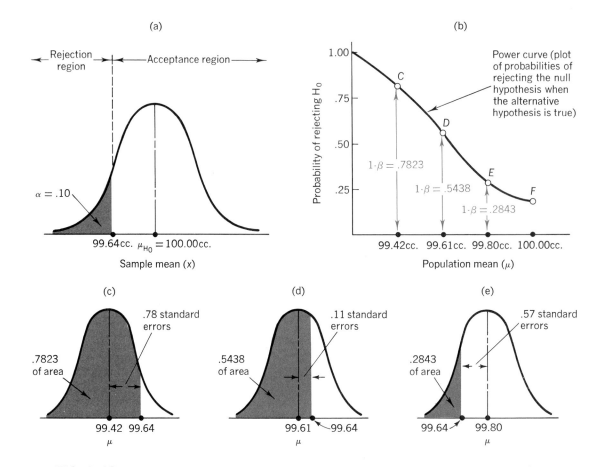

FIG. 9·12 **Left-tailed hypothesis test, associated power curve, and three values of μ**

Termination point
of the power curve

Using the same procedure at point E, we find the power of the test at $\mu = 99.80$ cc is .2843; this is illustrated as the colored area in e, Fig. 9·12. The values of $1 - \beta$ continue to decrease to the right of point E. How low do they get? As the population mean gets closer and closer to 100.00 cc, the power of the test $(1 - \beta)$ must get closer and closer to the probability of rejecting the null hypothesis when the population mean is exactly 100.00 cc. And we know *that* probability is nothing but the significance level of the test, in this case .10. Thus the curve terminates at point F, which lies at a height of .10 directly over the population mean.

Interpreting
the power curve

What does our power curve in b, Fig. 9·12, tell us? Just that as the shipment becomes less satisfactory (as the doses in the shipment become smaller), our test is more powerful (it has a greater probability of recognizing that the shipment is unsatisfactory). It also shows us, however, that because of sampling error, when the dosage is only slightly less than 100.00 cc the power of the test to recognize this situation is quite low. Thus if having *any* dosage below 100.00 cc is completely unsatisfactory, the test we have been discussing is not appropriate.

32. See problem 30, p. 314. Compute the power of the test for $\mu = 9.5$, 10.0, and 10.5 grams.

33. See problem 31, p. 314. Compute the power of the test for $\mu = 15.4$, 15.7, and 16.0 cards.

34. In problem 30, what happens to the power of the test for $\mu = 9.5$, 10.0, and 10.5 grams if the significance level is changed to .01?

35. In problem 31, what is the effect on the power of the test for $\mu = 15.4$, 15.7, and 16.0 cards of raising the significance level to 20 percent?

6 HYPOTHESIS TESTING OF PROPORTIONS— LARGE SAMPLES

Two-tailed tests of proportions

In this section, we'll apply what we have learned about tests concerning means to tests for *proportions* (that is, the proportion of occurrences in a population). But before we apply it, we'll review the important conclusions we made about proportions in Chapter 8. First, remember that the binomial is the theoretically correct distribution to use in dealing with proportions, since the data are discrete, not continuous. As the sample size increases, the binomial distribution approaches the normal in its characteristics, and we can use the normal distribution to approximate the sampling distribution. Specifically, the sample size needs to equal at least 30 *and np and nq each need to be at least 5* before we can use the normal distribution as a substitute for the binomial.

Dealing with proportions

Consider, as an example, a large university that is evaluating the promotability of its nonfaculty employees; that is, the proportion of them whose ability, training, and supervisory experience qualify them for promotion to the next highest level of management. The Office of Career Development tells the vice-chancellor for administration that 80 percent, or .8, of the employees in the university are "promotable." The vice-chancellor assembles a special committee to assess the promotability of all the employees. This committee conducts in-depth interviews with 150 employees and finds that in their judgment, only 70 percent of the sample are qualified for promotion.

Setting up the problem symbolically

$p_{H_0} = .8 \leftarrow$ hypothesized value of the population proportion of successes (judged promotable, in this case)

$q_{H_0} = .2 \leftarrow$ hypothesized value of the population proportion of failures (judged not promotable)

$n = 150 \leftarrow$ sample size

$\overline{p} = .7 \leftarrow$ sample proportion of promotables

$\overline{q} = .3 \leftarrow$ sample proportion judged not promotable

The vice-chancellor wants to test at the .05 significance level the hypothesis that .8 of its employees are promotable:

H_0: $p = .8 \leftarrow$ null hypothesis: 80 percent of the employees are promotable
H_1: $p \neq .8 \leftarrow$ alternative hypothesis: the proportion of promotable employees is not 80 percent
$\alpha = .05 \leftarrow$ level of significance for testing the hypothesis

---

Calculating
the standard error
of the proportion

To begin, we can calculate the standard error of the proportion, using the hypothesized value of p_{H_0} and q_{H_0} in Equation 8·4:

$$\sigma_{\bar{p}} = \sqrt{\frac{p_{H_0} q_{H_0}}{n}} \qquad [8·4]$$

$$= \sqrt{\frac{(.8)(.2)}{150}}$$

$$= \sqrt{.0010666}$$

$$= .0327 \leftarrow \text{standard error of the proportion}$$

Illustrating
the problem

In this instance, the vice-chancellor wants to know whether the true proportion is larger *or* smaller than the hypothesized proportion. Thus, a two-tailed test of a proportion is appropriate, and we have shown it graphically in Fig. 9·13. The significance level corresponds to the two colored regions, each containing .025 of the area. The acceptance region of .95 is illustrated as two areas of .475 each. Since n is larger than 30 and np and nq are each larger than 5, we can use the normal approximation of the binomial distribution. From Appendix Table 1, we can determine that the appropriate z value for .475 of the area under the curve is 1.96. Thus, the limits of the acceptance region are:

Determining
the limits of the
acceptance region

$$p_{H_0} + 1.96\sigma_{\bar{p}} = .8 + 1.96(.0327)$$

$$= .8 + .0641$$

$$= .8641 \leftarrow \text{upper limit}$$

$$p_{H_0} - 1.96\sigma_{\bar{p}} = .8 - 1.96(.0327)$$

$$= .8 - .0641$$

$$= .7359 \leftarrow \text{lower limit}$$

FIG. 9·13
**Two-tailed
hypothesis test of
proportion at
.05 level of
significance**

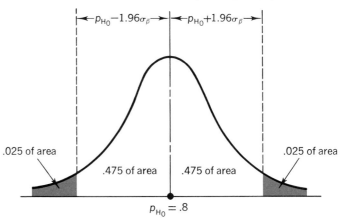

.025 of area

.475 of area   .475 of area

.025 of area

$p_{H_0} = .8$

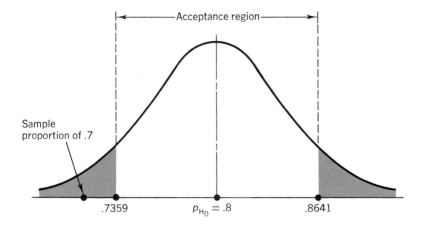

FIG. 9·14
Two-tailed
hypothesis test of
a proportion at .05
significance level,
showing
acceptance region
and sample
proportion

Figure 9·14 illustrates these two limits of the acceptance region, .8641 and .7359, as well as our sample proportion, .7. We can see that our sample proportion does *not* lie within the acceptance region. Therefore, in this case the vice-chancellor should reject the null hypothesis and conclude that there is a significant difference between the Office of Career Development's hypothesized proportion of promotable employees (.8) and the observed proportion of promotable employees in the sample. From this, they should infer that the true proportion of promotable employees in the entire university is not 80 percent.

Interpreting the results

One-tailed tests of proportions

A one-tailed test of a proportion is conceptually equivalent to a one-tailed test of a mean, as can be illustrated with this example. A member of a public interest group concerned with environmental pollution asserts at a public hearing that "fewer than 60 percent of the industrial plants in this area are complying with air pollution standards." Attending this meeting is an official of the Environmental Protection Agency who believes that 60 percent of the plants are complying with the standards; she decides to test that hypothesis at the .02 significance level.

H_0: $p = .6$ ← null hypothesis: the proportion of plants complying with air pollution
  standards is .6
H_1: $p < .6$ ← alternative hypothesis: the proportion complying with the standards
  is less than .6
$\alpha = .02$ ← level of significance for testing the hypothesis

---

The official makes a thorough search of the records in her office. She samples 60 plants from a population of over 10,000 plants and finds that 33 are complying with air pollution standards. Is the assertion by the member of the public interest group a valid one?

Setting up
the problem
symbolically

We begin by summarizing the case symbolically:

$p_{H_0} = .6 \leftarrow$ hypothesized value of the population proportion that are complying
with air pollution standards

$q_{H_0} = .4 \leftarrow$ hypothesized value of the population proportion that are not
complying and thus polluting

$n = 60 \leftarrow$ sample size

$\bar{p} = 33/60$ or $.55 \leftarrow$ sample proportion complying

$\bar{q} = 27/60$ or $.45 \leftarrow$ sample proportion polluting

---

Calculating
the standard error
of the proportion

Next, we can calculate the standard error of the proportion using the hypothesized population proportion as follows:

$$\sigma_{\bar{p}} = \sqrt{\frac{p_{H_0} q_{H_0}}{n}} \qquad [8 \cdot 4]$$

$$= \sqrt{\frac{(.6)(.4)}{60}}$$

$$= \sqrt{.004}$$

$$= .0632 \leftarrow \text{standard error of the population}$$

Illustrating
the problem

This is a one-tailed test: the EPA official wonders only whether the actual proportion is less than .6. Specifically, this is a left-tailed test. In order to reject the null hypothesis that the true proportion of plants in compliance is 60 percent, the EPA representative must accept the alternative hypothesis that fewer than .6 have complied. In Fig. 9·15, we have shown this hypothesis test graphically.

Determining
the limit of the
acceptance region

Since n is larger than 30, and since np and nq are each over 5, we can use the normal approximation of the binomial distribution. The appropriate z value from Appendix Table 1 for .48 of the area under the curve is 2.05. Thus, we can calculate the limit of the acceptance region as follows:

$$p_{H_0} - 2.05\sigma_{\bar{p}} = .6 - 2.05(.0632)$$

$$= .6 - .13$$

$$= .47 \leftarrow \text{lower limit}$$

FIG. 9·15
**One-tailed
hypothesis test at
.02 level of
significance**

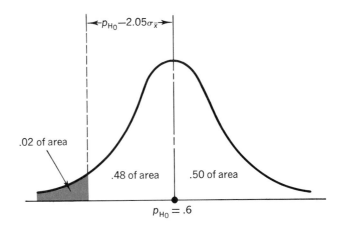

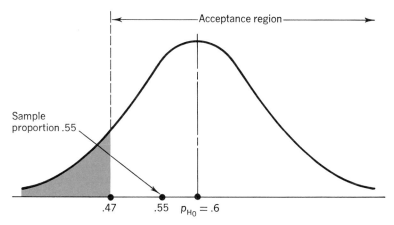

FIG. 9·16

One-tailed (left-tailed) hypothesis test at .02 significance level, showing acceptance region and sample proportion

Figure 9·16 illustrates the limit of the acceptance region, .47, and the sample proportion, .55, (33/60). Looking at this figure, we can see that the sample proportion lies within the acceptance region. Therefore, the EPA official should accept the null hypothesis that the true proportion of complying plants is .6. Although the observed sample proportion is below .6, *it is not significantly below* .6; that is, it is not far enough below .6 to make us accept the assertion by the member of the public interest group.

Interpreting the results

EXERCISES

36. For the treatment of a certain illness, the standard medication prescribed produces adverse side effects in 15 percent of the patients who take it. An experimental drug, recently tested on 75 patients suffering from the disorder, produced side effects in 18.7 percent of the patients who took it. At the .05 level of significance, is there evidence that the experimental drug leads to a higher incidence of side effects than the standard drug?

37. From a total of 8000 juveniles committed to detention schools in the most recent 5-year period, 300 were sampled to determine what proportion were girls. This sample showed 37 percent of those sent to the state detention schools were girls. A similar study made 5 years ago showed the percentage of girls in detention schools to be 32 percent. At a significance level of .10, has there been a significant change in the proportion of girls sent to detention schools?

38. A political candidate has consistently been favored by 58 percent of the population polled during the months preceding a general election. From a sample of 500 voters taken during the final week of campaigning, however, the proportion of voters in favor of this candidate was 54 percent.

    a) At the .10 level of significance, should the candidate believe that her support has changed?

    b) Reconsider the question in part a at the .05 level of significance.

39. A biology student became intrigued with the study of genetics and decided to do an extra project. According to his calculations for the offspring of a certain cross of fruit flies, .08 or 8 percent should exhibit certain phenotypic characteristics. Having performed an experiment to check his calculations, he observed that 49 of the 500 offspring exhibited the specified characteristics. Using a 5 percent level of significance, does he have reason to believe that his estimate of 8 percent is too low?

40. A government agency under the Department of Health, Education, and Welfare is concerned about the problem of malnutrition in young children. Though most people are aware of the problem for children of poor families, one study set out to find out about the incidence of children in middle class homes who were not receiving the proper nourishment. From a similar study made a few years ago, the percentage of children who showed severe malnutrition was 4 percent. In a recent screening of 5,000 middle class children throughout the country, 235 were found to show signs of severe malnutrition. At a 2 percent level of significance, should the agency be concerned about an increase in the proportion of undernourished children?

7 HYPOTHESIS TESTING OF MEANS UNDER DIFFERENT CONDITIONS

When to use the *t* distribution

When we estimated confidence intervals in Chapter 8, we learned that the difference in size between large and small samples is important when the population standard deviation σ is unknown and must be estimated from the sample standard deviation. If the sample size n is 30 or less and σ is not known, we should use the t distribution. The appropriate t distribution has $n - 1$ degrees of freedom. These rules apply to hypothesis testing, too.

Two-tailed tests of means using the t distribution

Setting up the problem symbolically

A personnel specialist of a major corporation is recruiting a large number of employees for an overseas assignment. During the testing process, management asks how things are going, and she replies, "Fine. I think the average score on the aptitude test will be 90." When management reviews 20 of the test results compiled, it finds that the mean score is 84, and the standard deviation of this score is 11.

$$\mu_{H_0} = 90 \leftarrow \text{hypothesized value of the population mean}$$
$$n = 20 \leftarrow \text{sample size}$$
$$\overline{x} = 84 \leftarrow \text{sample mean}$$
$$s = 11 \leftarrow \text{sample standard deviation}$$

If management wants to test her hypothesis at the .10 level of significance, what is the procedure?

$$H_0: \mu = 90 \leftarrow \text{null hypothesis: the true population mean score is 90}$$
$$H_1: \mu \neq 90 \leftarrow \text{alternative hypothesis: the mean score is not 90}$$
$$\alpha = .10 \leftarrow \text{level of significance for testing this hypothesis}$$

Calculating the standard error of the mean

Since the population standard deviation is not known, we must estimate it using the sample standard deviation and Equation 8·1:

$$\hat{\sigma} = s \tag{8·1}$$
$$= 11$$

Now we can compute the standard error of the mean. Since we are using $\hat{\sigma}$, an estimate of the population standard deviation, the standard error of the mean will also be an estimate. We can use Equation 8·6 as follows:

$$\hat{\sigma}_{\bar{x}} = \frac{\hat{\sigma}}{\sqrt{n}}$$ [8·6]

$$= \frac{11}{\sqrt{20}}$$

$$= \frac{11}{4.47}$$

$= 2.46 \leftarrow$ estimated standard error of the mean

Figure 9·17 illustrates this problem graphically. Since management is interested in knowing whether the true mean score is *larger or smaller* than the hypothesized score, a *two-tailed* test is the appropriate one to use. The significance level of .10 is shown in Fig. 9·17 as the two colored areas, each containing .05 of the area under the t distribution. Since the sample size is 20, the appropriate number of degrees of freedom is 19; that is, 20 − 1. Therefore, we look in the t distribution table, Appendix Table 2, under the .10 column until we reach the 19 degrees of freedom row. There we find the t value 1.729.

Illustrating
the problem

Determining
the limits of the
acceptance region

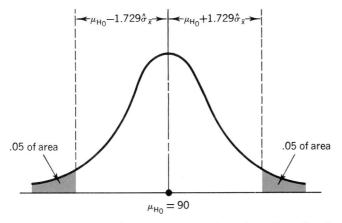

FIG. 9·17
**Two-tailed test of
hypothesis at
.10 level of
significance using
the t distribution**

This value is the appropriate one to use in calculating the limits of the acceptance region:

$$\mu_{H_0} + 1.729\hat{\sigma}_{\bar{x}} = 90 + 1.729(2.46)$$
$$= 90 + 4.25$$
$$= 94.25 \leftarrow \text{upper limit}$$

$$\mu_{H_0} - 1.729\hat{\sigma}_{\bar{x}} = 90 - 1.729(2.46)$$
$$= 90 - 4.25$$
$$= 85.75 \leftarrow \text{lower limit}$$

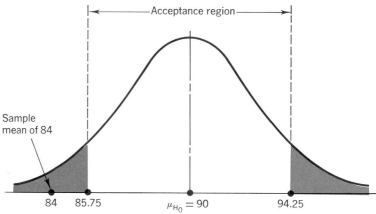

FIG. 9·18
Two-tailed hypothesis test at .10 level of significance, showing acceptance region and the sample mean

Sample mean of 84

84   85.75 $\mu_{H_0} = 90$ 94.25

Interpreting the results

Figure 9·18 illustrates these two limits of the acceptance region, 94.25 and 85.75, and sample mean, 84. From this figure, we can see that the sample mean lies outside the acceptance region. Therefore, management should reject the null hypothesis (the personnel specialist's assertion that the true mean score of the employees being tested is 90).

One-tailed tests of means using the t distribution

One difference from the z tables

The procedure for a one-tailed hypothesis test using the *t* distribution is the same conceptually as for a one-tailed test using the normal distribution and the *z* table. Performing such one-tailed tests may cause some difficulty, however. Notice that the column headings in Appendix Table 2 represent the *area in both tails combined.* Thus, they are appropriate to use in a two-tailed test with *two* rejection regions.

Using the t tables for one-tailed tests

If we use the *t* distribution for a one-tailed test, we wish to determine the area located in only one tail. So to find the appropriate *t* value for a one-tailed test at a significance level of .05 with 12 degrees of freedom, we would look in Appendix Table 2 under the .10 column opposite the 12 degrees of freedom row. The answer in this case is 1.782. This is true because the .10 column represents .10 of the area under the curve contained in *both tails combined,* and so it also represents .05 of the area under the curve contained in each of the tails separately.

EXERCISES

41. Given a sample mean of 19.1, a sample standard deviation of 4, and a sample of size 25, test the hypothesis that the value of the population mean is 17, against the alternative that it is greater than 17. Use the .01 significance level.

42. If a sample of 10 observations reveals a sample mean of 12 and a sample variance of 1.96, test the hypothesis that the population mean is 13, against the alternative that it is some other value. Use the .05 level of significance.

43. A sample of 16 observations has a mean of 210.5 and a standard deviation of 30. Test the hypothesis that in the entire population of 120 elements, the mean is 200, against the alternative that it is greater than 200, using the .01 level of significance.

44. For a sample of 50 taken from a population of 2,000, the sample mean is 105.1 and the sample standard deviation is 21.5. Using the .05 level of significance, test the hypothesis that the true population mean is 102, against the alternative that it is some other value.

45. An experimental psychologist has been using a new maze for conditioning rats. The 160 rats that he taught to run the maze averaged 14.1 trials before they gave a perfect performance, with a standard deviation of 4 trials. Long experience with a population of rats trained to run a similar maze showed that they averaged 12.6 trials before a perfect performance. At the .05 level of significance, should the psychologist conclude that the new maze is harder for the rats to learn?

46. The present best-selling remedy for headaches is reported to bring relief in 15 minutes. In a pilot study of 9 individuals, scientists at a pharmaceutical company found that their new formula brought relief in an average of 13.5 minutes, with a standard deviation of 1.2 minutes. At the .025 level of significance, is there reason to believe that the average relief time for the new medication is shorter than that for the old?

47. A television documentary on overeating claimed that Americans are 16 pounds overweight on average. To test this claim, 9 randomly selected individuals were examined, and the average excess weight was found to be 18 pounds with a standard deviation of 4 pounds. At the .05 level of significance, is there reason to believe the claim of 16 pounds to be in error?

8 HYPOTHESIS TESTING FOR DIFFERENCES BETWEEN MEANS AND PROPORTIONS

In many decision-making situations, people need to determine whether the parameters of two populations are alike or different. A company may want to test, for example, whether its female employees receive lower salaries than its male employees for the same work. A training director may wish to determine whether the proportion of promotable employees at one government installation is different from that at another. A drug manufacturer may need to know whether a new drug causes one reaction in one group of experimental animals but a different reaction in another group.

Comparing two populations

In each of these examples, decision makers are concerned with the parameters of two populations. In these situations, they are not as interested in the actual value of the parameters as they are in the *relation between* the values of the two parameters, that is, how these parameters differ. *Do* female employees earn less than male employees for the same work? *Is* the proportion of promotable employees at one installation different from that at another? *Did* one group of experimental animals react differently from the other? In this section, we shall introduce methods by which these questions can be answered, using hypothesis testing procedures.

Sampling distribution for the difference between two population parameters—basic concepts

A new way to generate a sampling distribution

In Chapter 7, we introduced the concept of the sampling distribution of the mean as the foundation for the work we would do in estimation and hypothesis testing. For a quick review of the sampling distribution of the mean, you may refer to Fig. 7·1 on page 239.

Since we now wish to study two populations, not just one, the sampling distribution of interest is the *sampling distribution of the difference between sample means.* Figure 9·19 may help us conceptualize this particular sampling distribution. At the top of this figure, we have drawn two populations, identified as Population 1 and Population 2. These two have means of μ_1 and μ_2 and standard deviations of σ_1 and σ_2, respectively. Beneath each population, we show the sampling distribution of the mean for that population. At the bottom of the figure is the sampling distribution of the difference between the sample means.

Deriving the sampling distribution of the difference between sample means

The two theoretical sampling distributions of the mean in Fig. 9·19 are each made up of all the possible samples of a given size that can be drawn from the corresponding population distribution. Now, suppose we take a random sample from the distribution of Population 1 and another random sample from the distribution of Population 2. If we then subtract the two sample means, we get:

$$\bar{x}_1 - \bar{x}_2 \leftarrow \text{difference between sample means}$$

Parameters of this sampling distribution

This difference will be positive if $\bar{x}_1$ is larger than $\bar{x}_2$, and negative if $\bar{x}_2$ is greater than $\bar{x}_1$. By constructing a distribution of *all* the possible sample differences of $\bar{x}_1 - \bar{x}_2$, we end up with the sampling distribution of the difference between sample means, which is shown at the bottom of Fig. 9·19.

The *mean of the sampling distribution of the difference between sample means* is symbolized $\mu_{\bar{x}_1 - \bar{x}_2}$ and is equal to $\mu_{\bar{x}_1} - \mu_{\bar{x}_2}$, which, as we saw in Chapter 7, is the same as $\mu_1 - \mu_2$. If $\mu_1 = \mu_2$, then $\mu_{\bar{x}_1} - \mu_{\bar{x}_2} = 0$.

The standard deviation of the distribution of the difference between the sample means is called the *standard error of the difference between two means* and is calculated using this formula:

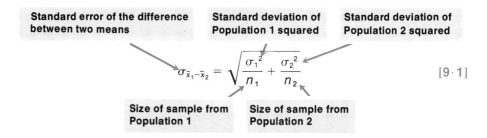

| Standard error of the difference between two means | Standard deviation of Population 1 squared | Standard deviation of Population 2 squared |

$$\sigma_{\bar{x}_1 - \bar{x}_2} = \sqrt{\frac{\sigma_1^2}{n_1} + \frac{\sigma_2^2}{n_2}}$$ [9·1]

| Size of sample from Population 1 | Size of sample from Population 2 |

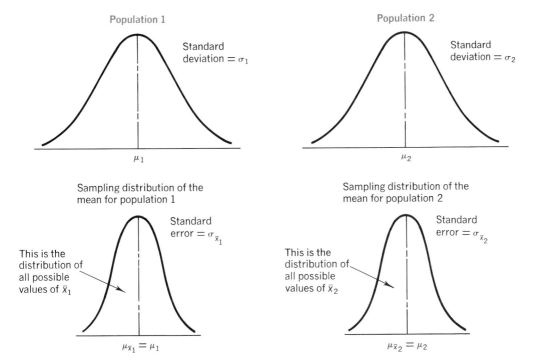

Population 1

Standard deviation $= \sigma_1$

μ_1

Population 2

Standard deviation $= \sigma_2$

μ_2

Sampling distribution of the mean for population 1

This is the distribution of all possible values of $\bar{x}_1$

Standard error $= \sigma_{\bar{x}_1}$

$\mu_{\bar{x}_1} = \mu_1$

Sampling distribution of the mean for population 2

This is the distribution of all possible values of $\bar{x}_2$

Standard error $= \sigma_{\bar{x}_2}$

$\mu_{\bar{x}_2} = \mu_2$

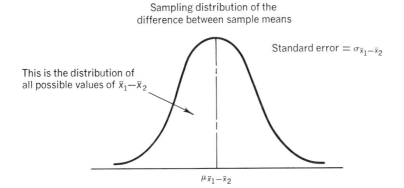

Sampling distribution of the difference between sample means

This is the distribution of all possible values of $\bar{x}_1 - \bar{x}_2$

Standard error $= \sigma_{\bar{x}_1 - \bar{x}_2}$

$\mu_{\bar{x}_1 - \bar{x}_2}$

FIG. 9·19

Basic concepts of population distributions, sampling distributions of the mean, and the sampling distribution of the difference between sample means

If the two population standard deviations are *not* known, we can *esti-mate* the standard error of the difference between two means. We can use the same method of estimating the standard error that we have used before by letting sample standard deviations estimate the population standard deviations as follows:

How to estimate the standard error of this sampling distribution

$$\hat{\sigma} = s \qquad\qquad [8\cdot1]$$

Sample standard deviations

Therefore, the formula for the estimated standard error of the difference between two means becomes:

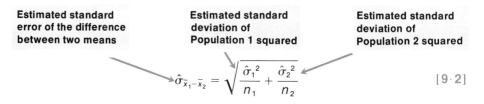

As the following examples show, depending on the sample sizes, we shall use different estimates for $\hat{\sigma}_1$ and $\hat{\sigma}_2$ in Equation 9·2.

Two-tailed tests for difference between means (large sample sizes)

Setting up
the problem
symbolically

When both sample sizes are greater than 30, this example illustrates how to do a two-tailed test of a hypothesis of the difference between two means. A Department of Labor statistician is asked to determine whether the hourly wages of semiskilled workers are the same in two cities. The statistician takes simple random samples of hourly earnings in both cities. The results of this survey are presented in Table 9·2. Suppose that the department wants to test the hypothesis at the .05 level that there is no difference between hourly wages for semiskilled workers in the two cities:

H_0: $\mu_1 = \mu_2$ ← null hypothesis: there is no difference
H_1: $\mu_1 \neq \mu_2$ ← alternative hypothesis: a difference exists
$\alpha = .05$ ← level of significance for testing this hypothesis

TABLE 9·2
Data from sample survey of hourly wages

City	Mean hourly earnings from sample	Standard deviation of sample	Size of sample
Apex	$6.95	$.40	200
Eden	7.10	.60	175

Calculating
the standard error
of the difference
between two means

Since the Department of Labor is interested only in whether the means are *or* are not equal, this is a two-tailed test.

The standard deviations of the two populations are not known. Therefore, our first step is to estimate them as follows:

$$\hat{\sigma}_1 = s_1 \quad \hat{\sigma}_2 = s_2$$
$$= \$.40 \quad = \$.60 \qquad [8·1]$$

Now the estimated standard error of the difference between two means can be determined by:

$$\hat{\sigma}_{\bar{x}_1 - \bar{x}_2} = \sqrt{\frac{\hat{\sigma}_1^2}{n_1} + \frac{\hat{\sigma}_2^2}{n_2}}$$

$$= \sqrt{\frac{(.40)^2}{200} + \frac{(.60)^2}{175}}$$

$$= \sqrt{\frac{.16}{200} + \frac{.36}{175}}$$

$$= \sqrt{.0028}$$

$$= \$.053 \leftarrow \text{estimated standard error}$$

[9·2]

We can illustrate this hypothesis test graphically. In Fig. 9·20 the significance level of .05 corresponds to the two colored areas, each of which contains .025 of the area. The acceptance region contains two equal areas of .475 each. Since both samples are large, we can use the normal distribution. From Appendix Table 1, we can determine the appropriate z value for .475 of the area under the curve to be 1.96. Now we can calculate the limits of the acceptance region:

Illustrating the problem

Determining the limits of the acceptance region

$$0 + 1.96\hat{\sigma}_{\bar{x}_1 - \bar{x}_2} = 0 + 1.96(\$.053)$$
$$= 0 + .1039$$
$$= \$.1039 \leftarrow \text{upper limit}$$

The hypothesized difference between the two population means is zero.

$$0 - 1.96\hat{\sigma}_{\bar{x}_1 - \bar{x}_2} = 0 - 1.96(\$.053)$$
$$= 0 - .1039$$
$$= -\$.1039 \leftarrow \text{lower limit}$$

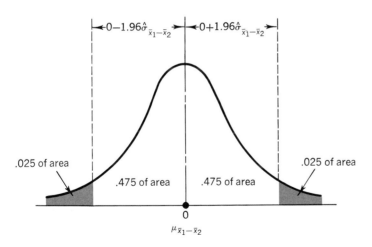

$\leftarrow 0-1.96\hat{\sigma}_{\bar{x}_1-\bar{x}_2} \rightarrow | \leftarrow 0+1.96\hat{\sigma}_{\bar{x}_1-\bar{x}_2} \rightarrow |$

.025 of area

.475 of area   .475 of area

.025 of area

0

$\mu_{\bar{x}_1-\bar{x}_2}$

FIG. 9·20
Two-tailed hypothesis test of the difference between two means at the .05 level of significance

Figure 9·21 illustrates these two limits of the acceptance region ($.1039 and −$.1039) and indicates the difference between the sample means. It is calculated:

Interpreting the result

FIG. 9·21
**Two-tailed
hypothesis test of
the difference
between two
means at .05 level
of significance,
showing
acceptance region
and difference
between sample
means**

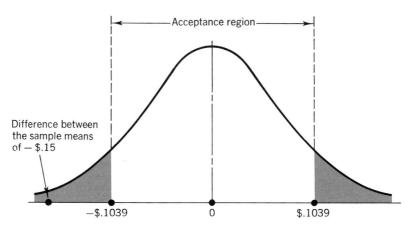

$$\text{Difference} = \bar{x}_1 - \bar{x}_2 \text{ (From Table 9·2)}$$
$$= \$6.95 - \$7.10$$
$$= -\$0.15$$

Figure 9·21 demonstrates that the difference between the two sample means lies outside the acceptance region. Thus, we reject the null hypothesis of no difference and conclude that the population means (the average semiskilled wages in these two cities) differ.

One-tailed tests for difference between means (small sample sizes)

**Setting up
the problem
symbolically**

The procedure for a one-tailed test of the difference between means is conceptually like that for the one-tailed tests of means we have already discussed. The only major difference will be in how we compute the estimated standard error of the difference between the two means. Suppose that the city of San Antonio has been investigating two education programs for increasing the sensitivity of its police officers to the needs of its large Spanish-speaking population. The original program consisted of several informal question-and-answer sessions with leaders of the Spanish-speaking community. Over the past few years, a program involving formal classroom contact with professional psychologists and sociologists has been developed. The new program is considerably more expensive and the mayor's office wants to know at the .05 level of significance whether this expenditure has resulted in greater sensitivity. Let's test the following:

$H_0: \mu_1 = \mu_2 \leftarrow$ null hypothesis: there is no difference in sensitivity levels achieved by the two programs
$H_1: \mu_1 > \mu_2 \leftarrow$ alternative hypothesis: the new program results in higher sensitivity levels
$\alpha = .05 \leftarrow$ level of significance for testing this hypothesis

Table 9·3 contains the data resulting from a sample of the officers trained in both programs. Because only limited data are available for the

Program sampled	Mean sensitivity after this program	Number of officers observed	Estimated standard deviation of sensitivity after this program
Formal	92%	12	15%
Informal	84%	15	19%

TABLE 9·3
Data from sample of two sensitivity programs

two programs, the population standard deviations are estimated from the data. The sensitivity level is measured as a percentage on a standard psychometric scale.

The mayor wishes to test whether the sensitivity achieved by the new program is *significantly higher* than that achieved under the older, more informal program. To reject the null hypothesis (a result that the mayor desires), the observed difference of sample means would need to fall sufficiently high in the *right* tail of the distribution. Then we would accept the alternative hypothesis that the new program leads to higher sensitivity levels and that the extra expenditures on this program are justified.

Our first task in performing the test is to calculate the standard error of the difference between the two means. Since the population standard deviations are not known, we must use Equation 9·2:

$$\hat{\sigma}_{\bar{x}_1-\bar{x}_2} = \sqrt{\frac{\hat{\sigma}_1{}^2}{n_1} + \frac{\hat{\sigma}_2{}^2}{n_2}} \qquad [9·2]$$

In the previous example, where the sample sizes were large (both greater than 30) we used Equation 8·1 and estimated $\hat{\sigma}_1{}^2$ by $s_1{}^2$, and $\hat{\sigma}_2{}^2$ by $s_2{}^2$. Now, with small sample sizes, that procedure is not appropriate. If we can assume that the unknown population variances are equal (and this assumption can be tested using a method discussed in Section 6 of the next chapter) we can continue. If we cannot assume that $\sigma_1{}^2 = \sigma_2{}^2$, then the problem is beyond the scope of this text.

Assuming for the moment that $\sigma_1{}^2 = \sigma_2{}^2$, how can we estimate the common variance σ^2? If we use either $s_1{}^2$ or $s_2{}^2$, we get an unbiased estimator of σ^2, but we don't use all of the information available to us since we ignore one of the samples. Instead we use a weighted average of $s_1{}^2$ and $s_2{}^2$, and the weights are the numbers of degrees of freedom in each sample. This weighted average is called a "pooled estimate" of σ^2. It is given by

Estimating σ^2 with small sample sizes

$$s_p{}^2 = \frac{(n_1 - 1)s_1{}^2 + (n_2 - 1)s_2{}^2}{n_1 + n_2 - 2} \qquad [9·3]$$

Plugging this into Equation 9·2 and simplifying gives us

$$\hat{\sigma}_{\bar{x}_1-\bar{x}_2} = s_p \sqrt{\frac{1}{n_1} + \frac{1}{n_2}} \qquad [9·4]$$

When we want to test hypotheses about differences of population means, and we have small samples but equal population variances, we use Equation 9·4 to estimate the standard error of the difference between the two means. Then as you might have guessed, the test is based on the t distribution. The appropriate number of degrees of freedom is $(n_1 - 1) + (n_2 - 1)$, or $n_1 + n_2 - 2$, which is the denominator in Equation 9·3.

Applying these results to our sensitivity example,

$$s_p{}^2 = \frac{(n_1-1)s_1{}^2 + (n_2-1)s_2{}^2}{n_1 + n_2 - 2} \qquad [9·3]$$

$$= \frac{(12-1)(15)^2 + (15-1)(19)^2}{12 + 15 - 2}$$

$$= \frac{11(225) + 14(361)}{25}$$

$$= 301.160$$

Taking square roots on both sides, we get $s_p = \sqrt{301.160}$ or 17.354, and so

$$\hat{\sigma}_{\bar{x}_1-\bar{x}_2} = s_p \sqrt{\frac{1}{n_1} + \frac{1}{n_2}} \qquad [9·4]$$

$$= 17.354 \sqrt{\frac{1}{12} + \frac{1}{15}}$$

$$= 17.354(.387)$$

$$= 6.721$$

Illustrating
the problem

In Fig. 9·22, a graphic illustration of this hypothesis test, the significance level of .05 is represented by the colored region at the right of the distribution. Since both samples are less than 30, the t distribution with $12 + 15 - 2 = 25$ degrees of freedom is the appropriate sampling distribution. The t value for .05 of the area under the curve is 1.708 according to Appendix Table 2. The value of $\mu_{\bar{x}_1-\bar{x}_2}$, the mean of the sampling distribution of the

FIG. 9·22

**Right-tailed
hypothesis test of
the difference
between two
means at the .10
level of
significance**

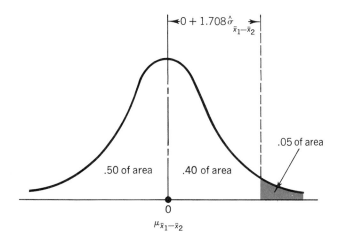

$\leftarrow 0 + 1.708\hat{\sigma}_{\bar{x}_1-\bar{x}_2} \rightarrow$

.05 of area

.50 of area   .40 of area

0

$\mu_{\bar{x}_1-\bar{x}_2}$

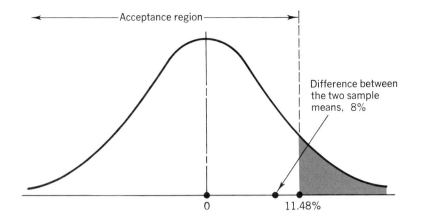

Acceptance region

Difference between
the two sample
means, 8%

0   11.48%

FIG. 9·23
**One-tailed test of
the difference
between two
means at the .05
level of
significance,
showing
acceptance region
and the difference
between the
sample means**

hypothesized difference between the two sensitivity means, is equal to zero. Thus, the calculation for determining the limit of the acceptance region is:

Determining the limits of the acceptance region

$$0 + 1.708\hat{\sigma}_{\bar{x}_1 - \bar{x}_2} = 0 + 1.708(6.721)$$
$$= 0 + 11.48$$
$$= 11.48\% \leftarrow \text{upper limit}$$

In Fig. 9·23, we have illustrated this limit of the acceptance region and the difference between the two sample sensitivities ($92\% - 84\% = 8\%$). We can see in Fig. 9·23 that the difference between the two sample means lies within the acceptance region. Thus, we accept the null hypothesis that there is no difference between the sensitivities achieved by the two programs. The city's expenditures on the formal instructional program have not produced significantly higher sensitivities.

Interpreting the result

Testing differences between means with dependent samples

In the last two examples our samples were chosen *independently* of each other. In the wage example, the samples were taken in two different cities. In the sensitivity example, samples were taken of policemen who had gone through two different training programs. Sometimes, however, it will make sense to take samples that are not independent of each other. Often the use of such *dependent* (or *paired*) samples will enable us to perform a more precise analysis, because they will allow us to control for extraneous factors. With dependent samples, we still follow the same basic procedure we have followed in all our hypothesis testing. The only differences are that we will use a different formula for the estimated standard error of the sample differences and that we will require that both samples be of the same size.

Conditions under which paired samples aid analysis

A health spa has advertised a weight reducing program and has claimed that the average participant in the program loses at least seventeen pounds. A somewhat overweight executive is interested in the program but is skeptical about the claims and asks for some hard evidence. The spa allows him to select randomly the records of 10 participants and record their weights

TABLE 9·4
**Weights before
and after a
reducing program**

Before	189	202	220	207	194	177	193	202	208	233
After	170	179	203	192	172	161	174	187	186	204

The overweight executive wants to test at the 5 percent significance level the claimed average weight loss of at least 17 pounds. Formally, we may state this problem:

H_0: $\mu_1 - \mu_2 = 17$ ← null hypothesis: average weight loss is only 17 pounds

H_1: $\mu_1 - \mu_2 > 17$ ← alternative hypothesis: average weight loss exceeds 17 pounds

$\alpha = .05$ ← level of significance

before and after the program. These data are recorded in Table 9·4. Here we have two samples (a *before* sample and an *after* sample) which are clearly dependent on each other, since the same ten people have been observed twice.

Conceptual
understanding
of differences

What we are really interested in is not the weights before and after but only their *differences.* **Conceptually, what we have is *not two samples* of before and after weights, but rather *one sample* of weight losses.** If the population of weight losses has a mean μ_ℓ, we can restate our hypothesis as:

H_0: $\mu_\ell = 17$

H_1: $\mu_\ell > 17$

Now we compute the individual losses, their mean and standard deviation, and proceed exactly as we did when testing hypotheses about a single mean. The computations are done in Table 9·5.

We use Equation 8·1 to estimate the unknown population standard deviation:

$$\hat{\sigma} = s$$
$$= 4.40$$
[8·1]

and now we can estimate the standard error of the mean:

$$\hat{\sigma}_{\bar{x}} = \frac{\hat{\sigma}}{\sqrt{n}}$$
[8·6]

$$= \frac{4.40}{\sqrt{10}}$$

$$= \frac{4.40}{3.16}$$

$$= 1.39 \leftarrow \text{estimated standard error of the mean}$$

Before	After	Loss x	Loss squared x^2
189	170	19	361
202	179	23	529
220	203	17	289
207	192	15	225
194	172	22	484
177	161	16	256
193	174	19	361
202	187	15	225
208	186	22	484
233	204	29	841

TABLE 9·5
Finding the mean weight loss and its standard deviation

$$\Sigma x = 197 \qquad\qquad \Sigma x^2 = 4{,}055$$

$$\bar{x} = \frac{\Sigma x}{n} \quad [3\cdot2] \qquad\qquad s = \sqrt{\frac{\Sigma x^2}{n-1} - \frac{n\,\bar{x}^2}{n-1}} \quad [4\cdot12]$$

$$= \frac{197}{10} \qquad\qquad\qquad = \sqrt{\frac{4055}{9} - \frac{10(19.7)^2}{9}}$$

$$= 19.7 \qquad\qquad\qquad\quad = \sqrt{19.34}$$

$$\qquad\qquad\qquad\qquad\qquad = 4.40$$

Figure 9·24 illustrates this problem graphically. Since we want to know if the mean weight loss *exceeds* 17 pounds, an upper-tailed test is appropriate. The .05 significance level is shown in Fig. 9·24 as the colored area under the t distribution. We use the t distribution because the sample size is only 10; the appropriate number of degrees of freedom is 9, $(10 - 1)$. Appendix Table 2 gives the t value of 1.833.

We use this t value to calculate the upper limit of the acceptance region

$$\mu_{H_0} + 1.833\,\hat{\sigma}_{\bar{x}} = 17 + 1.833(1.39)$$
$$= 17 + 2.55$$
$$= 19.55 \text{ pounds} \leftarrow \text{upper limit}$$

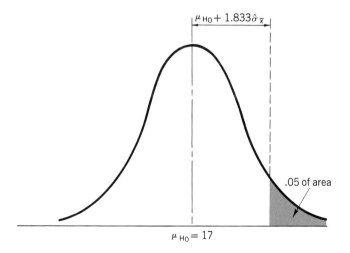

FIG. 9·24
One-tailed hypothesis test at .05 level of significance

335

FIG. 9·25

One-tailed
hypothesis test at
.05 level of
significance,
showing
acceptance region
and sample mean

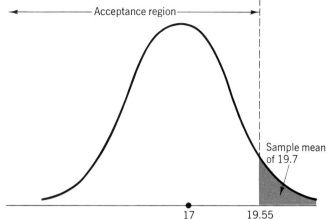

Interpreting
the results

Figure 9·25 illustrates the acceptance region and the sample mean 19.7. We see that the sample mean lies outside the acceptance region, so the executive can reject the null hypothesis and conclude that the claimed weight loss in the program is legitimate.

How does the paired difference test differ?

Let's see how this *paired difference* test differs from a test of the difference of means of *two independent* samples. Suppose that the data in Table 9·4 represent two independent samples of 10 individuals *entering* the program and *another* 10 randomly selected individuals *leaving* the program. The means and variances of the two samples are given in Table 9·6.

TABLE 9·6

Before and after means and variances

Population	Size	Mean	Variance
Before	10	202.5	253.61
After	10	182.8	201.96

Since the sample sizes are small, we use Equation 9·3 to get a pooled estimate of σ^2 and Equation 9·4 to estimate $\hat{\sigma}_{\bar{x}_1 - \bar{x}_2}$:

$$s_p^2 = \frac{(n_1 - 1)s_1^2 + (n_2 - 1)s_2^2}{n_1 + n_2 - 2} \qquad [9\cdot3]$$

$$= \frac{(10 - 1)(201.96) + (10 - 1)(253.61)}{10 + 10 - 2}$$

$$= \frac{1817.64 + 2282.49}{18}$$

$$= 227.79 \leftarrow \text{estimate of common population variance}$$

$$\hat{\sigma}_{\bar{x}_1 - \bar{x}_2} = s_p \sqrt{\frac{1}{n_1} + \frac{1}{n_2}} \qquad [9\cdot4]$$

$$= \sqrt{227.79} \sqrt{\frac{1}{10} + \frac{1}{10}}$$

$$= 15.09 \sqrt{0.2}$$

$$= 15.09(0.45)$$

$$= 6.79$$

The appropriate test is now based on the t distribution with 18 degrees of freedom $(10 + 10 - 2)$. With a significance level of .05, the appropriate t value from Appendix Table 2 is 1.734, so the upper limit of the acceptance region is

$$17 + 1.734\hat{\sigma}_{\bar{x}_1 - \bar{x}_2} = 17 + 1.734(6.79)$$
$$= 17 + 11.77$$
$$= 28.77 \text{ pounds}$$

The observed difference of the sample means is

$$\bar{x}_1 - \bar{x}_2 = 202.5 - 182.8$$
$$= 19.7 \text{ pounds}$$

so this test will *not* reject H_0.

Why did these two tests give such different results? In the paired sample test, the sample standard deviation of the individual differences was relatively small so 19.7 pounds was significantly larger than the hypothesized weight loss of 17 pounds. With independent samples, however, the estimated standard deviation of the difference between the means depended on the standard deviations of the before weights and the after weights. Since both of these were relatively large, $\hat{\sigma}_{\bar{x}_1 - \bar{x}_2}$ was also large, and thus 19.7 was not significantly larger than 17. The paired sample test controlled this initial and final variability in weights by looking only at the individual changes in weights. Because of this it was better able to detect the significance of the weight loss. *Explaining differing results*

We conclude this section with two examples showing when to treat 2 samples of equal size as dependent or independent.

1. An agricultural extension service wishes to determine whether a new hybrid seed corn has a greater yield than an old standard variety. If the service asks 10 farmers to record the yield of an acre planted with the new variety and asks another 10 farmers to record the yield of an acre planted with the old variety, the two samples are independent. If however it asks 10 farmers to plant one acre with each variety and record the results, then the samples are dependent, and the paired difference test is appropriate. In the latter case, differences due to fertilizer, insecticide, rainfall, etc., are controlled, because each farmer treats his two acres identically. Thus any differences in yield can be attributed solely to the variety planted.

2. The director of the secretarial pool at a large university central administrative office wants to determine whether typing speed depends upon the kind of typewriter used by a secretary. If she tests 7 secretaries using electric typewriters and 7 using manual typewriters, she should treat her samples as independent. If she tests the same 7 secretaries twice (once on each type of machine), then the two samples are dependent. In the paired difference test, differences among the secretaries are eliminated as a contributing factor, and the differences in typing speeds can be attributed to the different types of machines. *Should we treat samples as dependent or independent?*

Two-tailed tests for difference between proportions

Setting up
the problem
symbolically

Consider the case of a drug research organization that is testing two new compounds intended to reduce blood pressure levels. The compounds are administered to two different sets of laboratory animals. In group one, 71 of 100 animals tested respond to drug 1 with lower blood pressure levels. In group two, 58 of 90 animals tested respond to drug 2 with lower blood pressure levels. The organization wants to test whether there is a difference between the efficacies of these two drugs at the .05 level. How should we proceed with this problem?

$\bar{p}_1 = .71 \leftarrow$ sample proportion of successes with drug 1
$\bar{q}_1 = .29 \leftarrow$ sample proportion of failures with drug 1
$n_1 = 100 \leftarrow$ sample size for testing drug 1
$\bar{p}_2 = .644 \leftarrow$ sample proportion of successes with drug 2
$\bar{q}_2 = .356 \leftarrow$ sample proportion of failures with drug 2
$n_2 = 90 \leftarrow$ sample size for testing drug 2

$H_0: p_1 = p_2 \leftarrow$ null hypothesis: there is no difference between these two drugs
$H_1: p_1 \neq p_2 \leftarrow$ alternative hypothesis: there is a difference between them
$\alpha = .05 \leftarrow$ level of significance for testing this hypothesis

Calculating
the standard error
of the difference
between
two proportions

As in our previous examples, we can begin by calculating the standard deviation of the sampling distribution we are using in our hypothesis test. In this example, the binomial distribution is the correct sampling distribution. We want to find the *standard error of the difference between two proportions;* therefore, we should recall the formula for the *standard error of the proportion:*

$$\sigma_{\bar{p}} = \sqrt{\frac{pq}{n}} \qquad [8 \cdot 4]$$

Using this formula and the same form we previously used in Equation 9·1 for the standard error of the difference between two *means,* we get:

$$\sigma_{\bar{p}_1 - \bar{p}_2} = \sqrt{\frac{p_1 q_1}{n_1} + \frac{p_2 q_2}{n_2}} \qquad [9 \cdot 5]$$

How to estimate
this standard error

To test the two compounds, we do not know the population parameters p_1, p_2, q_1 and q_2, and thus we need to estimate them from the sample statistics $\bar{p}_1, \bar{p}_2, \bar{q}_1$, and $\bar{q}_2$. In this case, we might suppose that the practical formula to use would be:

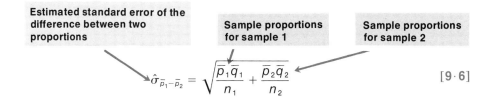

Estimated standard error of the difference between two proportions

Sample proportions for sample 1

Sample proportions for sample 2

$$\hat{\sigma}_{\bar{p}_1 - \bar{p}_2} = \sqrt{\frac{\bar{p}_1 \bar{q}_1}{n_1} + \frac{\bar{p}_2 \bar{q}_2}{n_2}} \qquad [9 \cdot 6]$$

But think about this a bit more. After all, if we hypothesize that there is *no difference* between the two population proportions, then our best estimate of the overall population proportion of successes is probably the *combined* proportion of successes in both samples, that is:

Best estimate of the overall proportion of successes in the population if the 2 proportions are hypothesized to be equal	=	$\dfrac{\textbf{Number of successes in sample 1} + \textbf{Number of successes in sample 2}}{\textbf{Total size of both samples}}$

And in the case of the two compounds, we use this equation with symbols rather than words:

$$\hat{p} = \frac{(n_1)(\bar{p}_1) + (n_2)(\bar{p}_2)}{n_1 + n_2} \qquad [9\cdot7]$$

$$= \frac{(100)(.71) + (90)(.644)}{100 + 90}$$

$$= \frac{71 + 58}{190}$$

$= .6789 \leftarrow$ estimate of the overall proportion of successes in the combined populations using combined proportions from both samples ($\hat{q}$ would be $1 - .6789 = .3211$)

Now we can appropriately modify Equation 9·6 using the values $\hat{p}$ and $\hat{q}$ from Equation 9·7.

Estimated standard error of the difference between two proportions using combined estimates

Estimates of the population proportions using combined proportions from both samples

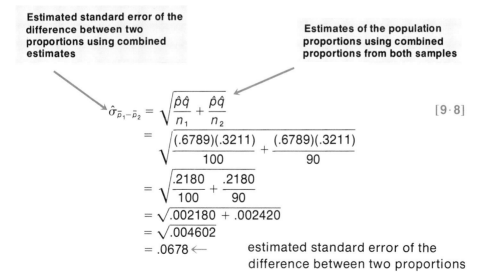

$$\hat{\sigma}_{\bar{p}_1-\bar{p}_2} = \sqrt{\frac{\hat{p}\hat{q}}{n_1} + \frac{\hat{p}\hat{q}}{n_2}} \qquad [9\cdot8]$$

$$= \sqrt{\frac{(.6789)(.3211)}{100} + \frac{(.6789)(.3211)}{90}}$$

$$= \sqrt{\frac{.2180}{100} + \frac{.2180}{90}}$$

$$= \sqrt{.002180 + .002420}$$

$$= \sqrt{.004602}$$

$= .0678 \leftarrow$ estimated standard error of the difference between two proportions

What did we save by using Equation 9·8 instead of Equation 9·6? In Equation 9·8 we needed only *one* value for $\hat{p}$ and *one* value for $\hat{q}$; thus we avoided some of the calculations involved in the use of Equation 9·6.

Illustrating the problem

Figure 9·26 illustrates this hypothesis test graphically. Since the management of the research organization wants to know whether there is a difference between the two compounds, this is a two-tailed test. The significance level of .05 corresponds to the colored regions in the figure. Both samples are large enough to justify using the normal distribution to approximate the binomial. From Appendix Table 1, we can determine that the appropriate z value for .475 of the area under the curve is 1.96. We can calculate the two limits of the acceptance region as follows:

Determining the limits of the acceptance region

The hypothesized difference between the 2 proportions is zero

$$\begin{aligned} 0 + 1.96\hat{\sigma}_{\bar{p}_1-\bar{p}_2} &= 0 + 1.96(.0678) \\ &= 0 + .1329 \\ &= .1329 \leftarrow \text{upper limit} \end{aligned}$$

$$\begin{aligned} 0 - 1.96\hat{\sigma}_{\bar{p}_1-\bar{p}_2} &= 0 - 1.96(.0678) \\ &= 0 - .1329 \\ &= -.1329 \leftarrow \text{lower limit} \end{aligned}$$

FIG. 9·26
Two-tailed hypothesis test of the difference between two proportions at the .05 level of significance

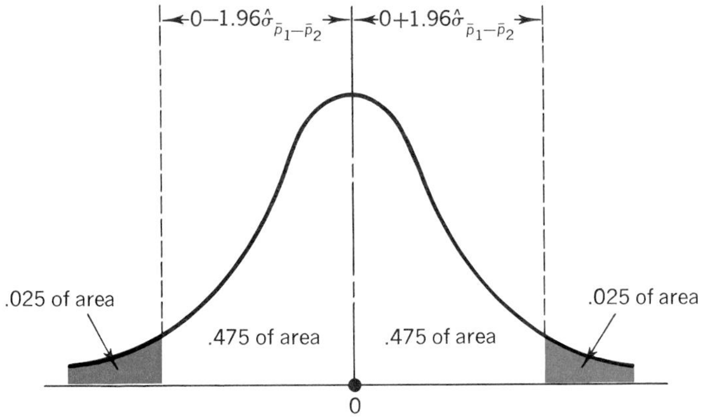
0−1.96σ̂ p̄1−p̄2
0+1.96σ̂ p̄1−p̄2
.025 of area
.475 of area
.475 of area
.025 of area
0

Figure 9·27 illustrates these two limits of the acceptance region, .1329 and −.1329. It also indicates the differences between the sample proportions, calculated as

$$\begin{aligned} \text{Difference} &= \bar{p}_1 - \bar{p}_2 \\ &= .71 - .644 \\ &= .066 \end{aligned}$$

Interpreting the results

We can see in Fig. 9·27 that the difference between the two sample proportions lies within the acceptance region. Thus, we accept the null hypothesis and conclude that these two compounds produce effects on blood pressure that are *not* different.

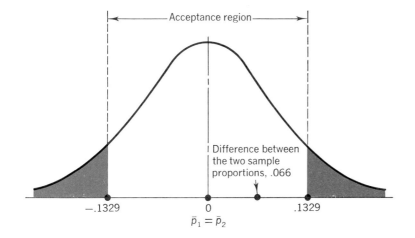

FIG. 9·27
Two-tailed
hypothesis test of
the difference
between two
proportions at the
.05 level of
significance,
showing
acceptance region
and the difference
between sample
proportions

Acceptance region

Difference between
the two sample
proportions, .066

$-.1329$ 0 $.1329$

$\bar{p}_1 = \bar{p}_2$

One-tailed tests for difference between proportions

Conceptually, the one-tailed test for the difference between two popu- Setting up
the problem
symbolically
lation proportions is similar to a one-tailed test for the difference between
two means. Suppose that for tax purposes, a city government has been
using two methods of listing property. The first requires the property owner
to appear in person before a tax lister, but the second permits the property
owner to mail in a tax form. The city manager thinks the personal appearance method produces far fewer mistakes than the mail-in method. She
authorizes an examination of 50 personal appearance listings and 75 mail-
in listings. Ten percent of the personal appearance forms contain errors;
13.3 percent of the mail-in forms contain them. The result of her sample can
be summarized:

$\bar{p}_1 = .10 \leftarrow$ proportion of personal appearance forms with errors
$\bar{q}_1 = .90 \leftarrow$ proportion of personal appearance forms without errors
$n_1 = 50 \leftarrow$ sample size of personal appearance forms
$\bar{p}_2 = .133 \leftarrow$ proportion of mail-in forms with errors
$\bar{q}_2 = .867 \leftarrow$ proportion of mail-in forms without errors
$n_2 = 75 \leftarrow$ sample size of mail-in forms

The city manager wants to test at the .15 level of significance the hypothesis
that the personal appearance method produces a lower proportion of
errors. What should she do?

$H_0: p_1 = p_2 \leftarrow$ null hypothesis: there is no difference between the two methods
$H_1: p_1 < p_2 \leftarrow$ alternative hypothesis: the personal appearance method has a lower
proportion of errors than the mail-in method
$\alpha = .15 \leftarrow$ level of significance for testing the hypothesis

Calculating
the standard error
of the difference
between
two proportions

To estimate the *standard error of the difference between two propor-tions,* we first use the combined proportions from both samples to estimate the overall proportion of successes:

$$\hat{p} = \frac{(n_1)(\bar{p}_1) + (n_2)(\bar{p}_2)}{n_1 + n_2} \qquad [9\cdot7]$$

$$= \frac{(50)(.10) + (75)(.133)}{50 + 75}$$

$$= \frac{5 + 10}{125}$$

$$= .12 \leftarrow \text{estimate of the overall proportion of successes in the population using combined proportions from both samples}$$

Now this answer can be used to calculate the standard error of the differ-ence between the two proportions, using Equation 9·8:

$$\hat{\sigma}_{\bar{p}_1-\bar{p}_2} = \sqrt{\frac{\hat{p}\hat{q}}{n_1} + \frac{\hat{p}\hat{q}}{n_2}} \qquad [9\cdot8]$$

$$= \sqrt{\frac{(.12)(.88)}{50} + \frac{(.12)(.88)}{75}}$$

$$= \sqrt{\frac{.10560}{50} + \frac{.10560}{75}}$$

$$= \sqrt{.002112 + .001408}$$

$$= \sqrt{.00352}$$

$$= .0593 \leftarrow \text{estimated standard error of the difference between two proportions using combined estimates}$$

Figure 9·28 illustrates this hypothesis test. Since the city manager wishes to test whether the personal appearance listing is better than the mailed-in listing, the appropriate test is a one-tailed test. Specifically, it is a *left-tailed* test because to reject the null hypothesis, the test result must fall

FIG. 9·28
One-tailed
hypothesis test of
the difference
between two
proportions at the
.15 level of
significance

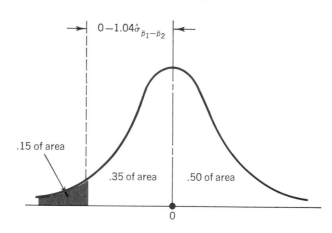

$$0 - 1.04\hat{\sigma}_{\bar{p}_1-\bar{p}_2}$$

.15 of area

.35 of area

.50 of area

0

in the colored portion of the left tail, indicating that *significantly fewer* errors exist in the personal appearance forms. This colored region in Fig. 9·28 corresponds to the .15 significance level.

Determining the limit of the acceptance region

With samples of this size, we can use the standard normal distribution and Appendix Table 1 to determine the appropriate *z* value for .35 of the area under the curve. We can use this value, 1.04, to calculate the one limit of the acceptance region:

The hypothesized difference between the 2 population proportions is zero $\longrightarrow$

$$0 - 1.04\hat{\sigma}_{\bar{p}_1-\bar{p}_2} = 0 - 1.04(.0593)$$
$$= 0 - .0617$$
$$\longleftarrow = -.0617 \quad \text{lower limit}$$

Interpreting the results

We have illustrated this limit to the acceptance region and the difference between the two sample proportions (.10 − .133 = −.033) in Fig. 9·29. This figure shows us that the difference between the two sample proportions lies well within the acceptance region, and the city manager should accept the null hypothesis that there is no difference between the two methods of tax listing. Therefore, if mailed-in listing is considerably less expensive to the city, the city manager should consider increasing the use of this method.

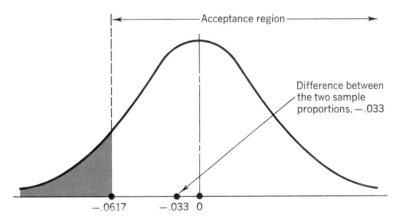

FIG. 9·29
One-tailed hypothesis test of the difference between two proportions at the .15 level of significance, showing acceptance region and the difference between the sample proportions

EXERCISES

48. Two independent samples of observations were collected. For the first sample of 36 elements, the mean was 240 and the standard deviation 14. The second sample of 49 elements had a mean of 230 and a standard deviation of 10.

    a) Compute the standard error of the difference between the two means.

    b) Test the hypothesis that the two samples are from populations with the same mean at the .05 level of significance.

49. In order to compare the performance of two training methods, samples of individuals from each of the methods were checked. For the six individuals from training method 1, the mean efficiency score was 35, with a variance of 40. For the eight individuals in training method 2, the mean efficiency score was 27, with a variance of 45.

   a) Compute the standard error of the difference between the two means.

   b) Test whether the efficiency scores by individuals from the two training methods may be concluded to be equal at the .01 level of significance.

50. The following two samples are physiological measures taken of patients before and after the administration of some medication.

   a) Find the mean change in the measure after the medication.

   b) Find the standard deviation of the change and the standard error of the mean.

   c) Test for a significant difference between the measures before and after the medication at the .05 level of significance.

Subject	1	2	3	4	5	6	7	8
Before	97	106	106	95	102	111	115	104
After	113	113	101	119	111	122	121	106

51. From a group of 1,000 men polled, 850 supported a certain bond issue. Of 500 women surveyed, 400 supported the issue.

   a) Calculate $\hat{p}$.

   b) Compute the standard error of the difference between the two proportions.

   c) Test the hypothesis that the proportions of men and women supporting the issue are equal, at the .01 level of significance.

52. Brenda Carson, a developmental psychologist, was curious about the ability of children to communicate new ideas to each other. Specifically, she wondered if there were certain things that children could teach one another as well as adults could teach these things to children. She ran a study in which adults taught some children a certain concept, and other children taught a peer the same concept. Of the 40 children taught by adults, the average score on a measure of their understanding was 20.2 with a standard deviation of 2.5. The 56 children instructed by other children had an average score of 21.0 with a standard deviation of 3.1. At a significance level of 10 percent, is there a significant difference in the understanding communicated by the adult and child teachers?

53. Two research laboratories have independently produced drugs that provide relief to arthritis sufferers. The first drug was tested on a group of 100 arthritis victims and produced an average of 8.5 hours of relief with a standard deviation of 2 hours. The second drug was tested on 75 arthritis victims, producing an average of 7.8 hours of relief with a standard deviation of 1.5 hours. At the .02 level of significance, does the first drug provide a significantly longer period of relief?

54. Two different areas of a city are being considered as sites for day-care centers. Of 150 households surveyed in one section, the proportion in which the mother worked full-time was .44. In the other section, 38 percent of the 100 households surveyed had mothers working at full-time jobs. At the .05 level of significance, is there a significant difference in the proportion of working mothers in the two areas of the city?

55. A comparative psychologist is currently studying the effects of different types of reinforcement used to promote learning in animals. It has been fairly well established that human infants respond and learn more when given attention and affection, such as holding and cuddling. To test the effectiveness of this type of reinforcement on animals, the psycholo-

gist records the number of trials leading to a criterion performance of a task accompanied by two types of reinforcement: either food pellets or handling and petting are rewards for correct responses. Of the 40 animals rewarded with food, the average number of trials leading to criterion performance is 28.8 with a standard deviation of 1.35 trials. Among the 45 animals rewarded with affection, the mean number of trials leading to criterion performance is 27.7 with a standard deviation of 1.49 trials. At the .05 level of significance, is a significant difference in learning time achieved by these two different rewards?

56. A coal-fired power plant is comparing two different systems for pollution abatement. The first system has reduced the emission of pollutants to acceptable levels 63 percent of the time, as determined from 200 air samples. The second (and more expensive) system has reduced the emission of pollutants to acceptable levels 79 percent of the time, as determined from 300 air samples. At the .10 level of significance, can management conclude that the more expensive system is no better than the inexpensive system?

57. Scientists often use artifacts found at archeological sites to make inferences about the political, cultural, and economic history of a society. In one study, the average gold content of Roman coins minted at two different times was examined to see if it was the same at the two mintings, or if, as happens from time to time in other societies, the gold content had changed. A chemical analysis was performed on 10 coins from the early first century and 8 coins minted in the late first century. The data below are the percentages of gold in the two sets of coins.

Century	Percent of gold in coins									
Early first	6.2	5.8	7.1	6.5	6.7	7.0	6.6	6.3	6.9	6.0
Late first	4.9	5.2	5.4	5.8	5.9	6.1	6.3	6.5		

Does it appear that coins minted at two different times during the first century contain different amounts of gold? (Use the .05 level of significance to do the test.)

58. A consumer research organization routinely selects several car models a year and tests their claims regarding safety, mileage and comfort. In one study of two similar subcompact models manufactured by two different automakers, the average gas mileage for 7 cars of make A was 21 miles per gallon with a standard deviation of 5.8. For 9 cars of make B, the average gas mileage was 26 miles per gallon with a standard deviation of 5.3 miles per gallon. Test the hypothesis that the average gas mileage for cars of make B is greater than the average gas mileage for cars of make A at the .05 level of significance.

59. Is the amount of responsibility for an action ascribed in relation to the severity of the consequences of that action? That question was the basis of a study of responsibility in which subjects read a description of an accident on an interstate highway. The consequences, in terms of cost and injury, were described as either very minor or serious. A questionnaire tested the subjects' comprehension of the facts in the story and asked them to rate the degree of responsibility that should be placed on the main figure in the story. Below are these ratings for the mild consequences group and the severe consequences group. High ratings indicate more responsibility attributed to the primary actor. Test the hypothesis that severe consequences lead to greater attribution of responsibility at the .01 level of significance.

Consequences	Degree of responsibility							
Mild	4	5	3	3	4	1	2	6
Severe	4	5	4	6	7	8	6	5

60. Circulatory problems in older people are major contributors to hospitalization and death through hardening of the arteries, heart disease, and stroke. A medication being developed

to help improve circulation and thus prevent or relieve such diseases was recently tested on a small number of elderly patients. Each patient's mean circulation time was measured before they were given the vasodilator and then again after they had been on the medication for three months. The data are given below. Test the hypothesis that the medication reduces mean circulation time. Use the .05 level of significance.

Patient	1	2	3	4	5	6	7	8	9	10
Before	12	14	12	13	15	13	14	13.5	12	12.5
After	9	13	14	10	12	11	13	10	11	13

61. Two different food rations given to swine were compared for the gain in weight they produced. Two animals from each of 9 litters of swine were randomly selected and one member of each pair was randomly assigned to each ration. The gains in weight recorded at the end of 30 days are given below. Test the hypothesis that average weight gains with the two rations are different. Use the .10 level of significance.

Litter	1	2	3	4	5	6	7	8	9
Ration 1	46	54	49	39	42	48	51	55	44
Ration 2	53	52	49	42	51	50	49	60	43

62. Many companies subscribe to music systems that play softly and are thought to produce a relaxing, comfortable work setting, leading to greater productivity. Mr. Kingpin, however, is not convinced of these benefits, especially for certain departments that he feels involve jobs requiring greater concentration. He assigns his personnel manager to run the music system for one week, to measure the productivity of the six employees in one of these departments, and to measure the same six employees' productivity in a week when the music is not piped in. The data given below are the productivity measures for the two week-long periods.

Employee	1	2	3	4	5	6
Week with music	142	136	158	145	150	148
Week without music	139	138	150	145	145	142

Test the hypothesis that when the music is piped in, the mean productivity for these employees is different from the mean productivity when the music is not piped in. Use the .05 level of significance.

9 TERMS INTRODUCED IN CHAPTER 9

- **ALPHA (α)** The probability of a Type I error.

- **ALTERNATIVE HYPOTHESIS** The conclusion we accept when the data fail to support the null hypothesis.

- **BETA (β)** The probability of a Type II error.

- **DEPENDENT SAMPLES** Samples drawn from two populations in such a way that the elements were not chosen independently of one another, in order to allow a more precise analysis or to control for some extraneous factors.

- **HYPOTHESIS** An assumption or speculation we make about a population parameter.

- **LOWER-TAILED TEST** A one-tailed hypothesis test in which a sample value significantly below the hypothesized population value will lead us to reject the null hypothesis.

- **NULL HYPOTHESIS** The hypothesis, or assumption, about a population parameter we wish to test, usually an assumption of the status quo.

- **ONE-TAILED TEST** A hypothesis test in which there is only one rejection region; i.e., we are concerned only with whether the observed value deviates from the hypothesized value in one direction.

- **PAIRED DIFFERENCE TEST** A hypothesis test of the difference between the sample means of two dependent samples.
- **POWER CURVE** A graph of the values of the power of a test for each value of μ, or other population parameter, for which the alternative hypothesis is true.
- **POWER OF THE HYPOTHESIS TEST** The probability of rejecting the null hypothesis when it is false; i.e., a measure of how well the hypothesis test is working.
- **SIGNIFICANCE LEVEL** A value indicating the percentage of sample values that is outside certain limits, assuming the null hypothesis is correct; i.e., the probability of rejecting the null hypothesis when it is true.

- **TWO-TAILED TEST** A hypothesis test in which the null hypothesis is rejected if the sample value is significantly higher or lower than the hypothesized value of the population parameter; a test involving two rejection regions.
- **TYPE I ERROR** Rejecting a null hypothesis when it is true.
- **TYPE II ERROR** Accepting a null hypothesis when it is false.
- **UPPER-TAILED TEST** A one-tailed hypothesis test in which a sample value significantly above the hypothesized population value will lead us to reject the null hypothesis.

10 EQUATIONS INTRODUCED IN CHAPTER 9

p. 326:

$$\sigma_{\bar{x}_1 - \bar{x}_2} = \sqrt{\frac{\sigma_1^2}{n_1} + \frac{\sigma_2^2}{n_2}}$$

[9·1]

This formula enables us to derive the standard deviation of the distribution of the difference between the sample means, that is, *the standard error of the difference between two means*. To do this, we take the square root of the value equal to the sum of Population 1's standard deviation squared and divided by its sample size and of Population 2's standard deviation squared and divided by its sample size.

p. 328:

$$\hat{\sigma}_{\bar{x}_1 - \bar{x}_2} = \sqrt{\frac{\hat{\sigma}_1^2}{n_1} + \frac{\hat{\sigma}_2^2}{n_2}}$$

[9·2]

If the two population standard deviations are unknown, we can use this formula to derive the *estimated* standard error of the difference between two means. We can use this equation after we have used the two sample standard deviations and Equation 8·1 to determine the estimated standard deviations of Population 1 and Population 2. ($\hat{\sigma} = s$)

p. 331:

$$s_p^2 = \frac{(n_1 - 1)s_1^2 + (n_2 - 1)s_2^2}{n_1 + n_2 - 2}$$

[9·3]

With this formula we can get a "pooled estimate" of σ^2. It uses a weighted average of s_1^2 and s_2^2, where the weights are the numbers of degrees of freedom in each sample. Use of this formula assumes that $\sigma_1^2 = \sigma_2^2$ (that the unknown population variances are equal). We use this formula when testing for the differences between means in situations with small sample sizes (less than 30).

p. 331:

$$\hat{\sigma}_{\bar{x}_1 - \bar{x}_2} = s_p \sqrt{\frac{1}{n_1} + \frac{1}{n_2}}$$

[9·4]

With the "pooled estimate" of σ^2 we obtained from Equation 9·3, we put this value into Equation 9·2 and simplify the expression. This gives us a formula to estimate the standard error of the difference between sample means when we have small samples (less than 30) but equal population variances.

p. 338:

$$\sigma_{\bar{p}_1 - \bar{p}_2} = \sqrt{\frac{p_1 q_1}{n_1} + \frac{p_2 q_2}{n_2}}$$

[9·5]

This is the formula to use to derive the standard error of the difference between two *proportions.* The symbols p_1 and p_2 represent the proportion of successes in Population 1 and Population 2, respectively, and q_1 and q_2 are the proportion of failures in Populations 1 and 2, respectively.

p. 338:

$$\hat{\sigma}_{\bar{p}_1 - \bar{p}_2} = \sqrt{\frac{\overline{p}_1 \overline{q}_1}{n_1} + \frac{\overline{p}_2 \overline{q}_2}{n_2}}$$

[9·6]

If the population parameters p and q are unknown, we can use the sample statistics $\overline{p}$ and $\overline{q}$ and this formula to *estimate* the standard error of the difference between two proportions.

p. 339:

$$\hat{p} = \frac{(n_1)(\overline{p}_1) + (n_2)(\overline{p}_2)}{n_1 + n_2}$$

[9·7]

Because the null hypothesis assumes that there is *no difference* between the two population proportions, it would be more appropriate to modify Equation 9·6 and to use the combined proportions from both samples to estimate the overall proportion of successes in the combined populations. Equation 9·7 combines the proportions from both samples. Notice that the value of $\hat{q}$ is equal to $1 - \hat{p}$.

p. 339:

$$\hat{\sigma}_{\bar{p}_1 - \bar{p}_2} = \sqrt{\frac{\hat{p}\hat{q}}{n_1} + \frac{\hat{p}\hat{q}}{n_2}}$$

[9·8]

Now we can substitute the results of Equation 9·7 both $\hat{p}$ and $\hat{q}$, into Equation 9·6 and get a more correct version of Equation 9·6. This new equation, 9·8, gives us the *estimated* standard error of the difference between the two proportions using combined estimates from both samples.

11 CHAPTER REVIEW EXERCISES

63. For the following situations, state the null and alternative hypotheses.

   a) A researcher wishes to test whether a certain enrichment class leads to test scores greater than the population average of 85 points.

b) An airlines employee wishes to determine if the average height of stewardesses is at least 66 inches.

c) A university official wishes to determine if average enrollment for the past 10 years is significantly different from a hypothesized value of 12,500.

64. Health Electronics, Inc., a manufacturer of pacemaker batteries, specifies that the life of each battery is equal to or greater than 28 months. If scheduling for replacement surgery for the batteries is to be based upon this claim, explain to the management of this company the consequences of Type I and Type II errors.

65. In the test of a hypothesis that the average score on a test is 110 versus the hypothesis that it is not equal to 110, one sample's estimate of the population parameter is 98 and another sample produces a sample statistic value of 122. In this situation, is one of these sample values more apt to lead us to accept the null hypothesis? Why or why not?

66. As part of a study of the reasons for lower scores made by minorities on standardized tests of basic skills, a survey was made to determine the amount of time children spent viewing television each week. A sample of 36 minority children showed an average weekly viewing time of 42 hours with a standard deviation of 5 hours. Another sample of 36 white children showed an average weekly viewing time of 40 hours with a standard deviation of 3. At the .05 level of significance, is there a significant difference between the number of hours these two groups of children spend watching television each week?

67. A university librarian suspects that the average number of books checked out to each student per visit has changed recently. In the past, an average of 3.5 books were checked out. However, a sample of 20 students averaged 4.2 books per visit, with a standard deviation of 1.8 books. At the .05 level of significance, has the average checkout changed?

68. Two lots of baby roosters were given two different hormones when they were one day old; and the weight of their combs was measured at 90 days. Of the 80 chicks receiving hormone A, the mean weight of their combs was 64.3 grams, with a standard deviation of .5. Combs of the 60 chicks given hormone C averaged 64.1 grams, with a standard deviation of .25. At the .05 level of significance, can we assume that there is no significant difference in the effects of the two hormones on the comb weights of roosters?

69. In an attempt to win an election against an incumbent senator who advocates an expanded welfare program, a candidate has charged that at least 20 percent of the present recipients are ineligible. The incumbent senator, upon hearing this charge, sponsors a study in which a random selection of 800 welfare recipients are carefully interviewed. This study finds that only 15 percent were not legally eligible. At a 5 percent level of significance, should the challenger's statements be accepted or rejected?

70. In response to criticism concerning lost mail, the U.S. Postal Service maintains that its historic loss rate of .3 percent or less has not changed. Concerned with this controversy, a government committee sponsors an investigation in which a total of 5,000 pieces of mail are mailed from various parts of the country. This mailing results in a total of 19 pieces not reaching their destination. At the .05 level of significance, is the U.S. Postal Service losing significantly more mail than its historic rate?

71. What is the probability that we are rejecting a true hypothesis when we reject the hypothesized value because of the following:

a) The sample statistic differs from it by more than 1.5 standard errors in either direction.

b) The value of the sample statistic is more than 2 standard errors above it.

c) The value of the sample statistic is more than 1 standard error below it.

72. If we wish to accept the null hypothesis 80 percent of the time when it is correct, how many standard errors around the hypothesized value should be used in determining whether to

reject it, based on sample information? How many for 90 percent certainty of accepting the null hypothesis when it is true?

73. Federal environmental statutes applying to a particular nuclear power plant specify that recycled water must, on the average, be no warmer than 82° F (28°C) before it can be released back into the river. From 100 samples, the average temperature of the recycled water was found to be 84° F (29°C). If the population standard deviation is 47.2° F (4°C), should the plant be cited for exceeding the limitations of the statute at a significance level of .04?

74. State inspectors, investigating charges that a Louisiana soft drink bottling company under-fills its product, have sampled 100 bottles and found the average contents to be 31.8 fluid ounces. The bottles are advertised to contain 32 fluid ounces. If the population standard deviation is 2 fluid ounces, should we reject at the 5 percent significance level the claim that the average is at least 32 fluid ounces?

75. Ms. Lacock, a first grade teacher, assumes that children in Head Start programs practice reading readiness skills. She wishes to see if, on their first reading readiness test, the Head Start children score better than those who have not been in the program. Scores of both groups are given below. Test the hypothesis that the average reading readiness score for Head Start children is greater than that for children who have not been in Head Start programs. Use the .025 level of significance.

No head start	Head start	No head start	Head start
64	68	86	84
65	75	72	88
92	95	85	82
81	89	74	85
91	78	75	
69	94	79	
89	79	82	
70	91		

76. A research assistant in the chemistry lab has just prepared 200 vials of a certain chemical compound for testing with other chemicals. Each vial is supposed to have 64 cc of the compound, with a standard deviation of 4 cc, in order to mix properly and provide accurate results for the experiments. A sample of 36 of the vials is checked and shows an average of 64.8 cc. The lab chemist wants to know if, at the .02 level of significance, he can assume that the vials contain 64 cc.

77. A biologist has been trying to improve an insecticide he developed to kill 50 percent of a certain type of insect. He sprayed his original mixture on an area infested by 300 insects, and 43 percent perished. After testing the new formula, he sprayed an area containing 400 of the insects, and 51 percent of this sample was killed. At the .05 level of significance, can the biologist conclude that the new formula is better than the original mixture?

78. An educational publishing firm, promoting a line of materials for left-handed children, claims that no more than 14 percent of elementary-school children are left-handed and need these special materials. To test this claim, the equipment supply officer at a private school surveyed 150 of the school's 1,200 children. This study showed that 17 of the 150 children were left-handed. If the supply manager is willing to run a 10 percent risk of rejecting the firm's claim when in fact it is true, what should the school officer conclude?

79. In a medical study on the treatment of skin infections, a researcher hypothesizes that the addition of vitamin A to patients' diets will produce more rapid healing than will a similar addition of vitamin D to their diets. Two groups of patients suffering from skin infections

were given these two different vitamin supplements and the length of time it took their infections to heal was recorded. Below are the healing times (in days) for the two groups.

Type of vitamin supplement

Vitamin A	6	8	9	10	12	9
Vitamin D	4	5	4	8	6	9

Is the difference between the average healing times for the two groups different at the .05 level of significance?

80. A clinical psychologist and a physician recently opened a clinic for people who want to lose large amounts of weight (over 20 pounds). Part of their plan involves large initial weight loss so that the person has incentive to stay on the program. This initial period is spent at a camp, where, the clinic claims, patients lose an average of 11 pounds in two weeks. To test this claim, 14 of the 300 persons who have attended the camp were surveyed, and the average number of pounds lost was found to be 9 with a sample standard deviation of 5. At the .05 level of significance, is the claim of an 11 pound weight loss in two weeks reasonable?

81. In studying insomnia, doctors have often felt that a significant part of the problem is psychological. In a well-controlled study of chronic insomniacs, 7 subjects were given placebos for 3 weeks, and they were monitored to determine how long it took them to fall asleep. For another 3-weeks' period, they were given sleeping pills, and the length of time it took them to fall asleep was again measured. The 3-weeks' period during which they received either placebos or sleeping pills was randomly assigned. The data below are the average number of minutes it took them to go to sleep during the two 3-weeks' periods.

Subject	1	2	3	4	5	6	7
Placebo	15.00	25.50	22.25	14.50	28.00	10.00	20.50
Sleeping pill	12.00	20.00	25.75	18.25	24.00	12.50	17.00

At the .05 level of significance, test the hypothesis that the average amount of time it takes an insomniac to fall asleep after taking a sleeping pill is different from the average time after taking a placebo.

82. A sample mean of 2.7 is found from a set of 25 observations. For the entire population of 200 elements, the standard deviation is known to be .4. Test whether we can accept a hypothesized value of 3 for the population mean or not, using the .10 level of significance, or if we must conclude that the true mean is some other value.

83. A chemist developing insect repellents wishes to know whether a new formula leads to greater protection from insect bites than given by the most popular product on the market. Sixteen volunteers were used in the experiment, and each had one arm sprayed with the old product and one arm sprayed with the new formula. Then each subject placed his arms in two insect chambers filled with equal numbers of mosquitoes, gnats, and other biting insects. The number of bites received on each arm was recorded and is given below. Test the hypothesis that there is a difference in the amount of protection provided by the two insect repellents. Use the .05 level of significance.

Subject	Old formula	New formula	Subject	Old formula	New formula
1	5	3	9	2	5
2	2	1	10	6	2
3	5	5	11	5	1
4	4	1	12	7	2
5	3	1	13	1	1
6	6	3	14	3	2
7	2	4	15	4	1
8	4	2	16	1	4

84. A company was recently criticized for not paying women as much as men working in the same positions. It claims that its average salary paid to all employees is $12,500. From a random sample of 36 women in the company, the average salary was calculated to be $11,900. If the population standard deviation is known to be $900 for these jobs, determine whether or not we could reasonably (within 2 standard deviations) expect to find $11,900 as the sample mean if, in fact, the company's claim is true.

85. An experimental psychologist working with rats has been using a new maze in conditioning them. The 49 rats taught to run the maze average 12 trials before giving a perfect performance. Long experience with a population of rats trained to run a similar maze showed that they made 11 trials before a perfect performance, on the average, with a standard deviation of 3.5 trials. At a significance level of .10, should the psychologist conclude that the new maze is harder for the rats to learn?

86. A hospital administrator opposes carpeting all rooms, claiming that the level of airborne bacteria in the rooms is not high enough to warrant the expenditure. Regulations require that the levels not exceed an average of 8 colonies of bacteria per cubic foot of air in the rooms. An inspector samples 15 of the 120 rooms to check the administrator's claim. The average level of airborne bacteria for the sample is found to be 10 colonies per cubic foot of air, and the sample standard deviation is found to be 3 colonies per cubic foot of air. At the .10 level of significance, is there sufficient reason to reject the claim that the levels of airborne bacteria in the hospital rooms satisfy the regulations?

87. A medical student doing an internship in obstetrics wondered whether the sexes differ in weight at birth. To examine the question, the intern kept a record of the weights of all the babies born during one week. Below are the weights in ounces of the babies at birth, according to their sex.

Weights at birth

Boys		Girls	
110	118	108	119
115	122	106	103
99	106	114	128
125	102	95	117
109	100	110	130
92	103	101	121
113	129		

Test the hypothesis that there are weight differences between the sexes at birth at the .05 level of significance.

88. Refer to Exercise 26. Compute the power of the test for $\mu = 116$, 117, and 117.6.

89. A university administrator hypothesized that 15 percent of the students admitted last year attended private high schools. If the observed proportion is .17 for a sample of 200 of 2,000 entering students, test whether we can accept his hypothesis as correct at the .10 level of significance or if we must conclude that some other value is more appropriate.

90. In Exercise 27, what would be the power of the test for $\mu = 12,500$; 12,000; and 11,500 if the significance level were changed to .05.

91. A fortune teller claims she can determine the sex of an unborn child with 80 percent accuracy several months before birth. As a test, she predicts the sex of 40 unborn children and is correct in 28 of the predictions. Does this evidence support the fortune teller's claim, or may we conclude that her percentage accuracy is less than 80 percent? Use the .05 level of significance.

92. In Exercise 26, what would be the power of the test for $\mu = 116$, 117, and 117.6 if the significance level were changed to .01?

93. Given that 55 of 1,000 failed a test in which the proportion that was hypothesized to fail was .04, test the hypothesis that the true proportion is .04 versus the alternative that it is greater than .04 at the .01 level of significance.

94. If an election worker hypothesized that 48 percent of the electorate would vote in the general election, and the observed proportion of voters casting their ballots was .45, should we reject her hypothesis at the .05 level of significance? There are 5,000 voters registered in the town and 10 percent were sampled.

95. Refer to Exercise 27. Compute the power of the test for $\mu = 12,500$; 12,000; and 11,500.

12 CHAPTER CONCEPTS TEST

Answer true *or* false. *Answers are in the back of the book.*

1. In hypothesis testing, we assume that some population parameter takes on a particular value before we sample. This assumption to be tested is called an alternative hypothesis.

2. Assuming a given hypothesis about a population mean is correct, then the percentage of a sample means that could fall outside certain limits from this hypothesized mean is called the significance level.

3. In hypothesis testing, the appropriate probability distribution to use is always the normal distribution.

4. If we were to make a Type I error, we would be rejecting a null hypothesis when it is really true.

5. A paired difference test is appropriate when the two samples being tested are dependent samples.

6. A one-tailed test for the difference between means may be undertaken when the sample sizes are either large or small and the procedures are similar. The only difference is that when sample sizes are large, we employ a normal distribution, whereas the t distribution is used when sample sizes are small.

7. If our null and alternative hypotheses are H_0: $\mu = 80$, and H_1: $\mu < 80$, it is appropriate to use a left-tailed test.

8. Suppose a hypothesis test is to be made regarding the difference in means between two populations, and our sample sizes are large. If we do not know the actual standard deviations of the two populations, we can use the sample standard deviations as estimates.

9. The value $1 - \beta$ is known as the power of the test.

10. If we took two independent samples and performed a hypothesis test to evaluate significant differences in their means, we would find the results very similar to a paired difference test performed on the same two samples.

10

Chi-square and analysis of variance

1. Introduction, 356

2. Chi-square as a test of independence, 357

3. Chi-square as a test of goodness of fit, 369

4. Analysis of variance, 374

5. Inferences about a population variance, 388

6. Inferences about two population variances, 394

7. Terms introduced, 399

8. Equations introduced, 400

9. Chapter review exercises, 403

10. Chapter concepts test, 407

354

Because of the rising number of complaints about the United States Postal Service, the Postmaster General has initiated an investigation of mail service between New York and Chicago. One of the goals of this investigation is to speed the delivery of first class mail.
To get a quick idea of current deliveries, one of the investigators followed nine randomly selected letters going between the two cities and recorded these hours between drop off and delivery: 50, 45, 27, 66, 43, 96, 45, 92, and 69.
Using the methods developed in this chapter, we can give the Postmaster General not only estimated information about *all* letter movement between New York and Chicago, but also information about how accurate that estimate is and what uncertainty is associated with it.

Uses of the
chi-square test

Function of
analysis
of variance

Inferences about
population
variances

1 INTRODUCTION In the last chapter, we learned how to test hypotheses using data from either one or two samples. We used one-sample tests to determine whether a mean or a proportion was significantly different from a hypothesized value. In the two-sample tests, we examined the difference between either two means or two proportions, and we tried to learn whether this difference was significant.

Suppose we have proportions from five populations instead of only two. In this case, the methods for comparing proportions described in Chapter 9 do not apply; we must use the *chi-square test,* the subject of the first portion of this chapter. Chi-square tests enable us to test whether more than two population proportions can be considered equal.

Actually, chi-square tests allow us to do a lot more than just test for the equality of several proportions. If we classify a population into several categories with respect to two attributes (for example, age and political affiliation) we can then use a chi-square test to determine if the two attributes are independent of each other.

Decision makers also encounter situations in which it is useful to test for the equality of more than two population means. Again, we cannot apply the methods introduced in Chapter 9 because they are limited to testing for the equality of only two means. The *analysis of variance,* discussed in the fourth section of this chapter, will enable us to test whether more than two population means can be considered equal.

It is clear that we will not always be interested in means and proportions. There are many decision making situations where we will be concerned about the variability in a population. Section 5 of this chapter shows how to use the chi-square distribution to form confidence intervals and test hypotheses about a population variance. In the last section, we show that hypotheses comparing the variances of two populations can be tested using the *F* distribution.

EXERCISES

1. Why do we use a chi-square test?

2. Why do we use analysis of variance?

3. What type of statistical test could be used in the following situations?

   a) We wish to know whether the average weight gain of cattle is significantly affected under 3 different feed mixtures.

   b) A research group is interested in determining whether there is a significant difference between the purchasing habits of men and women.

   c) We need to compare the differences in the proportions of voters favoring each of 4 candidates.

4. To make comparisons in the following groups, what type of statistical test is appropriate or what probability distribution is used?

   a) Percentage of the labor force from each age group: 16–23, 24–31, 32–39, 40–47, 48–55, and 56 and over.

   b) Average income of these age groups: 16–23, 24–31, 32–39, 40–47, 48–55, 56 and over.

   c) Average income of men and women, aged 16–56.

   d) Amount of dispersion in the earnings of men and women, aged 16–56.

5. To help keep straight which distribution or technique is used, complete the following table with either the name of a distribution or the technique involved. The row classification refers to the number of parameters involved in a test, and the column classification refers to the type of parameter involved. Some cells may not have an entry, while others may have more than one possible entry.

Number of parameters involved	Type of parameter μ	σ	p
1			
2			
3 or more			

2 CHI-SQUARE AS A TEST OF INDEPENDENCE

Many times, decision makers need to know whether the differences they observe among several sample proportions are significant or only due to chance. Suppose the campaign manager for a presidential candidate studies three geographically different regions and finds that 35 percent, 42 percent, and 51 percent of those voters surveyed in the three regions, respectively, recognize the candidate's name. If this difference is significant, the manager may conclude that location will affect the way the candidate should act. But if the difference is not significant (that is, if the manager concludes that the difference is solely due to chance), then he may decide that the place they choose to make a particular policy-making speech will have no effect on its reception. To run the campaign successfully, then, the manager needs to determine whether location and acceptance are dependent or independent.

Simple differences among proportions: dependent or independent?

Contingency tables

Suppose that in 4 regions the Veteran's Administration samples its hospital employees' attitudes toward job performance reviews. Respondents are given a choice between the present method (two reviews a year) and a proposed new method (quarterly reviews). Table 10·1, which illustrates the

Describing a contingency table

response to this question from the sample polled, is called a *contingency table*. A table such as this is made up of rows and columns, that is:

TABLE 10·1
Sample response concerning review schedules for veteran's administration hospital employees

	Northeast	Southeast	Central	West Coast		Total
Number who prefer present method	68	75	57	79	=	279
Number who prefer new method	32	45	33	31	=	141
Total employees sampled in each region	**100**	**120**	**90**	**110**	=	**420**

Notice that the 4 columns in Table 10·1 provide one basis of classification—geographical regions—and that the 2 rows classify the information another way: preference for review methods. Table 10·1 is called a "2 × 4 contingency table," because it consists of 2 rows and 4 columns. We describe the dimensions of a contingency table by first stating the number of rows and then the number of columns. The "total" column and the "total" row are not counted as part of the dimensions.

Observed and expected frequencies

Stating
the hypothesis

Suppose we now symbolize the true proportions of the total population of employees who prefer the present plan as:

$p_N \leftarrow$ proportion in Northeast who prefer present plan
$p_S \leftarrow$ proportion in Southeast who prefer present plan
$p_C \leftarrow$ proportion in Central region who prefer present plan
$p_W \leftarrow$ proportion in West Coast region who prefer present plan

Using these symbols, we can state the null and alternative hypotheses as follows:

$H_0: p_N = p_S = p_C = p_W \leftarrow$ null hypothesis
$H_1: p_N, p_S, p_C,$ and p_W are not all equal $\leftarrow$ alternative hypothesis

If the null hypothesis is true, we can combine the data from the four samples and then estimate the proportion of the total work force (the total population) that prefers the present review method:

Combined proportion who prefer present method assuming the null hypothesis of no difference is true	$=$	$\dfrac{68 + 75 + 57 + 79}{100 + 120 + 90 + 110}$			
	$=$	$\dfrac{279}{420}$			
	$=$	$.664$			

Obviously, if the value .664 estimates the population proportion expected to prefer the present compensation method, then .336 ($= 1 - .664$) is the estimate of the population proportion expected to prefer the proposed new method. Using .664 as the *estimate* of the population proportion who prefer the present review method, and .336 as the *estimate* of the population proportion who prefer the new method, we can estimate the number of sampled employees in each region whom we would expect to prefer each of the review methods. These calculations are done in Table 10·2.

Determining expected frequencies

	Northeast	Southeast	Central	West Coast
Total number sampled	100	120	90	110
Estimated proportion who prefer present method	$\times$ .664	$\times$ .664	$\times$ .664	$\times$ .664
Number *expected* to prefer present method	66	80	60	73
Total number sampled	100	120	90	110
Estimated proportion who prefer new method	$\times$ .336	$\times$ .336	$\times$ .336	$\times$ .336
Number *expected* to prefer new method	34	40	30	37

**TABLE 10·2
Proportion of sampled employees in each region expected to prefer the two review methods**

Table 10·3 combines all the information from Tables 10·1 and 10·2. It illustrates both the actual, or observed, frequency of the employees sampled who prefer each type of compensation method and the theoretical, or

Comparing expected and observed frequencies

	Northeast	Southeast	Central	West Coast
Frequency preferring present method				
Observed (actual) frequency	68	75	57	79
Expected (theoretical) frequency	66	80	60	73
Frequency preferring new method				
Observed (actual) frequency	32	45	33	31
Expected (theoretical) frequency	34	40	30	37

**TABLE 10·3
Comparison of observed and expected frequencies of sampled employees**

expected, frequency of sampled employees preferring each type of method. Remember that the expected frequencies, those in color, were estimated from our combined proportion estimate.

Reasoning
intuitively about
chi-square tests

To test the null hypothesis, $p_N = p_S = p_C = p_W$, we must compare the frequencies that were observed (the black ones in Table 10·3) with the frequencies we would expect if the null hypothesis is true (those in color). If the sets of observed and expected frequencies are nearly alike, we can reason intuitively that we will accept the null hypothesis. If there is a large difference between these frequencies, we may intuitively reject the null hypothesis, and conclude that there are significant differences in the proportions of employees in the four regions preferring the new method.

The chi-square statistic

Calculating
the chi-square
statistic

To go beyond our intuitive feelings about the observed and expected frequencies, we can use the chi-square statistic, which is calculated this way:

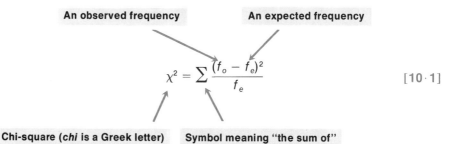

An observed frequency   **An expected frequency**

$$\chi^2 = \sum \frac{(f_o - f_e)^2}{f_e}$$

[10·1]

Chi-square (*chi* is a Greek letter)   **Symbol meaning "the sum of"**

This formula says that chi-square, or χ^2, is the sum we will get if we:

1. subtract f_e from f_o for each of the eight boxes, or cells, of Table 10·3
2. square each of the differences
3. divide each squared difference by f_e
4. sum all eight of the answers

Numerically, the calculations are easy to do using a table such as Table 10·4, which shows the steps.

Interpreting
the chi-square
statistic

The answer of 3.032 is the value for chi-square in our problem comparing preferences for review methods. If this value were as large as 20, it would indicate a substantial difference between our observed values and our expected values. A chi-square of zero, on the other hand, indicates that the observed frequencies exactly match the expected frequencies. The value of chi-square can never be negative, since the differences between the observed and expected frequencies are always *squared.*

TABLE 10·4
Calculation of
χ^2 (chi-square)
statistic from
data in Table 10·3

f_o	f_e	Step 1 $f_o - f_e$	Step 2 $(f_o - f_e)^2$	Step 3 $\dfrac{(f_o - f_e)^2}{f_e}$
68	66	2	4	.0606
75	80	−5	25	.3125
57	60	−3	9	.1500
79	73	6	36	.4932
32	34	−2	4	.1176
45	40	5	25	.6250
33	30	3	9	.3000
31	37	−6	36	.9730

Step 4 $\Sigma \dfrac{(f_o - f_e)^2}{f_e} = 3.032 \leftarrow \chi^2$ (chi-square)

The chi-square distribution

If the null hypothesis is true, then the sampling distribution of the chi-square statistic, χ^2, can be closely approximated by a continuous curve known as a *chi-square distribution.* As in the case of the t distribution, there is a different chi-square distribution for each different number of degrees of freedom. Figure 10·1 indicates the three different chi-square distributions that would correspond to 1, 5, and 10 degrees of freedom. For very small numbers of degrees of freedom, the chi-square distribution is severely skewed to the right. As the number of degrees of freedom increases, the curve rapidly becomes more symmetrical until the number reaches large values, at which point the distribution can be approximated by the normal.

Describing a chi-square distribution

The chi-square distribution is a probability distribution. Therefore, the total area under the curve in each chi-square distribution is 1.0. Like the t distribution, so many different chi-square distributions are possible that it is not practical to construct a table that illustrates the areas under the curve for all possible degrees of freedom. Instead, Appendix Table 5 illustrates only the areas in the tail most commonly used in significance tests using the chi-square distribution.

Finding probabilities when using a chi-square distribution

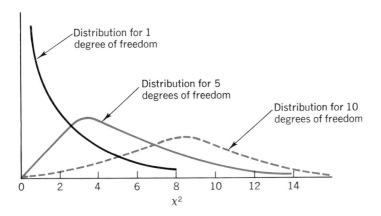

FIG. 10·1
Chi-square
distributions of 1,
5, and 10 degrees
of freedom

361

Determining degrees of freedom

Calculating
degrees of freedom

To use the chi-square test, we must calculate the number of degrees of freedom in the contingency table by applying Equation 10·2:

$$\text{Number of degrees of freedom} = (\text{Number of rows} - 1)(\text{Number of columns} - 1) \quad [10\cdot2]$$

Let's examine the appropriateness of this equation. Suppose we have a 3×4 contingency table like the one in Fig. 10·2. We know the row and column totals that are designated RT_1, RT_2, RT_3, and CT_1, CT_2, CT_3, CT_4. As we discussed in Chapter 8, the number of degrees of freedom is equal to the number of values that we can freely specify.

FIG. 10·2

A 3 × 4 contingency table illustrating determination of the number of degrees of freedom

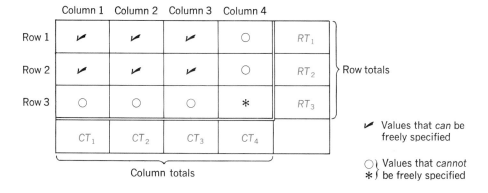

Look now at the first row of the contingency table in Fig. 10·2. Once we specify the first three values in that row (denoted by checks in the figure) the fourth value in that row (denoted by a circle) is already determined; we are not free to specify it because we know the row total.

Likewise, in the second row of the contingency table in Fig. 10·2, once we specify the first three values (denoted again by checks), the fourth value is determined and cannot be freely specified. We have denoted this fourth value by a circle.

Turning now to the third row, we see that its first entry is determined *because we already know the first two entries in the first column and the column total;* again we have denoted this entry with a circle. We can apply this same reasoning to the second and third entries in the third row, both of which have been denoted by a circle too.

Turning finally to the last entry in the third row (denoted by a star) we see that we cannot freely specify its value because we have already determined the first two entries in the fourth column. By counting the number of checks in the contingency table in Fig. 10·2, you can see that the number of values we are free to specify is 6 (the number of checks). This is equal to 2×3, or (the number of rows $-$ 1) times (the number of columns $-$ 1).

This is exactly what we have in Equation 10·2. Table 10·5 illustrates the row-and-column dimensions of three more contingency tables and indicates the appropriate degrees of freedom in each case.

Contingency table	Number of rows (r)	Number of columns (c)	(r − 1)	(c − 1)	Degrees of freedom (r − 1)(c − 1)
A	3	4	3 − 1 = 2	4 − 1 = 3	(2)(3) = 6
B	5	7	5 − 1 = 4	7 − 1 = 6	(4)(6) = 24
C	6	9	6 − 1 = 5	9 − 1 = 8	(5)(8) = 40

TABLE 10·5
Determination of degrees of freedom in three contingency tables

Using the chi-square test

Returning to our example of job review preferences of hospital employees, we use the chi-square test to determine whether attitude about reviews is independent of geographical region. If the VA wants to test the null hypothesis at the .10 level of significance, our problem can be summarized:

$H_0: p_N = p_S = p_C = p_W$ ← null hypothesis
$H_1: p_N, p_S, p_C,$ and p_W are *not* all equal ← alternative hypothesis
$\alpha = .10$ ← level of significance for testing this hypothesis

Stating the problem symbolically

Since our contingency table (Table 10·1) has 2 rows and 4 columns, the appropriate number of degrees of freedom is:

Calculating degrees of freedom

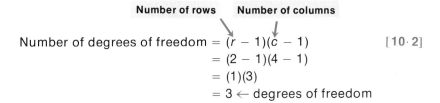

$$\text{Number of degrees of freedom} = (r − 1)(c − 1) \qquad [10·2]$$
$$= (2 − 1)(4 − 1)$$
$$= (1)(3)$$
$$= 3 \leftarrow \text{degrees of freedom}$$

Figure 10·3 illustrates a chi-square distribution for 3 degrees of freedom, showing the significance level in color. In Appendix Table 5, we can look under the .10 column and move down to the 3 degrees of freedom row.

Illustrating the hypothesis test

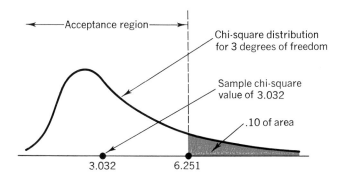

FIG. 10·3
Chi-square hypothesis test at .10 level of significance, showing acceptance region and sample chi-square value of 3.032

Interpreting
the results

There we find the value of the chi-square statistic 6.251. We can interpret this to mean that with 3 degrees of freedom, the region to the right of a chi-square value of 6.251 contains .10 of the area under the curve. Thus, the acceptance region for the null hypothesis in Fig. 10·3 goes from the left tail of the curve to the chi-square value of 6.251.

As we can see from Fig. 10·3, the sample chi-square value of 3.032, which we calculated in Table 10·4, falls within the acceptance region. Therefore, we accept the null hypothesis that there is no difference between the attitudes about job reviews in the four geographical regions. In other words, we conclude that attitude about performance reviews is independent of geography.

Contingency tables with more than two rows

Are hospital
stay and
insurance
coverage
independent?

Senator Ned Cannady has long been a proponent of a federally underwritten system of National Health Insurance. Opponents of the system argue that it would be too costly to implement, particularly since the existence of such a system would, among other effects, tend to encourage people to spend more time in hospitals. The senator believes that this argument is not valid and that lengths of stays in hospitals are independent of the types of health insurance that people have. He asked Donna McClish, his staff statistician, to check the matter out. Donna collected data on a random sample of 660 hospital stays and summarized it in Table 10·6.

TABLE 10·6
Hospital stay data classified by the type of insurance coverage and length of stay

| | | Days in hospital | | | |
		<5	5–10	>10	Total
Fraction of costs	<25%	40	75	65	180
covered by	25–50%	30	45	75	150
insurance	>50%	40	100	190	330
	Total	110	220	330	660

Cells

Table 10·6 gives observed frequencies in the nine different lengths of stay and the types of insurance categories (or "cells") into which we have divided the sample. Donna wishes to test the hypothesis

H_0 : length of stay and type of insurance
are independent
H_1 : length of stay depends on type of insurance
$\alpha = .01 \leftarrow$ level of significance for testing this hypothesis

Finding expected
frequencies

We will use a chi-square test, so we first have to find the expected frequencies for each of the nine cells. Let's demonstrate how to find them by looking at the cell that corresponds to stays of less than 5 days and insurance covering less than 25 percent of costs.

A total of 180 of the 660 stays in the sample had insurance covering less than 25 percent of costs. So we can use the figure 180/660 to *estimate* the proportion in the population having insurance covering less than 25 percent of the costs. Similarly, 110/660 *estimates* the proportion of all hospital stays that last fewer than 5 days. If length of stay and type of insurance really are independent, we can use Equation 5·4 to *estimate* the proportion in the first cell (less than 5 days and less than 25 percent coverage).

Estimating the proportions in the cells

We let:

A = the event "a stay corresponds to
someone whose insurance covers
less than 25 percent of the costs," and
B = the event "a stay lasts less than
5 days"

then,

$$P(\text{first cell}) = P(A \text{ and } B)$$
$$= P(A) \times P(B) \qquad [5 \cdot 4]$$
$$= \left(\frac{180}{660}\right)\left(\frac{110}{660}\right)$$
$$= 1/22$$

Since 1/22 is the expected *proportion* in the first cell, the expected *frequency* in that cell is:

$$(1/22)(660)$$
$$= 30 \text{ observations}$$

In general, we can calculate the expected frequency for any cell with Equation 10·3:

Calculating the expected frequencies for the cells

$$f_e = \frac{RT \times CT}{n} \qquad [10 \cdot 3]$$

where:

f_e = the expected frequency in a given cell
RT = the row total for the row containing that cell
CT = the column total for the column containing that cell
n = the total number of observations

Now we can use Equations 10·3 and 10·1 to compute all of the expected frequencies and the value of the chi-square statistic. The computations are done in Table 10·7.

TABLE 10·7
Calculation of expected frequencies and chi-square from data in Table 10·6

Row	Column	f_0	f_e	$= \dfrac{RT \times CT}{n}$	$f_0 - f_e$	$(f_0 - f_e)^2$	$\dfrac{(f_0 - f_e)^2}{f_e}$
1	1	40	30	$\dfrac{180 \times 110}{660}$	10	100	3.333
1	2	75	60	$\dfrac{180 \times 220}{660}$	15	225	3.750
1	3	65	90	$\dfrac{180 \times 330}{660}$	−25	625	6.944
2	1	30	25	$\dfrac{150 \times 110}{660}$	5	25	1.000
2	2	45	50	$\dfrac{150 \times 220}{660}$	−5	25	0.500
2	3	75	75	$\dfrac{150 \times 330}{660}$	0	0	0.000
3	1	40	55	$\dfrac{330 \times 110}{660}$	−15	225	4.091
3	2	100	110	$\dfrac{330 \times 220}{660}$	−10	100	0.909
3	3	190	165	$\dfrac{330 \times 330}{660}$	25	625	3.788

$$[10 \cdot 1] \quad \sum \frac{(f_0 - f_e)^2}{f_e} = 24.315 \leftarrow \chi^2 \text{ chi-square}$$

Figure 10·4 illustrates a chi-square distribution with 4 degrees of freedom, showing the .01 significance level in color. Appendix Table 5 (in the .01 column and the 4 degrees of freedom row) tells Donna that for her problem, the region to the right of a chi-square value of 13.277 contains .01 of

FIG. 10·4
Chi-square hypothesis test at .01 level of significance showing acceptance region and sample chi-square value of 24.315

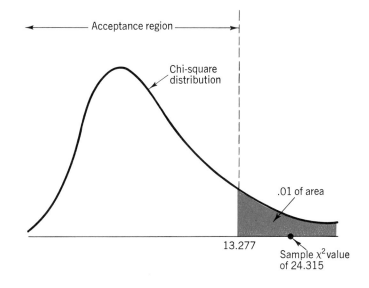

Acceptance region

Chi-square distribution

.01 of area

13.277

Sample χ^2 value of 24.315

the area under the curve. Thus, the acceptance region for the null hypothesis in Fig. 10·4 goes from the left tail of the curve to the chi-square value of 13.277.

As Fig. 10·4 shows Donna, the sample chi-square value of 24.315 she calculated in Table 10·7 is not within the acceptance region. Thus Donna must reject the null hypothesis and inform Senator Cannady that the evidence does not support his belief that length of hospital stay and insurance coverage are independent of each other.

Interpreting the results of the test

Precautions about using the chi-square test

To use a chi-square hypothesis test, we must have a sample size large enough to guarantee the similarity between the theoretically correct distribution and our sampling distribution of χ^2, the chi-square statistic. When the expected frequencies are too small, the value of χ^2 will be overestimated and will result in too many rejections of the null hypothesis. To avoid making incorrect inferences from χ^2 hypothesis tests, follow the general rule that an expected frequency of less than 5 in one cell of a contingency table is too small to use.* When the table contains more than one cell with an expected frequency of less than 5, we can combine these in order to get an expected frequency of 5 or more. But in doing this we reduce the number of categories of data and will gain less information from the contingency table.

Use large sample sizes

This rule will enable us to use the chi-square hypothesis test properly, but unfortunately, each test can only reflect (and not improve) the quality of the data we feed into it. So far, we have rejected the null hypothesis if the difference between the observed and expected frequencies, that is, the computed chi-square value, is too large. In the case of the job review preferences, we would reject the null hypothesis at a .10 level of significance if our chi-square value was 6.251 or more. But if the chi-square value was zero, we should be careful to question whether *absolutely no difference exists between observed and expected frequencies.* If we have strong feelings that some difference ought to exist, we should examine either the way the data were collected or the manner in which measurements were taken, or both, to be certain that existing differences had not been obscured or missed in collecting sample data.

Use carefully collected data

Experiments with the characteristics of peas led the monk Gregor Mendel to propose the existence of genes. Mendel's experimental results were astoundingly close to those predicted by his theory. Some time later, statisticians looked at Mendel's "pea data," performed a chi-square test, and concluded that chi-square was too small; that is, Mendel's reported experimental data was so close to what was expected that they could only conclude that he had fudged the data.

Mendel's pea data

*Statisticians have developed correction factors that, in some cases, allow us to use cells with expected frequencies of less than 5. The derivation and use of these correction factors are beyond the scope of this book.

6. Given the following dimensions for contingency tables, how many degrees of freedom will the chi-square statistic for each have?

   a) 2 rows, 5 columns   d) 5 rows, 5 columns

   b) 3 rows, 4 columns   e) 3 rows, 6 columns

   c) 4 rows, 6 columns

7. A presidential candidate is concerned that her support may be unevenly distributed throughout the country. In a survey in which the country was divided into four geographic regions, a random sampling of 100 voters in each region was polled with the following results:

	Region				
	A	**B**	**C**	**D**	**Total**
Support	47	52	43	49	**191**
Do not support	53	48	57	51	**209**
Total	**100**	**100**	**100**	**100**	**400**

   Develop a table of observed and expected frequencies (similar to Table 10·3) for this problem.

8. For problem 7:

   a) Calculate χ^2, using a frequency table similar to Table 10·4.

   b) State the null and alternative hypotheses.

   c) Using a .05 level of significance, should the null hypothesis be rejected?

9. To determine whether different occupational groups held different attitudes toward some proposed social legislation currently being considered by Congress, a political science researcher asked 4 groups: did they favor, oppose, or were they undecided? The results of this survey were:

	Occupational group				
	Doctors	**Lawyers**	**Businesspeople**	**Scientists**	**Total**
Favor	25	40	47	46	**158**
Oppose	69	51	74	57	**251**
Undecided	36	29	19	37	**121**
Total	**130**	**120**	**140**	**140**	**530**

   Calculate a table of observed and expected frequencies for this problem.

10. For problem 9:

    a) State the null and the alternative hypotheses.

    b) Calculate χ^2.

    c) At the .10 significance level, should the null hypothesis be rejected?

11. An anthropologist interested in the differences in family organization within three cultures surveyed a group of families from an Asian tribe, an African tribe, and a South American tribe. Each family was classified according to whether the family was father-dominant or mother-dominant in decisions regarding the children. The results of the survey were:

	Asian	African	South American	Total
Mother-dominant	7	10	8	**25**
Father-dominant	10	18	9	**37**
Total	**17**	**28**	**17**	**62**

Do the three cultures have the same family structure at the 10 percent significance level?

12. A sociologist studying family size wondered whether it was related to the level of educational achievement attained by the parents; that is, whether people with high educational levels have the same size families as those with less formal education. A survey questioned couples in the age group from 35 to 50 who had reportedly completed their families. The number of children was recorded along with the highest level of education attained by either parent. The results are shown in the following table:

Level of educational achievement

Number of children	Professional or postgraduate	College graduate	High school graduate	Did not complete high school	Total
6+	6	13	14	17	**50**
3–5	12	16	8	8	**44**
1–2	38	40	11	6	**95**
0	21	22	9	13	**65**
Total	**77**	**91**	**42**	**44**	**254**

At the .05 level of significance, does the number of children in the family differ according to the level of education of the parents?

3 CHI-SQUARE AS A TEST OF GOODNESS OF FIT: TESTING THE APPROPRIATENESS OF A DISTRIBUTION

In the previous section of this chapter, we used the chi-square test to decide whether to accept a null hypothesis that was a hypothesis of independence between two variables. In our example, these two variables were (1) attitude toward job performance reviews and (2) geographical region.

Function of a goodness-of-fit test

The chi-square test can also be used to decide whether a particular probability distribution, such as the binomial, Poisson, or normal, is the *appropriate* distribution. This is an important ability because as decision makers using statistics, we will need to choose a certain probability distribution to approximate the distribution of the data we happen to be considering. We will need the ability to question how far we can go from the assumptions that underlie a particular distribution before we must conclude that this distribution is no longer applicable. The chi-square test enables us to ask this question and to test whether there is a significant difference between an observed frequency distribution and a theoretical frequency distribution. In this manner, we can determine the *goodness of fit* of a theoretical distribution (that is, how well it fits the distribution of data that we have actually observed). Thus we can determine whether we should believe that the observed data constitute a sample drawn from the hypothesized theoretical distribution.

Calculating observed and expected frequencies

Suppose that a certain newspaper requires that college seniors who are seeking positions with the paper be interviewed by three different editors. This enables the paper to obtain a consensus evaluation of each candidate. Each editor gives the candidate either a positive or a negative rating. Table 10·8 contains the interview results of the last 100 candidates.

TABLE 10·8
Interview results
of 100 candidates

Possible positive ratings from 3 interviews	Number of candidates receiving each of these ratings
0	18
1	47
2	24
3	11
	100

For manpower planning purposes, the director of recruitment for this newspaper thinks that the interview process can be approximated by a binomial distribution with $p = .40$, that is, with a 40 percent chance of any candidate receiving a positive rating on any one interview. If the director wants to test this hypothesis at the .20 level of significance, how should he proceed?

Symbolical
statement
of the problem

H_0: A binomial distribution with $p = .40$
is a good description of the   ← null hypothesis
interview process
H_1: A binomial distribution with $p = .40$
is *not* a good description of the   ←alternative hypothesis
interview process
$\alpha = .20$ ← level of significance for   testing this hypothesis

Calculating
the binomial
probabilities

To solve this problem, we must determine whether the discrepancies between the observed frequencies and those we would expect (if the binomial distribution is the proper model to use) are actually due to chance. We can begin by determining what the binomial probabilities would be for this interview situation. For three interviews, we would find the probability of success in the Cumulative Binomial Distribution Table (Appendix Table 3) by looking for the column labeled $n = 3$ and $p = .40$. The results are summarized in Table 10·9.

TABLE 10·9
Binomial
probabilities for
interview problem

Possible positive ratings from 3 interviews	Binomial probabilities of these outcomes	
0	$1.0 \quad - .7840 =$	.2160
1	$.7840 - .3520 =$	.4320
2	$.3520 - .0640 =$	.2880
3		.0640
		1.0000

Now we can use the theoretical binomial probabilities of the outcomes to compute the expected frequencies. By comparing these expected frequencies with our observed frequencies using the χ^2 test, we can examine the extent of the difference between them. Table 10·10 lists the observed frequencies, the appropriate binomial probabilities from Table 10·9, and the expected frequencies for the sample of 100 interviews.

Determining the expected frequencies

TABLE 10·10

Observed frequencies, appropriate binomial probabilities, and expected frequencies for interview problem

Possible positive ratings from 3 interviews	Observed frequency of candidates receiving these ratings	Binomial probability of possible outcomes		Number of candidates interviewed		Expected frequency of candidates receiving these ratings
0	18	.2160	×	100	=	22
1	47	.4320	×	100	=	43
2	24	.2880	×	100	=	29
3	11	.0640	×	100	=	6
	100	1.0000				100

Calculating the chi-square statistic

To compute the chi-square statistic for this problem, we can use Equation 10·1:

$$\chi^2 = \sum \frac{(f_0 - f_e)^2}{f_e} \qquad [10·1]$$

and the format we introduced in Table 10·4. This process is illustrated in Table 10·11.

Observed frequency f_0	Expected frequency f_e	$f_0 - f_e$	$(f_0 - f_e)^2$	$\frac{(f_0 - f_e)^2}{f_e}$
18	22	−4	16	.7273
47	43	4	16	.3721
24	29	−5	25	.8621
11	6	5	25	4.1667

$$\sum \frac{(f_0 - f_e)^2}{f_e} = 6.1282 \leftarrow \chi^2$$

**TABLE 10·11
Calculation of χ^2 statistic from interview data listed in Table 10·10**

Determining degrees of freedom in a goodness-of-fit test

Before we can calculate the appropriate number of degrees of freedom for a chi-square goodness-of-fit test, we must count the number of classes (symbolized k) for which we have compared the observed and expected frequencies. Our interview problem contains 4 such classes: 0, 1, 2, and 3 positive ratings. Thus we begin with 4 degrees of freedom. Yet since the

First count the number of classes

four observed frequencies must sum to 100, the total number of observed frequencies we can freely specify is only $k - 1$ or 3. The fourth is determined, because the total of the four has to be 100.

Then subtract degrees of freedom lost from estimating population parameters

To solve a goodness-of-fit problem, we may be forced to impose additional restrictions on the calculations of the degrees of freedom. Suppose we are using the chi-square test as a goodness-of-fit test to determine whether a normal distribution fits a set of observed frequencies. If we have 6 classes of observed frequencies ($k = 6$), then we would conclude that we have only $k - 1$ or 5 degrees of freedom. If, however, we also have to use the sample mean as an estimate of the population mean, we will have to subtract an additional degree of freedom, which leaves us with only 4. And third, if we have to use the sample standard deviation to estimate the population standard deviation, we will have to subtract *one more* degree of freedom, leaving us with 3. Our general rule in these cases is **first employ the ($k - 1$) rule and then subtract an additional degree of freedom for each population parameter that has to be estimated from the sample data.**

In the interview example, we have 4 classes of observed frequencies. As a result, $k = 4$, and the appropriate number of degrees of freedom is $k - 1$ or 3. We are not required to estimate any population parameter, so we need not reduce this number further.

Using the chi-square goodness-of-fit test

Calculating the limit of the acceptance region

In the interview problem, the newspaper desires to test the hypothesis of goodness of fit at the .20 level of significance. In Appendix Table 5, then, we must look under the .20 column and move down to the row labeled 3 degrees of freedom. There we find that the value of the chi-square statistic is 4.642. We can interpret this value as follows: with three degrees of freedom, the region to the right of a chi-square value of 4.642 contains .20 of the area under the curve.

Illustrating the problem

Interpreting the results

Figure 10·5 illustrates a chi-square distribution for 3 degrees of freedom, showing in color a .20 level of significance. Notice that the acceptance region for the null hypothesis (the hypothesis that the sample data came from a binomial distribution with $p = .4$) extends from the left tail to the chi-square value of 4.642. Obviously, the sample chi-square value of 6.1282 falls

FIG 10·5
Goodness-of-fit test at the .20 level of significance, showing acceptance region and sample chi-square value of 6.1282

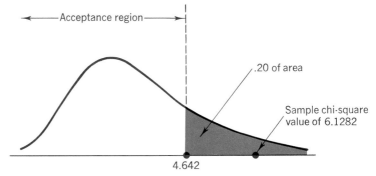

outside of this acceptance region. Therefore, we reject the null hypothesis and conclude that the binomial distribution with $p = .4$ fails to provide a good description of our observed frequencies.

EXERCISES

13. It is suggested that a certain variable may be described by the binomial distribution with 5 trials and a probability of success on each trial of .3. Given the following observed frequency distribution of success, can we conclude that the distribution does in fact follow the suggested distribution? Use the .05 significance level.

Number of successes/day	0	1	2	3	4	5
Frequency of the number of successes	20	65	42	14	6	3

14. At the .10 level of significance, can we conclude that the following distribution follows a Poisson distribution with $\lambda = 2$?

Number of arrivals/hour	0	1	2	3	4	5
Number of hours	10	19	31	26	11	3

15. Below is a table of observed frequencies along with the frequencies to be expected under a normal distribution.

   a) Calculate the chi-square statistic.

   b) Can we conclude that this distribution does in fact follow a normal distribution, using the .05 level of significance?

	Score				
	51–60	*61–70*	*71–80*	*81–90*	*91–100*
Observed frequency	3	10	44	50	13
Expected frequency	2	17	50	41	10

16. Below is an observed frequency distribution. Using a normal distribution with $\mu = 2.44$ and $\sigma = .4$:

   a) Find the probability of falling in each class.

   b) From part a, compute the expected frequency of each category.

   c) Calculate the chi-square statistic.

   d) At the .10 level of significance, does this distribution seem to be well-described by the suggested normal distribution?

Observed value of the variable	less than 1.8	1.8–2.19	2.2–2.59	2.6–2.99	3.0 and above
Observed frequency	3	17	33	22	5

17. A cognitive psychologist interested in problem solving has made a pilot test of subjects who were asked to solve a number of distinct types of problems. She has chosen 3 problems for use in a more extensive study of strategies used in problem solving. From past data, she believes that each subject has a 40 percent chance of solving each problem in the time limit and that performance on the 3 problems is independent, owing to the problems' requiring different cognitive skills. A sample of the results from 200 subjects showed that the number of subjects solving 0, 1, 2, or 3 problems, were 46, 73, 58, and 23, respectively. At the .10 level of significance, are these numbers of problems solved per subject well described by a binomial distribution with $p = .4$?

18. A chemical extraction plant processes sea water to collect sodium chloride and magnesium. From scientific analysis, sea water is known to contain sodium chloride, magnesium, and other elements in the ratio 62:4:34. A sample of 200 tons of sea water has resulted in 130 tons of sodium chloride and 6 tons of magnesium. Are these data consistent with the scientific model at the .05 level of significance?

19. Dennis Barry, a hospital administrator, has examined past records from 300 randomly selected eight-hour shifts to determine the frequency with which the hospital treats fractures. The number of days in which 0, 1, 2, 3, 4, 5, or 6 or more patients with broken bones were treated was 25, 45, 63, 71, 48, 26, and 22, respectively. At the .05 level of significance, can we reasonably believe that the incidence of broken bone cases follows a Poisson distribution with $\lambda = 3$?

20. A large city fire department calculates that for any given precinct, during any given eight-hour shift, there is a 30 percent chance of receiving at least one fire alarm. Here is a random sampling of 60 days:

Number of shifts during which alarms were received	1	2	3
Number of days	27	11	6

At the .05 level of significance, do these fire alarms follow a binomial distribution? (*Hint:* combine the last two groups so that all expected frequencies will be greater than 5).

21. A physician studying respiratory ailments has compiled the following table, classifying 150 patients by their maximal breathing capacity (MBC). Maximal breathing capacity refers to the greatest volume of air a person can inhale and exhale in one minute.

MBC (liters/minute)	50 or less	51–100	101–150	150 or more
Number of persons in each group	13	53	62	22

Before the individual MBC's were combined into these groups, the sample mean and standard deviation were calculated to be 106.3 liters/minute and 36.5 liters/minute respectively.

a) What is the probability (using a normal distribution with $\mu = 106.3$ and $\sigma = 36.5$) that a person's MBC will be less than 50.5 liters/minute; between 50.5 and 100.5 liters/minute; between 100.5 and 150.5 liters/minute; greater than 150.5 liters/minute?

b) Using the probabilities in part a, find the expected frequencies for the 150 patients' MBC's.

c) At the .05 level of significance, does the observed distribution follow the normal distribution found in part b?

22. The election supervisor is keeping track of the arrival of voters at representative precincts, to see how many election workers are needed to handle the flow. In a sample of 800 five-minute time periods, there were 36, 117, 194, 167, 138, 94, and 54 periods in which 0, 1, 2, 3, 4, 5, or 6 or more voters arrived at a precinct, respectively. Are these data consistent at the .05 level of significance with a Poisson distribution with $\lambda = 3$?

4 ANALYSIS OF VARIANCE Earlier in this chapter, we used the chi-square test to examine the difference between more than two sample proportions and to make inferences about whether such samples are drawn from populations each having the same proportion. In this section, we will learn a technique known as *analysis of variance (often abbreviated ANOVA),* which will enable us to test for the significance of the difference between more than two sample *means.* Using analysis of variance, we will be able to make inferences about whether our samples are drawn from populations having the same mean.

Analysis of variance will be useful in such situations as comparing the mileage achieved by five different brands of gasoline, testing which of four different training methods produces the fastest learning record, or comparing the first-year earnings of the graduates of half a dozen different schools of law or medicine. In each of these cases, we would compare the means of more than two samples.

Statement of the problem

Suppose that the director of training of a large state extension service agricultural laboratory is trying to evaluate three different methods that have been used for several years to train people to analyze soil. Method 1 assigns a new employee to an experienced analyst for individual guidance within the lab. With method 2, all new analysts work in a training room separate from the lab and are supervised by a trainer. Method 3 uses training films and programmed manuals. In this method, new analysts spend several days studying the films, then work independently, using the training manuals. The director wants to determine whether there is any difference in effectiveness among the three methods.

Method 1	Method 2	Method 3	
		18	
15	22	24	
18	27	19	
19	18	16	
22	21	22	
11	17	15	
85	105	114	
$\div 5$	$\div 5$	$\div 6$	
$17 = \bar{x}_1$	$21 = \bar{x}_2$	$19 = \bar{x}_3 \leftarrow$ sample means	
$n_1 = 5$	$n_2 = 5$	$n_3 = 6 \leftarrow$ sample sizes	

**TABLE 10·12
Daily
production of
16 new soil
sample analysts**

Calculating
the grand mean

After completion of the training period, the laboratory's statistical staff chooses 16 new analysts assigned at random to the three training methods.* Counting the samples analyzed per day by these 16 trainees, the staff has summarized the data and calculated the mean production of the trainees (see Table 10·12). Now if we wish to determine the *grand mean,* or $\bar{\bar{x}}$ (the mean for the entire group of 16 trainees), we can use one of two methods:

---

*Although in real practice 16 trainees would not constitute an adequate statistical sample, we have limited the number here to be able to demonstrate the basic techniques of analysis of variance and to avoid tedious calculations.

1. $\bar{\bar{x}} = \dfrac{15+18+19+22+11+22+27+18+21+17+18+24+19+16+22+15}{16}$

$= \dfrac{304}{16}$

$= 19 \leftarrow$ grand mean using all the data

2. $\bar{\bar{x}} = (5/16)(17) + (5/16)(21) + (6/16)(19)$

$= \dfrac{304}{16}$

$= 19 \leftarrow$ grand mean as a weighted average of the sample means, using the relative sample sizes as the weights

Statement of the hypothesis

Stating the problem symbolically

In this case, our reason for using analysis of variance is to decide whether these three samples (a *sample* is the performance of the analysts trained by any one method) were drawn from populations (a *population* is the total number of analysts trained by any one method) having the same means. Because we are testing the effectiveness of the three training methods, we must determine whether the three samples, represented by the sample means $\bar{x}_1 = 17$, $\bar{x}_2 = 21$, and $\bar{x}_3 = 19$, could have been drawn from populations having the same mean. A formal statement of the null hypothesis we wish to test would be:

$H_0: \mu_1 = \mu_2 = \mu_3 \leftarrow$ null hypothesis
$H_1: \mu_1, \mu_2,$ and μ_3 are *not* all equal $\leftarrow$ alternative hypothesis

Interpreting the results

If we can conclude from our test that the sample means do *not* differ significantly, we can infer that the choice of training method does not influence the productivity of the analysts. On the other hand, if we find a difference among the sample means that is too large to attribute to chance sampling error, we can infer that the method used in training *does* influence the productivity of the analysts. In that case, we would adjust our training program accordingly.

Analysis of variance: basic concepts

Assumptions made in analysis of variance

In order to use analysis of variance, we must assume that each of the samples is drawn from a normal population and that each of these populations has the same variance σ^2. If, however, the sample sizes are large enough, we do not need the assumption of normality.

In our training methods problem, our null hypothesis states that the three populations have the same means. If this hypothesis is true, classifying the data into three columns in Table 10·12 is unnecessary, and the entire set of 16 measurements of productivity can be thought of as a sample from one population. This overall population also has a variance of σ^2.

Analysis of variance is based on a comparison of two different estimates of the variance, σ^2, of our overall population. In this case, we can calculate one of these two estimates by examining the variance among the three sample means, which are 17, 21, and 19. The other estimate of the population variance is determined by the variation within the three samples themselves, that is (15, 18, 19, 22, 11), (22, 27, 18, 21, 17), and (18, 24, 19, 16, 22, 15). Then we compare these two estimates of the population variance. Since both are estimates of σ^2, they should be approximately equal in value *when the null hypothesis is true.* If the null hypothesis is *not* true, these two estimates will differ considerably. The three steps in analysis of variance, then, are:

1. Determine one estimate of the population variance from the variance *among the sample means.*

2. Determine a second estimate of the population variance from the variance *within the samples.*

3. Compare these two estimates. If they are approximately equal in value, *accept* the null hypothesis.

Steps in analysis of variance

In the remainder of this section, we shall learn how to calculate these two estimates of the population variance, how to compare these two estimates, and how to make a hypothesis test and interpret the results. As we learn how to do these computations, however, keep in mind that all are based on the concepts we have presented in this section.

Calculating the variance among the sample means

Step 1 in analysis of variance indicates that we must obtain one estimate of the population variance from the variance among the three sample means. In statistical language, this estimate is called the *between-column variance.*

Finding the first estimate of the population variance

In Chapter 4, we used Equation 4·11 to calculate the sample variance:

Sample variance

$$s^2 = \frac{\Sigma(x - \bar{x})^2}{n - 1} \qquad [4 \cdot 11]$$

Now, because we are working with three sample means and a grand mean, let's substitute $\bar{x}$ for x, $\bar{\bar{x}}$ for $\bar{x}$, and k (the number of samples) for n to get a formula for the variance among the sample means:

First find the variance among sample means

Variance among sample means

$$s_{\bar{x}}^2 = \frac{\Sigma(\bar{x} - \bar{\bar{x}})^2}{k - 1} \qquad [10 \cdot 4]$$

Next, we can return for a moment to Chapter 7, where we defined the standard error of the mean as the standard deviation of all possible samples

Then find the
population variance
using this variance
among sample means

of a given size. The formula to derive the standard error of the mean is Equation 7·1:

Standard error of the mean (standard deviation of all possible sample means of a given size)

Population standard deviation

$$\sigma_{\bar{x}} = \frac{\sigma}{\sqrt{n}} \qquad \text{← Square root of the sample size}$$

$$[7 \cdot 1]$$

Let's simplify this equation by cross multiplying the terms and then squaring both sides in order to change the population standard deviation, σ, into the population variance, σ^2:

Population variance $\rightarrow \sigma^2 = \sigma_{\bar{x}}^2 \times n$ $\qquad [10 \cdot 5]$

Standard error squared (this is the variance among the sample means)

For our soil analysts problem, we do not have all the information we need to use this equation to find σ^2. Specifically, we do not know $\sigma_{\bar{x}}^2$. We could however, calculate the variance among the 3 sample means $s_{\bar{x}}^2$, using Equation 10·4. So why not substitute $s_{\bar{x}}^2$ for $\sigma_{\bar{x}}^2$ in Equation 10·5 and calculate an estimate of the population variance? This will give us:

$$\hat{\sigma}^2 = s_{\bar{x}}^2 \times n = \frac{\Sigma n(\bar{x} - \bar{\bar{x}})^2}{k - 1}$$

There is a slight difficulty in using this equation as it stands. In Equation 7·1, n represents the sample size, but *which* sample size should we use when the different samples have different sizes? We solve this problem with Equation 10·6, where each $(\bar{x}_j - \bar{\bar{x}})^2$ is multiplied by its own appropriate n_j.

$$\hat{\sigma}^2 = \frac{\Sigma n_j(\bar{x}_j - \bar{\bar{x}})^2}{k - 1} \qquad [10 \cdot 6]$$

where:

$\hat{\sigma}^2 =$ our first estimate of the population variance based on the variance among the sample means (the *between column variance*)
$n_j =$ the size of the jth sample
$\bar{x}_j =$ the sample mean of the jth sample
$\bar{\bar{x}} =$ the grand mean
$k =$ the number of samples

Now we can use Equation 10·6 and the data from Table 10·12 to calculate the between-column variance. Table 10·13 shows how to make these calculations.

TABLE 10·13 Calculation of the between-column variance

n	$\bar{x}$	$\bar{\bar{x}}$	$\bar{x} - \bar{\bar{x}}$	$(\bar{x} - \bar{\bar{x}})^2$	$n(\bar{x} - \bar{\bar{x}})^2$
5	17	19	$17 - 19 = -2$	$(-2)^2 = 4$	$5 \times 4 = 20$
5	21	19	$21 - 19 = 2$	$(2)^2 = 4$	$5 \times 4 = 20$
6	19	19	$19 - 19 = 0$	$(0)^2 = 0$	$6 \times 0 = 0$

$$\hat{\sigma}^2 = \frac{\Sigma n_j(\bar{x}_j - \bar{\bar{x}})^2}{k - 1} = \frac{40}{3 - 1}$$ [10·6] $\Sigma n_j(\bar{x}_j - \bar{\bar{x}})^2 \rightarrow \overline{40}$

$$= \frac{40}{2}$$

$= 20 \leftarrow$ the between-column variance

Calculating the variances within the samples

Finding the second estimate of the population variance

The second estimate of the population variance is based on the variance within the samples. In statistical terms, this can be called the *within-column variance.* Our soil analysts problem has three samples of 5 or 6 items each. We can calculate the variance within each of these three samples using Equation 4·11:

Sample variance

$$s^2 = \frac{\Sigma(x - \bar{x})^2}{n - 1}$$ [4·11]

Since we have assumed that the variances of our three populations are the same, we could use any one of the three sample variances (s_1^2 or s_2^2 or s_3^2) as the second estimate of the population variance. Statistically, we can get a better estimate of the population variance by using a weighted average of all three sample variances. The general formula for this second estimate of σ^2 is:

$$\hat{\sigma}^2 = \Sigma \left(\frac{n_j - 1}{n_T - k} \right) s_j^2$$ [10·7]

where:
$\hat{\sigma}^2 = $ our second estimate of the population variance based on the variances within the samples (the *within-column variance*)
$n_j = $ the size of the jth sample
$s_j^2 = $ the sample variance of the jth sample
$k = $ the number of samples
$n_T = \Sigma n_j = $ the total sample size

Using all the information at our disposal

This formula uses all the information that we have at our disposal, not just a portion of it. Had there been 7 samples instead of 3 we would have taken a weighted average of all 7. The weights used in Equation 10·7 will be explained shortly on pages 382-83. Table 10·14 illustrates how to calculate this second estimate of the population variance using the variances within all three of our samples.

TABLE 10·14 Calculation of variances within the samples and the within-column variance

Training method 1 Sample mean: $\bar{x} = 17$		Training method 2 Sample mean: $\bar{x} = 21$		Training method 3 Sample mean: $\bar{x} = 19$	
$x - \bar{x}$	$(x - \bar{x})^2$	$x - \bar{x}$	$(x - \bar{x})^2$	$x - \bar{x}$	$(x - \bar{x})^2$
$15 - 17 = -2$	$(-2)^2 = 4$	$22 - 21 = 1$	$(1)^2 = 1$	$18 - 19 = -1$	$(-1)^2 = 1$
$18 - 17 = 1$	$(1)^2 = 1$	$27 - 21 = 6$	$(6)^2 = 36$	$24 - 19 = 5$	$(5)^2 = 25$
$19 - 17 = 2$	$(2)^2 = 4$	$18 - 21 = -3$	$(-3)^2 = 9$	$19 - 19 = 0$	$(0)^2 = 0$
$22 - 17 = 5$	$(5)^2 = 25$	$21 - 21 = 0$	$(0)^2 = 0$	$16 - 19 = -3$	$(3)^2 = 9$
$11 - 17 = -6$	$(-6)^2 = 36$	$17 - 21 = -4$	$(-4)^2 = 16$	$22 - 19 = 3$	$(3)^2 = 9$
	$\Sigma(x - \bar{x})^2 = 70$		$\Sigma(x - \bar{x})^2 = 62$	$15 - 19 = -4$	$(-4)^2 = 16$
	$\dfrac{\Sigma(x - \bar{x})^2}{n - 1} = \dfrac{70}{5 - 1}$		$\dfrac{\Sigma(x - \bar{x})^2}{n - 1} = \dfrac{62}{5 - 1}$		$\Sigma(x - \bar{x})^2 = 60$
	$= \dfrac{70}{4}$		$= \dfrac{62}{4}$		$\dfrac{\Sigma(x - \bar{x})^2}{n - 1} = \dfrac{60}{6 - 1}$
					$= \dfrac{60}{5}$
sample variance $\rightarrow s_1^2 = 17.5$		sample variance $\rightarrow s_2^2 = 15.5$		sample variance $\rightarrow s_3^2 = 12.0$	

And:

$$\hat{\sigma}^2 = \Sigma \left(\frac{n_j - 1}{n_T - K} \right) s_j^2 = (4/13)(17.5) + (4/13)(15.5) + (5/13)(12.0) \quad [10·7]$$

$$= \frac{192}{13} \qquad \text{Second estimate of the population}$$
variance based on the variances within

$$= 14.769 \quad \leftarrow \text{the samples (the within-column variance)}$$

The F hypothesis test:
computing and interpreting the F statistic

Finding the *F* ratio, a statistic of the *F* distribution

We can compare these two estimates of the population variance by computing their ratio, called *F* as follows:

$$F = \frac{\text{First estimate of the population variance}}{\text{Second estimate of the population variance}} \quad [10·8]$$
based on the variance among the sample means
based on the variances within the samples

If we substitute the statistical shorthand for the numerator and denominator of this ratio, Equation 10·8 becomes:

$$F = \frac{\text{Between-column variance}}{\text{Within-column variance}} \quad [10·9]$$

Now we can find the F ratio for the soil analysts problem with which we have been working:

$$F = \frac{\text{Between-column variance}}{\text{Within-column variance}} \qquad [10 \cdot 9]$$

$$= \frac{20}{14.769}$$

$$= 1.354 \leftarrow F\ ratio$$

Now that we have found it, how can we interpret this F ratio of 1.354? First, let's examine the denominator, which is based on the variance within the samples. The denominator is a good estimate of σ^2 (the population variance) whether the null hypothesis is true or not. What about the numerator? If the null hypothesis that the three methods of training have equal effects is true, then the numerator, or the variation among the sample means of the three methods, is also a good estimate of σ^2 (the population variance). As a result, **the denominator and numerator should be about equal if the null hypothesis is true.** The nearer the F ratio comes to one, then, the more we are inclined to accept the null hypothesis. Conversely, as the F ratio becomes larger, we will be more inclined to reject the null hypothesis and accept the alternative (that a difference does exist in the effects of the three training methods).

Interpreting the F ratio

Shortly, we shall learn a more formal way of deciding when to accept or reject the null hypothesis. But even now, you should understand the basic logic behind this F *statistic.* When populations are not the same, the between-column variance (which was derived from the variance among the sample means) will tend to be larger than the within-column variance (which was derived from the variances within the samples), and the value of F will tend to increase. This will lead us to reject the null hypothesis.

The F distribution

Like other statistics we have studied, if the null hypothesis is true, then the F statistic has a particular sampling distribution. Like the t and chi-square distributions, this F distribution is actually a whole family of distributions, three of which are shown in Fig. 10·6. Notice that each is identified by a *pair* of degrees of freedom, unlike the t and chi-square distributions, which have only one value for the number of degrees of freedom. **The first number refers to the number of degrees of freedom in the numerator of the F ratio; the second, to the degrees of freedom in the denominator.**

Describing an F distribution

As we can see in Fig. 10·6, the F distribution has a single mode. The specific shape of an F distribution depends upon the number of degrees of freedom in both the numerator and the denominator of the F ratio. But in general, the F distribution is skewed to the right and tends to get more symmetrical as the number of degrees of freedom in the numerator and denominator increase.

FIG 10·6
Three F distributions (first value in parentheses = number of degrees of freedom in the numerator of the F ratio; second = number of degrees of freedom in the denominator)

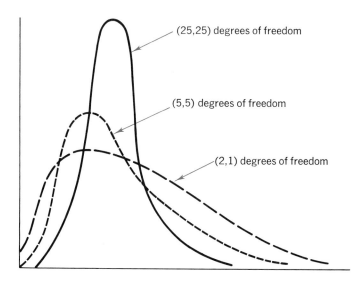

- (25,25) degrees of freedom
- (5,5) degrees of freedom
- (2,1) degrees of freedom

Using the F distribution: degrees of freedom

Calculating degrees of freedom

As we have mentioned, each *F* distribution has a pair of degrees of freedom, one for the numerator of the *F* ratio and the other for the denominator. How can we calculate both of these?

Finding the numerator degrees of freedom

First, let's think about the numerator, the between-column variance. In Table 10-13, we used three values of $(\bar{x} - \bar{\bar{x}})^2$ one for each sample, to calculate $\Sigma(\bar{x} - \bar{\bar{x}})^2$. Once we knew two of these $(\bar{x} - \bar{\bar{x}})^2$ values, the third was *automatically determined* and could not be freely specified. Thus, one degree of freedom is lost when we calculate the between-column variance, and the number of degrees of freedom for the numerator of the *F* ratio is always one fewer than the number of samples. The rule, then, is:

$$\text{Number of degrees of freedom in } \textit{numerator} \text{ of the } F \text{ ratio} = (\text{Number of samples} - 1) \quad [10 \cdot 10]$$

Finding the denominator degrees of freedom

Now, what of the denominator? Look at Table 10-14 for a moment. There we calculated the variances within the samples, and we used all three samples. For the *j*th sample, we used n_j values of $(x - \bar{x})$ to calculate the $(x - \bar{x})^2$ for that sample. Once we knew all but one of these $(x - \bar{x})$ values, the last was *automatically determined* and could not be freely specified. Thus we lost one degree of freedom in the calculations for *each* sample, leaving us with 4, 4, and 5 degrees of freedom in the samples. Since we had three samples, we were left with $4 + 4 + 5 = 5 + 5 + 6 - 3 = 13$ degrees of freedom. We can state the rule like this:

$$\text{Number of degrees of freedom in denominator of the } F \text{ ratio} = \Sigma(n_j - 1) = n_T - k \quad [10 \cdot 11]$$

where:

$$n_j = \text{the size of the } j\text{th sample}$$
$$k = \text{the number of samples}$$
$$n_T = \Sigma n_j = \text{the total sample size}$$

Now we can see that the weight assigned to $s_j{}^2$ in Equation 10·7 was just its fraction of the total number of degrees of freedom in the denominator of the F-ratio.

Using the F table

To do F hypothesis tests, we shall use an F table in which the columns represent the number of degrees of freedom for the numerator and the rows represent the degrees of freedom for the denominator. Separate tables exist for each level of significance.

Suppose we are testing a hypothesis at the .01 level of significance, using the F distribution. Our degrees of freedom are 8 for the numerator and 11 for the denominator. In this instance, we would turn to Appendix Table 6. In the body of that table, the appropriate value for 8 and 11 degrees of freedom is 4.74. If our calculated value of F exceeds this table value of 4.74, we would reject the null hypothesis. If not, we would accept it.

Testing the hypothesis

We can now test our hypothesis that the three different training methods produce identical results, using the material we have developed to this point. Let's begin by reviewing how we calculated the F ratio:

Finding the F statistic and the degrees of freedom

$$F = \frac{\text{First estimate of the population variance}}{\text{Second estimate of the population variance}} \quad [10\cdot 8]$$
$$\text{based on the variances within the samples}$$

$$= \frac{20}{14.769}$$

$$= 1.354 \leftarrow F \text{ statistic}$$

Next, calculate the number of degrees of freedom in the numerator of the F ratio, using Equation 10·10 as follows:

$$\begin{array}{ll} \text{Number of degrees of freedom} \\ \text{in numerator of the } F \text{ ratio} \end{array} = (\text{Number of samples} - 1) \quad [10\cdot 10]$$

$$= 3 - 1$$

$$= 2 \leftarrow \begin{array}{l} \text{degrees of freedom} \\ \text{in the numerator} \end{array}$$

And we can calculate the number of degrees of freedom in the denominator of the F ratio by use of Equation 10·11:

$$\text{Number of degrees of freedom in denominator of the } F \text{ ratio} = \Sigma(n_j - 1) = n_T - k \qquad [10·11]$$

$$= (5 - 1) + (5 - 1) + (6 - 1)$$
$$= 16 - 3$$
$$= 13 \leftarrow \text{degrees of freedom in the denominator}$$

Calculating
the limit of the
acceptance region

Suppose the director of training wants to test at the .05 level the hypothesis that there is no difference between the three training methods. We can look in Appendix Table 6 for 2 degrees of freedom in the numerator and 13 in the denominator. The value we find there is 3.81. Figure 10·7 shows this hypothesis test graphically. The colored region represents the level of significance. The table value of 3.81 sets the upper limit of the acceptance region. Since the calculated value for F of 1.354 lies within the acceptance region, we would accept the null hypothesis and conclude that, according to the sample information we have, there is no difference in the effects of the three training methods on analyst productivity.

Interpreting
the results

**FIG. 10·7
Hypothesis test at
the .05 level of
significance, using
the F distribution
and showing
the acceptance
region and the
calculated F value**

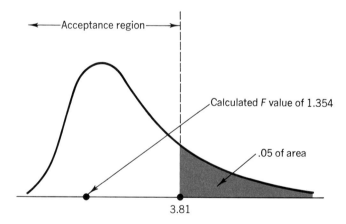

Precautions about using the F test

Use large
sample sizes

As we stated earlier, our sample sizes in this problem are too small for us to be able to draw valid inferences about the effectiveness of the various training methods. We chose small samples so that we could explain the logic of analysis of variance without tedious calculations. In actual practice, our methodology would be the same, but our samples would be larger.

Control all factors
but the one
being tested

In our example, we have assumed the absence of many factors that might have affected our conclusions. We accepted as given, for example, the fact that all the new analysts we sampled had the same demonstrated aptitude for learning—which may or may not be true. We assumed that all the instructors of the three training methods had the same ability to teach and to manage, which may not be true. And we assumed that the laboratory's statistical staff collected the data on productivity during work periods that were similar in terms of time of day, day of the week, time of the year, and so on. To be able to make significant decisions based on analysis of variance, we need to be certain that all these factors are effectively controlled.

Finally, notice that we have discussed only *one-way*, or one factor, analysis of variance. Our problem examined the effect of the type of training method on analyst productivity, nothing else. Had we wished to measure the effect of two factors, such as the training program and the age of the analysts we would need the ability to use two-way analysis of variance, a statistical method best saved for more advanced textbooks.

A test for
one factor only

EXERCISES

23. A study compared effects of 4 tranquilizer drugs on vision. Below are the visual scores for 5 individuals in each drug group.

Drug 1	77	86	80	88	84
Drug 2	95	92	88	91	89
Drug 3	72	77	68	82	75
Drug 4	80	84	79	70	82

a) Compute the mean score for each group and then determine the grand mean.

b) Calculate the variance among the sample means.

c) Estimate the population variance using the between-column variance (Equation 10·6).

d) Estimate the population variance using the within-column variance computed from the variances within the samples.

e) Calculate the F ratio. At the .05 level of significance, do the drugs produce differential effects on vision?

24. Three training methods were compared to see if they led to greater productivity after training. Below are productivity measures for individuals trained by each method.

Method 1	36	26	31	20	34	25
Method 2	40	29	38	32	39	34
Method 3	32	18	23	21	33	27

At the .05 level of significance, do the three training methods lead to different levels of productivity?

25. For the five groups of observations below, test the hypothesis that their mean scores are all the same. Use the .01 level of significance.

Group 1	15	17	14	11		
Group 2	12	10	13	17	14	
Group 3	10	14	13	15	12	
Group 4	14	9	7	10	8	7
Group 5	13	12	9	14	10	9

26. Given the measurements on the 4 samples below, can we conclude that they come from populations having the same mean value? Use the .05 level of significance.

Sample 1	17	22	25	29	30	
Sample 2	29	18	20	19	30	21
Sample 3	13	14	20	18	27	16
Sample 4	21	28	20	22	18	

27. A forestry major decided to study how different amounts of rainfall affected the heights of trees. To examine this, he planted 4 groups of trees. Each group consisted of 5 trees of the same type, age, and size, and they were planted in chambers provided with different amounts of simulated rainfall. At the end of the specified period of time, he measured the heights of the trees.

Achieved heights (inches)

Group 1	Group 2	Group 3	Group 4
36	29	31	36
34	34	35	28
37	34	32	34
35	36	33	32
33	32	39	30

a) Calculate the mean height, $\bar{x}$, for each group; then determine the grand mean, $\bar{\bar{x}}$.

b) Calculate the variance *among* the sample means.

c) Using Equation 10·6, estimate the population variance (the between-column variance).

d) Calculate the variances *within* the samples and estimate the population variance based upon these variances (the within-column variance).

e) Calculate the F ratio. At the .05 level of significance, do the four different rainfall amounts produce trees with the same height?

28. How will a small lesion in a rat's brain affect its performance on a task that requires discrimination? To answer this question, an experimenter has designed a study introducing a lesion into a symmetrical part of rats' brains. Either the right side or the left side may be used. The experimenter decides to run 3 groups of rats: one with the lesion on the right, one with the lesion on the left, and a control group with no lesion. Recorded are the number of trials required for each rat to attain a certain criterion level of performance.

Left lesion	65	79	73	55	68	74
Right lesion	60	64	57	75	62	56
Control	61	54	74	59	46	

a) Calculate the average number, $\bar{x}$, of trials it took the rats in each group to reach criterion performance level and then determine the grand mean, $\bar{\bar{x}}$.

b) Calculate the variance among the sample means.

c) Estimate the population variance by the between-column variance.

d) Calculate the variances within the samples and estimate the population variance based upon these variances (the within-column variance).

e) Calculate the F ratio. At the .05 level of significance, do the three groups take the same number of trials to reach the criterion level of performance?

29. It has often been noted that there seem to be relatively more suicide attempts during the Christmas holiday season, perhaps because people feel especially lonely then. Along the same lines, one psychologist wondered if the number of admissions to mental hospitals

during the holiday season was greater than the number of admissions in weeks before or after the holiday. So for the past six years, the number of mental hospital admissions in the state during the months of November, December, and January were noted. They are listed in the table below.

Number of admissions

November	42	36	58	54	37	47
December	51	38	45	32	47	46
January	37	29	35	42	31	33

At the .05 level of significance, is the number of admissions the same during these three months?

30. A research company has designed four different systems to clean up oil spills. The following table contains the results, measured by how much surface area (in meters^2) is cleared in one hour. The data were found by testing each method in five trials. Are the four systems equally effective at the .05 level of significance?

System A	55	60	58	61	54
System B	47	53	54	49	52
System C	63	59	58	64	63
System D	51	56	54	59	54

31. As part of research on the effects of smoking on living organisms (and more specifically, humans), small animals are exposed to smoke; their survival times are recorded. One study placed paramecia in 5 smoke chambers through which smoke was drawn from cigarettes with 5 different charcoal filters. At the .01 level of significance, is the survival time the same for all 5 filters? (Shown are the survival times, in minutes, for the paramecia in the 5 smoke chambers.)

Chamber 1	24	11	19	27	15	16	22	32	17
Chamber 2	29	35	37	26	45	26	29	35	38
Chamber 3	30	28	29	32	22	17	23	29	11
Chamber 4	16	14	5	19	21	17	11	26	9
Chamber 5	21	16	19	15	16	28	23	29	17

32. In some prisons, inmates are rewarded with tokens that can be exchanged for various items or privileges. It occurred to a criminologist that the character of the items or privileges offered might influence the behavior of the inmates, motivating them to earn more or fewer tokens. To explore this possibility, inmates were assigned to one of three conditions: tokens would be exchangeable for (1) privileges, (2) luxury items such as cigarettes and magazines, and (3) something unspecified. At the end of one month, the number of tokens each subject had earned was recorded and is presented in the following table. At the .05 level of significance, is there any difference in the number of tokens earned by subjects in the three conditions?

For privileges	41	53	54	55	43
For luxuries	45	51	48	43	39
For something unspecified	34	44	46	45	51

33. Helping people lose weight, one therapist tried three different counseling techniques and kept records to compare their effectiveness. Each subject was weighed at the beginning of the study and assigned a goal—the number of pounds to be lost in a given period. At the end of the study, subjects were weighed, in order to record the percent of the goal each

had attained. At the .05 level of significance, can we consider the three methods of counseling equally effective, as illustrated by the percentage of goal lost in the following table?

Percent of desired weight loss attained

Supportive counseling	85	84	79	84	88	
Aversive therapy	83	94	96	85	87	90
Behavior modification	74	76	73	81		

5 INFERENCES ABOUT A POPULATION VARIANCE

In chapters 8 and 9, we learned how to form confidence intervals and test hypotheses about one or two population means or proportions. Earlier in this chapter, we used chi-square and F tests to make inferences about more than two means or proportions. But we are not always interested in means and proportions. In many situations, responsible decision makers have to make inferences about the variability in a population. In order to schedule the labor force at harvest time, a peach grower needs to know not only the mean time to maturity of the peaches, but also their variance around that mean. A sociologist investigating the effect of education on earning power wants to know if the incomes of college graduates are more variable than those of high school graduates. Precision instruments used in laboratory work must be quite accurate on the average; but in addition, repeated measurements should show very little variation. In this section, we shall see how to make inferences about a single population variance. The next section looks at problems involving the variances of two populations.

Need to make decisions about variability in a population

The distribution of the sample variance

Let's go back now to the Postmaster General's problem that opened this chapter. Recall that an investigator followed 9 letters from New York to Chicago, to estimate the standard deviation in time of delivery. Table 10·15 repeats the data and computes $\bar{x}$, s^2, and s. As we saw in Chapter 8, we use s to estimate σ.

We can tell the Postmaster General that the *population* standard deviation, as estimated by the *sample* standard deviation, is approximately 23 hours. But she also wants to know how accurate that estimate is and what uncertainty is associated with it. In other words, she wants a confidence interval, not just a point estimate of σ. In order to find such an interval, we must know the sampling distribution of s. It is traditional to talk about s^2 rather than s, but this will cause us no trouble, since we can always go from s^2 and σ^2 to s and σ by taking square roots; and we can go in the other direction by squaring.

Determining the uncertainty attached to estimates of the population standard deviation

If the population variance is σ^2, then the statistic

$$\chi^2 = \frac{(n-1)s^2}{\sigma^2}$$

[10·12]

TABLE 10·15
**Delivery time
(in hours) for
letters going
between New York
and Chicago**

Time x	$\bar{x}$	$x - \bar{x}$	$(x - \bar{x})^2$
50	59	−9	81
45	59	−14	196
27	59	−32	1024
66	59	7	49
43	59	−16	256
96	59	37	1369
45	59	−14	196
90	59	31	961
69	59	10	100

$\Sigma x = 531$ $\qquad\qquad\qquad \Sigma(x - \bar{x})^2 = 4232$

$$\bar{x} = \frac{\Sigma x}{n} = \frac{531}{9} \quad [3\cdot2]$$
$$= 59 \text{ hours}$$

$$s^2 = \frac{\Sigma(x - \bar{x})^2}{n - 1} = \frac{4232}{8} \quad [4\cdot11]$$
$$= 529 \text{ hours squared}$$
$$s = \sqrt{s^2} = \sqrt{529} \quad [4\cdot12]$$
$$= 23 \text{ hours}$$

has a chi-square distribution with $n - 1$ degrees of freedom. This result is exact if the population is normal; but even for samples from non-normal populations, it is frequently a good approximation. We can now use the chi-square distribution to form confidence intervals and test hypotheses about σ^2.

Confidence intervals for the population variance

Suppose we want a 95 percent confidence interval for the variance in our mail delivery problem. Figure 10·8 shows how to begin constructing this interval.

Constructing a confidence interval for a variance

We locate two points on the χ^2 distribution, χ_U^2 cuts off .025 of the area in the upper tail of the distribution, and χ_L^2 cuts off .025 of the area in the lower tail. (For a 99 percent confidence interval, we would put .005 of the area in each tail, and similarly for other confidence levels.) The values of χ_L^2 and χ_U^2 can be found in Appendix Table 5. In our mail problem, with $9 - 1 = 8$ degrees of freedom, $\chi_L^2 = 2.180$, and $\chi_U^2 = 17.535$.

Upper and lower limits for the confidence interval

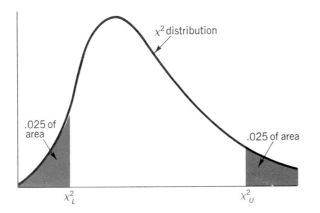

FIG. 10·8
**Constructing a
confidence interval
for σ^2**

Now Equation 10·12 gives χ^2 in terms of s^2, n, and σ^2. To get a confidence interval for σ^2, we solve Equation 10·12 for σ^2.

$$\sigma^2 = \frac{(n-1)s^2}{\chi^2} \qquad [10·13]$$

and then our confidence interval is given by

$$\sigma_L^2 = \frac{(n-1)s^2}{\chi_U^2} \leftarrow \text{lower confidence limit}$$
$$\qquad [10·14]$$
$$\sigma_U^2 = \frac{(n-1)s^2}{\chi_L^2} \leftarrow \text{upper confidence limit}$$

Notice that since χ^2 appears in the denominator in Equation 10·13, we can use χ_U^2 to find σ_L^2 and χ_L^2 to find σ_U^2. Continuing with the Postmaster General's problem, we see she can be 95 percent confident that the population variance lies between 241.35 and 1941.28 hours squared:

$$\sigma_L^2 = \frac{(n-1)s^2}{\chi_U^2} = \frac{8(529)}{17.535} = 241.35$$
$$\qquad [10·14]$$
$$\sigma_U^2 = \frac{(n-1)s^2}{\chi_L^2} = \frac{8(529)}{2.180} = 1941.28$$

So a 95 percent confidence interval for σ would be from $\sqrt{241.35}$ to $\sqrt{1941.28}$ hours; that is, from 15.54 to 44.06 hours.

A two-tailed test of a variance

Testing
hypotheses about
a variance: two-tailed
tests

A statistics professor has given careful thought to the design of examinations. In order for him to be reasonably certain that an exam does a good job of distinguishing the differences in achievement shown by the students, the standard deviation in scores on the examination cannot be too small. On the other hand, if the standard deviation is too large, there will tend to be a lot of very low scores, which is bad for student morale. Past experience has led the professor to believe that a standard deviation of about 13 points on a 100-point exam indicates that the exam does a good job of balancing these two objectives.

The professor just gave an examination to his class of 41 freshmen and sophomores. The mean score was 72.7, and the sample standard deviation was 15.7. Does this exam meet his goodness criterion? We can summarize the data:

$\sigma_{H_0} = 13$ $\leftarrow$ hypothesized value of the population
standard deviation
$s = 15.7 \leftarrow$ sample standard deviation
$n = 41$ $\leftarrow$ sample size

If the professor uses a significance level of .10 in testing his hypothesis, we can symbolically state the problem:

Statement of
the problem

$H_0 : \sigma = 13 \leftarrow$ null hypothesis: the true standard
deviation is 13 points

$H_1 : \sigma \neq 13 \leftarrow$ alternative hypothesis: the true standard
deviation is not 13 points

$\alpha = .10 \leftarrow$ level of significance for testing this
hypothesis

The first thing we do is to use Equation 10·12 to calculate the χ^2 statistic:

$$\chi^2 = \frac{(n-1)s^2}{\sigma^2}$$

[10·12] Calculating
the χ^2 statistic

$$= \frac{40(15.7)^2}{(13)^2}$$

$$= 58.34$$

This statistic has a χ^2 distribution with $n - 1$ ($= 40$ in this case) degrees of freedom. We will accept the null hypothesis if χ^2 is neither too big nor too small. From the χ^2 distribution table (Appendix Table 5) we can see that the appropriate χ^2 values for .05 of the area to lie in each tail of the curve are 26.509 and 55.759. These two limits of the acceptance region and the observed sample statistic ($\chi^2 = 58.34$) are shown in Figure 10·9. We see that the sample value of χ^2 is not in the acceptance region, so the professor should reject the null hypothesis; this exam does not meet his goodness criterion.

Interpreting
the results

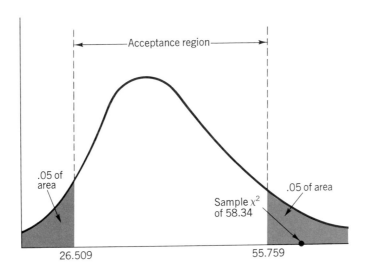

FIG. 10·9
**Two-tailed
hypothesis test at
.05 level of
significance
showing
acceptance region
and sample χ^2**

A one-tailed test of a variance

Testing hypotheses
about a variance:
one-tailed tests

Precision Analytics, Inc. manufactures a wide line of precision instruments and has a fine reputation in the field for the quality of its instruments. In order to preserve that reputation, it maintains strict quality control on all of its output. It will not release an analytic balance for sale, for example, unless that balance shows a variability significantly below one microgram (at $\alpha = .01$) when weighing quantities of about 500 grams. A new balance has just been delivered to the quality control division from the production line.

The new balance is tested by using it to weigh the same 500 gram standard weight 30 different times. The sample standard deviation turns out to be 0.73 micrograms. Should this balance be sold? We summarize the data and state the problem:

$\sigma_{H_0} = 1$ ← hypothesized value of the population standard deviation

$s = 0.73$ ← sample standard deviation

$n = 30$ ← sample size

$H_0 : \sigma = 1$ ← null hypothesis: the true standard deviation is one microgram

$H_1 : \sigma < 1$ ← alternative hypothesis: the true standard deviation is less than one microgram

$\alpha = .01$ ← level of significance for testing this hypothesis

We begin by using Equation 10·12 to calculate the χ^2 statistic:

$$\chi^2 = \frac{(n-1)s^2}{\sigma^2} \qquad [10\cdot12]$$

$$= \frac{29(.73)^2}{(1)^2}$$

$$= 15.45$$

We will reject the null hypothesis and release the balance for sale, if this statistic is sufficiently small. From Appendix Table 5, we see that with 29 degrees of freedom (30 − 1), the value of χ^2 which leaves an area of .01 in the lower tail of the curve is 14.257. The acceptance region and the observed value of χ^2 are shown in Fig. 10·10. We see that we cannot reject the null hypothesis. The balance should be returned to the production line for adjusting.

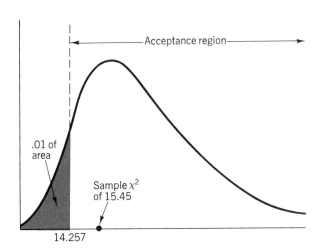

FIG. 10·10
One-tailed hypothesis test at .01 significance level showing acceptance region and sample χ^2

Acceptance region

.01 of area

Sample χ^2 of 15.45

14.257

EXERCISES

34. A sample of 16 observations from a normal distribution has a mean of 32.5 and a variance of 16.9. Construct a 95 percent confidence interval for the true population variance.

35. The standard deviation of a distribution is hypothesized to be 310. If an observed sample of 10 yields a sample standard deviation of 220, should we reject the null hypothesis that the true standard deviation is 310? Use the .05 level of significance.

36. Based on past data, the variance of a population is hypothesized to be 48. If a sample of 15 observations yields a variance of 55, should we reject the hypothesized value and conclude that the variance has increased? Use the .10 significance level.

37. Given a sample variance of 224 from a set of 12 observations, construct a 90 percent confidence interval for the population variance.

38. A developmental psychologist feels that the attention span of 6-year-olds is surely greater than that of 5-year-olds, but he does not expect the variability in attention span to differ for the two age groups. In previous studies, it has been shown that the average attention span of 5-year-olds for a particular type of task is 20 minutes with a variance of 56 minutes squared. For a group of twenty 6-year-olds, the average attention span at the same type of task is 30 minutes with a sample variance of 28 minutes squared. Does the variability in attention span appear to differ at the two age levels? Test the hypothesis at the .05 level.

39. A teacher, giving a standardized achievement test to her class, reads in the scoring manual that the population variance for scores on the test is 45 points squared, from a large standardization group employed in the development of the test. The sample variance of scores for the 24 students in her class is 25 points squared. At the .05 significance level, is there evidence to indicate that her class is less variable in their scores than the national standardization population used?

40. In checking its cars for adherence to emissions standards set by the government, an automaker measured emissions of 25 cars. The average number of particles of pollutants emitted was found to be within the required levels, but the sample variance was 54. Find the 95 percent confidence interval for the variance in emission particles for these cars.

41. Dr. Rhodes is studying a certain illness that has always shown a wide amount of variability in recovery time. Based on previous work, he has found that the variance in recovery time is 84 days squared. After developing a new treatment that he believes will reduce the vari-

ance as well as the mean of recovery times, he decides to try it out. Using the treatment on a group of 15 patients recently diagnosed as suffering from the illness, he finds the sample variance of their recovery times to be 28 days squared. Is he justified in claiming that his new treatment reduces the variance in recovery time? Test the hypothesis at the .05 level of significance.

6 INFERENCES ABOUT TWO POPULATION VARIANCES

In Chapter 9, we saw several situations in which we wanted to compare the means of two different populations. Recall that we did this by looking at the *difference* of the means of two samples drawn from those populations. Here, we want to compare the variances of two populations. However, rather than looking at the *difference* of the two sample variances, it turns out to be more convenient if we look at their *ratio*. The next two examples show how this is done.

Comparing the variances of two populations

A one-tailed test of two variances

A prominent sociologist at a large midwestern university believes that incomes earned by college graduates show much greater variability than the earnings of those who did not attend college. In order to test out this theory, she dispatches two research assistants to Chicago to look at the earnings of these two populations. The first assistant takes a random sample of 21 college graduates and finds that their earnings have a sample standard deviation of $s_1 = \$17,000$. The second assistant samples 25 nongraduates and obtains a standard deviation in earnings of $s_2 = \$7,500$. The data of our problem can be summarized as follows:

Statement of the problem

$$s_1 = 17,000 \leftarrow \text{standard deviation of first sample}$$
$$n_1 = 21 \quad \leftarrow \text{size of first sample}$$
$$s_2 = 7500 \quad \leftarrow \text{standard deviation of second sample}$$
$$n_2 = 25 \quad \leftarrow \text{size of second sample}$$

Since the sociologist theorizes that the earnings of college graduates are *more* variable than those of people not attending college, a one-tailed test is appropriate. She wishes to verify her theory at the .01 level of significance. We can formally state her problem:

Why a one-tailed test is appropriate

$$H_0 : \sigma_1^2 = \sigma_2^2 \ (\text{or } \sigma_1^2/\sigma_2^2 = 1) \leftarrow \text{null hypothesis: the two variances are the same}$$

$$H_1 : \sigma_1^2 > \sigma_2^2 \ (\text{or } \sigma_1^2/\sigma_2^2 > 1) \leftarrow \text{alternative hypothesis: earnings of college graduates have more variance}$$

$$\alpha = .01 \leftarrow \text{level of significance for testing this hypothesis}$$

We know that s_1^2 can be used to estimate σ_1^2, and s_2^2 can be used to estimate σ_2^2. If the alternative hypothesis is true, we would expect that s_1^2 will be greater than s_2^2 (or, equivalently, that s_1^2/s_2^2 will be greater than one). But how much greater must s_1^2 be in order for us to be able to reject the null hypothesis? To answer this question, we must know the distribution of s_1^2/s_2^2. If we assume that the two populations are reasonably well described by normal distributions, then the ratio

$$F = s_1^2/s_2^2$$ [10·15]   Description of
the F-statistic

has an F-distribution with $n_1 - 1$ degrees of freedom in the numerator and $n_2 - 1$ degrees of freedom in the denominator.

In the earnings problem, we calculate the sample F-statistic:

$$\begin{aligned} F &= s_1^2/s_2^2 \\ &= \frac{(17000)^2}{(7500)^2} \\ &= \frac{289{,}000{,}000}{56{,}250{,}000} \\ &= 5.14 \end{aligned}$$ [10·15]

For 20 degrees of freedom (21 − 1) in the numerator and 24 degrees of freedom (25 − 1) in the denominator, Appendix Table 6 tells us that the critical value that separates the acceptance and rejection regions is 2.74. Figure 10·11 shows the acceptance region and the observed F-statistic of 5.14. Our sociologist rejects the null hypothesis, and the sample data support her theory.   Interpreting
the results

A word of caution about the use of Appendix Table 6 is necessary at this point. You will notice that the table gives values of the F-statistic that are appropriate for only *upper-tailed* tests. Contrast this with Appendix Table 5, which gives values appropriate for both upper- and lower-tailed tests. How can we handle alternative hypotheses of the form   Handling lower-
tailed tests
in Table 6

FIG. 10·11
**One-tailed
hypothesis test at
.01 level of
significance
showing the
acceptance region
and the sample F-
statistic**

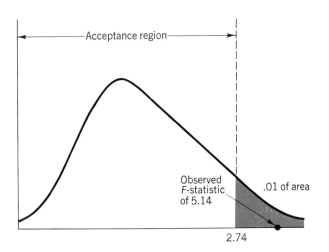

Acceptance region

Observed
F-statistic
of 5.14

.01 of area

2.74

$\sigma_1^2 < \sigma_2^2$ (or $\sigma_1^2/\sigma_2^2 < 1$)? This is easily done if we notice that $\sigma_1^2/\sigma_2^2 < 1$ is equivalent to $\sigma_2^2/\sigma_1^2 > 1$. Thus, all we need to do is calculate the ratio s_2^2/s_1^2 which also has an F-distribution (but with $n_2 - 1$ numerator degrees of freedom and $n_1 - 1$ denominator degrees of freedom), and then we can use Appendix Table 6. There is another way to say the same thing: when-ever you are doing a one-tailed test of two variances, number the popu-lations so that the alternative hypothesis has the form

$$H_1 : \sigma_1^2 > \sigma_2^2 \text{ (or } \sigma_1^2/\sigma_2^2 > 1)$$

and then proceed as we did in the earnings example.

A two-tailed test of two variances

Finding the
critical value
in a two-tailed
test

The procedure for a two-tailed test of two variances is similar to that for a one-tailed test. The only problem arises in finding the critical value in the lower tail. This is related to the problem about lower-tailed tests discussed in the last paragraph, and we will resolve it in the same way.

One criterion in evaluating oral anesthetics for use in general dentistry is the variability in the length of time between injection and complete loss of sensation in the patient. (This is called the effect delay time.) A large pharmaceutical firm has just developed two new oral anesthetics, which it will market under the names Oralcaine and Novasthetic. From similarities in the chemical structure of the two compounds, it has been predicted that they should show the same variance in effect delay time. Sample data from tests of the two compounds (which controlled other variables such as age and weight) are given in Table 10·16.

TABLE 10·16
Effect delay times for two anesthetics

Anesthetic	Sample size n	Sample variance (seconds squared) s^2
Oralcaine	31	1296
Novasthetic	41	784

The company wants to test at a 2 percent significance level whether the two compounds have the same variance in effect delay time. Symbolically, the hypotheses are:

Statement
of the
problem

$H_0 : \sigma_1^2 = \sigma_2^2 \text{(or } \sigma_1^2/\sigma_2^2 = 1) \leftarrow$ null hypothesis: the two variances
are the same
$H_1 : \sigma_1^2 \neq \sigma_2^2 \text{(or } \sigma_1^2/\sigma_2^2 \neq 1) \leftarrow$ alternative hypothesis: the two
variances are different
$\alpha = .02 \leftarrow$ significance level of the test

To test these hypotheses, we again use Equation 10·15

$$F = s_1^2/s_2^2 \qquad \text{[10·15]}$$
$$= 1296/784$$
$$= 1.65$$

This statistic comes from an F-distribution with $n_1 - 1$ degrees of freedom in the numerator (30 in this case) and $n_2 - 1$ degrees of freedom in the denominator (40 in this case). Let us use the notation

$$F(n, d, \alpha)$$

to denote that value of F with n numerator degrees of freedom, d denominator degrees of freedom, and an area of α in the upper tail. In our problem, the acceptance region extends from $F(30,40,.99)$ to $F(30,40,.01)$ as illustrated in Fig. 10·12.

We can get the value of $F(30,40,.01)$ directly from Appendix Table 6; it is 2.20. However, the value of $F(30,40,.99)$ is not in the table. Now $F(30,40,.99)$ will correspond to a *small* value of s_1^2/s_2^2, but to a *large* value of s_2^2/s_1^2, which is just the reciprocal of s_1^2/s_2^2. Given the discussion on page 396 about lower-tailed tests, we might suspect that

$$F(n, d, \alpha) = \frac{1}{F(d, n, 1 - \alpha)} \qquad \text{[10·16]}$$

and this turns out to be true. We can use this equation to find $F(30,40,.99)$:

$$F(30,40,.99) = \frac{1}{F(40,30,.01)} \qquad \text{[10·16]}$$
$$= \frac{1}{2.30}$$
$$= 0.43$$

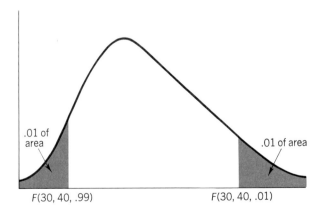

FIG. 10·12

**Two-tailed test of
hypothesis at the
.02 significance
level**

.01 of
area

.01 of area

F(30, 40, .99)

F(30, 40, .01)

FIG. 10·13
Two-tailed
hypothesis test at
.02 level of
significance
showing
acceptance region
and the sample *F-*
statistic

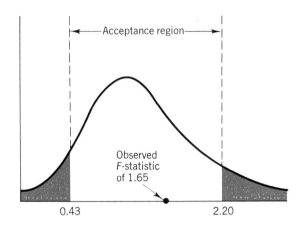

Interpreting
the results

In Fig. 10·13, we have illustrated the acceptance region for this hypothesis test and the observed value of *F*. We see there that the null hypothesis is accepted, so we conclude that the observed difference in the sample variances of effect delay times for the two anesthetics is not statistically significant.

EXERCISES

42. For the following two samples, test the hypothesis that their variances are equal at the .05 level of significance.

Sample 1	38	27	34	25	28	
Sample 2	34	37	35	40	30	34

Can we conclude that their variances are equal, or must we accept the alternative that Sample 1 has a larger variance?

43. From a sample of 16 observations, the estimate of the standard deviation of a population was found to be 8.2. From another sample of 12 observations, the estimate was found to be 4.8. Can we accept the hypothesis that the two samples come from populations with equal variances, or must we conclude that the variance of the second sample is smaller? Use the .05 level of significance.

44. Two normally distributed populations are being compared to see if they have the same variance. A sample of 15 observations from population 1 had a variance of 146, while a sample of 18 observations from population 2 had a variance of 124. Can we accept the null hypothesis of equal variances, or does the evidence appear to indicate that the first sample's variance is larger? Use the .05 significance level.

45. For two populations thought to have the same variance, the following information was found. A sample of 12 from population 1 exhibited a sample variance of 1.96, while a sample of 10 from population 2 had a variance of 3.64.

a) Calculate the *F* ratio for the test of equality of variances.

b) Find the critical *F* value for the upper tail, using the .05 significance level.

c) Find the corresponding *F* value for the lower tail.

d) State the conclusion of your test.

46. In our study of comparisons between the means of two groups, it was noted that the most common form of the two-group t-test for the difference between two means assumes that the population variance for the two groups is the same. One experimenter, using a control condition and an experimental condition in his study of drug reactions, wished to verify that this assumption held; i.e., that the treatment administered affected only the mean, not the variance of the variable under study. From his data he calculated the variance of the experimental group to be 27.8 and that of the control group to be 18.6. The experimental group had 35 subjects, and the control group had 32. Can he proceed to use the t-test, which assumes equal variances for the two groups?

47. Ben Morris, a dedicated football fan, is taking a statistics course this semester and has become enthralled in using his knowledge to compare his team with their opponents. This week, he wants to see whether there is a difference between the variance in weight of the home team and that of the visiting team. Weights of the 11 members of the starting offensive lineup for both teams are shown in the table. Calculate the sample variances and test the hypothesis that the variance in weight is the same for the two teams. Use the .10 level of significance.

| Home team | 191 | 189 | 183 | 201 | 193 | 198 | 244 | 218 | 208 | 225 | 238 |
| Visitors | 172 | 196 | 180 | 185 | 188 | 223 | 215 | 244 | 245 | 252 | 212 |

48. A sociologist suspects that the variability of attitudes toward abortion may be different for males and females, since males are less directly affected by the situation. Using a standard instrument developed to assess attitudes toward the issue, he surveys 25 male and 25 female undergraduates. He found that the sample variance of response measures for females was 12.4, and that for males the sample variance was 5.2. Does there appear to be a difference in the variability of responses for members of the two sexes? Use the .10 level of significance to test the hypothesis.

49. A professor in anthropology and a colleague in sociology were discussing the issue of family size. The anthropologist felt that American families vary in size more than those of other countries do, because Americans have diverse national, cultural, and religious backgrounds. The sociologist, however, felt that because social pressures to conform are so strongly felt in this country, people tend to adopt the values of their contemporaries, and thus variation in family size is less here. To test their conflicting ideas, they took random samples of 65 families from the census records of the United States and from similar records in Germany, a modern but older and less diverse country than the U.S. They found that the variance in family size among the U.S. sample was 9.6, while the sample variance for the German group was 4.2. Is the difference between the variances in family size significant at the .05 level? Who was right, the sociologist or the anthropologist?

7 TERMS INTRODUCED IN CHAPTER 10

• ANALYSIS OF VARIANCE (ANOVA) A statistical technique used to test the equality of 3 or more sample means and thus make inferences as to whether the samples come from populations having the same mean.

• BETWEEN-COLUMN VARIANCE An estimate of the population variance derived from the variance among the sample means.

• CHI-SQUARE DISTRIBUTION A family of probability distributions, differentiated by their degrees of freedom, used to test a number of different hypotheses about variances, proportions, and distributional goodness of fit.

• CONTINGENCY TABLE A table having R rows and C columns. Each row corresponds to a level of one variable; each column, to a level

- **GOODNESS-OF-FIT TEST** A statistical test for determining whether there is a significant difference between an observed frequency distribution and a theoretical probability distribution hypothesized to describe the observed distribution.

- **EXPECTED FREQUENCIES** The frequencies we would expect to see in a contingency table or frequency distribution if the null hypothesis is true.

- **F-DISTRIBUTION** A family of distributions differentiated by two parameters (df-numerator, df-denominator), used primarily to test hypotheses regarding variances.

- **F-RATIO** A ratio used in the analysis of variance, among other tests, to compare the magnitude of two estimates of the population variance to determine if the two estimates are approximately equal; in ANOVA, the ratio of between-column variance to within-column variance is used.

- **GRAND MEAN** The mean for the entire group of subjects from all the samples in the experiment.

- **TEST OF INDEPENDENCE** A statistical test of proportions or frequencies, to determine if membership in categories of one variable is different as a function of membership in the categories of a second variable.

- **WITHIN-COLUMN VARIANCE** An estimate of the population variance based on the variances within the k samples, using a weighted average of the k sample variances.

8 EQUATIONS INTRODUCED IN CHAPTER 10

p. 360:

$$\chi^2 = \sum \frac{(f_0 - f_e)^2}{f_e}$$

[10·1]

This formula says that the *chi-square statistic* (χ^2) is equal to the sum (Σ) we will get if we

1. subtract the expected frequencies, f_e, from the observed frequencies, f_0, for each category of our contingency table
2. square each of the differences
3. divide each squared difference by f_e
4. sum all the results of step 3

Numerically, the calculations are easy to accomplish using a table. A large value for chi-square indicates a great difference between the observed and expected values. A chi-square of zero indicates that the observed frequencies exactly match the expected frequencies. The value of chi-square can never be negative because the differences between observed and expected frequencies are always squared.

p. 362:

Number of degrees of freedom = [Number of rows − 1] [Number of columns − 1]

[10·2]

To calculate number of *degrees of freedom for a chi-square test,* multiply the number of rows (less one) times the number of columns (less one).

p. 365:

$$f_e = \frac{RT \times CT}{n} \qquad [10\cdot3]$$

With this formula, we can calculate the expected frequency for any cell within a contingency table. RT is the row total for the row containing the cell, CT is the column total for the column containing the cell and n is the total number of observations.

p. 377:

$$s_{\bar{x}}^2 = \frac{\Sigma(\bar{x} - \bar{\bar{x}})^2}{k - 1} \qquad [10\cdot4]$$

To calculate the *variance among the sample means,* use this formula and a format such as Table 10·13:

1. subtract the grand mean, $\bar{\bar{x}}$, from the sample mean, $\bar{x}$, for each sample taken
2. square each of the differences
3. sum all the answers
4. divide the total by the number of samples (less one)

p. 378:

$$\sigma^2 = \sigma_{\bar{x}}^2 \times n \qquad [10\cdot5]$$

The *population variance* is equal to the product of the square of the standard error of the mean and the sample size.

p. 378:

$$\hat{\sigma}^2 = \frac{\Sigma n_j(\bar{x}_j - \bar{\bar{x}})^2}{k - 1} \qquad [10\cdot6]$$

One estimate of the population variance (the between-column variance) can be obtained by using this equation. We obtain this equation by first substituting $s_{\bar{x}}^2$ for $\sigma_{\bar{x}}^2$ in Equation 10·5, and then by weighting each $(\bar{x}_j - \bar{\bar{x}})^2$ by its own appropriate sample size (n_j).

p. 379:

$$\hat{\sigma}^2 = \Sigma \left(\frac{n_j - 1}{n_T - k}\right) s_j^2 \qquad [10\cdot7]$$

A second estimate of the population variance (the within-column variance) can be obtained from this equation. This equation uses a weighted average of all the sample variances. In this formulation, $n_T = \Sigma n_j$, the total sample size.

p. 380:

$$F = \frac{\text{First estimate of the population variance based on the variance among the sample means}}{\text{Second estimate of the population variance based on the variances within the samples}} \qquad [10\cdot8]$$

This ratio is the way we can compare the two estimates of the population variance, which we calculated in Equations 10·6 and 10·7. In a hypothesis test based on an F distribution, we are more likely to accept the null hypothesis if this *F ratio* or *F statistic* is near to the value of one. As the *F* ratio increases, the more likely it is that we will reject the null hypothesis.

p. 380:
$$F = \frac{\text{Between-column variance}}{\text{Within-column variance}}$$
[10·9]

This restates Equation 10·8, using statistical shorthand for the numerator and denominator of the *F* ratio.

p. 382:
$$\text{Number of degrees of freedom in numerator} = [\text{Number of samples} - 1] \text{ of the } F \text{ ratio}$$
[10·10]

To do an analysis of variance, we calculate the number of *degrees of freedom in the between-column variance* (the numerator of the *F* ratio) by subtracting one from the number of samples collected.

$$\text{Number of degrees of freedom in denominator of the } F \text{ ratio} = \Sigma(n_j - 1)$$

p. 382:
$$= n_T - k$$
[10·11]

We use this equation to calculate the number of degrees of freedom in the denominator of the *F* ratio. This turns out to be the total sample size, n_T, minus the number of samples *k*.

p. 388:
$$\chi^2 = \frac{(n-1)s^2}{\sigma^2}$$
[10·12]

With a population variance of σ^2, the χ^2 statistic given by this equation has a chi-square distribution with $n - 1$ degrees of freedom. This result is exact if the population is normal, but even in samples from non-normal populations, frequently it is still a good approximation.

p. 390:
$$\sigma^2 = \frac{(n-1)s^2}{\chi^2}$$
[10·13]

To get a confidence interval for σ^2, we solve Equation 10·12 for σ^2.

p. 390:
$$\sigma_L^2 = \frac{(n-1)s^2}{\chi_U^2} \leftarrow \text{lower confidence limit}$$
$$\sigma_U^2 = \frac{(n-1)s^2}{\chi_L^2} \leftarrow \text{upper confidence limit}$$
[10·14]

These formulas give the lower and upper confidence limit for a confidence interval for σ^2. (Notice that since χ^2 appears in the denominator, we use χ_U^2 to find σ_L^2, and χ_L^2 to find σ_U^2.)

$$F = \frac{s_1^2}{s_2^2}$$

[10·15]

This ratio has an F distribution with $n_1 - 1$ degrees of freedom in the numerator and $n_2 - 1$ degrees of freedom in the denominator. (This assumes that the two populations are reasonably well described by normal distributions.)

9 CHAPTER REVIEW EXERCISES

50. A consumer group concerned with health care costs decided to do a survey of doctors in a large city, to determine how much variation there was in charges for a standard medical checkup. For a sample of 20 doctors they found an average charge of $95 for a complete physical exam, excluding lab work, with a standard deviation of $8. Find a 90 percent confidence interval for the variance in cost of a medical checkup in this city.

51. For the contingency table below, calculate the observed and expected frequencies and the chi-square statistic. Test the appropriate hypothesis at the .10 significance level.

Attitude toward social legislation

Occupation	Favor	Neutral	Oppose
Blue-collar	18	12	36
White-collar	11	15	42
Professional	24	8	32

52. A Lexington bank is considering the acquisition of a fully automated, tellerless branch banking system. To test whether the public's acceptance of this innovation would be affected by income level, the bank has installed experimental units in three shopping centers, each catering to one of three general economic groups—lower, middle, or upper class. The results of this survey were the following:

Frequency of responses

	Low	Middle	Upper	Total
Approve: Observed	30	45	23	98
Expected	31	41	26	
Disapprove: Observed	30	35	27	92
Expected	29	39	24	

a) What are the null and alternative hypotheses for this problem?

b) Calculate the value of χ^2.

c) Using a .05 level of significance, should the null hypothesis be rejected?

53. What probability distribution is used to do each of the following types of statistical tests?

a) Comparing 2 population proportions.

b) Value of a single population variance.

c) Comparing 3 or more population means.

d) Comparing 2 population means from small, dependent samples.

54. What probability distribution is used to do each of the following types of statistical tests?

 a) Comparing the means of 2 small samples from populations with unknown variances.

 b) Comparing 2 population variances.

 c) Value of a single population mean based on large samples.

 d) Comparing 3 or more population proportions.

55. A farmer decided to plant 3 varieties of alfalfa to see if different varieties produced different seed yields. The following table contains 6 samples of the seed yields (in hundreds of lbs per acre) for each of the 3 varieties of alfalfa.

Seed yields (hundreds of lbs/acre)

Variety 1	6.5	7.2	6.8	6.9	6.4	7.3
Variety 2	4.9	5.3	4.8	4.6	5.9	5.0
Variety 3	6.1	5.9	5.8	6.1	6.0	5.7

At the .01 level of significance, do the three varieties of alfalfa produce the same seed yields?

56. A department of transportation's survey of a city's three main thoroughfares measured the volume of traffic passing a checkpoint on each street during five-minute time intervals. The following table shows the number of vehicles at randomly selected periods for each street. At the .05 level of significance, is the average volume of traffic equal on the three streets?

Highway 1	30	45	26	44	18	38	42	29
Highway 2	24	33	31	16	31	13	12	25	27
Highway 3	35	47	43	46	27	31	21

57. The director of a county system of technical schools, questioning the quality of education at the three member institutions, randomly selected six students from the electrical programs of each school and administered a test to each. From the following student scores on this exam, can the director conclude that the three programs are equal, using an F test at the .05 level of significance?

School 1	85	71	78	89	74	95
School 2	65	77	84	75	71	96
School 3	72	86	77	76	84	85

58. For the following contingency table:

 a) Construct a table of observed and expected frequencies.

 b) Calculate the chi-square statistic.

 c) State the null and alternative hypotheses.

 d) Using a .05 level of significance, should the null hypothesis be rejected?

	Income level		
Church attendance	*Low*	*Middle*	*High*
Never	28	52	16
Occasional	25	66	14
Regular	18	73	8

59. For the following contingency table:

 a) Construct a table of observed and expected frequencies.

 b) Calculate the chi-square statistic.

c) State the null and alternative hypothesis.

d) Using a .01 level of significance, should the null hypothesis be rejected?

	Age group			
Type of car driven	16–21	22–30	31–45	45+
Sports car	10	15	12	8
Compact	5	7	6	8
Midsize	12	14	20	25
Full size	8	12	21	25

60. As part of a federal air traffic study at a local airport, a record was made of the number of transient aircraft arrivals during 250 half-hour time intervals. The table below presents the observed number of periods in which there were 0, 1, 2, 3, or 4 or more arrivals, as well as the expected number of such periods if arrivals per half hour have a Poisson distribution with $\lambda = 2$. At the .05 level of significance, does this Poisson distribution describe the observed arrivals?

Number of observed arrivals (per half hour)	0	1	2	3	4 or more
Number of periods observed	47	56	71	44	32
Number of periods expected (Poisson, $\lambda = 2$)	34	68	68	45	35

61. There has been some sociological evidence that women as a group are more variable than men in their attitudes and beliefs. A large private research organization has conducted a survey of men's attitudes on a certain issue and found the standard deviation on this attitude scale to be 15 points. A sociologist gave the same scale to a group of 30 women and found that the sample variance was 360 points. At the .05 significance level, is there reason to believe that women do indeed show greater variability on this attitude scale?

62. A social psychologist has tested 150 subjects to construct an attitude scale measuring feelings toward the women's movement. She presents a number of statements varying in their favorability toward the movement and the subjects respond either "agree" or "disagree." The final attitude score or measure for each subject is the number of statements agreed with. She thinks that attitudes reflected by her scale should follow a normal distribution. Using the sample mean and sample standard deviation as parameters for the normal distribution, the psychologist constructed the following table. Do the data in this table confirm the conclusion, at the .025 level of significance, that attitudes as measured by this scale follow a normal distribution?

Number of items agreed with	10 or fewer	11–12	13–14	15–16	17–18	19+
Number of subjects in each group	8	27	53	48	26	4
Number of subjects in normal distribution	14	26	41	36	22	11

63. Psychologists have often wondered about the effects of stress and anxiety on test performance. An aptitude test was given to two randomly chosen groups of 18 college students, one group in a nonstressful situation and the other in a stressful situation. The experimenter expects the stress treatment to increase the variance of scores on the test because he feels some students perform better under stress while others experience adverse reactions to stress. The variances computed for the two groups are $s_1^2 = 22.8$ for the nonstress group and $s_2^2 = 78.5$ for the stress group. Was his hypothesis confirmed? Use the .05 level of significance to test the hypothesis.

405

64. A study was made of the possible relationship between the number of years of preschool experience children had before entering first grade and their ability to get along with one another and behave in class. Children were classified with regard to the type and extent of their preschool experience and their behavior in first grade. The results of this study were:

Prior experience

	Nursery school + kindergarten	Kindergarten only	Nursery school only	No experience	Total
Poorly behaved	24	42	38	29	133
Satisfactorily behaved	70	41	45	47	203
Well behaved	36	27	27	29	119
Total	130	110	110	105	455

At the .025 level of significance, is preschool experience a significant factor in the behavior of children in first grade?

65. In the development of new drugs for the treatment of anxiety, it is important to check the drugs' effects on various motor functions, one of which is driving. The Confab Pharmaceutical Company is testing 4 different tranquilizing drugs for their effects on driving skill. Subjects take a simulated driving test, and their score reflects their errors. The more severe errors lead to higher scores. The results of these tests produced the following table:

Drug 1	230	258	239	241	
Drug 2	285	276	263	274	
Drug 3	215	232	204	247	226
Drug 4	241	253	237	246	210

At a .05 level of significance, do the four drugs affect driving skill differentially?

66. In test theory, which is concerned with the construction of reliable and valid measuring instruments for a variety of uses, there is a concept known as parallel tests. Parallel tests measure the same attribute or ability, so that individuals should score about the same on both of two parallel tests. One of the characteristics of parallel tests is that they have the same variance in their scores. A professor of elementary statistics teaches two sections of the course, which meet on consecutive days. He has constructed two tests that he believes are parallel, to be sure that nobody on the second day will know any of the questions on the test, even though the two tests cover the same material. After giving the tests, he computes the basic statistics, including the two variances. For the first test given to a class of 32, $s^2 = 396$; and for the second test given to 30 students, $s^2 = 344$. Can he feel secure in the assumption that the variances are equal, based on these groups representing random samples from the same population? Use the .10 level of significance to test the hypothesis.

67. Andrea Johnson, a social psychologist studying attitude change is searching for the most effective type of persuasion in modifying attitudes toward an institution or group. She measures subjects' attitudes before and after a persuasive manipulation involving either a film, a lecture, or role-playing activity. The data used are composite measures of change in attitude toward the target group.

Film	46	48	52	43	47	51
Lecture	38	43	39	45	36	43
Role-playing	45	44	47	46	46	39

At the .05 level of significance, can the researcher conclude that the three methods of inducing attitude change are equally effective?

Answer true *or* false. *Answers are in the back of the book.*

1. Analysis of variance may be used to test whether the means of more than two populations can be considered equal.

2. Analysis of variance is based upon a comparison of two estimates of the variance of the overall population which contains all samples.

3. When comparing the variances of two populations, it is convenient to look at the difference in the sample variances, just as we looked at the difference in sample means to make inferences about population means.

4. When using the chi-square distribution as a test of independence, the number of degrees of freedom is related to both the number of rows and the number of columns in the contingency table.

5. Chi-square may be used as a test to decide whether a particular distribution closely approximates a sample from some population. We refer to such tests as goodness-of-fit tests.

6. If samples are taken from two populations that are both nearly normal, then the ratio of all possible sets of the two sample variances also is normally distributed.

7. When using a chi-square test, we must ensure an adequate sample size, so that we can avoid any tendency for the value of the chi-square statistic to be overestimated.

8. When testing hypotheses about a population's variance, we may form confidence intervals by using the chi-square distribution.

9. The specific shape of an *F*-distribution depends on the number of degrees of freedom in both the numerator and denominator of the *F*-ratio.

10. One convenient aspect of hypothesis testing using the *F*-statistic is that all such tests are upper-tailed tests.

11

Regression and correlation analysis

1. Introduction, 410

2. Estimation using the regression line, 415

3. Correlation analysis, 435

4. Making inferences about population parameters, 444

5. Using regression and correlation analysis: limitations, errors, and caveats, 449

6. Multiple regression and correlation analysis, 451

7. The computer and multiple regression, 461

8. Terms introduced, 465

9. Equations introduced, 466

10. Chapter review exercises, 469

11. Chapter concepts test, 473

The director of New Mexico's Human Services Division believes that the employment rate of educable mentally retarded high school graduates depends on the state's expenditure for special education. Data for the last 6 years look like this:

Year	Millions spent on special education	Employment rate in percent
1979	5	31
1978	11	40
1977	4	30
1976	5	34
1975	3	25
1974	2	20

The director wants an equation to use in predicting the annual employment rate from the amount budgeted for special education. Using the methods in this chapter, we can provide him with just such a decision-making tool and, in addition, tell him something about the accuracy he can expect in using it to make decisions.

1 INTRODUCTION Every day, people make personal and professional decisions that are based upon predictions of future events. To make these forecasts, they rely upon the relationship (intuitive and calculated) between what is already known and what is to be estimated. If decision makers can determine how the known is related to the future event, they can aid the decision-making process considerably. That is the subject of this chapter: how to determine the *relationship between variables.*

In Chapter 10, we used chi-square tests of independence to determine whether a statistical relationship existed between two variables. The chi-square test tells us if there is such a relationship, but it does not tell us what that relationship is. **Regression and correlation analysis will show us how to determine both the nature and the strength of a relationship between two variables.** We will learn to predict, with some accuracy, the value of an unknown variable based on past observations of that variable and others.

The term *regression* was first used as a statistical concept in 1877 by Sir Francis Galton. Galton made a study that showed that the height of children born to tall parents will tend to move back, or "regress," toward the mean height of the population. He designated the word *regression* as the name of the general process of predicting one variable (the height of the children) from another (the height of the parent). Later, statisticians coined the term *multiple regression* to describe the process by which several variables are used to predict another.

In *regression analysis,* we shall develop an *estimating equation,* that is, a mathematical formula that relates the known variables to the unknown variable. Then, after we have learned the pattern of this relationship, we can apply *correlation analysis* to determine the degree to which the variables are related. Correlation analysis, then, tells us how well the estimating equation actually describes the relationship.

Types of relationships

Regression and correlation analyses are based on the relationship, or association, between two (or more) variables. The known variable (or variables) is called the *independent* variable(s). The variable we are trying to predict is the *dependent* variable.

Scientists know, for example, that there is a relationship between the annual sales of aerosol spray cans and the quantity of fluorocarbons released into the atmosphere each year. If we studied this relationship, "the number of aerosol cans sold each year" would be the independent variable, and "the quantity of fluorocarbons released annually" would be the dependent variable.

Let's take another example. Economists might base their predictions of the annual gross national product, or GNP, on the final consumption spending within the economy. Thus, "the final consumption spending" is the independent variable, and "the GNP" would be the dependent variable.

In regression, we can have only one dependent variable in our estimating equation. However, we can use more than one independent variable. Often when we add independent variables, we improve the accuracy of our prediction. Economists, for example, frequently add a second independent variable, "the level of investment spending," to improve their estimate of the nation's GNP.

Our two examples of fluorocarbons and GNP are illustrations of direct associations between independent and dependent variables. As the independent variable increases, the dependent variable also increases. In like manner, we expect the sales of a company to increase as the advertising budget increases. We can graph such a *direct relationship,* plotting the independent variable on the *X*-axis and the dependent variable on the *Y*-axis. We have done this in *a,* Fig. 11·1. Notice how the line slopes up as *X* takes on larger and larger values. The slope of this line is said to be *positive* because *Y* increases as *X* increases.

Direct relationship between *X* and *Y*

Relationships can also be *inverse* rather than direct. In these cases, the dependent variable decreases as the independent variable increases. The government assumes that such an inverse association exists between a company's increased annual expenditures for pollution abatement devices and decreased pollution emissions. This type of relationship is illustrated in *b,* Fig. 11·1 and is characterized by a *negative* slope (the dependent variable *Y* decreases as the independent variable *X* increases).

Inverse relationship between *X* and *Y*

Frequently, we find a *causal* relationship between variables; that is, the independent variable "causes" the dependent variable to change. This is true in the antipollution example above. But in many cases, some other factor causes the change in both the dependent and the independent variables. We might be able to predict the sales of diamond earrings from the sale of new Cadillacs, but we could not say that one is caused by the other. Instead, we realize that the sales levels of both Cadillacs and diamond earrings are caused by another factor, such as the level of disposable income.

(a) Direct relationship

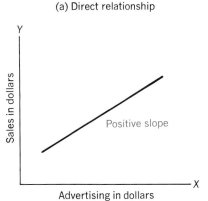

(b) Inverse relationship

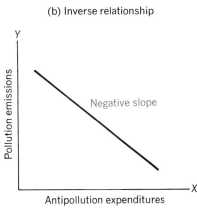

FIG. 11·1

Direct and inverse relationships between independent variable *X* and dependent variable *Y*

For this reason, it is important that you consider the relationships found by regression to be relationships of association but *not* necessarily of cause and effect. Unless you have specific reasons for believing that the values of the dependent variable are caused by the values of the independent variable(s), do not infer causality from the relationships you find by regression.

Scatter diagrams

The first step in determining whether there is a relationship between two variables is to examine the graph of the observed (or known) data. This graph, or chart, is called a *scatter diagram.*

A scatter diagram can give us two types of information. Visually, we can look for patterns that indicate the variables are related. Then, if the variables are related, we can see what kind of line, or estimating equation, describes this relationship.

We are going to develop and use a specific scatter diagram. Suppose a university admissions director asks us to determine whether any relationship exists between a student's scores on an entrance examination and that student's cumulative grade point average (GPA) upon graduation. The administrator has accumulated a random sample of data from the records of the university. This information is recorded in Table 11·1.

TABLE 11·1
Student scores on entrance examinations and cumulative grade point averages at graduation

Student	A	B	C	D	E	F	G	H
Entrance examination scores (100 = maximum possible score)	74	69	85	63	82	60	79	91
Cumulative GPA (4.0 = A)	2.6	2.2	3.4	2.3	3.1	2.1	3.2	3.8

To begin, we should transfer the information in Table 11·1 to a graph. Since the director wishes to use examination scores to predict success in college, we have placed the cumulative GPA (the dependent variable) on the vertical or *Y*-axis and the entrance examination score (the independent variable) on the horizontal or *X*-axis. Figure 11·2 shows the completed scatter diagram.

FIG. 11·2
Scatter diagram of student scores on entrance examinations plotted against cumulative grade point averages

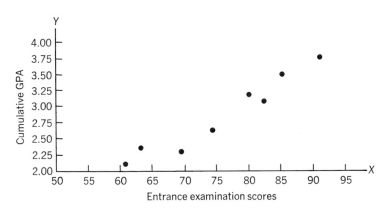

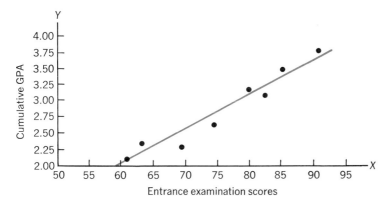

At first glance we can see why we call this a scatter diagram. The pattern of points results from the fact that each pair of data from Table 11·1 has been recorded as a single point. When we view all these points together, we can visualize the relationship that exists between the two variables. As a result, we can draw, or "fit," a straight line through our scatter diagram to represent the relationship. We have done this in Fig. 11·3. It is common to try to draw these lines so that an equal number of points lie on either side of the line.

Drawing, or "fitting," a straight line to a scatter diagram

In this case, the line drawn through our data points represents a direct relationship because Y increases as X increases. Because the data points are relatively close to this line, we can say that there is a high degree of association between the examination scores and the cumulative GPAs. In Fig. 11·3, we can see that the relationship described by the data points is well described by a straight line. Thus, we can say that it is a *linear* relationship.

Interpreting our straight line

The relationship between the X and Y variables can also take the form of a curve. Statisticians call such a relationship *curvilinear.* The employees of many industries, for example, experience what is called a "learning curve"; that is, as they produce a new product, the time required to produce one unit is reduced by some fixed proportion as the total number of units doubles. One such industry is aviation. Manufacturing time per unit for a new aircraft tends to decrease by 20 percent each time the total number of completed new planes doubles. Figure 11·4 illustrates the curvilinear relationship of this "learning curve" phenomenon.

Curvilinear relationships

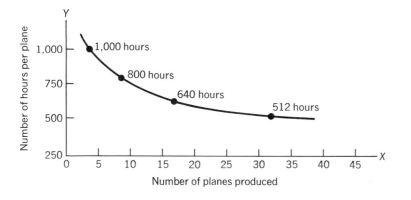

Review of possible
relationships

The direction of the curve can indicate whether the curvilinear relationship is direct or inverse. The curve in Fig. 11·4 describes an inverse relationship because Y decreases as X increases.

To review the relationships possible in a scatter diagram, examine the graphs in Fig. 11·5. Graphs a and b show direct and inverse linear relationships. Graphs c and d are examples of curvilinear relationships that demonstrate direct and inverse associations between variables, respectively. Graph e illustrates an inverse linear relationship with a wide pattern of points. This wider scattering indicates that there is a lower degree of association between the independent and dependent variables than there is in graph b. The pattern of points in graph f seems to indicate that there is no relationship between the two variables; therefore, knowledge of the past concerning one variable will not allow us to predict future occurrences of the other.

FIG. 11·5
Possible relationships between X and Y in scatter diagrams

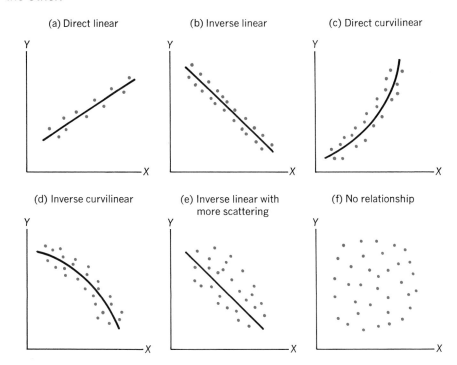

(a) Direct linear   (b) Inverse linear   (c) Direct curvilinear

(d) Inverse curvilinear   (e) Inverse linear with more scattering   (f) No relationship

EXERCISES

1. What is regression analysis?
2. In regression analysis, what is an estimating equation?
3. What is the purpose of correlation analysis?
4. Define direct and inverse relationships.
5. To what does the term *causal relationship* refer?
6. Explain the difference between linear and curvilinear relationships.

7. Explain why and how we construct a scatter diagram.

8. What is multiple regression analysis?

9. Construct a scatter diagram for the following set of data. Is there a relationship between the two variables? If so, is it linear or curvilinear, direct or inverse?

X	40	58	33	65	80	80	56	30	33	90	72
Y	15	14	12	20	26	26	14	12	12	30	22

10. For each of the following scatter diagrams, indicate whether a relationship exists and if so, whether it is direct or inverse and linear or curvilinear.

    (a)   (b)   (c)

11. Construct a scatter diagram for the following set of data and indicate the type of relationship.

X	22	20	16	10	18	18	15	6	21	16	13	3	17	15	11	5
Y	7	15	27	34	9	24	26	33	12	20	30	37	14	24	29	38

2 ESTIMATION USING THE REGRESSION LINE

Calculating the regression line using an equation

In the scatter diagrams we have used to this point, the *regression lines* were put in place by fitting the lines visually among the data points. In this section, we shall learn how to calculate the regression line somewhat more precisely using an equation that relates the two variables mathematically. Here, we examine only linear relationships involving two variables. We shall deal with relationships among more than two variables in Section 6 of this chapter.

The equation for a straight line where the dependent variable Y is determined by the independent variable X is: Equation for a straight line

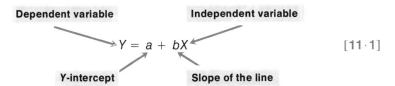

$$Y = a + bX \qquad\qquad [11 \cdot 1]$$

Using this equation we can take a given value of X and compute the value of Y. The a is called the "Y-intercept" because its value is the point at which the regression line crosses the Y-axis, that is, the vertical axis. The b in Equation $11 \cdot 1$ is the "slope" of the line. It represents how much each change of the independent variable X changes the dependent variable Y. Both a Interpreting the equation and *b* are numerical *constants* since, for any given straight line, their value does not change.

Calculating *Y*
from *X*
using the equation
for a straight line

Suppose we know that *a* is 3 and *b* is 2. Let us determine what *Y* would be for an *X* equal to 5. When we substitute the values for *a*, *b*, and *X* in Equation 11·1, we find the corresponding value of *Y* to be:

$$Y = a + bX \qquad [11·1]$$
$$= 3 + 2(5)$$
$$= 3 + 10$$
$$= 13 \leftarrow \text{value for } Y \text{ given } X = 5$$

Using the estimating equation for a straight line

Finding the values
for *a* and *b*

How can we find the values of the numerical constants, *a* and *b*? To illustrate this process, let's use the straight line in Fig. 11·6.

Visually, we can find *a* (the *Y*-intercept) by locating the point where the line crosses the *Y*-axis. In Fig. 11·6, this happens where *a* = 3.

To find the slope of the line, *b*, we must determine how the dependent variable, *Y*, changes as the independent variable, *X*, changes. We can begin by picking two points on the line in Fig. 11·6. Now, we must find the values of *X* and *Y* (the *coordinates*) of both points. We can call the coordinates of our first point (X_1, Y_1) and those of the second point (X_2, Y_2). By examining Fig. 11·6, we can see that $(X_1, Y_1) = (1,5)$ and $(X_2, Y_2) = (2,7)$. At this point, then, we can calculate the value of *b*, using this equation:

$$b = \frac{Y_2 - Y_1}{X_2 - X_1} \qquad [11·2]$$
$$= \frac{7 - 5}{2 - 1}$$
$$= \frac{2}{1}$$
$$= 2 \leftarrow \text{slope of the line}$$

Writing and using
the equation for
a straight line

In this manner, we can learn the values of the numerical constants, *a* and *b*, and write the equation for a straight line. The line in Fig. 11·6 can be described by Equation 11·1 where *a* = 3 and *b* = 2. Thus:

$$Y = a + bX \qquad [11·1]$$

and: $$Y = 3 + 2X \quad \text{in Fig. } 11·6.$$

Using this equation, we can determine the corresponding value of the dependent variable for any value of *X*. Suppose we wish to find the value of *Y* when *X* = 7. The answer would be:

$$Y = a + bX \qquad [11 \cdot 1]$$
$$= 3 + 2(7)$$
$$Y = 3 + 14$$
$$= 17$$

FIG. 11·6
**Straight line with a
positive slope, with
Y-intercept and
two points on the
line designated**

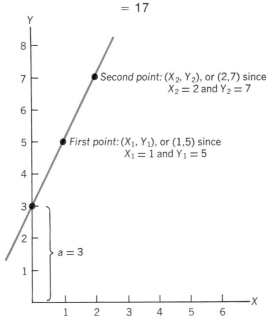

If you substitute more values for X into the equation, you will notice that Y increases as X increases. Thus, the relationship between the variables is *direct,* and the slope is *positive.*

Now consider the line in Fig. 11·7 (p. 418). We see that it crosses the Y-axis at 6. Therefore, we know that $a = 6$. If we select the two points where $(X_1, Y_1) = (0,6)$ and $(X_2, Y_2) = (1,3)$ we will find that the slope of the line is -3:

$$b = \frac{Y_2 - Y_1}{X_2 - X_1} \qquad [11 \cdot 2]$$
$$= \frac{3 - 6}{1 - 0}$$
$$= -\frac{3}{1}$$
$$= -3$$

*Direct relationship;
positive slope*

Notice that when b is negative, the line represents an *inverse* relationship, and the slope is *negative* (Y decreases as X increases). Now, with the numerical values of a and b determined, we can substitute them into the general equation for a line:

*Determining
the equation for
the line*

$$Y = a + bX \qquad [11 \cdot 1]$$
$$Y = 6 + (-3)X$$
$$Y = 6 - 3X$$

FIG. 11·7
**Straight line with
negative slope**

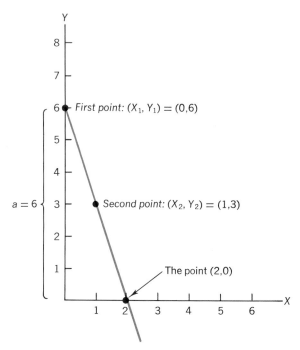

Assume we wish to find the value of the dependent variable that corresponds to $X = 2$. Substituting into the above equation, we get:

$$Y = 6 - (3)(2)$$
$$= 6 - 6$$
$$= 0$$

Thus, when $X = 2$, Y must equal 0. If we refer to the line in Fig. 11·7, we can see that the point (2,0) does lie on the line.

The method of least squares

Now that we have seen how to determine the equation for a straight line, let's think about how we can calculate an equation for a line that is drawn through the middle of a set of points in a scatter diagram. How can we "fit" a line mathematically if none of the points lie on the line? To a statistician, the line will have a "good fit" if it *minimizes the error* between the estimated points on the line and the actual observed points that were used to draw it.

Before we proceed, we need to introduce a new symbol. So far, we have used Y to represent the individual values of the observed points measured along the Y-axis. Now we should begin to use $\hat{Y}$ *(Y-hat)* to symbolize the individual values of the *estimated* points, that is, those points that lie on the estimating line. Accordingly, we shall write the equation for the estimating line as:

$$\hat{Y} = a + bX$$

[11·3]

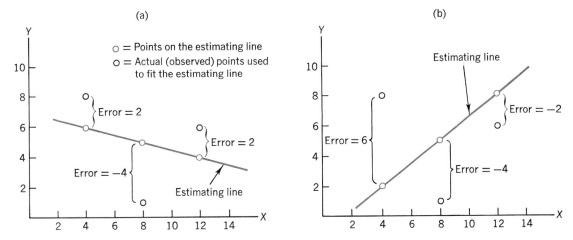

FIG. 11·8 **Two different estimating lines fitted to the same three observed data points, showing errors in both cases**

In Fig. 11·8, we have two estimating lines that have been fitted to the same set of three data points. These three given, or observed, data points are shown in black. Two very different lines have been drawn to describe the relationship between the two variables. Obviously, we need a way to decide which of these lines gives us a better fit.

Which line fits best

One way we can "measure the error" of our estimating line is to *sum* all the individual differences, or errors, between the estimated points shown in color and the observed points shown in black. In Table 11·2, we have calculated the individual differences between the corresponding Y and $\hat{Y}$, and then we have found the sum of them.

Using total error to determine best fit

A quick visual examination of the two estimating lines in Fig. 11·8 reveals that the line in graph *a* fits the three data points better than the line in graph *b*.* However, our process of summing the individual differences in Table 11·2 indicates that both lines describe the data equally well (the total error in both cases is zero). Thus, we must conclude that this process for calculating the error is not a reliable way to judge the goodness of fit of an estimating line.

The problem with adding the individual errors is the canceling effect of the positive and negative values. From this, we might deduce that the proper criterion for judging the goodness of fit would be to add the *absolute values*

Using absolute value of error to measure best fit

TABLE 11·2
Summing the errors of the two estimating lines in Fig. 11·8

Graph a $Y - \hat{Y}$		Graph b $Y - \hat{Y}$	
$8 - 6 =$	2	$8 - 2 =$	6
$1 - 5 =$	-4	$1 - 5 =$	-4
$6 - 4 =$	2	$6 - 8 =$	-2
	0 ← total error		0 ← total error

*We can reason that this is so by noticing that whereas both estimating lines miss the second and third points (reading from left to right) by an equal distance, the line in graph *a* misses the first point by considerably less than the line in graph *b*.

419

TABLE 11·3	Graph a		Graph b	
Summing the absolute values of the errors of the two estimating lines in Fig. 11·8	$\|Y - \hat{Y}\|$		$\|Y - \hat{Y}\|$	
	$\|8 - 6\| = 2$		$\|8 - 2\| = 6$	
	$\|1 - 5\| = 4$		$\|1 - 5\| = 4$	
	$\|6 - 4\| = 2$		$\|6 - 8\| = 2$	
	8 ← total absolute error		12 ← total absolute error	

(the values without their algebraic signs) of each error. We have done this in Table 11·3. (The symbol for absolute value is two parallel vertical lines, $|\ \ |$.) Since the total absolute error in graph a is smaller than the total absolute error in graph b, and since we are looking for the "minimum absolute error," we have confirmed our intuitive impression that the estimating line in graph a is the better fit.

On the basis of this success, we might conclude that minimizing the sum of the absolute values of the error is the best criterion for finding a good fit. But before we feel too comfortable with it, we should examine a different situation.

In Fig. 11·9, we again have two identical scatter diagrams with two different estimating lines fitted to the three data points. In Table 11·4, we have added the absolute values of the errors and found that the estimating line in graph a is a better fit than the line in graph b. Intuitively, however, it appears that the line in graph b is the better fit line because it has been moved vertically to take the middle point into consideration. Graph a, on the other hand, seems to ignore the middle point completely. So we would probably discard this second criterion for finding the best fit. Why? **The sum of the absolute values does not stress the *magnitude* of the error.**

Giving more weight to farther points; squaring the error

It seems reasonable that the farther away a point is from the estimating line, the more serious is the error. We would rather have several small absolute errors than one large one, as we saw in the last example. **In effect, we want to find a way to "penalize" large absolute errors, so that we can**

FIG. 11·9

Two different estimating lines fitted to the same three observed data points, showing errors in both cases

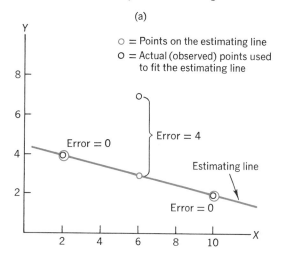

(a)

○ = Points on the estimating line
⊙ = Actual (observed) points used to fit the estimating line

Error = 0
Error = 4
Estimating line
Error = 0

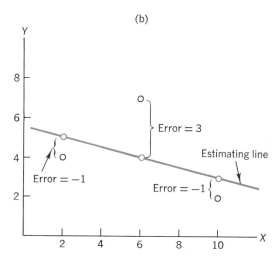

(b)

Error = 3
Estimating line
Error = −1
Error = −1

TABLE 11·4
**Summing the
absolute values
of the errors
of the two
estimating lines
in Fig. 11·9**

Graph a $\|Y - \hat{Y}\|$	Graph b $\|Y - \hat{Y}\|$
$\|4 - 4\| = 0$	$\|4 - 5\| = 1$
$\|7 - 3\| = 4$	$\|7 - 4\| = 3$
$\|2 - 2\| = \underline{0}$	$\|2 - 3\| = \underline{1}$
4 ← total absolute error	5 ← total absolute error

avoid them. We can accomplish this if we *square* the individual errors before we add them. Squaring each term accomplishes two purposes:

1. It magnifies, or penalizes, the larger errors.
2. It cancels the effect of the positive and negative values (a negative error squared is still positive).

Since we are looking for the estimating line that minimizes the sum of the squares of the errors, we call this the *least squares method.*

Let's apply the least squares criterion to the problem in Fig. 11·9. After we have organized the data and summed the squares in Table 11·5, we can see that, as we thought, the estimating line in graph b is the better fit.

Using least squares as a measure of best fit

TABLE 11·5
Applying the least squares criterion to the estimating lines

Graph a $(Y - \hat{Y})^2$	Graph b $(Y - \hat{Y})^2$
$(4 - 4)^2 = (0)^2 = 0$	$(4 - 5)^2 = (-1)^2 = 1$
$(7 - 3)^2 = (4)^2 = 16$	$(7 - 4)^2 = (3)^2 = 9$
$(2 - 2)^2 = (0)^2 = \underline{0}$	$(2 - 3)^2 = (-1)^2 = \underline{1}$
16 ← sum of the squares	11 ← sum of the squares

Using the criterion of least squares, we can now determine whether one estimating line is a better fit than another. But for a set of data points through which we could draw an infinite number of estimating lines, how can we tell when we have found *the best-fitting line?*

Statisticians have derived two equations we can use to find the slope and the Y-intercept of the best-fitting regression line. The first formula calculates the slope:

Finding best-fitting least squares line mathematically

$$b = \frac{\Sigma XY - n\overline{X}\overline{Y}}{\Sigma X^2 - n\overline{X}^2}$$

[11·4]

Slope of the least squares regression line

where:

b = slope of the best-fitting estimating line
X = values of the independent variable
Y = values of the dependent variable
$\overline{X}$ = mean of the values of the independent variable
$\overline{Y}$ = mean of the values of the dependent variable
n = number of data points (that is, the number of the pairs of values for the independent and dependent variables)

The second formula calculates the *Y*-intercept of the line whose slope we calculated using Equation 11·4:

Intercept of the
least squares
regression line

$$a = \overline{Y} - b\overline{X}$$ [11·5]

where:

> a = *Y*-intercept
> b = slope from Equation 11·4
> $\overline{Y}$ = mean of the values of the dependent variable
> $\overline{X}$ = mean of the values of the independent variable

With these two equations, we can find the best-fitting regression line for any two-variable set of data points.

Using the least squares method in two problems

Suppose the director of the Chapel Hill Sanitation Department is interested in the relationship between the age of a garbage truck and the annual repair expense he should expect to incur. In order to determine this relationship, the director has accumulated information concerning four of the trucks the city presently owns (Table 11·6).

TABLE 11·6
**Annual truck
repair expenses**

Truck number	Age of truck in years (X)	Repair expense during last year in hundreds of $ (Y)
101	5	7
102	3	7
103	3	6
104	1	4

Example of the
least squares
method

The first step in calculating the regression line for this problem is to organize the data as outlined in Table 11·7. This allows us to substitute directly into Equations 11·4 and 11·5 in order to find the slope and the *Y*-intercept of the best-fitting regression line.

With the information in Table 11·7, we can now use the equations for the slope (Equation 11·4) and the *Y*-intercept (Equation 11·5) to find the numerical constants for our regression line. The slope is:

Finding the value
of *b*

$$b = \frac{\Sigma XY - n\overline{X}\overline{Y}}{\Sigma X^2 - n\overline{X}^2}$$ [11·4]

$$= \frac{78 - (4)(3)(6)}{44 - (4)(3)^2}$$

$$= \frac{78 - 72}{44 - 36}$$

$$= \frac{6}{8}$$

$$= .75 \leftarrow \text{the slope of the line}$$

422

TABLE 11·7

Calculation of inputs for Equations 11·4 and 11·5

Trucks (n = 4) (1)	Age (X) (2)	Repair expense (Y) (3)	XY (2) × (3)	X² (3)²
101	5	7	35	25
102	3	7	21	9
103	3	6	18	9
104	1	4	4	1
	$\Sigma X = 12$	$\Sigma Y = 24$	$\Sigma XY = 78$	$\Sigma X^2 = 44$

$$\bar{X} = \frac{\Sigma X}{n} \qquad [3·2]$$

$$= \frac{12}{4}$$

$$= 3 \leftarrow \text{mean of the values of the independent variable}$$

$$\bar{Y} = \frac{\Sigma Y}{n} \qquad [3·2]$$

$$= \frac{24}{4}$$

$$= 6 \leftarrow \text{mean of the values of the dependent variable}$$

And the Y-intercept is:

Finding the value of a

$$a = \bar{Y} - b\bar{X} \qquad [11·5]$$

$$= 6 - (.75)(3)$$

$$= 6 - 2.25$$

$$= 3.75 \leftarrow \text{the Y-intercept}$$

Now, to get the estimating equation that describes the relationship between the age of a truck and its annual repair expense, we can substitute the values of a and b in the general equation for a straight line.

Getting the estimating equation

$$\hat{Y} = a + bX \qquad [11·3]$$
$$\hat{Y} = 3.75 + .75X$$

Using this estimating equation (which we could plot as a regression line if we wished), the Sanitation Department director can estimate the annual repair expense, given the age of his equipment. If, for example, the city has a truck that is four years old, the director could use the equation to predict the annual repair expense for this truck as follows:

Using the estimating equation

$$\hat{Y} = 3.75 + .75(4)$$

$$= 3.75 + 3$$

$$= 6.75 \leftarrow \text{expected annual repair expense of } \$675.00$$

Thus, the city might expect to spend about $675 annually in repairs on a 4-year-old truck.

TABLE 11·8
Annual
employment rate
and special
education
expenses

Year	Expenditures for special education (X) ($ millions)	Annual employment rate (Y)
1979	$ 5	31%
1978	11	40
1977	4	30
1976	5	34
1975	3	25
1974	2	20

Another example

Now we can solve the chapter opening problem concerning the relationship between money spent on special education and employment rates of educable mentally retarded high school graduates. Table 11·8 presents the information for the preceding 6 years. With this, we can determine the regression equation describing the relationship.

Again, we can facilitate the collection of the necessary information if we perform the calculations in a table such as Table 11·9.

With this information, we are ready to find the numerical constants a and b for the estimating equation. The value of b is:

Finding b

$$b = \frac{\Sigma XY - n\overline{X}\overline{Y}}{\Sigma X^2 - n\overline{X}^2} \qquad [11\cdot4]$$

$$= \frac{1000 - (6)(5)(30)}{200 - (6)(5)^2}$$

$$= \frac{1000 - 900}{200 - 150}$$

$$= \frac{100}{50}$$

$$= 2 \leftarrow \text{the slope of the line}$$

Finding a

And the value for a is:

$$a = \overline{Y} - b\overline{X} \qquad [11\cdot5]$$

$$= 30 - (2)(5)$$

$$= 30 - 10$$

$$= 20 \leftarrow \text{the Y-intercept}$$

So we can substitute these values for a and b into Equation 11·3 and get:

Determining the
estimating equation

$$\hat{Y} = a + bX \qquad [11\cdot3]$$
$$\hat{Y} = 20 + 2X$$

Using the
estimating equation
to predict

Using this estimating equation, the division can predict what the annual employment rate will be from the amount budgeted for special education. If the division spends $8 million for special education in 1980, it can expect to see approximately 36 percent of its educable mentally retarded high school graduates getting jobs that year.

$$\hat{Y} = 20 + 2(8)$$
$$= 20 + 16$$
$$= 36\% \leftarrow \text{expected annual employment rate}$$

TABLE 11·9
Calculation of inputs for Equations 11·4 and 11·5

Year (n = 6)	Expenditures for special education (X)	Annual employment rate (Y)	XY	X²
1979	5	31	155	25
1978	11	40	440	121
1977	4	30	120	16
1976	5	34	170	25
1975	3	25	75	9
1974	2	20	40	4
	$\Sigma X = 30$	$\Sigma Y = 180$	$\Sigma XY = 1,000$	$\Sigma X^2 = 200$

$$\overline{X} = \frac{\Sigma X}{n} \quad [3\cdot2]$$
$$= \frac{30}{6}$$
$$= 5 \leftarrow \text{mean of the values of the independent variable}$$
$$\overline{Y} = \frac{\Sigma Y}{n} \quad [3\cdot2]$$
$$= \frac{180}{6}$$
$$= 30 \leftarrow \text{mean of the values of the dependent variable}$$

Estimating equations are not perfect predictors. In Fig. 11·10, which plots the points found in Table 11·8, the 36 percent estimate of employment for 1980 is only that—an estimate. Even so, the regression line does give us an idea of what to expect for the coming year.

Shortcoming of estimating equation

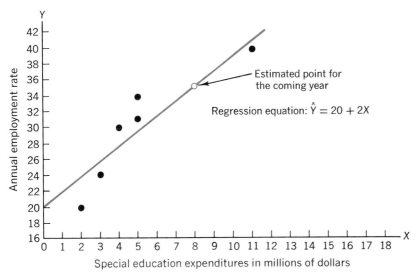

**FIG. 11·10
Scattering of points around regression line**

Regression equation: $\hat{Y} = 20 + 2X$

Estimated point for the coming year

Annual employment rate

Special education expenditures in millions of dollars

Checking the estimating equation

Checking the
estimating equation:
one way

Now that we know how to calculate the regression line, we can learn how to check our work. A crude way to verify the accuracy of the estimating equation is to examine the graph of the sample points. As we can see from the previous problem, the regression line in Fig. 11·10 does appear to follow the path described by the sample points.

Another way
to check the
estimating equation

A more sophisticated method comes from one of the mathematical properties of a line fitted by the method of least squares; that is, the individual positive and negative errors must sum to zero. Using the information from Table 11·9, check to see whether the sum of the errors in the last problem is equal to zero. This is done in Table 11·10.

Since the sum of the errors in Table 11·10 does equal zero, and since the regression line appears to "fit" the points in Fig. 11·10, we can be reasonably certain that we have not committed any serious mathematical mistakes in determining the estimating equation for this problem.

TABLE 11·10
**Calculating the
sum of the
individual
errors
in Table 11·9**

Y	$-$	$\hat{Y}$ (that is, 20 + 2X)		Individual error
31	$-$	[20 + (2)(5)]	$=$	1
40	$-$	[20 + (2)(11)]	$=$	-2
30	$-$	[20 + (2)(4)]	$=$	2
34	$-$	[20 + (2)(5)]	$=$	4
25	$-$	[20 + (2)(3)]	$=$	-1
20	$-$	[20 + (2)(2)]	$=$	-4
				0 ← total error

The standard error of estimate

Measuring the
reliability of the
estimating equation

The next process we need to learn in our study of regression analysis is how to measure the reliability of the estimating equation that we have developed. We alluded to this topic when we introduced scatter diagrams. There, we realized intuitively that a line must be more accurate as an estimator when the data points lie close to the line (as in graph *a* of Fig. 11·11) than when the points are farther away from the line (as in graph *b* of Fig. 11·11).

Definition and use
of standard error
of estimate

To measure the reliability of the estimating equation, statisticians have developed the *standard error of estimate.* This standard error is symbolized s_e and is similar to the standard deviation (which we first examined in Chapter 4) in that both are measures of dispersion. You will recall that the standard deviation is used to measure the dispersion of a set of observations about the mean. **The standard error of estimate, on the other hand, measures the variability, or scatter, of the observed values around the regression line.**
Even so, you will see the similarity between the standard error of estimate and the standard deviation if you compare Equation 11·6, which defines the

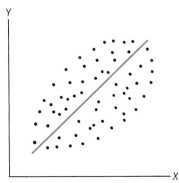

(a) This regression line is a more accurate estimator of the relationship between X and Y

(b) This regression line is a less accurate estimator of the relationship between X and Y

FIG. 11·11

Contrasting degrees of scattering of data points and the resulting effect on accuracy of the regression line

standard error of estimate, with Equation 4·12, which defines the standard deviation:

Standard error of estimate

$$s_e = \sqrt{\frac{\Sigma(Y - \hat{Y})^2}{n - 2}}$$

[11·6]

where:

Y = values of the dependent variable

$\hat{Y}$ = estimated values from the estimating equation which correspond to each Y value

n = number of data points with which we are fitting the regression line

Equation for calculation of standard error of estimate

Notice that in Equation 11·6 the sum of the squared deviations is divided by n − 2 and not by n. This happens because we have lost two degrees of freedom in estimating the regression line. We can reason that since the values of a and b were obtained from a sample of data points, we lose two degrees of freedom when we use these points to estimate the regression line.

n − 2 as the divisor in Equation 11·6

Now let's refer again to our earlier example of the Sanitation Department director who related the age of his trucks to the amount of annual repairs. On page 422 we found the estimating equation in that situation to be:

$$\hat{Y} = 3.75 + .75X$$

where X is the age of the truck and $\hat{Y}$ is the estimated amount of annual repairs (in hundreds of dollars).

To calculate s_e for this problem, we must first determine the value of $\Sigma(Y - \hat{Y})^2$, that is, the numerator of Equation 11·6. We have done this in Table 11·11, using (3.75 + .75X) for $\hat{Y}$ wherever it was necessary. Since

Calculating the standard error of estimate

X (1)	Y (2)	Ŷ (that is, 3.75 + 75X) (3)	Individual error (Y − Ŷ) (2) − (3)	(Y − Y)² [(2) − (3)]²
5	7	3.75 + (.75)(5)	7 − 7.5 =−.5	.25
3	7	3.75 + (.75)(3)	7 − 6.0 = 1.0	1.00
3	6	3.75 + (.75)(3)	6 − 6.0 = 0.0	.00
1	4	3.75 + (.75)(1)	4 − 4.5 =−.5	.25

$$\Sigma(Y - \hat{Y})^2 = \mathbf{1.50} \leftarrow \textbf{sum of squared errors}$$

$\Sigma(Y - \hat{Y})^2$ is equal to 1.50, we can now use Equation 11·6 to find the standard error of estimate:

$$s_e = \sqrt{\frac{\Sigma(Y - \hat{Y})^2}{n - 2}} \qquad [11·6]$$

$$= \sqrt{\frac{1.50}{4 - 2}}$$

$$= \sqrt{\frac{1.50}{2}}$$

$$= \sqrt{.75}$$

$$= .866 \leftarrow \text{standard error of estimate of \$86.60}$$

Using a short-cut method to calculate the standard error of estimate

To use Equation 11·6, we must do the tedious series of calculations outlined in Table 11·11. For every value of Y, we must compute the corresponding value of $\hat{Y}$. Then we must substitute these values into the expression $\Sigma(Y - \hat{Y})^2$.

Fortunately, we can eliminate some of the steps in this task by using the short cut provided by Equation 11·7; that is,

A quicker way to calculate s_e

$$s_e = \sqrt{\frac{\Sigma Y^2 - a\Sigma Y - b\Sigma XY}{n - 2}} \qquad [11·7]$$

where:

X = values of the independent variable
Y = values of the dependent variable
a = Y-intercept from Equation 11·5
b = slope of the estimating equation from Equation 11·4
n = number of data points

This equation is a short cut because, when we first organized the data in this problem so that we could calculate the slope and the Y-intercept (Table

except one—the value of $\Sigma \hat{Y}^2$. Table 11·12 is a repeat of Table 11·7 with the Y^2 column added.

Now we can refer to Table 11·12 and our previous calculations of a and b in order to calculate s_e using the short-cut method:

$$s_e = \sqrt{\frac{\Sigma Y^2 - a\Sigma Y - b\Sigma XY}{n - 2}} \qquad [11·7]$$

$$= \sqrt{\frac{150 - (3.75)(24) - (.75)(78)}{4 - 2}}$$

$$= \sqrt{\frac{150 - 90 - 58.5}{2}}$$

$$= \sqrt{\frac{1.5}{2}}$$

$$= \sqrt{.75}$$

$$= .866 \leftarrow \text{standard error of } \$86.60$$

This is the same result as the one we obtained using Equation 11·6, but think of how many steps we saved!

TABLE 11·12
Calculation of inputs for Equation 11·7

Trucks $n = 4$ (1)	Age X (2)	Repair expense Y (3)	XY (2) × (3)	X² (2)²	Y² (3)²
101	5	7	35	25	49
102	3	7	21	9	49
103	3	6	18	9	36
104	1	4	4	1	16
	$\Sigma X = 12$	$\Sigma Y = 24$	$\Sigma XY = 78$	$\Sigma X^2 = 44$	$\Sigma Y^2 = 150$

Interpreting the standard error of estimate

As was true of the standard deviation, the larger the standard error of estimate, the greater the scattering (or dispersion) of points around the regression line. Conversely, if $s_e = 0$, we expect the estimating equation to be a "perfect" estimator of the dependent variable. In that case, all the data points should lie directly on the regression line, and no points would be scattered around it.

Interpreting and using the standard error of estimate

We shall use the standard error of estimate as a tool in the same way that we can use the standard deviation. That is to say, assuming that the observed points are normally distributed around the regression line, we can expect to find 68 percent of the points within $\pm 1s_e$ (or plus-and-minus one standard error of estimate), 95.5 percent of the points within $\pm 2s_e$, and 99.7 percent of the points within $\pm 3s_e$. Figure 11·12 illustrates these "bounds"

Using s_e to form bounds around the regression line

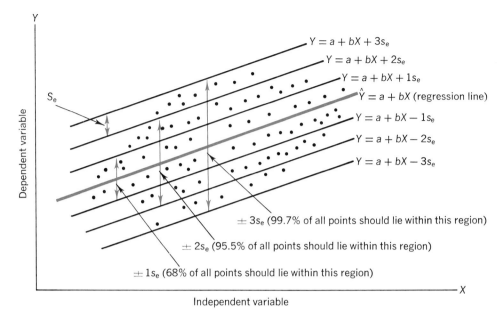

FIG. 11·12
$\pm 1 s_e$, $\pm 2 s_e$, and
$\pm 3 s_e$ **bounds**
around the
regression line

around the regression line. Another thing to notice in Fig. 11·12 is that the standard error of estimate is measured along the Y-axis, rather than perpendicularly from the regression line.

Assumptions we
make in use of s_e

At this point, we should state the assumptions we are making, because shortly, we shall make some probability statements based on these assumptions. Specifically, we have assumed that:

1. The observed values for Y are normally distributed around each estimated value of $\hat{Y}$
2. The variance of the distributions around each possible value of $\hat{Y}$ is the same.

If this second assumption were not true, then the standard error at one point on the regression line could differ from the standard error at another point on the line.

Approximate prediction intervals

Using s_e
to generate
prediction
intervals

One way to view the standard error of estimate is to think of it as the statistical tool we can use to make a probability statement about the interval around an estimated value of $\hat{Y}$, within which the actual value of Y lies. We can see, for instance, in Fig. 11·12 that we can be 95.5 percent certain that the actual value of Y will lie within 2 standard errors of the estimated value of $\hat{Y}$. We call these intervals around the estimated $\hat{Y}$ *approximate prediction*

intervals. They serve the same function as the confidence intervals did in
Chapter 8.

431
Section 2
ESTIMATION USING
REGRESSION LINE

Now, applying the concept of approximate prediction intervals to the
Sanitation Department director's repair expenses, we know that the estimating equation used to predict the annual repair expense is:

$$\hat{Y} = 3.75 + .75X$$

Applying
prediction
intervals

And we know that if the department has a 4-year-old truck, we predict it will
have an annual repair expense of $675:

$$\hat{Y} = 3.75 + .75(4)$$
$$= 3.75 + 3.00$$
$$= 6.75 \leftarrow \text{expected annual repair expense of } \$675$$

Finally, you will recall that we calculated the standard error of estimate to
be $s_e = .866$, ($86.60). We can now combine these two pieces of information
and say that we are roughly 68 percent confident that the actual repair
expense will be within ± 1 standard error of estimate from $\hat{Y}$. We can calculate the upper and lower limits of this prediction interval as follows:

One standard error
prediction
intervals

$$\hat{Y} + 1s_e = \$675 + (1)(\$86.60)$$
$$= \$761.60 \leftarrow \text{upper limit of prediction interval}$$

and:

$$\hat{Y} - 1s_e = \$675 - (1)(\$86.60)$$
$$= \$588.40 \leftarrow \text{lower limit of prediction interval}$$

If, instead, we say that we are roughly 95.5 percent confident that the actual
repair expense will be within plus-and-minus 2 standard errors of estimate
from $\hat{Y}$, we would calculate the limits of this new prediction interval like this:

Two standard error
prediction
intervals

$$\hat{Y} + 2s_e = \$675 + (2)(\$86.60)$$
$$= \$848.20 \leftarrow \text{upper limit}$$

and:

$$\hat{Y} - 2s_e = \$675 - (2)(\$86.60)$$
$$= \$501.80 \leftarrow \text{lower limit}$$

Keep in mind that statisticians apply the prediction intervals of the normal distribution (68 percent for $1s_e$, 95.5 percent for $2s_e$, and 99.7 percent
for $3s_e$) *only* to large samples, that is, where $n > 30$. In this problem, our
sample size is too small ($n = 4$). Thus, *our conclusions are inaccurate.* But
the method we have used nevertheless demonstrates the principle involved
in prediction intervals.

n is too small
to use the
normal distribution

If we wish to avoid the inaccuracies caused by the size of the sample, we need to use the *t* distribution. Recall that the *t* distribution is appropriate when *n* is less than 30 and the population standard deviation is unknown. We meet both these conditions, since $n = 4$ and s_e is an estimate rather than the known population standard deviation.

An example using
the t distribution
to calculate
prediction
intervals

Now suppose the Sanitation Department director wants to be roughly 90 percent certain that the annual truck repair expense will lie within the prediction interval. How should we calculate this interval? Since the *t* distribution table focuses on the probability that the parameter we are estimating will lie *outside* the prediction interval, we need to look in Appendix Table 2 under the 100% − 90% = 10% value column. Once we locate that column, we look for the row representing two degrees of freedom; since $n = 4$ and since we know we lose two degrees of freedom (in estimating the values of *a* and *b*) then $n - 2 = 2$. Here we find the appropriate *t* value to be 2.920.

Now we can make a more accurate calculation of our prediction interval limits as follows:

$$\hat{Y} + t(s_e) = \$675 + (2.920)(\$86.60)$$
$$= \$675 + \$252.87$$
$$= \$927.87 \leftarrow \text{upper limit}$$

and:

$$\hat{Y} - t(s_e) = \$675 - (2.920)(\$86.60)$$
$$= \$675 - \$252.87$$
$$= \$422.13 \leftarrow \text{lower limit}$$

So the director can be 90 percent certain that the annual repair expense on a four-year-old truck will lie between \$422.13 and \$927.87.

We stress again that the above prediction intervals are only *approximate*. In fact, statisticians can calculate the exact standard error for the prediction, s_p, using this formula:

$$s_p = s_e \sqrt{1 + \frac{1}{n} + \frac{(\overline{X} - X_0)^2}{\Sigma X^2 - n\overline{X}^2}}$$

where:

X_0 = the specific value of *X* at which we want to predict the value of *Y*

Notice that if we use this formula, s_p will be different for *each* value of X_0. In particular, if X_0 is *far* from $\overline{X}$, then s_p will be large because $(\overline{X} - X_0)^2$ will be large. If, on the other hand, X_0 is close to $\overline{X}$, and *n* is moderately large (greater than 10), then s_p will be close to s_e. This happens because $1/n$ will be small and $(\overline{X} - X_0)^2$ will be small. Therefore, the value under the square root sign will be close to 1, the square root will be even closer to 1, and s_e will be very close to s_p. This justifies our use of s_e to compute approximate prediction intervals.

EXERCISES

12. For the following set of data:

    a) Plot the scatter diagram.

    b) Develop the estimating equation that best describes the data.

    c) Predict Y for $X = 4, 9,$ and 12.

X	7	10	8	5	11	3	7	11	12	6
Y	2.0	3.0	2.4	1.8	3.2	1.5	2.1	3.8	4.0	2.2

13. Using the data in the table below:

    a) Plot the scatter diagram.

    b) Develop the estimating equation that best describes the data.

    c) Predict Y for $X = 12, 14,$ and 18.

X	20	11	15	10	17	19
Y	5	15	14	17	8	9

14. Given the following set of data:

    a) Find the best fitting line.

    b) Compute the standard error of estimate.

    c) Find a prediction interval (with a 95 percent confidence level) for the dependent variable given that X is 44.

X	56	48	42	58	40	39	50
Y	9.5	7.5	7.0	9.5	6.2	6.6	8.7

15. For the following set of data:

    a) Find the best fitting line.

    b) Calculate the standard error of estimate.

    c) Find a prediction interval (with 90 percent confidence level) for Y when X is 27.

X	25	18	32	21	35	28	30
Y	16	11	20	15	26	32	20

16. As part of a study of the effect of unemployment upon society, a University of Southern California sociologist would like to determine the relationship between unemployment and suicide attempts. The following results of a survey of 12 California cities presents the number of suicide attempts per 1,000 residents and the corresponding percentage of unemployment:

Unemployment	7.3	6.4	6.2	5.5	6.4	4.7	5.8	7.9	6.7	9.6	10.3	7.2
Number of suicide attempts per 1,000 residents	22	17	9	8	12	5	7	19	13	29	33	18

a) Develop the estimating equation that best describes these data.

b) Calculate the standard error of estimate for this relationship.

c) Find a prediction interval (with 95 percent confidence level) for the attempted suicide rate in a California city where unemployment is currently running at 6 percent.

17. A study by the Atlanta, Georgia, Department of Transportation on the effect of bus ticket prices upon the number of passengers produced the following results:

Ticket price (cents)	15	20	25	30	40	50
Passengers per 100 miles	440	430	450	370	340	370

a) Plot these data.

b) Develop the estimating equation that best describes these data.

c) Predict the number of passengers per mile if the ticket price were 35 cents.

18. A therapist who prescribes tranquilizing drugs for some of his patients was concerned about the relationship between the dosage prescribed and the patient's ability to perform normal tasks and to solve problems requiring reasoning, since this might affect job performance. He administered the following dosage proportions to 8 patients of similar age and weight and then gave them a series of standard reasoning problems to solve. Below are the dosages and the scores obtained on the reasoning task.

Dosage	5	10	10	15	15	20	20	25
Score on reasoning task	58	41	45	27	26	12	16	3

a) Plot these data.

b) Develop the equation that best describes the relationship between dosage amount and score on a reasoning task.

c) Predict the score on a reasoning task that we could expect to find if we administered 18 units of the tranquilizer.

19. Investigating the effects of noise on a person's mood, level of tension, and nervousness, a psychologist first interviewed subjects to ensure that they were calm and relaxed at the beginning of the experiment. Then he placed them in rooms with varying levels of noise for half an hour. Following this, he gave them a questionnaire designed to measure their mood and level of anxiety. The following table shows the index of their degree of arousal or nervousness and the level of noise to which they were exposed (5.0 = low noise level; 10.0 = very loud noise level).

Noise level	7.0	6.5	5.5	6.0	8.0	8.5	6.0	6.5
Degree of arousal	23	38	45	36	16	18	39	41

a) Plot these data.

b) Develop an estimating equation that describes these data.

c) Predict the degree of arousal that we might expect when the noise level is 7.25.

20. Anne Morgan, a physician at Tulane Medical School, believes that there is a relationship between a man's age and his physical stamina. In order to discover this relationship, she has compiled the following information for 10 randomly selected men.

Age	42	27	36	25	22	39	57	19	33	30
Number of minutes maintained in strenuous physical task	2	7	5	9	10	4	4	8	6	5

a) Plot these data.

b) Develop the equation that best describes the relationship between age and physical stamina.

c) How long might a 30-year-old man be expected to maintain strenuous physical activity of the sort in this task?

3 CORRELATION ANALYSIS

Correlation analysis is the statistical tool that we can use to describe *the degree to which one variable is linearly related to another.* Frequently, correlation analysis is used in conjunction with regression analysis to measure how well the regression line explains the variations of the dependent variable, *Y*. Correlation can also be used by itself, however, to measure the degree of association between two variables.

What correlation analysis does

Statisticians have developed two measures for describing the correlation between two variables: the *coefficient of determination* and the *correlation coefficient.* Introducing these two measures of association is the purpose of this section.

Two measures that describe correlation

The coefficient of determination

The coefficient of determination is the primary way we can measure the extent, or strength, of the association that exists between two variables, *X* and *Y*. Since we have used a sample of points to develop regression lines, we refer to this measure as the *sample coefficient of determination.*

The sample coefficient of determination is developed from the relationship between two kinds of variation: the variation of the *Y* values in a data set around

Developing the sample coefficient of determination

1. the fitted regression line
2. their own mean

The term *variation* in both these cases is used in its usual statistical sense to mean "the sum of a group of squared deviations." Using this definition, then, it is reasonable to express the variation of the *Y* values around the regression line with this equation:

$$\text{Variation of the } Y \text{ values around the regression line} = \Sigma(Y - \hat{Y})^2 \qquad [11\cdot8]$$

And the second variation, that of the *Y* values around their own mean, is determined by:

$$\text{Variation of the } Y \text{ values around their own mean} = \Sigma(Y - \overline{Y})^2 \qquad [11\cdot9]$$

435

One minus the ratio between these two variations is the sample coefficient of determination, which is symbolized r^2:

$$r^2 = 1 - \frac{\Sigma(Y - \hat{Y})^2}{\Sigma(Y - \bar{Y})^2} \qquad [11\cdot10]$$

The next two sections will show you that r^2, as defined by Equation $11\cdot10$, is a measure of the degree of linear association between X and Y.

An intuitive interpretation of r^2

Consider two extreme ways in which the variables X and Y can be related. In Table $11\cdot13$ every observed value of Y lies on the estimating line, as can be proven visually by Fig. $11\cdot13$. This is *perfect correlation.*

TABLE 11·13
Illustration of
perfect
correlation
between
two variables,
X and Y

Data point	Value of X	Value of Y
1st	1	4
2nd	2	8
3rd	3	12
4th	4	16
5th	5	20
6th	6	24
7th	7	28
8th	8	32

$$\Sigma Y = 144$$
$$\bar{Y} = \frac{144}{8}$$
$$= 18 \quad \leftarrow \text{mean of the values of } Y$$

FIG. 11·13
Perfect correlation
between X and Y:
every data point
lies on the
regression line

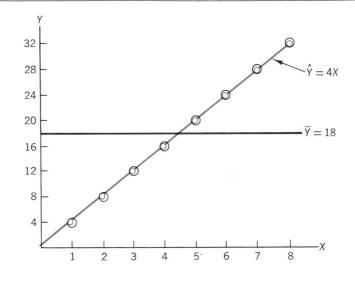

The estimating equation appropriate for this data is easy to determine. Estimating equation appropriate for perfect correlation Since the regression line passes through the origin, we know that the Y-intercept is zero; and since Y increases by 4 every time X increases by 1, the slope must equal 4. Thus, the regression line is:

$$\hat{Y} = 4X$$

Now, to determine the sample coefficient of determination for the regression line in Fig. 11·13, we first calculate the numerator of the fraction in Equation 11·10:

Variation of the Y values
around the regression line $= \Sigma(Y - \hat{Y})^2$ [11·8]

$$= \Sigma(0)^2$$
$$= 0$$

Since every Y value is on the regression line, the difference between Y and $\hat{Y}$ is zero in each case

Then we can find the denominator of the fraction:

Variation of the Y values
around their own mean $= \Sigma(Y - \bar{Y})^2$ [11·9]

Determining sample coefficient of determination for perfect correlation

$$
\begin{aligned}
&= (\ 4 - 18)^2 = (-14)^2 = 196 \\
&+ (\ 8 - 18)^2 = (-10)^2 = 100 \\
&+ (12 - 18)^2 = (-\ 6)^2 = \ \ 36 \\
&+ (16 - 18)^2 = (-\ 2)^2 = \ \ \ \ 4 \\
&+ (20 - 18)^2 = (\ \ \ 2)^2 = \ \ \ \ 4 \\
&+ (24 - 18)^2 = (\ \ \ 6)^2 = \ \ 36 \\
&+ (28 - 18)^2 = (\ \ 10)^2 = 100 \\
&+ (32 - 18)^2 = (\ \ 14)^2 = \underline{196} \\
& 672 \leftarrow \Sigma(Y - \bar{Y})^2
\end{aligned}
$$

With these values to substitute into Equation 11·10, we can find that the sample coefficient of determination is equal to +1:

$$r^2 = 1 - \frac{\Sigma(Y - \hat{Y})^2}{\Sigma(Y - \bar{Y})^2} \qquad [11·10]$$

$$= 1 - \frac{0}{672}$$

$$= 1 - 0$$

$$= 1 \leftarrow \text{ sample coefficient of determination}$$
$$\text{when there is perfect correlation}$$

The value of r^2 is equal to $+1$, then, whenever the regression line is a perfect estimator.

A second extreme way in which the variables X and Y can be related is that the points could lie at equal distances on both sides of a horizontal regression line, as is pictured in Fig. 11·14. The data set here consists of 8 points, all of which have been recorded in Table 11·14.

FIG. 11·14
Zero correlation between X and Y: same values of Y appear for different values of X

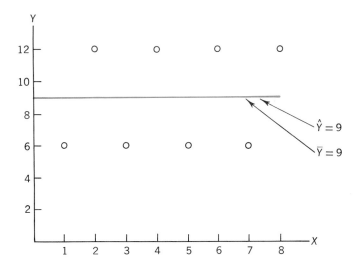

TABLE 11·14
Illustration of zero correlation between two variables X and Y

Data point	Value of X	Value of Y
1st	1	6
2nd	2	12
3rd	3	6
4th	4	12
5th	5	6
6th	6	12
7th	7	6
8th	8	12

$$\Sigma Y = 72$$
$$\bar{Y} = \frac{72}{8}$$
$$= 9 \leftarrow \text{mean of the values of } Y$$

Determining sample coefficient of determination for zero correlation

From Fig. 11·14, we can see that the least squares regression line appropriate for this data is of the form $\hat{Y} = 9$. The slope of the line is *zero* because the same values of Y appear for all the different values of X. Both the Y-intercept and the mean of the Y values are equal to 9.

Now we'll compute the two variations using Equations 11·8 and 11·9, so that we can determine the sample coefficient of determination for this regression line. First, the variation of the Y values around the estimating line $\hat{Y} = 9$:

Variation of the Y values
around the regression line $= \Sigma(Y - \hat{Y})^2$ [11·8]

$$
\begin{aligned}
&= (6 - 9)^2 = (-3)^2 = 9 \\
&+ (12 - 9)^2 = (3)^2 = 9 \\
&+ (6 - 9)^2 = (-3)^2 = 9 \\
&+ (12 - 9)^2 = (3)^2 = 9 \\
&+ (6 - 9)^2 = (-3)^2 = 9 \\
&+ (12 - 9)^2 = (3)^2 = 9 \\
&+ (6 - 9)^2 = (-3)^2 = 9 \\
&+ (12 - 9)^2 = (3)^2 = \underline{9} \\
&\qquad\qquad\qquad\qquad\quad 72 \leftarrow \Sigma(Y - \hat{Y})^2
\end{aligned}
$$

Then, the variation of the Y values around the mean of 9:

Variation of the Y values
around their own mean $= \Sigma(Y - \overline{Y})^2$ [11·9]

$$
\begin{aligned}
&= (6 - 9)^2 = (-3)^2 = 9 \\
&+ (12 - 9)^2 = (3)^2 = 9 \\
&+ (6 - 9)^2 = (-3)^2 = 9 \\
&+ (12 - 9)^2 = (3)^2 = 9 \\
&+ (6 - 9)^2 = (-3)^2 = 9 \\
&+ (12 - 9)^2 = (3)^2 = 9 \\
&+ (6 - 9)^2 = (-3)^2 = 9 \\
&+ (12 - 9)^2 = (3)^2 = \underline{9} \\
&\qquad\qquad\qquad\qquad\quad 72 \leftarrow \Sigma(Y - \overline{Y})^2
\end{aligned}
$$

Substituting these two values into Equation 11·10, we see that the sample coefficient of determination is 0:

$$
\begin{aligned}
r^2 &= 1 - \frac{\Sigma(Y - \hat{Y})^2}{\Sigma(Y - \overline{Y})^2} \\
&= 1 - \frac{72}{72} \\
&= 1 - 1 \\
&= 0 \leftarrow \text{sample coefficient of determination}
\end{aligned}
$$ [11·10]

when there is no correlation

Thus, the value of r^2 is zero when there is no correlation.

In the problems most decision makers encounter, r^2 will lie somewhere between these two extremes of 1 and 0. Keep in mind, however, that an r^2 close to 1 indicates a strong correlation between X and Y, while an r^2 near 0 means there is little correlation between these two variables.

Interpreting r^2 values

One point that we must emphasize strongly is that r^2 measures only the strength of a linear relationship between two variables. For example, if we had a lot of X, Y points that all fell on the circumference of a circle but at randomly scattered places, clearly there would be a relationship among these points (they all lie on the same circle). But in this instance, if we computed r^2 it would turn out in fact to be close to zero, because the points do not have a *linear* relationship with each other.

Interpreting r^2 another way

Another way
to interpret the
sample coefficient
of determination

Statisticians also interpret the sample coefficient of determination by looking at the *amount of the variation in Y that is explained by the regression line.* To understand this meaning of r^2, consider the regression line (shown in color) in Fig. 11·15. Here, we have singled out one observed value of Y, shown as the upper black circle. If we use the mean of the Y values, $\bar{Y}$, to estimate this black-circled value of Y, then the *total deviation* of this Y from its mean would be $(Y - \bar{Y})$. Notice that if we used the regression line to estimate this black-circled value of Y, we would get a better estimate. However, even though the regression line accounts for, or explains, $(\hat{Y} - \bar{Y})$ of the total deviation, the remaining portion of the total deviation, $(Y - \hat{Y})$ still is *unexplained.*

Explained
and unexplained
deviation

But consider a whole set of observed Y values instead of only one value. The total variation, that is, the sum of the squared total deviations, of these points from their mean would be:

$$\Sigma(Y - \bar{Y})^2 \qquad [11 \cdot 9]$$

Explained
and unexplained
variation

and the *explained* portion of the total variation, or the sum of the squared explained deviations of these points from their mean, would be:

$$\Sigma(\hat{Y} - \bar{Y})^2$$

FIG. 11·15
Total deviation,
explained
deviation, and
unexplained
deviation for *one*
observed value of Y

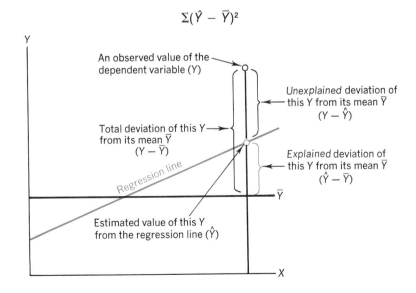

The *unexplained* portion of the total variation (the sum of the squared unexplained deviations) of these points from the regression line would be:

$$\Sigma(Y - \hat{Y})^2 \qquad [11 \cdot 8]$$

If we want to express the fraction of the total variation that remains unexplained, we would divide the unexplained variation, $\Sigma(Y - \hat{Y})^2$, by the total variation, $\Sigma(Y - \bar{Y})^2$ as follows:

$$\frac{\Sigma(Y - \hat{Y})^2}{\Sigma(Y - \bar{Y})^2} \leftarrow \text{fraction of the total variation that is unexplained}$$

And finally, if we subtract the fraction of the total variation that remains unexplained from one, we will have the formula for finding that fraction of the total variation of Y which *is* explained by the regression line. That formula is:

$$r^2 = 1 - \frac{\Sigma(Y - \hat{Y})^2}{\Sigma(Y - \bar{Y})^2} \qquad [11 \cdot 10]$$

the same equation that we have previously used to calculate r^2. It is in this sense, then, that r^2 measures how well X explains Y, that is, the degree of association between X and Y.

One final word about calculating r^2. To obtain r^2 using Equations $1 \cdot 8$, $11 \cdot 9$, and $11 \cdot 10$ requires a series of tedious calculations. To bypass these calculations, statisticians have developed a short-cut version, using values we would have determined already in the regression analysis. The formula is:

Short-cut method to calculate r^2

$$r^2 = \frac{a\Sigma Y + b\Sigma XY - n\bar{Y}^2}{\Sigma Y^2 - n\bar{Y}^2} \qquad [11 \cdot 11]$$

where:

r^2 = sample coefficient of determination
a = Y-intercept
b = slope of the best-fitting estimating line
X = values of the independent variable
Y = values of the dependent variable
$\bar{Y}$ = mean of the observed values of the dependent variable

To see why this formula is a short cut, apply it to an earlier regression relating special education expenditures to employment rate. In Table $11 \cdot 15$, we have repeated the columns from Table $11 \cdot 9$, page 425, adding a Y^2 column.

Applying the short-cut method

TABLE 11·15
Calculations of inputs for Equation 11·11

Year n = 6 (1)	Special education expense X (2)	Annual employment rate Y (3)	XY (2) × (3)	X² (2)²	Y² (3)²
1979	5	31	155	25	961
1978	11	40	440	121	1,600
1977	4	30	120	16	900
1976	5	34	170	25	1,156
1975	3	25	75	9	625
1974	2	20	40	4	400
	$\Sigma X = 30$	$\Sigma Y = 180$	$\Sigma XY = 1,000$	$\Sigma X^2 = 200$	$\Sigma Y^2 = 5,642$

$$\overline{Y} = \frac{180}{6}$$

$= 30 \leftarrow$ mean of the values of
the dependent variable

Recall that when we found the values for *a* and *b* on page 424, the regression line for this problem was described by:

$$\hat{Y} = 20 + 2X$$

Using this line and the information in Table 11·15, we can solve for r^2 as follows:

$$r^2 = \frac{a\Sigma Y + b\Sigma XY - n\overline{Y}^2}{\Sigma Y^2 - n\overline{Y}^2} \qquad [11\cdot11]$$

$$= \frac{(20)(180) + (2)(1000) - (6)(30)^2}{5642 - (6)(30)^2}$$

$$= \frac{3600 + 2000 - 5400}{5642 - 5400}$$

$$= \frac{200}{242}$$

$= .826 \leftarrow$ sample coefficient of determination

Interpreting r^2

Thus, we can conclude that the variations in the special education expenditures (the independent variable *X*) explain 82.6 percent of the variation in the annual employment rate (the dependent variable *Y*).

The coefficient of correlation

Sample coefficient
of correlation

The coefficient of correlation is the second measure that we can use to describe how well one variable is explained by another. When we are dealing with samples, the *sample coefficient of correlation* is denoted by r and is the square root of the sample coefficient of determination:

$$r = \sqrt{r^2}$$

[11·12]

When the slope of the estimating equation is positive, r is the positive square root, but if b is negative, r is the negative square root. Thus, **the sign of r indicates the direction of the relationship between the two variables X and Y.** If an inverse relationship exists, that is, if Y decreases as X increases, then r will fall between 0 and -1. Likewise, if there is a direct relationship (if Y increases as X increases), then r will be a value within the range of 0 to 1. Figure 11·16 illustrates these various characteristics of r.

The coefficient of correlation is more difficult to interpret than r^2. What does $r = .9$ mean? To answer that question, we must remember that $r = .9$ is the same as $r^2 = .81$. The latter tells us that 81 percent of the variation in Y is explained by the regression line. So we see that r is nothing more than the square root of r^2, and we cannot interpret its meaning directly.

Interpreting r

Now let's find the coefficient of correlation for our problem relating special education expenditures and employment rate. Since, in the previous section, we found that the sample coefficient of determination is $r^2 = .826$, we can substitute this value into Equation 11·12 and find that:

Calculating r for the special education problem

$$
\begin{aligned}
r &= \sqrt{r^2} \\
&= \sqrt{.826} \\
&= .909 \leftarrow \text{sample coefficient of correlation}
\end{aligned}
$$

[11·12]

(a) $r^2 = 1$ and $r = 1$

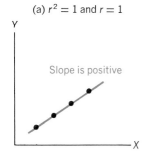

(b) $r^2 = 1$ and $r = -1$

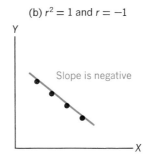

FIG. 11·16
Various characteristics of r, the sample coefficient of correlation

(c) $r^2 = .81$ and $r = .9$

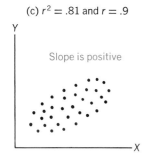

(d) $r^2 = .49$ and $r = -.7$

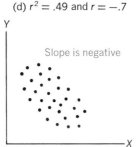

(e) $r^2 = 0$ and $r = 0$

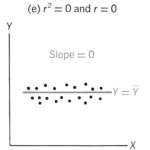

The relation between the two variables is direct and the slope is positive; therefore, the sign for *r* is positive.

EXERCISES

In the following exercises, calculate the sample coefficient of determination and the sample coefficient of correlation for the problems specified.

21. Problem 16, p. 433.

22. Problem 17, p. 434.

23. Problem 18, p. 434.

24. Problem 19, p. 434.

25. Problem 20, p. 434.

26. What type of correlation (positive, negative, or zero) should we expect from these variables?

    a) Intelligence of parents and intelligence of their children.

    b) Age at first marriage and number of years of education.

    c) Weight and blood pressure.

    d) High-school grade-point average and student's height.

4 MAKING INFERENCES ABOUT POPULATION PARAMETERS

Relationship of sample regression line and population regression line

So far, we have used regression and correlation analyses to relate two variables on the basis of sample information. But data from a sample represents only part of the total population. Because of this, we may think of our estimated sample regression line as an estimate of a true, but unknown population regression line of the form:

$$Y = A + BX \qquad [11 \cdot 13]$$

Recall our discussion of the Sanitation Department director who tried to use the age of a truck to explain the annual repair expense on it. That expense will probably consist of two parts:

1. Regular maintenance that does not depend on the age of the truck: tune-ups, oil changes, and lubrication. This expense is captured in the intercept term *A* in Equation 11·13 above.

2. Expenses for repairs due to aging: relining brakes, engine and transmission overhauls, and painting. Such expenses will tend to increase with the age of the truck, and they are captured in the *BX* term of the population regression line $Y = A + BX$ in Equation 11·13 above.

Of course, all the brakes of all the trucks will not wear out at the same time, and some of the trucks will probably never need engine overhauls. Because of this, the individual data points will probably not lie exactly on the population regression line. Some will be above it; some will fall below it. So, instead of satisfying

Why data points do not lie exactly on the regression line

$$Y = A + BX \qquad [11 \cdot 13]$$

the individual data points will satisfy the formula

$$Y = A + BX + e \qquad [11 \cdot 13a]$$

where e is a random disturbance from the population regression line. On the average, e equals zero because disturbances above the population regression line are canceled out by disturbances below the line. We can denote the standard deviation of these individual disturbances by σ_e. The standard error of estimate s_e, then, is an estimate of σ_e, the standard deviation of the disturbances.

Random disturbance e and its behavior

Let us look more carefully at Equations 11·13 and 11·13a. Equation 11·13a expresses the individual values of Y (in this case, annual repair expense) in terms of (1) the individual values of X (the age of the truck) and (2) the random disturbance (e). Since disturbances above the population regression line are cancelled out by those below the line, we know that the expected value of e is zero, and we see that if we had several trucks of the same age, X, we would expect the average annual repair expense on these trucks to be $Y = A + BX$. This shows us that the population regression line (Equation 11·13) gives the mean value of Y associated with each value of X.

Since our *sample* regression line, $\hat{Y} = a + bX$ (Equation 11·3), estimates the *population* regression line, $Y = A + BX$ (Equation 11·13), we should be able to use it to make inferences about the population regression line. In this section then, we shall make inferences about the slope B of the "true" regression equation (the one for the entire population) that are based upon the slope b of the regression equation estimated from a sample of values.

Making inferences about B from b

Slope of the population regression line

The regression line is derived from a sample and not from the entire population. As a result, we cannot expect the true regression equation, $Y = A + BX$ (the one for the entire population), to be exactly the same as the equation estimated from the sample observations, or $\hat{Y} = a + bX$. Even so, we can use the value of b, the slope we calculate from a sample, to test hypotheses about the value of the B, the slope of the regression line for the entire population.

Difference between true regression equation and one estimated from sample observations

The procedure for testing a hypothesis about *B* is similar to procedures discussed in Chapter 9, on hypothesis testing. To understand this process, return to the problem that related annual expenditures for special education to employment rate. On page 424, we pointed out that $b = 2$. The first step is to find some value for *B* to compare with $b = 2$.

Suppose that over an extended past period of time the slope of the relationship between *X* and *Y* was 2.1. To test if this were still the case, we could define the hypotheses as:

H_0: $B = 2.1$ ←Null hypothesis
H_1: *B* is not equal to 2.1 ←Alternative hypothesis

In effect, then, we are testing to learn whether current data indicate that *B* has changed from its historical value of 2.1.

To find the test statistic for *B*, it is necessary first to find the *standard error of the regression coefficient*. Here, the regression coefficient we are working with is *b*, so the standard error of this coefficient is denoted s_b. Equation 11·14 presents the mathematical formula for s_b:

$$s_b = \frac{s_e}{\sqrt{\Sigma X^2 - n\overline{X}^2}}$$

[11·14]

where:

s_b = standard error of the regression coefficient
s_e = standard error of estimate
X = values of the independent variable
$\overline{X}$ = mean of the values of the independent variable
n = number of data points

Finding upper
and lower limits
of the
acceptance region
for our hypothesis test

Once we have calculated s_b, we can use the *t* distribution with $n - 2$ degrees of freedom and the following equation to calculate the upper and lower limits of the acceptance region.

upper limit of acceptance region = $B + t(s_b)$
lower limit of acceptance region = $B - t(s_b)$

[11·15]

where:

t = appropriate *t* value (with $n - 2$ degrees of freedom) for the significance level of the test
B = actual slope hypothesized for the population
s_b = standard error of the regression coefficient

Of course, for a one-tailed test you would calculate only an upper or lower limit as appropriate.

$$s_e = \sqrt{\frac{\Sigma Y^2 - a\Sigma Y - b\Sigma XY}{n - 2}}$$

[11·7]   Calculating s_e

$$= \sqrt{\frac{5,642 - (20)(180) - (2)(1,000)}{6 - 2}}$$

$$= \sqrt{\frac{5,642 - 3,600 - 2,000}{4}}$$

$$= \sqrt{\frac{42}{4}}$$

$$= \sqrt{10.5}$$

$$= 3.24 \leftarrow \text{standard error of estimate}$$

Now we can determine the standard error of the regression coefficient:

$$s_b = \frac{s_e}{\sqrt{\Sigma X^2 - n\bar{X}^2}}$$

[11·14]   Finding s_b

$$= \frac{3.24}{\sqrt{200 - (6)(5)^2}}$$

$$= \frac{3.24}{\sqrt{50}}$$

$$= \frac{3.24}{7.07}$$

$$= .46 \leftarrow \text{standard error of the regression coefficient}$$

Suppose we have reason to test our hypothesis at the 10 percent level
of significance. Since we have 6 observations in our sample data, we know
that we have $n - 2$ or $6 - 2 = 4$ degrees of freedom. We look in Appendix
Table 2 under the 10 percent column and come down until we find the 4
degrees-of-freedom row. There, we see that the appropriate t value is 2.132.
Since we are concerned whether b (the slope of the sample regression line)
is significantly *different* from B (the hypothesized slope of the population
regression line), this is a two-tailed test and the limits of the acceptance
region are found using Equation 11·15.

Conducting the
hypothesis test

$$B + t(s_b) = 2.1 + 2.132\,(0.46)$$
$$= 3.081 \leftarrow \text{upper limit of acceptance region}$$
$$B - t(s_b) = 2.1 - 2.132\,(0.46)$$
$$= 1.119 \leftarrow \text{lower limit of acceptance region}$$

The slope of our regression line (b) is 2.0, which is inside the acceptance
region. Therefore, we accept the null hypothesis that B still equals 2.1. In

other words, there is not enough difference between b and 2.1 for us to conclude that B has changed from its historical value. Because of this we feel that each additional million dollars spent on special education still increases the employment rate by 2.1 percent, as it has in the past.

In addition to hypothesis testing, we can also construct a *confidence interval* for the value of B. In the same way that b is a point estimate of B, such confidence intervals are interval estimates of B. The problem we just completed, and for which we did a hypothesis test, will illustrate the process of constructing a confidence interval. There, we found that:

$$b = 2.0$$
$$s_b = 0.46$$
$$t = 2.132 \leftarrow 10\% \text{ level of significance and 4 degrees of freedom}$$

Confidence interval
for B

With this information, we can calculate confidence intervals like this:

$$b + t(s_b) = 2 + (2.132)(.46)$$
$$= 2 + .981$$
$$= 2.981 \leftarrow \text{upper limit}$$
$$b - t(s_b) = 2 - (2.132)(.46)$$
$$= 2 - .981$$
$$= 1.019 \leftarrow \text{lower limit}$$

Interpreting the
confidence interval

In this situation, then, we are 90 percent confident that the true value of B lies between 1.019 and 2.981; that is, each additional dollar spent on special education increases employment rate by some amount between 1.02 percent and 2.98 percent.

EXERCISES

27. From the data below, test the hypothesis that the population slope is 1.2 against the alternative that it is not equal to that value.

X	9	17	20	19	20	23
Y	23	35	29	33	43	32

28. In a regression problem with a sample of size 6, the slope was found to be .75 and the standard error of estimate 30.412. The quantity, $(\Sigma X^2 - n \bar{X}^2) = 240,083.3$

   a) Find the standard error of the regression coefficient.

   b) Construct a 90 percent confidence interval for the population slope.

29. From a data set based on 12 observations, the sample slope was found to be 2.4. If the standard error of the regression coefficient is .15, is there reason to believe that (at the .05 significance level) the slope has changed from its past value of 2.8?

30. For a sample of size 10, the slope was found to be .265 and the standard error of the regression coefficient was .02. Is there reason to believe that the slope has changed from its past value of .30? Use the .01 significance level.

31. Doctors wonder whether the older type contact lens and a newly developed one respond differently to light intensity. A patient was fitted with the new contact lenses and tested for the distance at which she could read printed material under different light intensities. From the following data, should the doctors conclude (at the .10 level of significance) that the slope of the relationship is significantly different from that of the old contact lenses, for which it is 2.8?

Light intensity	7.3	6.6	6.4	6.8	5.9
Reading distance (inches)	20	18	19	20	18

32. In 1969, a government health agency found that in a number of countries, the relationship between smokers and heart disease fatalities per 100,000 population had a slope of .08. A recent study of 12 countries produced a slope of .14 and a standard error of the regression coefficient of .02.

a) Construct a 95 percent confidence interval estimate of the slope of the true regression line. Does the result from this recent study indicate that the true slope has changed?

b) Construct a 99 percent confidence interval estimate of the slope of the true regression line. Does the result from this recent study indicate that the true slope has changed?

33. The Energy Research Administration specifies that the relationship between the number of cars in a train and a diesel engine's consumption of fuel oil has a slope of .046. A particular railroad company has compiled the operation records for 10 different train lengths. From this information, the slope of this relationship (train length versus fuel oil consumption) was calculated to be .061, and the standard error of the regression coefficient was determined to be .005.

a) Construct a 95 percent confidence interval of estimate of the slope of the true regression line. Does the information from this study indicate that the ERA claim is invalid?

b) Repeat the analysis from part a at the 99 percent level of confidence.

5 USING REGRESSION AND CORRELATION ANALYSIS: LIMITATIONS, ERRORS, AND CAVEATS

Regression and correlation analysis are statistical tools that, when properly used, can significantly help people make decisions. Unfortunately, they are frequently misused. As a result, decision makers often make inaccurate forecasts and less-than-desirable decisions. We'll mention the most common errors made in the use of regression and correlation in the hope that you will avoid them.

Misuse of regression and correlation

Extrapolation beyond the range of the observed data

A common mistake is to assume that the estimating line can be applied over any range of values. Hospital administrators can properly use regression analysis to predict the relationship between costs per bed and occupancy levels at various occupancy levels. Some administrators, however, incorrectly use the same regression equation to predict the costs per bed for occupancy levels that are significantly higher than those that were used to estimate the regression line. Although one relationship holds over the range of sample points, an entirely different relationship may exist for a

Specific limited range over which regression equation holds

different range. As a result, these people make decisions on one set of costs and find that the costs change drastically as occupancy increases (owing to things such as overtime costs and capacity constraints). Remember that **an estimating equation is valid only over the same range as the one from which the sample was taken initially.**

Cause and effect

Regression
and correlation
analysis
do not determine
cause and effect

Another mistake we can make when we use regression analysis is to assume that a change in one variable is "caused" by a change in the other variable. As we discussed earlier, **regression and correlation analyses can in no way determine cause and effect.** If we say that there is a correlation between students' grades in college and their annual earnings 5 years after graduation, we are *not* saying that one causes the other. Rather, both may be caused by other factors such as sociological background, parental attitudes, quality of teachers, effectiveness of the job-interviewing process, and economic status of parents—to name only a few potential factors.

Using past trends to estimate future trends

Conditions change
and invalidate the
regression equation

We must take care to reappraise the historical data we use to estimate the regression equation. Conditions can change and violate one or more of the assumptions on which our regression analysis depends. Earlier in this chapter we made the point that we assume that the variance of the disturbance e around the mean is constant. In many situations, however, this variance changes from year to year.

Values of variables
change over time

Another error that can arise from the use of historical data concerns the dependence of some variables on time. Suppose a firm uses regression analysis to determine the relationship between the number of employees and the production volume. If the observations used in the analysis extend back for several years, the resulting regression line may be too steep, because it may fail to recognize the effect of changing technology.

Misinterpreting the coefficients of correlation and determination

Misinterpreting r
and r^2

The coefficient of correlation is occasionally misinterpreted as a percentage. If $r = .6$, it is incorrect to state that the regression equation "explains" 60 percent of the total variation in Y. Instead, if $r = .6$, then r^2 must be $.6 \times .6 = .36$. Only 36 percent of the total variation is explained by the regression line.

The coefficient of determination is misinterpreted if we use r^2 to describe the percentage of the change in the dependent variable that is *caused* by a change in the independent variable. This is wrong because r^2 is a measure only of how well one variable describes another, *not* of how much of the change in one variable is caused by the other variable.

When applying regression analysis, people sometimes find a relationship between two variables that, in fact, have no common bond. Even though one variable does not "cause" a change in the other, they think that there must be some factor common to both variables. It might be possible, for example, to find a statistical relationship between a random sample of the number of miles per gallon consumed by eight different cars and the distance from earth to each of the other eight planets. But since there is absolutely no common bond between gas mileage and the distance to other planets, this "relationship" would be meaningless.

Relationships
that have
no common bond

EXERCISES

34. Explain why an estimating equation is valid over only the range of values used for its development.

35. Explain the difference between the coefficient of determination and the coefficient of correlation.

36. Why should we be cautious in using past data to predict future trends?

37. Why must we not attribute causality in a relationship even when there is strong correlation between the variables or events?

6 MULTIPLE REGRESSION AND
CORRELATION ANALYSIS
As we mentioned earlier, we may use more than one independent variable to estimate the dependent variable and, in this way, attempt to increase the accuracy of the estimate. This process is called multiple regression and correlation analysis. It is based on the same assumptions and procedures we have encountered using simple regression.

Using more than
one independent
variable
to estimate the
dependent variable

Consider the real estate agent who wishes to relate the number of houses the firm sells in a month to the amount of her monthly advertising. Certainly we can find a simple estimating equation that relates these two variables. Could we also improve the accuracy of our equation by including in the estimating process the number of salespersons she employs each month? The answer is yes. And now, since we want to use both the number of sales agents and the advertising expenditures to predict monthly house sales, we must use *multiple,* not simple, regression to determine the relationship.

The principal advantage of multiple regression is that it allows us to utilize more of the information available to us to estimate the dependent variable. Sometimes the correlation between two variables may be insufficient to determine a reliable estimating equation. Yet, if we add the data

Value of
multiple regression

from more independent variables, we may be able to determine an estimating equation that describes the relationship with greater accuracy.

Multiple regression and correlation analysis is a three-step process such as the one we used in simple regression. In this process, we must:

Steps in
multiple regression
and correlation

1. Describe the multiple regression equation.
2. Examine the multiple regression standard error of estimate.
3. Use multiple correlation analysis to determine how well the regression equation describes the observed data.

For convenience, we shall use only two independent variables in the problems we work in this section. Keep in mind, however, that the same process is applicable to any number of independent variables.

The multiple regression equation

A problem
to demonstrate
multiple regression

The Internal Revenue Service is trying to estimate the monthly amount of unpaid taxes discovered by its auditing division. In the past, the IRS estimated this figure on the basis of the expected number of field audit labor hours. In recent years, however, field audit labor hours have become an erratic predictor of the actual unpaid taxes. As a result, the IRS is looking for another factor with which it can improve the estimating equation.

The auditing division does keep a record of the number of hours during which the computer is used to detect unpaid taxes. Could we combine this information with the data on field audit labor hours and come up with a more accurate estimating equation for the unpaid taxes discovered each month? Table 11·16 presents this data for the last 10 months.

Appropriate symbols

In simple regression, X is the symbol used for the values of the independent variable. In multiple regression, we have more than one independent variable. So we shall continue to use X, but we shall add a subscript (for example, X_1, X_2) to distinguish between the independent variables we are using.

TABLE 11·16
Data from IRS auditing records during last 10 months

Month	X_1 Field audit labor hours (00's omitted)	X_2 Computer hours (00's omitted)	Y Actual unpaid taxes discovered (millions of dollars)
January	45	16	$29
February	42	14	24
March	44	15	27
April	45	13	25
May	43	13	26
June	46	14	28
July	44	16	30
August	45	16	28
September	44	15	28
October	43	15	27

In this problem X_1 will represent the number of field audit labor hours and X_2 the number of computer hours. The dependent variable, Y, will be the actual unpaid taxes discovered.

Defining the two independent variables

Recall that in simple regression, the estimating equation $\hat{Y} = a + bX$ describes the relationship between the two variables X and Y. In multiple regression, we must extend that equation, adding one term for each new variable. In symbolic form, Equation 11·16 is the formula we can use when we have two independent variables:

Estimating equation for multiple regression

$$\hat{Y} = a + b_1X_1 + b_2X_2 \qquad [11\cdot16]$$

where:

$\hat{Y}$ = estimated value corresponding to the dependent variable
a = Y-intercept
X_1 and X_2 = values of the two independent variables
b_1 and b_2 = slopes associated with X_1 and X_2, respectively

We can visualize the simple estimating equation as a line on a graph; similarly, we can picture a two-variable multiple regression equation as a plane, such as the one shown in Fig. 11·17. Here we have a 3-dimensional

Visualizing multiple regression

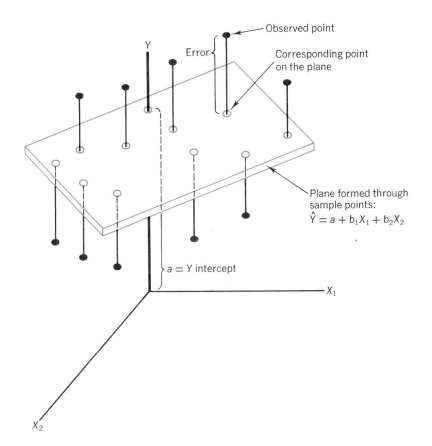

Observed point

Error

Corresponding point on the plane

Plane formed through sample points:
$\hat{Y} = a + b_1X_1 + b_2X_2$

$a = Y$ intercept

Y

X_1

X_2

FIG. 11·17
Multiple regression plane for ten data points

453

shape that possesses depth, length, and width. To get an intuitive feel for this 3-dimensional shape, visualize the intersection of the axes Y, X_1, and X_2 as one corner of a room.

Interpretation of
Fig. 11·17

Figure 11·17 is a graph of the ten sample points from Table 11·16 and the plane about which these points seem to cluster. Some points lie above the plane, and some fall below it—just as points lay above and below the simple regression line.

Using
the least squares
criterion to fit
a regression plane

Our problem is to decide which of the possible planes that we could draw will be the best fit. To do this, we shall again use the least squares criterion and locate the plane that minimizes the sum of the squares of the errors; that is, the distances from the points around the plane to the corresponding points *on* the plane. We use our data and the following three equations to determine the values of the numerical constants a, b_1, and b_2.

$$\Sigma Y = na + b_1 \Sigma X_1 + b_2 \Sigma X_2 \qquad [11 \cdot 17]$$

$$\Sigma X_1 Y = a \Sigma X_1 + b_1 \Sigma X_1^2 + b_2 \Sigma X_1 X_2 \qquad [11 \cdot 18]$$

$$\Sigma X_2 Y = a \Sigma X_2 + b_1 \Sigma X_1 X_2 + b_2 \Sigma X_2^2 \qquad [11 \cdot 19]$$

Solving Equations 11·17, 11·18, and 11·19 for a, b_1, and b_2 will give us the coefficients for the regression plane. Obviously, the best way to compute all the sums in these three equations is to use a table to collect and organize the necessary information, just as we did in simple regression. This we have done for the IRS problem in Table 11·17.

TABLE 11·17
Values for fitting least squares plane, where $n = 10$

Y (1)	X_1 (2)	X_2 (3)	X_1Y (2) × (1)	X_2Y (3) × (1)	X_1X_2 (2) × (3)	X_1^2 (2)^2	X_2^2 (3)^2	Y^2 (1)^2
29	45	16	1,305	464	720	2,025	256	841
24	42	14	1,008	336	588	1,764	196	576
27	44	15	1,188	405	660	1,936	225	729
25	45	13	1,125	325	585	2,025	169	625
26	43	13	1,118	338	559	1,849	169	676
28	46	14	1,288	392	644	2,116	196	784
30	44	16	1,320	480	704	1,936	256	900
28	45	16	1,260	448	720	2,025	256	784
28	44	15	1,232	420	660	1,936	225	784
27	43	15	1,161	405	645	1,849	225	729
272	441	147	12,005	4,013	6,485	19,461	2,173	7,428
↑	↑	↑	↑	↑	↑	↑	↑	↑
ΣY	ΣX_1	ΣX_2	$\Sigma X_1 Y$	$\Sigma X_2 Y$	$\Sigma X_1 X_2$	ΣX_1^2	ΣX_2^2	ΣY^2

$\overline{Y} = 27.2$
$\overline{X}_1 = 44.1$
$\overline{X}_2 = 14.7$

Now, using the information from Table 11·17 in Equations 11·17, 11·18, and 11·19, we get three equations in the three unknown constants $(a, b_1, \text{and } b_2)$, which we denote below as ①, ②, and ③.

$$272 = 10a + 441b_1 + 147b_2 \qquad ①$$
$$12{,}005 = 441a + 19{,}461b_1 + 6{,}485b_2 \qquad ②$$
$$4{,}013 = 147a + 6{,}485b_1 + 2{,}173b_2 \qquad ③$$

We can find the values for the three numerical constants by solving these three equations simultaneously, as follows:

Step 1. Multiply Equation ① by -441. Multiply Equation ② by 10. Add ① to ②. This eliminates a and produces Equation ④.

$$① \times (-441): -119{,}952 = -4410a - 194{,}481b_1 - 64{,}827b_2$$
$$② \times (10) \quad : \quad 120{,}050 = 4410a + 194{,}610b_1 + 64{,}850b_2$$
$$④: \quad 98 = \qquad\qquad 129b_1 + \qquad 23b_2$$

Step 2. Multiply Equation ① by -147 and Equation ③ by 10. Add ① to ③. This eliminates a and produces Equation ⑤.

$$① \times (-147): -39{,}984 = -1470a - 64{,}827b_1 - 21{,}609b_2$$
$$③ \times (10) \quad : \quad 40{,}130 = 1470a + 64{,}850b_1 + 21{,}730b_2$$
$$⑤: \quad 146 = \qquad\qquad 23b_1 + \qquad 121b_2$$

Step 3. Multiply Equation ④ by -23 and Equation ⑤ by 129. Add ④ to ⑤ to eliminate b_1. This produces Equation ⑥ which can be solved for b_2:

$$④ \times (-23) : -2{,}254 = -2{,}967b_1 - \qquad 529b_2$$
$$⑤ \times (129) : \quad 18{,}834 = 2{,}967b_1 + 15{,}609b_2$$
$$⑥: \quad 16{,}580 = \qquad\qquad 15{,}080b_2$$
$$b_2 = \frac{16{,}580}{15{,}080}$$

$$b_2 = 1.099$$

Step 4. Find the value of b_1 by substituting the value for b_2 into Equation ④:

$$④: 98 = 129b_1 + 23b_2$$
$$98 = 129b_1 + (23)(1.099)$$
$$98 = 129b_1 + 25.277$$
$$72.723 = 129b_1$$

$$.564 = b_1$$

Step 5. Substitute the values of b_1 and b_2 into Equation ① to determine the value of a:

$$\text{①:}\ 272 = 10a + 441b_1 + 147b_2$$
$$272 = 10a + (441)(.564) + (147)(1.099)$$
$$272 = 10a + 248.724 + 161.553$$
$$-138.277 = 10a$$

$$\boxed{-13.828 = a}$$

Step 6. Substitute the values of a, b_1, and b_2 into the general multiple regression equation (Equation 11·16). The resulting Equation ⑦ describes the relationship among the number of field audit labor hours, the number of computer hours, and the unpaid taxes discovered by the auditing division.

$$\hat{Y} = a + b_1 X_1 + b_2 X_2 \qquad\qquad [11\cdot16]$$
$$\text{⑦:}\ \hat{Y} = -13.828 + .564X_1 + 1.099X_2$$

The auditing division can use this equation monthly to estimate the amount of unpaid taxes it will discover.

Using the multiple regression equation to estimate

Suppose the IRS wants to increase its discoveries in the coming month. Since trained auditors are scarce, the IRS does not intend to hire additional personnel. The number of field audit labor hours, then, will remain at October's level of about 4,300 hours. But in order to increase its discoveries of unpaid taxes, the IRS expects to increase the number of computer hours to about 1,600. As a result:

$$X_1 = 43 \leftarrow 4,300 \text{ hours of field audit labor}$$
$$X_2 = 16 \leftarrow 1,600 \text{ hours of computer time}$$

Substituting these values into Equation ⑦ , we get:

$$\hat{Y} = -13.828 + .564X_1 + 1.099X_2 \qquad\qquad ⑦$$
$$= -13.828 + (.564)(43) + (1.099)(16)$$
$$= -13.828 + 24.252 + 17.584$$
$$= 28.008 \leftarrow \text{estimated discoveries of } \$28,008,000$$

Interpreting our estimate

Therefore, in the November forecast, the audit division can indicate that it expects about $28 million of discoveries for this combination of factors.

So far, we have referred to a as the Y-intercept and to b_1 and b_2 as the slopes of the multiple regression line. But to be more precise, we should say that these numerical constants are the *estimated regression coefficients.* The constant a is the value of $\hat{Y}$ (in this case, the estimated unpaid taxes) *if* both X_1 and X_2 happen to be zero. The coefficients b_1 and b_2 describe how changes in X_1 and X_2 affect the value of $\hat{Y}$. In Equation ⑦ , for example,

a, b_1, and b_2 are the estimated regression coefficients

we can hold the number of field audit labor hours, X_1, constant and change the number of computer hours, X_2. When we do, the value of $\hat{Y}$ will increase $1,099,000 for every additional 100 hours of computer time. Likewise, we can hold X_2 constant and find that, for every 100-hour increase in the number of field audit labor hours, $\hat{Y}$ increases by $564,000.

In multiple regression analysis, the regression coefficients become less reliable as the degree of correlation between the independent variables increases. If there is a high level of correlation between them, we have a problem that statisticians call *multicollinearity.*

457

Section 6
MULTIPLE
REGRESSION AND
CORRELATION
ANALYSIS

Definition
and effect of
multicollinearity

Multicollinearity might occur if we wished to estimate a firm's sales revenue, and we used both the number of salespersons employed and their salaries. Since the values associated with these two independent variables are highly correlated, we need to use only one set of them to make our estimate. In fact, adding a second variable that is correlated with the first distorts the values of the regression coefficients. Nevertheless, we can often predict Y well, even when multicollinearity is present.

It is important to exercise care in selecting the independent variables to estimate a dependent variable. To ensure that multicollinearity is kept to a minimum, do some simple correlations between pairs of independent variables before you set up the problem as a multiple regression.

Recognizing
multicollinearity

A measure of dispersion: the standard error of estimate for multiple regression

Now that we have determined the equation that relates our three variables, we need some measure of the dispersion around this multiple regression plane. In simple regression, the estimation becomes more accurate as the degree of dispersion around the regression line gets smaller. The same is true of the sample points around the multiple regression plane. To measure this variation, we shall again use the measure called the standard error of estimate:

Measuring
dispersion around
the multiple
regression plane
using the
standard error
of estimate

$$s_e = \sqrt{\frac{\Sigma(Y - \hat{Y})^2}{n - 3}}$$
[11·20]

where:
   Y = sample values of the dependent variable
   $\hat{Y}$ = corresponding estimated values from the regression equation
   n = number of elements in the sample

The denominator of this equation indicates that in multiple regression with two independent variables, the standard error has $n - 3$ degrees of freedom. This occurs because the degrees of freedom are reduced by the three numerical constants, a, b_1, and b_2, that have all been estimated from the same sample.

Short-cut method
to calculate s_e

Equation 11·20 is very tedious to calculate. Consequently, statisticians have devised a shortcut formula that uses the values we have already calculated in order to determine the estimating equation:

$$s_e = \sqrt{\frac{\Sigma Y^2 - a\Sigma Y - b_1 \Sigma X_1 Y - b_2 \Sigma X_2 Y}{n - 3}} \qquad [11\cdot21]$$

where:

Y = values of the dependent variable
X_1 and X_2 = values of the independent variables
n = number of elements in the sample
a, b_1 and b_2 = the regression coefficients determined through the use of Equations 11·17, 11·18, and 11·19

Calculating
the value of s_e

Now we can use the values we calculated for the three numerical constants and the information in Table 11·17 to find the standard error of estimate in our IRS problem, as follows:

$$s_e = \sqrt{\frac{\Sigma Y^2 - a\Sigma Y - b_1 \Sigma X_1 Y - b_2 \Sigma X_2 Y}{n - 3}} \qquad [11\cdot21]$$

$$= \sqrt{\frac{7428 - (-13.828)(272) - (.564)(12{,}005) - (1.099)(4013)}{10 - 3}}$$

$$= \sqrt{\frac{7428 + 3761.216 - 6770.82 - 4410.287}{7}}$$

$$= \sqrt{\frac{8.109}{7}}$$

$$= \sqrt{1.158}$$

$$= 1.076 \text{ million dollars} \leftarrow \text{standard error of estimate}$$

Forming a
confidence interval
around Y

As was the case in simple regression, we can use the standard error of the estimate and the t distribution to form an approximate *confidence interval* around our estimated value, $\hat{Y}$. In the unpaid tax problem, for 4,300 field audit labor hours and 1,600 computer hours, our $\hat{Y}$ is $28,008,000 estimated unpaid taxes discovered, and our s_e is $1,076,000. If we want to construct a 95 percent confidence interval around this estimate of $28,008,000, we look in Appendix Table 2 under the 5 percent column until we locate the $n - 3$ = 10 − 3 = 7 degrees-of-freedom row. The appropriate t value for our interval estimate is 2.365. Therefore, we can calculate the limits of our confidence interval like this:

$$\hat{Y} + t(s_e) = 28{,}008{,}000 + (2.365)(1{,}076{,}000)$$
$$= 28{,}008{,}000 + 2{,}544{,}740$$
$$= 30{,}552{,}740 \leftarrow \text{upper limit}$$

and:

459

Section 6
MULTIPLE
REGRESSION AND
CORRELATION
ANALYSIS

$$\hat{Y} - t(s_e) = 28{,}008{,}000 - (2.365)(1{,}076{,}000)$$
$$= 28{,}008{,}000 - 2{,}544{,}740$$
$$= 25{,}463{,}260 \leftarrow \text{lower limit}$$

With a confidence level as high as 95 percent, the auditing division can feel certain that the actual discoveries will lie in this large interval from $25,463,260 to $30,552,740. If the IRS wishes to use a lower confidence level, such as 90 percent, it can narrow the range of values in estimating the unpaid taxes discovered. As was true with simple regression, we can use the standard normal distribution, Appendix Table 1, to approximate the t distribution whenever our degrees of freedom (n minus the number of estimated regression coefficients) are greater than 30.

Interpreting the confidence interval

The coefficient of multiple determination

In our discussion of simple correlation analysis, we measured the strength of the relation between two variables using the sample coefficient of determination, r^2. This coefficient of determination is the fraction of the total variation of the dependent variable Y that is explained by the estimating equation.

Meaning of the coefficient of determination

Similarly, in multiple correlation we shall measure the strength of the relationship among three variables using the *coefficient of multiple determination, R^2* or its square root, R (the multiple coefficient of correlation). This coefficient of multiple determination is also the fraction that represents the proportion of the total variation of Y that is "explained" by the regression plane. We can calculate the value of R^2 using Equation 11·22:

Using the coefficient of multiple determination in multiple correlation

$$R^2 = \frac{a\Sigma Y + b_1 \Sigma X_1 Y + b_2 \Sigma X_2 Y - n\overline{Y}^2}{\Sigma Y^2 - n\overline{Y}^2} \qquad [11\cdot22]$$

where:

$$Y = \text{values of the dependent variable}$$
$$X_1 \text{ and } X_2 = \text{values of the independent variables}$$
$$\overline{Y} = \text{mean of the values of the dependent variable}$$
$$n = \text{number of elements in the sample}$$
$$a, b_1, \text{ and } b_2 = \text{the regression coefficients determined through}$$
$$\text{the use of Equations } 11\cdot17, 11\cdot18 \text{ and } 11\cdot19$$

In Table 11·17, we have already calculated the necessary values for Equation 11·22 if, in addition, we remember that $a = -13.828$, $b_1 = .564$, and $b_2 = 1.099$. So now let's find R^2 for our IRS problem to determine how well the two independent variables describe the dependent variable:

Calculating R^2

$$R^2 = \frac{a\Sigma Y + b_1\Sigma X_1 Y + b_2\Sigma X_2 Y - n\bar{Y}^2}{\Sigma Y^2 - n\bar{Y}^2} \qquad [11 \cdot 22]$$

$$= \frac{(-13.828)(272) + (.564)(12,005) + (1.099)(4013) - (10)(27.2)^2}{7428 - (10)(27.2)^2}$$

$$= \frac{-3761.216 + 6770.82 + 4410.287 - (10)(739.84)}{7428 - (10)(739.84)}$$

$$= \frac{7419.891 - 7398.4}{7428 - 7398.4}$$

$$= \frac{21.491}{29.6}$$

$$= .726 \leftarrow \text{coefficient of multiple determination}$$

Interpreting
an R^2 of .726

In this case, $R^2 = .726$ indicates that the number of field audit labor hours, together with the number of computer hours, explains 72.6 percent of the total variation in unpaid taxes discovered for our sample. While this may be an improvement over whatever estimating equation was previously used, the auditing division may wish to add still another independent variable in an attempt to improve the accuracy of their estimate even more.

EXERCISES

38. Given the following set of data:
    a) Calculate the multiple regression line.
    b) Predict Y when $X_1 = 4.8$ and $X_2 = 4.0$.

Y	X_1	X_2
34	5.0	5.0
29	4.2	4.5
43	8.5	10.0
12	1.4	2.5
35	3.6	5.0
27	1.3	3.0

39. For the following set of data:
    a) Calculate the multiple regression line.
    b) Predict Y for $X_1 = 36$ and $X_2 = 16$

Y	X_1	X_2
8	10	8
36	37	21
23	18	14
27	29	11
14	14	9
12	21	4

40. Calculate the coefficient of multiple determination for problem 38 above.

41. Jean Barker, a reading specialist, wants to develop an estimating equation to predict reading achievement for first graders. She believes that achievement is related to both the perceptual-motor development of a child and the degree to which he or she is above average in mental age. She has compiled the following information from five students.

Student	X_1 Perceptual-motor task score	X_2 Years difference between mental and chronological ages	Y Reading achievement score
1	100	1.05	71
2	103	1.21	82
3	112	1.26	91
4	104	1.35	98
5	116	1.47	104

a) Calculate the least squares equation that best relates these three variables.

b) If a child's perceptual-motor score is 120 and her mental age is 1.56 years above her chronological age, what score would you expect her to obtain on the reading achievement test at the end of the year?

42. A developer of food for pigs would like to determine what relationship exists among the age of a pig when it starts receiving a newly developed food supplement, the initial weight of the pig at the same time, and the amount it gains in a one-week period with the food supplement. The following information is the result of a study of eight piglets.

Piglet number	X_1 Initial weight (lbs)	X_2 Initial age (weeks)	Y Weight gain
1	39	8	7
2	52	6	6
3	48	7	7
4	46	12	10
5	61	9	9
6	34	6	4
7	25	10	3
8	55	4	4

a) Calculate the least squares equation that best describes these three variables.

b) How much might we expect a pig to gain in a week with the food supplement if it was 9 weeks old and weighed 48 pounds?

c) Calculate the standard error of estimate for these data and construct an approximate 90 percent confidence interval around the estimate in part b.

43. To what does the term *multicollinearity* refer?

44. How is the coefficient of multiple determination interpreted?

7 THE COMPUTER AND MULTIPLE REGRESSION

In this chapter, we have examined general techniques for developing the regression equation, measuring the associated error, and calculating the degree

Limitation of simplified problems and small samples

of association between variables. To do this, we have presented simplified problems and samples of small sizes. By using small samples, however, we have minimized the accuracy of the regression equations for the sake of expedience. Furthermore, we have limited our discussion of multiple regression to only two independent variables in order to avoid the tedious calculations of more advanced multiple regressions when many independent variables are involved.

Making use
of the computer

As decision makers, we shall deal with complex problems that will require larger samples and additional independent variables. To assist us in solving these more detailed problems, we will make use of a computer, which allows us to perform a large number of computations in a very small period of time.

Besides enabling us to calculate multiple regression equations relating many variables, some computer programs also allow us to develop *curvilinear regressions* to describe nonlinear relationships. Consult the "program library" of your data-processing facility to learn which statistical programs are available to you.

Demonstration of
multiple regression
using the computer

To demonstrate how a computer handles multiple regression analysis, take our IRS problem from the previous section. Suppose the auditing division adds to their model the information concerning rewards to informants. The IRS wishes to include this third independent variable, X_3, because it feels certain that there is some relationship between these payments and the unpaid taxes discovered. Information for the last ten months is recorded in Table 11·18.

TABLE 11·18

**Factors related
to the discovery
of unpaid taxes**

Month	Field audit labor hours (00's omitted) X_1	Computer hours (00's omitted) X_2	Rewards to informants (00's omitted) X_3	Actual unpaid taxes discovered (000,000's omitted) Y
January	45	16	71	29
February	42	14	70	24
March	44	15	72	27
April	45	13	71	25
May	43	13	75	26
June	46	14	74	28
July	44	16	76	30
August	45	16	69	28
September	44	15	74	28
October	43	15	73	27

Using STATPACK
to solve
multiple regression
problems

To solve this problem, the auditing division has used a computer multiple regression program called STATPACK. This particular program tells the user how to enter (and correct) the data. Actually, typing with one finger is about all the skill you need to use the STATPACK program. Of course, interpreting the answer requires the same level of understanding that we used to explain the 72.6 percent answer we had from our hand-computed, two-variable multiple regression analysis. As you can see from examining the computer printout, the addition of a third independent variable (that is, rewards to informants) increases the total sample variation we are able to explain from 72.6 percent up to 98.3 percent.

STATPACK output

The computer output below contains the input data and solution to the IRS problem, using *three* independent variables. To help you understand the output format, illustrative comments are added in color.

RUN ★★★STATPACK

STATPACK   3:55   01/06/77

ARE YOU A STATPACK EXPERT
?★ NO

THE RESPONSE 'SOS' MAY BE ENTERED IN ORDER TO GAIN ADDITIONAL INFORMATION ABOUT THE RESPONSE NEEDED BY STATPACK. SOS MAY BE ENTERED ONLY IF THE QUESTION IS FOLLOWED BY THE CHARACTERS ?★

SPECIFY THE NAMES OF THE INPUT AND OUTPUT FILES(FORM:IN,OUT)
?★ ★,★ ◀——Indicates that all input and output will be on the computer terminal keyboard.

WHAT ANALYSIS DO YOU WISH TO PERFORM
?★ MULTIPLE REGRESSION

HOW MANY ROWS IN YOUR DATA MATRIX ◀——How many data points?
?★ 10

HOW MANY COLUMNS ◀——How many variables (dependent and independent)?
?★ 4

NOW, ENTER EACH ROW
?45,16,71,29
?42,14,70,24
?44,15,72,27
?45,13,71,25
?43,13,75,26
?46,14,74,23
?44,16,76,(23) ◀——Entered incorrectly; will be corrected below.
?45,16,69,23
?44,15,74,23
?43,15,73,27

DO YOU WISH TO PRINT THE DATA JUST READ IN
?* NO

DO YOU WISH TO CHANGE SOME VALUES
?* YES ←——— We need to correct input mistake above.

TYPE EDIT CODE
?* SOS

THE FOLLOWING CODES SIGNIFY TYPES OF EDIT FEATURES..
  0 -   NO MORE EDIT
  1 -   REPLACE AN INDIVIDUAL VALUE
  2 -   REPLACE AN ENTIRE ROW
  3 -   ADD A ROW
  4 -   DELETE A ROW
  5 -   SORT DATA (DESCENDING)
  6 -   SORT DATA (ASCENDING)   The incorrect value will be replaced.
?* 1 ←

TYPE ROW NUMBER, COLUMN NUMBER, AND NEW VALUE
?7,4,30 ←—— Corrected value for July actual unpaid taxes discovered.

TYPE EDIT CODE
?* 0 ←——— No more changes to be made.

DO YOU WISH TO PRINT THE DATA MATRIX
?* YES ←——— To double-check that everything has been properly entered.
       45.000   16.000   71.000   29.000
       42.000   14.000   70.000   24.000
       44.000   15.000   72.000   27.000
       45.000   13.000   71.000   25.000
       43.000   13.000   75.000   26.000
       46.000   14.000   74.000   23.000
       44.000   16.000   76.000   30.000
       45.000   16.000   69.000   23.000
       44.000   15.000   74.000   23.000
       43.000   15.000   73.000   27.000

SPECIFY THE DEPENDENT VARIABLE
?* 4 ←——— Unpaid taxes discovered Y.

HOW MANY INDEPENDENT VARIABLES
?* 3

SPECIFY THESE VARIABLES
?1,2,3 ←——— Field audit labor hours, computer hours, and rewards to informants: X_1, X_2, and X_3.

464

VARIABLE	REG.COEF.	STD.ERROR COEF.
1	0.59697	0.03113
2	1.17634	0.03407
3	0.40511	0.04223

INTERCEPT	-45.79634
MULTIPLE CORRELATION	0.99167
STD. ERROR OF ESTIMATE	0.23613

(R^2 would equal $.99167^2$, or .983; this means that 98.3% of the total variation in unpaid taxes discovered is explained by the three independent variables used.)

8 TERMS INTRODUCED IN CHAPTER 11

• **COEFFICIENT OF CORRELATION** The square root of the coefficient of determination. Its sign indicates the direction of the relationship between two variables, direct or inverse.

• **COEFFICIENT OF DETERMINATION** A measure of the proportion of variation in Y, the dependent variable, that is explained by the regression line; i.e., by Y's relationship with the independent variable.

• **COEFFICIENT OF MULTIPLE CORRELATION, R** The square root of the coefficient of multiple determination.

• **COEFFICIENT OF MULTIPLE DETERMINATION, R^2** A measure of the proportion of the total variation in the dependent variable that is explained by the regression plane.

• **CORRELATION ANALYSIS** A technique to determine the degree to which variables are linearly related.

• **CURVILINEAR RELATIONSHIP** An association between two variables which is described by a curved line.

• **DEPENDENT VARIABLE** The variable we are trying to predict in regression analysis.

• **DIRECT RELATIONSHIP** A relationship between two variables such that, as the independent variable's value increases, so does the value of the dependent variable.

• **ESTIMATING EQUATION** A mathematical formula that relates the known variables to the unknown variable in regression analysis.

• **INDEPENDENT VARIABLES** The known variable, or variables, in regression analysis.

• **INVERSE RELATIONSHIP** A relationship between two variables such that, as the independent variable increases, the dependent variable decreases.

• **LEAST SQUARES METHOD** A technique for fitting a straight line through a set of points in such a way that the sum of the squared vertical distances from the n points to the line is minimized.

• **LINEAR RELATIONSHIP** A particular type of association between two variables that can be described mathematically by a straight line.

• **MULTICOLLINEARITY** A statistical problem sometimes present in multiple regression analysis in which the reliability of the regression coefficients is reduced, owing to a high level of correlation between the independent variables.

• **MULTIPLE REGRESSION** The statistical process by which several variables are used to predict another variable.

• **REGRESSION** The general process of predicting one variable from another by statistical means, using previous data.

• **REGRESSION LINE** A line fitted to a set of data points to estimate the relationship between two variables.

• **SCATTER DIAGRAM** A graph of points on a rectangular grid; the x- and y-coordinates of each point correspond to the two measurements made on some particular sample element, and the pattern of points illustrate the relationship between the two variables.

- **SLOPE** A constant for any given straight line, the value of which represents how much each change of the independent variable changes the dependent variable.
- **STANDARD ERROR OF ESTIMATE** A measure of the reliability of the estimating equation, indicating the variability of the observed points around the regression line; i.e., the extent to which observed values differ from their predicted values on the regression line.
- **STANDARD ERROR OF THE REGRESSION COEFFICIENT** A measure of the variability of sample regression coefficients around the true population regression coefficient.
- **Y-INTERCEPT** A constant for any given straight line, whose value represents the predicted value of the y-variable when the x-variable has a value of 0.

9 EQUATIONS INTRODUCED IN CHAPTER 11

p. 415:

$$Y = a + bX \qquad [11 \cdot 1]$$

This is the equation for *a straight line* where the dependent variable Y is "determined" by the independent variable X. The a is called the *Y-intercept* because its value is the point at which the line crosses the Y-axis (the vertical axis). The b is the *slope* of the line; that is, it tells how much each change of the independent variable X changes the dependent variable Y. Both a and b are numerical constants, since for any given straight line their value does not change.

p. 416:

$$b = \frac{Y_2 - Y_1}{X_2 - X_1} \qquad [11 \cdot 2]$$

To calculate the numerical constant b for any given line, find the value of the coordinates, X and Y, for two points that lie on the line. The coordinates of the first point are (X_1, Y_1) and the second point (X_2, Y_2). Remember that b is the slope of the line.

p. 418:

$$\hat{Y} = a + bX \qquad [11 \cdot 3]$$

In regression analysis, $\hat{Y}$ *(Y-hat)* symbolizes the individual Y values of the *estimated* points; that is, those points that lie on the estimating line. Accordingly, Equation 11·3 is the equation for the estimating line.

p. 421:

$$b = \frac{\Sigma XY - n\overline{X}\,\overline{Y}}{\Sigma X^2 - n\overline{X}^2} \qquad [11 \cdot 4]$$

The equation enables us to calculate the *slope of the best-fitting regression line* for any two-variable set of data points. We introduce two new symbols in this equation, $\bar{X}$ and $\bar{Y}$, which represent the means of the values of the independent variable and the dependent variable, respectively. In addition, this equation contains n, which, in this case, represents the number of data points with which we are fitting the regression line.

p. 422:
$$a = \bar{Y} - b\bar{X}$$
[11·5]

Using this formula, we can compute the *Y-intercept of the best-fitting regression line* for any two-variable set of data points.

p. 427:
$$s_e = \sqrt{\frac{\Sigma(Y - \hat{Y})^2}{n - 2}}$$
[11·6]

The *standard error of estimate, s_e,* measures the variability or scatter of the observed values around the regression line. In effect, it indicates the reliability of the estimating equation. The denominator is $n - 2$ because we lose two degrees of freedom (for the values a and b) in estimating the regression line.

p. 428:
$$s_e = \sqrt{\frac{\Sigma Y^2 - a\Sigma Y - b\Sigma XY}{n - 2}}$$
[11·7]

Since Equation 11·6 requires tedious calculations, statisticians have devised this *shortcut method for finding the standard error of estimate.* In calculating the values for b and a, we already calculated every quantity in Equation 11·7 except ΣY^2, which we can do easily.

p. 435:
Variation of the Y values around the regression line $= \Sigma(Y - \hat{Y})^2$
[11·8]

The variation of the Y values in a data set around the fitted regression line is one of two quantities from which the sample coefficient of determination is developed. Equation 11·8 shows how to measure this particular dispersion, which is the *unexplained* portion of the total variation.

p. 435:
Variation of the Y values around their own mean $= \Sigma(Y - \bar{Y})^2$
[11·9]

This formula measures the *total variation* of a whole set of Y values; that is, the dispersion of these Y values around their own mean.

p. 436:
$$r^2 = 1 - \frac{\Sigma(Y - \hat{Y})^2}{\Sigma(Y - \bar{Y})^2}$$
[11·10]

The *sample coefficient of determination, r^2,* gives the fraction of the total variation of Y that is explained by the regression line. It is an important measure of the degree of association between X and Y. If the value of r^2 is $+1$, then the regression line is a perfect estimator. If $r^2 = 0$, there is no correlation between X and Y.

p. 441:

$$r^2 = \frac{a\Sigma Y + b\Sigma XY - n\overline{Y}^2}{\Sigma Y^2 - n\overline{Y}^2}$$

[11·11]

This is a shortcut equation for calculating r^2.

p. 443:

$$r = \sqrt{r^2}$$

[11·12]

The *sample coefficient of correlation* is denoted by r and is found by taking the square root of the sample coefficient of determination. It is a second measure (in addition to r^2) we can use to describe how well one variable is explained by another. The sign of r indicates the direction of the relationship between the two variables X and Y.

p. 444:

$$Y = A + BX$$

[11·13]

Each *population regression line* is of the form in Equation 11·13, where A is the Y-intercept for the population, and B is the slope.

p. 445:

$$Y = A + BX + e$$

[11·13a]

Because all the individual points in a population do not lie on the population regression line, the *individual* data points will satisfy Formula 11·13a, where e is a random disturbance from the population regression line. On the average, e equals zero because disturbances above the population regression line are canceled out by disturbances below it.

p. 446:

$$s_b = \frac{s_e}{\sqrt{\Sigma X^2 - n\overline{X}^2}}$$

[11·14]

When we are dealing with a population, we can use this formula to find the *standard error of the regression coefficient, b*.

p. 446:

Upper limit of acceptance region $B + t(s_b)$
Lower limit of acceptance region $B - t(s_b)$

[11·15]

Once we have calculated s_b using Equation 11·14, we can determine the upper and lower limits of the acceptance region for a hypothesis test using this pair of equations.

p. 453:

$$\hat{Y} = a + b_1 X_1 + b_2 X_2$$

[11·16]

In multiple regression, this is the formula for the estimating equation that describes the relationship between three variables: Y, X_1, and X_2. Picture a two-variable multiple regression equation as a plane, rather than a line.

p. 454:

$$\Sigma Y = na + b_1\Sigma X_1 + b_2\Sigma X_2$$ [11·17]

$$\Sigma X_1 Y = a\Sigma X_1 + b_1\Sigma X_1{}^2 + b_2\Sigma X_1 X_2$$ [11·18]

[11·19]

$$\Sigma X_2 Y = a\Sigma X_2 + b_1\Sigma X_1 X_2 + b_2\Sigma X_2{}^2$$

Solving these three equations determines the values of the numerical constants a, b_1, and b_2 and thus the best-fitting multiple regression plane.

p. 457:

$$s_e = \sqrt{\frac{\Sigma(Y - \hat{Y})^2}{n - 3}}$$

[11·20]

To measure the variation around a multiple regression plane, use this equation to find the *standard error of estimate.* The standard error, in this case, has $n - 3$ degrees of freedom, owing to the three numerical constants that must be calculated from the data first (a, b_1, and b_2).

p. 458:

$$s_e = \sqrt{\frac{\Sigma Y^2 - a\Sigma Y - b_1\Sigma X_1 Y - b_2\Sigma X_2 Y}{n - 3}}$$

[11·21]

Equation 11·21 is a *shortcut formula for finding the standard error of estimate* in multiple regression problems.

p. 459:

$$R^2 = \frac{a\Sigma Y + b_1\Sigma X_1 Y + b_2\Sigma X_2 Y - n\bar{Y}^2}{\Sigma Y^2 - n\bar{Y}^2}$$

[11·22]

In multiple correlation, we can measure the strength of the relationship among three variables using the *coefficient of multiple determination,* or R^2. This coefficient of determination is the fraction that represents the proportion of the total variation of Y that is "explained" by the regression plane.

10 CHAPTER REVIEW EXERCISES

45. Melinda Wilde, an HEW economist, speculates about the relationship between a family's income and its expenditures for food. The following table presents the results of a survey of 8 randomly selected families:

Income ($\times$ $1,000)	8	12	9	24	13	37	19	16
Percent spent for food	36	25	33	15	28	19	20	22

   a) Develop an estimating equation that best describes these data.

   b) Calculate the standard error of estimate, s_e, for this relationship.

   c) Find an approximate 90 percent confidence interval for the percent of income spent on food by a family earning $25,000 annually.

46. For the following table of observed and predicted values of Y, compute the sample coefficient of determination and sample correlation coefficient.

Y	55	64	54	63	68	70	76	66	75	74
$\hat{Y}$	55.5	59.5	60.5	63.5	67.5	65.5	73.5	70.5	72.5	76.5

47. (Fill in the blanks.) Regression and correlation analysis deal with the _____ between variables. Regression analysis, through _____ equations, enables us to _____ an unknown variable from a set of known variables. The unknown variable is called the _____ variable; known variables are referred to as _____ variables. The correlation between two variables indicates the _____ of the linear relationship between

them and thus gives an idea of how well the _____ _____ in regression describes the relationship between the variables.

48. Calculate the sample coefficient of determination and sample correlation coefficient for problem 13, p. 433.

49. A University of Southern California sociologist, working on the link between unemployment and crime, surveyed twelve cities. Following are the number of robberies and assaults per 1,000 residents and the corresponding unemployment rate for each city.

Percent unemployment	7.3	6.4	6.2	5.5	6.4	4.7	5.8	7.9	6.7	9.6	10.3	7.2
Number of robberies and assaults per 1,000 residents	64	53	42	29	71	26	32	68	53	64	85	73

   a) Develop the best-fitting estimating equation that describes these data.

   b) Calculate the standard error of estimate for this relationship.

   c) The city of San Marino is currently suffering from 8.6 percent unemployment. Find a 90 percent confidence interval for San Marino's assault and robbery rate.

50. For each of the following pairs of plots, state which has a higher value of r, the correlation coefficient, and what the sign of r is:

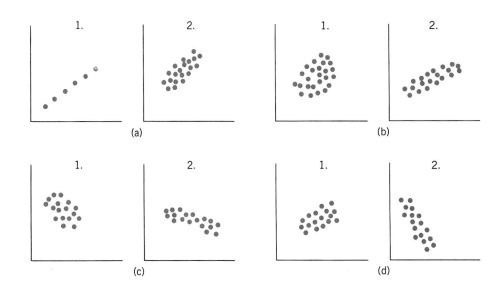

51. The manager of the Durham, N.C., water purification plant has compiled the data shown below to determine whether or not water usage has changed. These pairs of data are the volumes of water consumed and the corresponding number of households serviced for six recent months. Previous studies indicated that the relationship describing these two variables has a slope of 13. At the .10 level of significance, has the slope of this relationship increased?

Number of households (1,000s)	8.1	7.8	8.4	7.6	8.0	8.1
Water consumed (10 million liters)	94	83	97	85	89	92

52. Calculate the sample coefficient of determination and the sample correlation coefficient for problem 12 on page 433.

53. We should not extrapolate to predict values outside the range of the data used in construct-ing the regression line. The reason (choose one):

a) The relationship between the variables may not be the same for different values of the variables.

b) The independent variable may not have the causal effect on the dependent variable for these values.

c) The variables' values may change over time.

d) There may be no common bond to explain the relationship.

54. A Milwaukee survey of 10 city human resource agencies indicates that the relationship between the amount budgeted for job placement for the unemployed (in thousands of dollars) and the number of people successfully employed through this program has a slope of 1.5 and a standard error of the regression coefficient of .35. Does this information con-tradict, at the .10 level of significance, the agency's claim that 60,000 more jobs would be produced by an additional allocation of $25 million?

55. Unlike the coefficient of determination, the coefficient of correlation (choose one):

a) Indicates whether the slope of the regression line is positive or negative.

b) Measures the strength of association between the two variables more exactly.

c) Can never have an absolute value greater than 1.

d) Measures the percentage of variance explained by the regression line.

56. A USDA research team has investigated the relationship between the corn harvest and the average temperature during the growing season. Measurements over a period of years produced the following results:

Average temperature (°C)	19	23	25	24	26	21
Bushels per acre	66	74	72	76	78	72

a) Plot these data.

b) Develop the estimating equation that best describes these data.

57. A college math professor needs information on the relationship between a student's score on a mathematics aptitude test, his high school math preparation, and his success in the introductory calculus course she teaches. She asks students what math courses they have had in high school and weights them in a routine way, which yields an index between 2 and 12. The measure of course success is based on a point system of assignments and tests with the maximum value possible of 50. The following information was compiled for 6 randomly selected students.

Student	X_1 Math aptitude test score	X_2 High school preparation	Y Total points at end of course
A	84	7	36
B	74	5	28
C	89	8	39
D	78	7	30
E	92	10	45
F	70	3	22

a) Develop the estimating equation that best describes these data.

b) If a student scored 83 on the aptitude test and had a math preparation index score of 7, how many points might we expect for this person?

58. A horticulturist wants to determine the relationship that exists among the relative humidity in a greenhouse, the temperature in the greenhouse, and the rate of growth of plants of a certain variety. The following data were collected from 5 randomly selected greenhouses.

X_1	X_2	Y
Relative humidity (%)	Temperature (°C)	Growth rate (cm./wk.)
20	11	7
40	16	26
35	19	13
30	12	5
50	26	33

a) Develop the multiple regression equation that best describes these data.

b) If a greenhouse had a 30 percent relative humidity reading and a temperature of 15°C, what rate of growth might we expect plants of this type to show?

59. A team of 25 evaluators rated all ten of the high schools in the city last year. Each evaluator's ratings of several areas were averaged to yield a below average, average, good, or excellent rating. An administrator inquired about the relationship, if any, between these ratings and the dropout rates at the schools. The following table gives the number of good and excellent ratings for each school and the number of students who dropped out last year.

Number of good and excellent ratings	11	19	7	12	13	10	16	22	14	12
Number of dropouts	6	4	8	3	7	8	3	2	5	6

a) Plot these data.

b) Develop the estimating equation that best describes these data.

c) Predict the number of dropouts we might expect for a school if it had received 15 good or excellent ratings.

60. Sociologists studying various problems with large cities have been attracted by several relationships between a city's population and certain social problems. In one particular instance, they decided to determine what relationship might exist between the population of a city and its violent crimes. The following are the populations of several cities (in thousands) and the corresponding number of violent crimes per 1,000 residents for last year.

Population (thousands)	27	18	44	36	12	23	22
Number of violent crimes (per 1,000 people)	32	12	46	30	24	27	18

a) Develop an estimating equation that best describes these data.

b) Calculate the standard error of estimate for this relationship.

c) If a city had a population of 30,000, predict (with 95 percent confidence) the expected number of violent crimes per thousand population for the city.

61. Calculate the coefficient of multiple determination for problem 39, p. 460.

62. In a CAB study of airline operations, a survey of 12 companies disclosed that the relationship between the number of pilots employed and the number of planes in service has a slope of 3.5. Previous studies indicated that the slope of this relationship was 4.0. If the standard error of the regression coefficient has been calculated to be .20, is there reason to believe that, at the .01 level of significance, the true slope has changed?

472

Answer true *or* false. *Answers are in the back of the book.*

1. Regression analysis is used to describe how well an estimating equation describes the relationship being studied.

2. Given that the equation for a line is $Y = 26 - 24X$, we may say the relationship of Y to X is direct linear.

3. An r^2 value close to 0 indicates a strong correlation between X and Y.

4. Regression and correlation analysis are used to determine cause and effect relationships.

5. The sample coefficient of correlation, r, is nothing more than $\sqrt{r^2}$, and we cannot interpret its meaning directly as a percentage of some kind.

6. The standard error of estimate measures the variability of the observed values around the regression equation.

7. In multiple regression analysis, there could exist a high degree of correlation between the independent variables. This is a problem known as multicollinearity.

8. We may interpret the sample coefficient of determination as the amount of the variation in Y that is explained by the regression line.

9. Lines drawn on either side of the regression line at ± 1, ± 2 and ± 3 times the value of the standard error of estimate are called confidence lines.

10. The estimating equation is valid over only the same range as that given by the original sample data upon which it was developed.

Nonparametric methods

1. Introduction to nonparametric statistics, 476

2. The sign test for paired data, 478

3. A rank sum test: the Mann-Whitney *U* test, 484

4. One-sample runs tests, 490

5. Rank correlation, 496

6. Terms introduced, 506

7. Equations introduced, 506

8. Chapter review exercises, 508

9. Chapter concepts test, 512

474

Although the effect of air pollution on health is a complex problem, an international organization has decided to make a preliminary investigation of (1) average year-round quality of air and (2) the incidence of pulmonary-related diseases. A preliminary study ranked 11 of the world's major cities from 1 (worst) to 11 (best) in these two variables.

City	A	B	C	D	E	F	G	H	I	J	K
Air quality rank	4	7	9	1	2	10	3	5	6	8	11
Pulmonary disease rank	5	4	7	3	1	11	2	10	8	6	9

The health organization's data are different from any we have seen so far in this book: they do not give us the *variable* used to determine these ranks. (We don't know if the rank of pulmonary disease is a result of pneumonia, emphysema, or other illnesses per 100,000 population.) Nor do we know the *values* (whether City D has twice as much pollution as City K or 20 times as much).

If we knew the variables and their values, we could use the regression techniques of Chapter 11. Unfortunately, that is not the case; but even without any knowledge of either variables or values, we can use the techniques in this chapter to provide the health organization with help on its problem.

The majority of hypothesis tests discussed so far have made inferences about population *parameters*, such as the mean and the proportion. These parametric tests have used the parametric statistics of samples that came from the population being tested. To formulate these tests, we made restrictive assumptions about the populations from which we drew our samples. In each case in Chapter 9, for example, we assumed that our samples either were large or came from *normally distributed* populations. But populations are not always normal. And even if a goodness-of-fit test (Chapter 10) indicates that a population *is* approximately normal, we cannot always be sure we're right, because the test is not 100 percent reliable. Clearly, there are certain situations in which the use of the normal curve is not appropriate. For these cases, we need alternatives to the parametric statistics and the specific hypothesis tests we've been using so far.

1 INTRODUCTION TO NONPARAMETRIC STATISTICS

Fortunately, in recent times statisticians have developed useful techniques that do not make restrictive assumptions about the shape of population distributions. **These are known as *distribution-free* or, more commonly, *nonparametric* tests.** The hypothesis of a nonparametric test is concerned with something other than the value of a population parameter. A large number of these tests exist, but this chapter will examine only a few of the better known and more widely used ones:

1. The sign test for paired data, where positive or negative signs are substituted for quantitative values.
2. A rank sum test, often called the Mann-Whitney U Test, which can be used to determine whether two independent samples have been drawn from the same population. It uses more information than the sign test.
3. The one-sample runs test, a method for determining the randomness with which sampled items have been selected.
4. Rank correlation, a method for doing correlation analysis when the data are not available to use in numerical form, but when information is sufficient to rank the data first, second, third, and so forth.

Advantages of nonparametric methods

Nonparametric methods have a number of clear advantages over parametric methods:

1. They do not require us to make the assumption that a population is distributed in the shape of a normal curve or another specific shape.

2. Generally, they are easier to do and to understand. Most nonparametric tests do not demand the kind of laborious computations often required, for example, to calculate a standard deviation. A nonparametric test may ask us to replace numerical values with the order in which those values occur in a list, as has been done in Table 12·1. Obviously, dealing computationally with 1, 2, 3, 4, and 5 takes less effort than working with 13.33, 76.50, 101.79, 113.45, and 189.42.

Parametric value	113.45	189.42	76.50	13.33	101.79	TABLE 12·1 **Converting parametric values to nonparametric ranks**
Nonparametric rank	4	5	2	1	3	

3. Sometimes even formal ordering or ranking is not required. Often, all we can do is describe one outcome as "better" than another. When this is the case, or when our measurements are not as accurate as is necessary for parametric tests, we can use nonparametric methods.

Disadvantages of nonparametric methods

Two disadvantages accompany the use of nonparametric tests:

1. They ignore a certain amount of information. We have demonstrated how the values 1, 2, 3, 4, and 5 can replace the numbers 13.33, 76.50, 101.79, 113.45, and 189.42. Yet if we represent "189.42" by "5," we lose information that is contained in the value 189.42. Notice that in our ordering of the values 13.33, 76.50, 101.79, 113.45, and 189.42, the value 189.42 can become 1,189.42 and still be the fifth, or largest, value in the list. But if this list is a data set, we can learn more knowing that the highest value is 1,189.42 instead of 189.42 than we can by representing both of these numbers by the value 5.

2. They are often not as efficient or "sharp" as parametric tests. The estimate of an interval at the 95 percent confidence level using a nonparametric test may be twice as large as the estimate using a parametric test such as those in Chapter 8. When we use nonparametric tests, we make a trade-off: we lose sharpness in estimating intervals, but we gain the ability to use less information and to calculate faster.

EXERCISES

1. What is the difference between the kinds of questions answered by parametric tests and those answered by nonparametric tests?

2. The null hypothesis most often examined in nonparametric tests

   a) includes specification of a population's parameters.

   b) is used to evaluate some general population aspect.

   c) is very similar to that used in regression analysis.

   d) simultaneously tests more than two population parameters.

3. What are the major advantages of nonparametric methods over parametric methods?

4. What are the primary shortcomings of nonparametric tests?

5. Dr. Geof Sher is an internist with a large private practice. To make the best use of his office hours, his receptionist schedules appointments according to a precise schedule. There is no five-minute period unaccounted for, including telephone calls. Unfortunately, she has been underestimating the amount of time appointments will take and scheduling too many patients, resulting in long waits in the doctor's office. Though waiting periods may be short in the morning, as the day progresses and the doctor gets further and further behind, the

waits become longer. In assessing the problem, should the doctor assume that the successive waiting times are independent and normally distributed?

6. The County Mental Hospital is planning to paint the interior of the hospital this spring. Since certain colors and shades are reputed to be more soothing and relaxing than others, the board of directors has been considering different qualities of hue, saturation, and lightness. Samples of a broad range of combinations were labeled and distributed among patients and employees, whose preferences were recorded. The results follow:

Rank	Hue-saturation-lightness combination	Number of preferences
1	15	52
2	5	49
3	14	39
4	4	38
5	6	37
6	16	36
7	7	32
8	8	29
9	13	26
10	3	25
11	17	24
12	18	18
13	13	15
14	2	15
15	9	14
16	1	10
17	11	10
18	19	10
19	7	9

Will the directors sacrifice any real information by using the ranking test as its decision criterion? (Hint: you might graph the data.)

2 THE SIGN TEST FOR PAIRED DATA

One of the easiest nonparametric tests to use is the sign test. Its name comes from the fact that it is based on the direction (or the signs for pluses or minuses) of a pair of observations and not on their numerical magnitude.

Use sign test for paired data

Consider the result of a test panel of 40 college juniors evaluating the effectiveness of two types of classes: large lectures by full professors or small sections by graduate assistants. Table 12·2 lists the responses to this

TABLE 12·2
Evaluation by 40 students of 2 types of classes

Panel member number	1	2	3	4	5	6	7	8	9	10	11	12	13	14	15	16
Score for large lectures (1)	2	1	4	4	3	3	4	2	4	1	3	3	4	4	4	1
Score for small sections (2)	3	2	2	3	4	2	2	1	3	1	2	3	4	4	3	2
Sign of score (1) minus score (2)	−	−	+	+	−	+	+	+	+	0	+	0	0	0	+	−

478

request: "Indicate how you rate the effectiveness in transmitting knowledge of these two types of classes by giving them a number from 4 to 1. A rating of 4 is excellent, and 1 is poor." In this case, the sign test can help us hypothesize about whether students feel there is a difference between the effectiveness of the two types of classes.

Converting values to signs

We can begin, as we have in Table 12·2, by converting the evaluations of the two teaching methods into a sign. Here, a plus sign means the student prefers large lectures; a minus sign indicates a preference for small sections; and a zero represents a tie (no preference). If we count the bottom row of Table 12·2, we get these results:

Number of + signs	19
Number of − signs	11
Number of 0's	10
Total sample size	**40**

Stating the hypothesis

We are using the sign test to determine whether our panel can discern a real difference between the two types of classes. Since we are testing perceived differences, we shall exclude tie evaluations (0's). We can see that we have 19 plus signs and 11 minus signs, for a total of 30 usable responses. If there is no difference between the two types of classes p (the probability that the first score exceeds the second score) would be .5, and we would expect to get about 15 plus signs and 15 minus signs. We would set up our hypotheses like this:

Finding the sample size

$H_0: p = .5 \leftarrow$ null hypothesis: there is no difference
between the 2 types of classes
$H_1: p \neq .5 \leftarrow$ alternative hypothesis: there is a
difference between the 2 types of classes

If you look carefully at the hypotheses, you will see that the situation is similar to the fair coin toss that we discussed in Chapter 5. If we tossed a fair coin 30 times, p would be .5, and we would expect about 15 heads and 15 tails. In that case, we would use the binomial distribution as the appropriate sampling distribution. You may also remember that when the sample size n is at least 30 and np and nq are each at least 5, we can use the normal distribution to approximate the binomial. This is just the case with the

Choosing the distribution

17	18	19	20	21	22	23	24	25	26	27	28	29	30	31	32	33	34	35	36	37	38	39	40
1	2	2	4	4	4	4	3	3	2	3	4	3	4	3	1	4	3	2	2	2	1	3	3
3	2	3	3	1	4	3	3	2	2	1	1	1	3	2	2	4	4	3	3	1	1	4	2
−	0	−	+	+	0	+	0	+	0	+	+	+	+	+	−	0	−	−	−	+	0	−	+

results from our panel of college juniors. Thus, we can apply the normal distribution to our test of the two teaching methods.

$p_{H_0} = .5$ ← hypothesized proportion of the population who feel that both types of classes are the same

$q_{H_0} = .5$ ← hypothesized proportion of the population who feel the 2 types of classes are different ($q_{H_0} = 1 - p_{H_0}$)

$n = 30$ ← sample size

$\bar{p} = .633$ ← proportion of successes in the sample (19/30)

$\bar{q} = .367$ ← proportion of failures in the sample (11/30)

Testing a hypothesis of no difference

Calculating the standard error

Suppose the chancellor's office wants to test the hypothesis that there is no difference between student perception of the two types of classes at the .05 level of significance. We shall conduct this test using the methods we introduced in Chapter 9. The first step is to calculate the standard error of the proportion:

$$\sigma_{\bar{p}} = \sqrt{\frac{pq}{n}} \qquad [8\cdot4]$$

$$= \sqrt{\frac{(.5)(.5)}{30}}$$

$$= \sqrt{.00833}$$

$$= .091 \leftarrow \text{standard error of the proportion}$$

Illustrating the test

Since we want to know whether the true proportion is larger *or* smaller than the hypothesized proportion, this is a two-tailed test. Figure 12·1 illustrates this hypothesis test graphically. The two colored regions represent the .05 level of significance.

Because we are using the normal distribution in our test, we can determine from Appendix Table 1 that the z value for .475 of the area under the

FIG. 12·1
Two-tailed hypothesis test of a proportion at the .05 level of significance

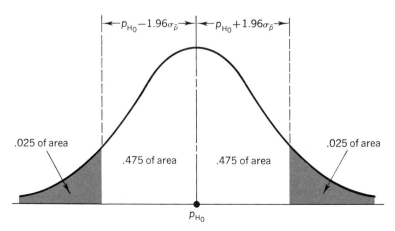

.025 of area

.475 of area   .475 of area

.025 of area

$p_{H_0} - 1.96\sigma_{\bar{p}}$ $p_{H_0} + 1.96\sigma_{\bar{p}}$

p_{H_0}

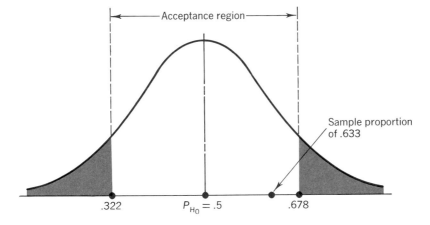

FIG. 12·2
Two-tailed
hypothesis test at
the .05 level of
significance,
illustrating the
acceptance region
and the sample
proportion

Acceptance region

Sample proportion
of .633

.322 $P_{H_0} = .5$ .678

curve is 1.96. Thus, we can calculate the limits of the acceptance region for the null hypothesis as follows:

$$p_{H_0} + 1.96\sigma_{\bar{p}} = .5 + (1.96)(.091)$$
$$= .5 + .178$$
$$= .678 \leftarrow \text{upper limit}$$

and:

$$p_{H_0} - 1.96\sigma_{\bar{p}} = .5 - (1.96)(.091)$$
$$= .5 - .178$$
$$= .322 \leftarrow \text{lower limit}$$

Figure 12·2 illustrates these two limits of the acceptance region, .322 and .678, and the sample proportion, .633. We can see that the sample proportion falls within the acceptance region for this hypothesis test. Therefore, the chancellor should accept the null hypothesis that students perceive no difference between the two types of classes.

A sign test such as this is quite simple to do and applies to both one-tailed and two-tailed tests. It is usually based on the binomial distribution. Remember, however, that we were able to use the normal approximation to the binomial as our sampling distribution because the sample size was large enough and np and nq were both greater than 5. If these conditions are not met, we must use the binomial instead.

EXERCISES

7. Compare the two sets of therapist's ratings below to see if shock treatments helped the patients who received them. A lower rating indicates a better mental state. Ratings were made before and after the treatments. Use the .05 level of significance.

Before	8	7	6	9	11	10	8	6	5	8	9	10
After	6	5	8	6	9	8	10	7	4	6	9	10

8. Use the sign test to see if there is a difference between the behavior scores given to children before and after medication for hyperactivity. Use the .10 significance level.

| Before | 32 | 35 | 33 | 36 | 44 | 41 | 32 | 39 | 31 | 47 | 30 | 29 |
| After | 36 | 37 | 34 | 40 | 40 | 42 | 40 | 42 | 33 | 46 | 29 | 35 |

9. Different methods of teaching reading have been popular with different educators for several years. Teachers at an elementary school decided to compare two of the more common methods over a three-months' period. In two first-grade classes, children were matched on similarity of intelligence and background. Teachers of the classes used the two methods; and at the end of the three-months' period, each child was given a reading achievement test. From the data, the teachers hoped to draw some conclusions about which method worked better, at least for children with backgrounds like those at this school. The scores of the 34 children in each class on the reading achievement test are presented below.

Method I	Method II	Method I	Method II
79	76	75	76
84	82	90	92
92	92	79	74
83	81	83	81
75	76	85	80
89	88	86	79
76	74	94	94
94	97	82	81
94	92	91	94
82	80	84	83
70	73	88	89
81	81	98	96
88	86	79	80
77	75	88	85
86	84	93	90
90	86	78	78
85	86	94	92

At a 5 percent significance level, does Method II surpass Method I in leading to better scores on the reading achievement test?

10. Because of the severity of recent winters, there has been talk of the earth slowly progressing toward another ice age. Some scientists hold different views, however, because the summers have brought extreme temperatures as well. One scientist suggested looking at the mean temperature for each month to see if it was lower than in the previous year. Another meteorologist at the government weather service argued that perhaps they should look as well at temperatures in the spring and fall months of the last two years, so that their conclusions would be based on other than extreme temperatures. In this way, he said, they could detect whether there appeared to be a general warming or cooling trend or just extreme temperatures in the summer and winter months. So 15 dates in the spring and fall were randomly selected, and the temperatures in the last two years were noted for a particular location with generally moderate temperatures. Following are the dates and corresponding temperatures for 1978 and 1979.

a) Is the meteorologist's reasoning as to the method of evaluation sound? Explain.

b) Using a sign test, determine whether the meteorologist can be 95 percent sure that 1979 was cooler than 1978, based on these data.

	Temperature (Fahrenheit)					
Date	1978	1979	Date	1978	1979	
Mar. 29	46°	44°	Oct. 12	58°	56°	
Apr. 4	51	53	May 31	73	75	
Apr. 13	48	48	Sept. 28	68	66	
May 22	67	64	June 5	74	71	
Oct. 1	55	52	June 17	80	82	
Mar. 23	46	50	Oct. 5	64	64	
Nov. 12	49	49	Nov. 28	53	49	
Sept. 30	63	60				

11. With the concern over radiation exposure and its relationship to incidence of cancer, city environmental specialists keep a close eye on the types of industry coming into the area and the degree to which they employ radiation in their production. An index of exposure to radioactive contamination has been developed and is used daily to determine if the levels are increasing or are higher under certain atmospheric conditions.

Environmentalists claim that radioactive contamination has increased in the last year because of new industry in the city. City administrators, however, claim that new, more stringent regulations on industry in the area have made levels lower than last year, even with new industry using radiation. To test their claim, records for 10 randomly selected days of the year have been checked, and the index of exposure to radioactive contamination has been noted. The following results were obtained.

Index of radiation exposure

1978	1.402	1.401	1.400	1.404	1.395	1.402	1.406	1.401	1.404	1.406
1979	1.440	1.395	1.398	1.404	1.393	1.400	1.401	1.402	1.400	1.403

Can the administrators be 92 percent sure that the levels of radioactive contamination have changed—or more specifically, that they have been reduced?

12. As part of the recent interest in population growth and the sizes of families, a population researcher examined a number of hypotheses concerning the family size that various people look upon as ideal. She suspected that variables of race, sex, age, and background might account for some of the different views. In one pilot sample, the researcher tested the hypothesis that women today think of an ideal family as being smaller than the ideal held by their mothers. She asked each of the participants in the pilot study to state the number of children she would choose to have or that she considered ideal. Responses were anonymous, to guard against the possibility that people would feel obligated to give a socially desirable answer. In addition, people of different backgrounds were included in the sample. Below are the responses of the mother-daughter pairs.

	Ideal family size												
Sample pair	A	B	C	D	E	F	G	H	I	J	K	L	M
Daughter	3	4	2	1	5	4	2	2	3	3	1	4	2
Mother	4	4	4	3	5	3	3	5	3	2	2	3	1

a) Can the researcher be 97 percent sure that the mothers and daughters do not have essentially the same ideal of family size?

b) Ignore momentarily the fact that n is not large enough for us to use the normal distribution to approximate the binomial. Determine if the researcher could conclude that the mothers do not have essentially the same family size preferences as their daughters by using the normal approximation to the binomial.

c) Assume that for each pair listed there were 10 more pairs who responded in an identical manner. Calculate the range of the proportion for which the researcher would conclude that there is no difference in the mothers and daughters. Is your conclusion changed?

d) Explain any differences in conclusions obtained in parts a, b, and c.

3 A RANK SUM TEST: THE MANN-WHITNEY U TEST

Mann-Whitney *U* test introduced

Rank sum tests are a whole family of tests. We shall concentrate on one member of this family, the Mann-Whitney *U* test, which will enable us to determine whether two independent samples have been drawn from the same population (or from two different populations having the same distribution). It uses *ranking* information rather than pluses and minuses and, therefore, is less wasteful of data than the sign test.

Approaching a problem using the Mann-Whitney U test

Suppose that the board of regents of a large eastern state university wants to test the hypothesis that the mean SAT scores of students at two branches of the state university are equal. The board keeps statistics on all students at all branches of the system. A random sample of 15 students from each branch has produced the data shown in Table 12·3.

TABLE 12·3 SAT scores for students at two state university branches

Branch A	1,000	1,100	800	750	1,300	950	1,050	1,250	1,400	850	1,150	1,200	1,500	600	775
Branch S	920	1,120	830	1,360	650	725	890	1,600	900	1,140	1,550	550	1,240	925	500

Ranking the items to be tested

To apply the Mann-Whitney *U* test to this problem, we begin by ranking all the scores in order from lowest to highest, indicating beside each the symbol of the branch.

Table 12·4 accomplishes this.

Next, let's learn the symbols used in a Mann-Whitney *U* test in the context of this problem:

Symbols for expressing the problem

n_1 = number of items in sample 1; that is, the number of students at Branch A

n_2 = number of items in sample 2; that is, the number of students at Branch S

R_1 = sum of the ranks of the items in sample 1: the sum from Table 12·4 of the ranks of all the Branch A scores

R_2 = sum of the ranks of the items in sample 2: the sum from Table 12·4 of the ranks of all the Branch S scores

Rank	Score	Branch	Rank	Score	Branch
1	500	S	16	1,000	A
2	550	S	17	1,050	A
3	600	A	18	1,100	A
4	650	S	19	1,120	S
5	725	S	20	1,140	S
6	750	A	21	1,150	A
7	775	A	22	1,200	A
8	800	A	23	1,240	S
9	830	S	24	1,250	A
10	850	A	25	1,300	A
11	890	S	26	1,360	S
12	900	S	27	1,400	A
13	920	S	28	1,500	A
14	925	S	29	1,550	S
15	950	A	30	1,600	S

TABLE 12·4
SAT scores ranked from lowest to highest

In this case, both n_1 and n_2 are equal to 15, but it is *not* necessary for both samples to be of the same size. Now in Table 12·5, we can reproduce the data from Table 12·3, adding the ranks from Table 12·4. Then we can total the ranks for each branch. As a result, we have all the values we need to solve this problem because we know that:

$$n_1 = 15$$
$$n_2 = 15$$
$$R_1 = 247$$
$$R_2 = 218$$

Branch A	Rank	Branch S	Rank
1,000	16	920	13
1,100	18	1,120	19
800	8	830	9
750	6	1,360	26
1,300	25	650	4
950	15	725	5
1,050	17	890	11
1,250	24	1,600	30
1,400	27	900	12
850	10	1,140	20
1,150	21	1,550	29
1,200	22	550	2
1,500	28	1,240	23
600	3	925	14
775	7	500	1
	247 ← total ranks		218 ← total ranks

TABLE 12·5
Raw data and rank list for SAT scores

Calculating the U statistic

Using the values for n_1 and n_2 and the ranked sums of R_1 and R_2, we can determine the *U statistic,* a measurement of the difference between the ranked observations of the two samples of SAT scores:

Defining
the U statistic

$$U = n_1 n_2 + \frac{n_1(n_1 + 1)}{2} - R_1 \qquad [12 \cdot 1]$$

$$= (15)(15) + \frac{(15)(16)}{2} - 247$$

$$= 225 + 120 - 247$$

$$= 98 \leftarrow U \text{ statistic}$$

If the null hypothesis that the $n_1 + n_2$ observations came from identical populations is true, then this U statistic has a sampling distribution with a mean of:

Mean of
the U statistic

$$\mu_U = \frac{n_1 n_2}{2} \qquad [12 \cdot 2]$$

$$= \frac{(15)(15)}{2}$$

$$= 112.5 \leftarrow \text{ mean of the } U \text{ statistic}$$

and a standard error of:

Standard
error of the
U statistic

$$\sigma_U = \sqrt{\frac{n_1 n_2 (n_1 + n_2 + 1)}{12}} \qquad [12 \cdot 3]$$

$$= \sqrt{\frac{(15)(15)(15 + 15 + 1)}{12}}$$

$$= \sqrt{\frac{6975}{12}}$$

$$= \sqrt{581.25}$$

$$= 24.1 \leftarrow \text{ standard error of the } U \text{ statistic}$$

Testing the hypothesis

Stating
the hypothesis

The sampling distribution of the U statistic can be approximated by the normal distribution when *both* n_1 and n_2 are larger than 10. Because our problem meets this condition, we can use the normal distribution and the z table to make our test. The board of regents wishes to test at the .15 level of significance the hypothesis that these samples were drawn from identical populations.

$H_0: \mu_1 = \mu_2 \leftarrow$ null hypothesis: there is no difference between the two populations, and so they have the same mean

$H_1: \mu_1 \neq \mu_2 \leftarrow$ alternative hypothesis: there is a difference between the two populations; in particular, they have different means

$\alpha = .15 \leftarrow$ level of significance for testing this hypothesis

The board of regents wants to know whether the mean SAT score for students at either of the two schools is better or worse than the other. Therefore, this is a two-tailed test of the hypothesis. Figure 12·3 illustrates this test graphically. The two colored areas represent the .15 level of significance. Since we are using the normal distribution as our sampling distribution in this test, we can determine from Appendix Table 1 that the appropriate *z* value for an area of .425 is 1.44. The two limits of the acceptance region can be calculated like this:

Illustrating the problem

Finding the limits of the acceptance region

$$\mu_U + 1.44\sigma_U = 112.5 + (1.44)(24.1)$$
$$= 112.5 + 34.7$$
$$= 147.2 \leftarrow \text{upper limit}$$

and:

$$\mu_U - 1.44\sigma_U = 112.5 - (1.44)(24.1)$$
$$= 112.5 - 34.7$$
$$= 77.8 \leftarrow \text{lower limit}$$

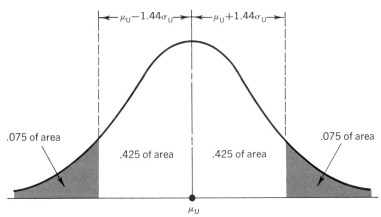

FIG. 12·3
Two-tailed test of hypothesis at the .15 level of significance

Figure 12·4 illustrates the limits of the acceptance region, 77.8 and 147.2, and the *U* value calculated earlier, 98. We can see that the sample *U* statistic does lie within the acceptance region. Thus, we would accept the null hypothesis of no difference and conclude that the distributions, and hence the mean SAT scores at the two schools, are the same.

Interpreting the results

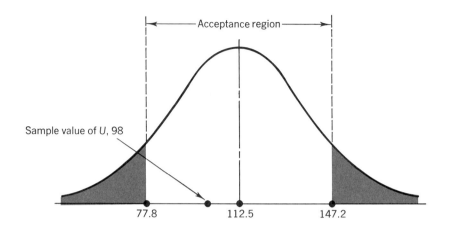

FIG. 12·4
Two-tailed hypothesis test at .15 level of significance, showing the acceptance region and the sample *U* statistic

Sample value of U, 98

Acceptance region

77.8   112.5   147.2

Special properties of the U test

Another way
to compute
the *U* statistic

The *U* statistic has a feature that enables users to save calculating time when the two samples under observation are of unequal size. We computed the value of *U* using Equation 12·1:

$$U = n_1 n_2 + \frac{n_1(n_1 + 1)}{2} - R_1 \qquad [12·1]$$

But just as easily, we could have computed the *U* statistic using the R_2 value like this:

$$U = n_1 n_2 + \frac{n_2(n_2 + 1)}{2} - R_2 \qquad [12·4]$$

The answer would have been 127 (which is just as far above the mean of 112.5 as 98 was below it). In this problem, we would have spent the same amount of time calculating the value of the *U* statistic using either Equation 12·1 or Equation 12·4. In other cases, when the number of items is larger in one sample than in the other, choose the equation that will require less labor. Regardless of whether you calculate *U* using Equation 12·1 or 12·4, you will come to the same conclusion. Notice that in this example, the answer 127 falls in the acceptance region just as 98 did.

Handling ties
in the data

What about *ties* that may happen when we rank the items for this test? For example, what if the two scores ranked 13 and 14 in Table 12·4 both had the value 920? In this case, we would find the average of their ranks (13 + 14)/2 = 13.5, and assign the result to both of them. If there were a three-way tie among the scores ranked 13, 14, and 15, we would average these ranks (13 + 14 + 15)/3 = 14, and use that value for all three items.

13. Test the hypothesis of no difference between the two populations, using the Mann-Whitney U test for the data below. Use the .05 level of significance.

Sample 1 26 25 38 33 42 40 44 26 43 35
Sample 2 44 30 34 47 35 46 35 47 48 34

14. Use the Mann-Whitney U test to determine if there is any difference between the two populations represented by the samples below. Use the .01 significance level. (Disregard the fact that we have fewer than 10 observations for n_1 and n_2.)

Group 1 89 90 92 81 76 88 85
Group 2 78 93 81 87 89 71 90 96 82

15. Test the hypothesis of no difference between the two populations represented by these samples. Use the rank sum test with a level of significance of .05. (Disregard the fact that we have fewer than 10 observations for n_1 and n_2.)

Sample 1 78 64 75 45 82 69 60
Sample 2 110 70 53 51 61 68

16. Use the Mann-Whitney U test to determine if there is any difference between the populations represented by the samples below. Use the .05 level of significance. (Disregard the fact that we have fewer than 10 observations for n_1 and n_2.)

Group 1 24 28 15 47 23 25 53 20
Group 2 22 12 30 16 26 14 18 21 16 18

17. A corn farmer changed the type of insecticide he was using, because a new one gave evidence of being more effective. As he began harvesting the crop, however, he felt that the yield was somewhat smaller than in past years. Since records for last year were easily available, he could compare the yields for all of the acres last year and the 10 acres that were harvested this year. Records show these yields with old and new insectides:

Yield for acres harvested (bushels)

Old insecticide	New insecticide	Old insecticide	New insecticide
992	965	966	956
945	1,054	889	900
938	912	972	
1,027	850	940	
892	796	873	
983	911	1,015	
1,014	877	1,016	
1,258	902	897	

Can the farmer be 95 percent sure that the new insecticide is reducing his yield?

18. Authorities for the Massachusetts Highway Department were considering purchasing a new ferry for the Martha's Vineyard crossing. The existing ferry held, on the average, 32 vehicles. The ferry under consideration was larger, but authorities doubted whether the additional space was really usable. Because of the many differing sizes of vehicles that typically used any ferry, it was impossible to estimate with accuracy whether the new ferry would be able to carry more vehicles. Relative capacities were particularly important because of a Massachusetts law requiring competitive bidding on all public contracts. The highway department had to know whether the ferry it was considering was compatible in size with a duplicate of the original (for which they had received a lower bid).

To aid in the decision, several tests were conducted on the existing ferry. Another state, which already had one of the new ferries in operation, conducted similar tests in response to the request of the Massachusetts Highway Department. The statistic that was recorded was the lengthwise footage for vehicles that were loaded on board. The data are presented below:

Lengthwise footage of vehicles loaded per trip (at capacity)

Existing ferry	453	438	447	449	452	450	439	445	446	454	451	448	442	447
Proposed ferry	458	459	450	448	459	457	462	439	448	454				

Use the Mann-Whitney U test to draw a conclusion about the capacity differential of the two ships at the 10 percent significance level.

19. In weather modification studies, days with certain weather conditions are selected as favorable for cloud seeding. In order to keep researchers as blind as possible, certain of the appropriate days are selected at random, and seeding is employed. On other appropriate days, clouds are not seeded. The researchers often do not know until the end of a series on which of the days the clouds were seeded. From a sample of days that were good for seeding, the data below give the amount of rainfall on days when seeding was done and on the days when it was not done.

Index of rainfall amounts

Unseeded days	18	21	23	15	19	26	17	18	22	20	18	21
Seeded days	22	17	15	23	25	22	26	24	16	17	23	21

Use a Mann-Whitney U test and a 9 percent significance level to decide if cloud seeding produced greater amounts of rainfall.

20. A question of interest to medical scientists as well as to some some social scientists is whether men or women have higher pain thresholds. Many have argued that women must have higher pain thresholds because they endure labor and childbirth, which are often very painful. Others contend that men are the ones who can endure pain better. To test this hypothesis, Judy Jenkins paid subjects to participate in shock experiments, in which the shocks were painful but harmless. Using various physiological as well as psychological instruments, subjects were monitored, and an index of their pain threshold was arrived at. Somehow, one of the assistants misplaced a portion of the data, and Judy was able to locate only the following information from records of the tests.

$$\sigma_U = 64.265077$$
$$\mu_U = 420$$
$$R_1 = 830$$

Judy also remembered that the sample size for men, n_2, had been 2 units larger than n_1.

Reconstruct a z value for the test and determine if the pain threshold can be assumed with a 5 percent level of significance to be the same for both men and women. Indicate also the values for n_1, n_2, and R_2.

Concept of randomness

4 ONE-SAMPLE RUNS TESTS
So far, we have assumed that the samples in our problems were randomly selected; that is, chosen without preference or bias. What if you were to notice recurrent patterns in a sample chosen by someone else? Suppose that applicants for advanced job training were to be selected without regard to sex from a large population. Using

the notation W = woman and M = man, you find that the first group enters in this order:

W, W, W, W, M, M, M, M, W, W, W, W, M, M, M, M

By inspection, you would conclude that although the total number of applicants is equally divided between the sexes, the order is not random. A random process would rarely list two items in alternating groups of four. Suppose now that the applicants begin to arrive in this order:

W, M, W, M, W, M, W, M, W, M, W, M, W, M, W, M

It is just as unreasonable to think that a random selection process would produce such an orderly pattern of men and women. In this case, too, the *proportion* of women to men is right, but you would be suspicious about the *order* in which they are arriving.

To allow us to test samples for the randomness of their order, statisticians have developed the *theory of runs.* **A run is a sequence of identical occurrences preceded and followed by different occurrences or by none at all.** If men and women enter as follows, the sequence will contain three runs:

The theory of runs

$$\underbrace{W,}_{\text{1st}} \underbrace{M, M, M, M,}_{\text{2nd}} \underbrace{W}_{\text{3rd}}$$

And this sequence contains six runs:

$$\underbrace{W, W, W,}_{\text{1st}} \underbrace{M, M,}_{\text{2nd}} \underbrace{W,}_{\text{3rd}} \underbrace{M, M, M, M,}_{\text{4th}} \underbrace{W, W, W, W,}_{\text{5th}} \underbrace{M}_{\text{6th}}$$

A *test of runs* would use the following symbols if it contained just two kinds of occurrences:

Symbols for expressing the problem

---

n_1 = number of occurrences of type 1
n_2 = number of occurrences of type 2
r = number of runs

Let's apply these symbols to a new pattern for the arrival of applicants:

M, W, W, M, M, M, M, W, W, W, M, M, W, M, W, W, M

In this case, the values of n_1, n_2, and r would be:

$n_1 = 8 \leftarrow$ number of women
$n_2 = 9 \leftarrow$ number of men
$r = 9 \leftarrow$ number of runs

A problem using a one-sample runs test

A manufacturer of breakfast cereal uses a machine to insert randomly one of two types of toys in each box. The company wants randomness so that every child in the neighborhood does not get the same toy. Testers choose samples of 60 successive boxes to see if the machine is properly mixing the two types of toys. Using the symbols A and B to represent the two types of toys, a tester reported that one such batch looked like this:

B, A, B, B, B, A, A, A, B, B, A, B, B, B,B, A, A, A, A, B,
A, B, A, A, B, B, B, A, A, B, A, A, A, A, A, B, B, A, B, B, A,
A, A, A, B, B, A, B, B, B, B, A, A, B, B, A, B, A, A, B, B

Stating the problem symbolically

The values in our test will be:

$n_1 = 29 \leftarrow$ number of boxes containing toy A
$n_2 = 31 \leftarrow$ number of boxes containing toy B
$r = 29 \leftarrow$ number of runs

The sampling distribution of the r statistic

The r statistic, the basis of a one-sample runs test

The *number of runs,* or *r,* is a statistic with its own special sampling distribution and its own test. Obviously, runs may be of differing lengths, and various numbers of runs can occur in one sample. Statisticians can prove that too many or too few runs in a sample indicate that something other than chance was at work when the items were selected. **A one-sample runs test, then, is based on the idea that too few or too many runs show that the items were not chosen randomly.** Statisticians used this test in 1970 to check the randomness of the draft lottery.

To derive the mean of the sampling distribution of the r statistic, use the following formula:

Mean and standard error of the r statistic

$$\mu_r = \frac{2n_1 n_2}{n_1 + n_2} + 1 \qquad [12\cdot5]$$

Applying this to the cereal company, the mean of the r statistic would be:

$$\mu_r = \frac{(2)(29)(31)}{29 + 31} + 1$$

$$= \frac{1798}{60} + 1$$

$$= 29.97 + 1$$

$$= 30.97 \leftarrow \text{mean of the r statistic}$$

The standard error of the r statistic can be calculated with this formidable-looking formula:

$$\sigma_r = \sqrt{\frac{2n_1 n_2(2n_1 n_2 - n_1 - n_2)}{(n_1 + n_2)^2(n_1 + n_2 - 1)}} \qquad [12\cdot 6]$$

For our problem, the standard error of the r statistic becomes:

$$\sigma_r = \sqrt{\frac{(2)(29)(31)(2 \times 29 \times 31 - 29 - 31)}{(29 + 31)^2(29 + 31 - 1)}}$$

$$= \sqrt{\frac{(1798)(1738)}{(60)^2(59)}}$$

$$= \sqrt{\frac{3,124,924}{212,400}}$$

$$= \sqrt{14.71}$$

$$= 3.84 \leftarrow \text{standard error of the } r \text{ statistic}$$

Testing the hypothesis

In the one-sample runs test, the sampling distribution of r can be closely approximated by the normal distribution if *either* n_1 or n_2 is larger than 20. Since our cereal company has a sample of 60 boxes, we can use the normal approximation. Management is interested in testing at the .20 level the hypothesis that the toys are randomly mixed, so the test becomes:

H_0:	In a one-sample runs test, no symbolic statement of	← null hypothesis: the toys are randomly mixed
H_1:	the hypothesis is appropriate.	← alternative hypothesis: the toys are not randomly mixed
	$\alpha = .20$	← level of significance for testing this hypothesis

Stating the hypothesis

Since too many *or* too few runs would indicate that the process by which the toys are inserted is not random, a two-tailed test is appropriate. Figure 12·5 illustrates this test graphically.

Illustrating the test

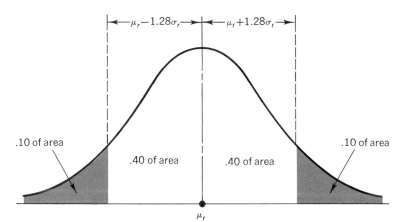

FIG. 12·5
Two-tailed hypothesis test at the .20 level of significance

$\leftarrow \mu_r - 1.28\sigma_r \rightarrow | \leftarrow \mu_r + 1.28\sigma_r \rightarrow$

.10 of area

.40 of area | .40 of area

.10 of area

μ_r

Finding the limits of the acceptance region	Because we can use the normal distribution, we can turn to Appendix Table 1 to find the appropriate z value for .40 of the area under the curve. We can then use this value, 1.28, to calculate the limits of the acceptance region:

$$\mu_r + 1.28\sigma_r = 30.97 + (1.28)(3.84)$$
$$= 30.97 + 4.92$$
$$= 35.89 \leftarrow \text{upper limit}$$

and:

$$\mu_r - 1.28\sigma_r = 30.97 - (1.28)(3.84)$$
$$= 30.97 - 4.92$$
$$= 26.05 \leftarrow \text{lower limit}$$

Interpreting the results	Both these limits to the acceptance region, 26.05 and 35.89, and the number of runs in the sample, 29, are shown in Fig. 12·6. There, we can see that the observed number of runs, 29, falls within the acceptance region. Therefore, management should accept the null hypothesis and conclude from this test that the toys are being inserted in the boxes in random order.

FIG. 12·6
Two-tailed hypothesis test at the .20 level of significance, showing the acceptance region and the observed number of runs

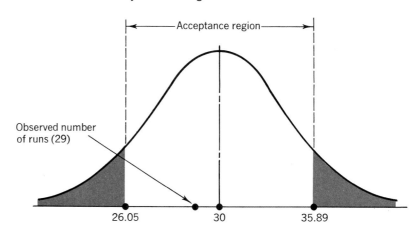

EXERCISES

21. Test for the randomness of the following sample using the .05 significance level.

A,B,B,B,B,A,B,A,B,B,A,A,B,A,A,A,B,B,B,A,B,B,A,A,A,A,B,A,B,B,A,A,A,B,A,A,B,A,B,B

22. A sequence of plants was inspected for disease. The sequence of healthy and infected plants was as follows:

H,H,H,H,I,H,I,I,I,H,H,I,I,H,H,H,H,H,H,H,H,I,I,I,I,I

Test for the randomness of the infection using the .10 significance level.

23. The following data represent the percentage of defective items turned out by a single machine for 25 consecutive days. Test the randomness of this sequence at the .05 level by designating values by whether they are above or below the median.

8.2	9.4	11.1	10.4	8.6
10.3	12.3	12.0	9.3	9.7
8.9	10.0	11.8	9.9	10.9
9.4	8.4	10.1	12.2	11.9
10.3	11.4	8.8	7.4	11.2

24. The following is the order of male and female arrivals at a theater box office before a show. Test this sequence for randomness at the .01 level of significance.

M,F,M,F,M,M,M,F,F,M,F,M,F,M,F,M,M,M,M,F,M,F,M,F,M,M,F,F,F,M,F, M,F,M,F,M,M,F,M,M,F,M,M,M,M,F,M,F,M,M

25. A restaurant owner has noticed over the years that older couples appear to eat earlier than young couples at his quiet, romantic restaurant. He suspects that perhaps it is because of children having to be left with babysitters and also because the older couples may retire earlier at night. One night he decided to keep a record of couples' arrivals at the restaurant. He noted whether each couple was over or under 30. His notes are reproduced below. (A = 30 and older; B = younger than 30.)

(5:30 P.M.) A, A, A, A, A, A, B, A, A, A, A, A, A, B, B, B, A, B, B, B, B, B, A, B, B, B, B, B, B, A (10 P.M.)

At a 5 percent level of significance, was the restaurant owner correct in his thought that the age of his customers at different dining hours is less than random?

26. Kathy Summers, the new principal at an elementary school, was concerned about the number of children being sent to her office as behavior problems each day. She had suspected for some time that the lunch and playground schedules set up before she came were not the best ones. Specifically, she thought that the younger children (grades 1, 2, and 3) had shorter attention spans and therefore became restless earlier in the day and misbehaved first, getting worse as the day went on. With longer attention spans, the older children (grades 4, 5, and 6) became fidgety later in the day. Kathy felt that if this hypothesis were true, she could rearrange lunchroom and playground schedules to give younger children lunch earlier, then more class time, and a play period as a later break. This would mean delaying lunches for older children and giving them their play periods either right before lunch or right afterward. If her hypothesis were false, she would run the risk of breaking up younger children's days too much and creating more problems with older children by expecting too much of them. Kathy recorded the grade levels of all children sent to her office one day, in order of their arrival.

Grade of child by order of arrival

1, 2, 3, 1, 4, 5, 3, 1, 2, 3, 1, 3, 6, 2, 3, 6, 2, 2, 3, 5, 4, 6, 4, 5, 1, 3, 4, 5, 5, 6, 4, 5, 2, 3, 5, 6, 4, 3, 2, 5, 4, 6

a) At a 2 percent level of significance, does Kathy have a valid hypothesis that the behavior problems of children of different ages are not random?

b) Is her hypothesis appropriate for the decision she wishes to make about rescheduling lunch and play periods?

27. Martha Thomas, a biochemist working in a large medical laboratory, is in charge of all the chemical analyses done in the medical complex where it is located. Accuracy and thoroughness are her responsibility. The lab employs a number of technicians and other biochemists to do some analyses and use machines that have been developed for certain types of analysis. Typically, each week Martha randomly chooses completed analyses before they are reported and conducts tests to ensure that they have been done correctly and thoroughly. Martha's assistant, Kim Nathan, randomly chooses 49 analyses per week

from the completed and filed ones of each day, and Martha does the reanalyses. Martha wanted to make certain that the selection process was a random one, so she could provide assurances that the machine analyses and those done by hand were both periodically checked. She arranged to have lab personnel place a special mark on the back of the records, so that they could be identified. Since Kim was unaware of the mark, the randomness of the test would not be affected. Kim completed her sample with the following results.

Sample of chemical analyses for one week

1-*Done by machines*; 2-*Done by hand*

1, 1, 1, 1, 1, 1, 1, 1, 1, 2, 1, 1, 1, 1, 1, 1, 1, 1, 1, 2,
1, 1, 1, 1, 1, 1, 1, 1, 2, 1, 1, 1, 1, 1, 1, 1, 1, 1, 2,
1, 1, 1, 1, 1, 1, 1, 1, 1

a) At a 1 percent significance level, can you conclude that the sample was random?

b) If the sample was distributed as below, would the sample be random?

1, 1, 1, 1, 1, 1, 1, 1, 1, 1, 1, 1, 1, 1, 1, 1, 1, 1, 1, 1,
1, 1, 1, 1, 1, 1, 1, 1, 1, 1, 1, 1, 1, 1, 1, 1, 1, 1, 1, 1,
1, 1, 1, 1, 1, 2, 2, 2, 2

c) Since machine analyses are much faster than those done by hand, and since a number of the tests are possible to do by machine, there are about three times as many machine analyses per week as hand analyses. Is there statistical evidence in part a to support the belief that somewhere in the sampling process there is something less than randomness occurring? If so, what is the evidence?

d) Does the conclusion you reached in part c lead you to any new conclusions about the one-sample runs test, particularly in reference to your answer for part a?

Function of the rank correlation coefficient

5 RANK CORRELATION

Chapter 11 introduced us to the notion of correlation and to the correlation coefficient, a measure of the closeness of association between two variables. Often in correlation analysis, information is not available in the form of numerical values like those we used in the problems of that chapter. But if we can assign rankings to the items in each of the two variables we are studying, a *rank correlation coefficient* can be calculated. This is a meausre of the correlation that exists between the two sets of ranks, a measure of the degree of association between the variables that we would not have been able to calculate otherwise.

Advantage of using rank correlation

A second reason for learning the method of rank correlation is to be able to simplify the process of computing a correlation coefficient from a very large set of data for each of two variables. To prove how tedious this can be, try expanding one of the correlation problems in Chapter 11 by a factor of 10 and performing the necessary calculations. Instead of having to do these calculations, we can compute a measure of association that is based on the *ranks* of the observations, *not the numerical values* of the data. This measure is called the Spearman rank correlation coefficient, in honor of the statistician who developed it in the early 1900s.

By working a couple of examples, we can learn how to calculate and interpret this measure of the association between two ranked variables. First, consider Table 12·6, which lists 5 persons and compares the academic rank they achieved in college with the level they have attained in a certain organization 10 years after graduation. The value of 5 represents the highest rank in the group; the rank of 1, the lowest.

Listing the ranked variables

Student	College rank	Organization rank 10 years later
John	4	4
Margaret	3	3
Debbie	1	1
Steve	2	2
Lisa	5	5

TABLE 12·6
Comparison of the ranks of 5 students

Using the information in Table 12·6, we can calculate a coefficient of rank correlation between success in college and organization level achieved 10 years later. All we need is Equation 12·7 and a few computations. Equation 12·7 is:

Calculating the rank correlation coefficient

$$r_s = 1 - \frac{6\Sigma d^2}{n(n^2 - 1)} \qquad [12·7]$$

where: r_s = coefficient of rank correlation (Notice that the subscript s, from Spearman, distinguishes this r from the one we calculated in Chapter 11.)

n = number of paired observations

Σ = notation meaning "the sum of"

d = difference between the ranks for each pair of observations

The computations are easily done in tabular form, as we show in Table 12·7. Therefore, we have all the information we need to find the rank correlation coefficient for this problem:

$$r_s = 1 - \frac{6\Sigma d^2}{n(n^2 - 1)} \qquad [12·7]$$

$$= 1 - \frac{6(0)}{5(25 - 1)}$$

$$= 1 - \frac{0}{120}$$

$$= 1 \leftarrow \text{rank correlation coefficient}$$

TABLE 12·7   Generating information to compute the rank correlation coefficient

Student	College rank (1)	Organization rank (2)	Difference between the 2 ranks (1) − (2)	Difference squared [(1) − (2)]²
John	4	4	0	0
Margaret	3	3	0	0
Debbie	1	1	0	0
Steve	2	2	0	0
Lisa	5	5	0	0
				$\Sigma d^2 = 0 \leftarrow$ sum of the squared differences

Explaining values
of the
rank correlation
coefficient

As we learned in Chapter 11, this correlation coefficient of 1 shows that there is a perfect association or *perfect correlation* between the two variables. This verifies our intuitive reaction to the fact that the college and organization ranks for each person were identical.

Computing another
rank correlation
coefficient

One more example should make us feel comfortable with the coefficient of rank correlation. Table 12·8 illustrates 5 more individuals, but this time the ranks in college and in an organization 10 years later seem to be extreme opposites. We can compute the difference between the ranks for each pair of observations, find d^2, and then take the sum of all the d^2's. Substituting these values into Equation 12·7, we find a rank correlation coefficient of −1:

$$r_s = 1 - \frac{6\Sigma d^2}{n(n^2 - 1)} \qquad [12\cdot7]$$

$$= 1 - \frac{6(40)}{5(25 - 1)}$$

$$= 1 - \frac{240}{120}$$

$$= 1 - 2$$

$$= -1 \leftarrow \text{rank correlation coefficient}$$

TABLE 12·8   Generating data to compute rank correlation coefficient

Student	College rank (1)	Organization rank (2)	Difference between the 2 ranks (1) − (2)	Difference squared [(1) − (2)]²
Roy	5	1	4	16
David	1	5	−4	16
Jay	3	3	0	0
Charlotte	2	4	−2	4
Kathy	4	2	2	4
				$\Sigma d^2 = 40 \leftarrow$ sum of the squared differences

In Chapter 11, we learned that a correlation coefficient of −1 represents *perfect inverse correlation.* And that is just what happened in our case: the people who did the best in college wound up 10 years later in the lowest ranks of an organization. Now let's apply these ideas.

Interpreting the results

Solving a problem using rank correlation

Rank correlation is a useful technique for looking at the connection between air quality and the evidence of pulmonary-related diseases that we discussed in our chapter opening problem. Table 12·9 reproduces the data found by the health organization studying the problem. In the same table, we also do some of the calculations needed to find r_s.

TABLE 12·9
Ranking of eleven cities

City	Air-quality rank (1)	Pulmonary-disease rank (2)	Difference between the 2 ranks (1) − (2)	Difference squared [(1) − (2)]²
A	4	5	−1	1
B	7	4	3	9
C	9	7	2	4
D	1	3	−2	4
E	2	1	1	1
F	10	11	−1	1
G	3	2	1	1
H	5	10	−5	25
I	6	8	−2	4
J	8	6	2	4
K	11	9	2	4
	Best rank = 11			$\Sigma d^2 = 58$ ← sum of the squared
	Worst rank = 1			differences

Using the data in Table 12·9 and Equation 12·7, we can find the rank correlation coefficient for this problem:

Finding the rank correlation coefficient

$$r_s = 1 - \frac{6\Sigma d^2}{n(n^2 - 1)} \qquad [12\cdot7]$$

$$= 1 - \frac{6(58)}{11(121 - 1)}$$

$$= 1 - \frac{348}{11(120)}$$

$$= 1 - \frac{348}{1320}$$

$$= 1 - .264$$

$$= .736 \leftarrow \text{rank correlation coefficient}$$

A correlation coefficient of .736 suggests a substantial positive association between average air quality and the occurrence of pulmonary disease, at

Interpreting the results

least in the eleven cities sampled; that is, high levels of pollution go with high incidence of pulmonary disease.

How can we test this value of .736? We can apply the same methods we used to test hypotheses in Chapter 9. In performing such tests on r_s we are trying to avoid the error of concluding that an association exists between two variables if, in fact, no such association exists in the population from which these two samples were drawn; that is, if the *population* rank correlation coefficient, ρ_s (*rho-sub-s*), is really equal to zero.

For small values of n (n less than 30), the distribution of r_s is not normal, and unlike other small sample statistics we have encountered, it is not appropriate to use the t distribution for testing hypotheses about the rank correlation coefficient. Instead, we use Appendix Table 7 to determine the acceptance and rejection regions for such hypotheses. In our current problem, suppose that the health organization wants to test, at the .05 level of significance, the null hypotheses that there is zero correlation in the ranked data of *all* cities in the world. Our problem then becomes:

$H_0: \rho_s = 0 \leftarrow$ null hypothesis: there is no correlation in
the ranked data of the population
$H_1: \rho_s \neq 0 \leftarrow$ alternative hypothesis: there is a correlation
in the ranked data of the population
$\alpha = .05 \leftarrow$ level of significance for testing this
hypothesis

A two-tailed test is appropriate; so we look at Appendix Table 7 in the row for $n = 11$ and the column tor a significance level of .05. There we find that the critical values for r_s are $\pm.6091$; that is, the upper limit of the acceptance region is .6091, and the lower limit of the acceptance region is $-.6091$.

Figure 12·7 shows the limits of the acceptance region and the rank correlation coefficient we calculated from the air quality sample. From this figure, we can see that the rank correlation coefficient lies outside the acceptance region. Therefore, we would reject the null hypothesis of no correlation and conclude that there is an association between air quality levels and the incidence of pulmonary disease in the world's cities.

FIG. 12·7

Two-tailed hypothesis test, using Appendix Table 7 at the .05 level of significance, showing the acceptance region and the sample rank correlation coefficient

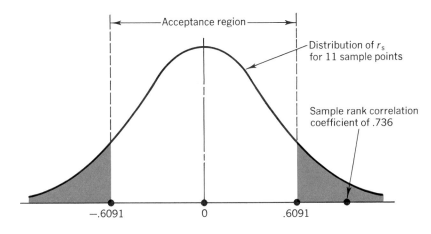

If the sample size is greater than 30, we can no longer use Appendix Table 7. However, when n is greater than 30, the sampling distribution of r_s is approximately normal, with a mean of zero and a standard deviation of $1/\sqrt{n-1}$. Thus, the standard error of r_s is:

The appropriate distribution for values of n greater than 30

$$\sigma_{r_s} = \frac{1}{\sqrt{n-1}} \qquad [12\cdot8]$$

and we can use Appendix Table 1 to find the appropriate z values for testing hypotheses about the population rank correlation.

As an example of hypothesis testing of rank correlation coefficients when n is greater than 30, consider the case of a social scientist who tries to determine whether bright people tend to choose spouses who are also bright. He randomly chooses 32 couples and tests to see if there is a significant rank correlation in the IQ's of the couples. His data and computations are given in Table 12·10 (p. 502).

Example with n greater than 30

Using the data in Table 12·10 and Equation 12·7, we can find the rank correlation coefficient for this problem:

$$
\begin{aligned}
r_s &= 1 - \frac{6\Sigma d^2}{n(n^2 - 1)} \qquad [12\cdot7]\\
&= 1 - \frac{6(1,043.5)}{32(1,024 - 1)}\\
&= 1 - \frac{6,261}{32,736}\\
&= 1 - .191\\
&= .809 \leftarrow \text{rank correlation coefficient}
\end{aligned}
$$

If the social scientist wishes to test his hypothesis at the .01 level of significance, his problem can be stated:

$H_0: \rho_s = 0 \leftarrow$ null hypothesis: there is no rank correlation in the population; that is, husband and wife intelligence is randomly mixed

$H_1: \rho_s > 0 \leftarrow$ alternative hypothesis: the population rank correlation is positive; that is, bright people choose bright spouses

$\alpha = .01 \leftarrow$ level of significance for testing this hypothesis

An upper-tailed test is appropriate. From Appendix Table 1, we find that the appropriate z value for the .01 level of significance is 2.33. Figure 12·8 (p. 503) illustrates this hypothesis test graphically; we show there the colored region in the upper tail of the distribution that corresponds to the .01 level of significance.

Stating the hypothesis

Couple (1)	Husband's IQ (2)	Wife's IQ (3)	Husband's rank (4)	Wife's rank (5)	Difference between ranks (4) − (5)	Difference squared [(4) − (5)]²
1	95	95	8	4.5	3.5	12.25
2	103	98	20	8.5	11.5	132.25
3	111	110	26	23	3	9.00
4	92	88	4	2	2	4.00
5	150	106	32	18	14	196.00
6	107	109	24	21.5	2.5	6.25
7	90	96	3	6	−3	9.00
8	108	131	25	32	−7	49.00
9	100	112	17.5	25.5	−8	64.00
10	93	95	5.5	4.5	1	1.00
11	119	112	29	25.5	3.5	12.25
12	115	117	28	30	−2	4.00
13	87	94	1	3	−2	4.00
14	105	109	21	21.5	− .5	0.25
15	135	114	31	27	4	16.00
16	89	83	2	1	1	1.00
17	99	105	14.5	16.5	−2	4.00
18	106	115	22.5	28	−5.5	30.25
19	126	116	30	29	1	1.00
20	100	107	17.5	19	− 1.5	2.25
21	93	111	5.5	24	−18.5	342.25
22	94	98	7	8.5	− 1.5	2.25
23	100	105	17.5	16.5	1	1.00
24	96	103	10	15	−5	25.00
25	99	101	14.5	13	1.5	2.25
26	112	123	27	31	−4	16.00
27	106	108	22.5	20	2.5	6.25
28	98	97	12.5	7	5.5	30.25
29	96	100	10	11.5	−1.5	2.25
30	98	99	12.5	10	2.5	6.25
31	100	100	17.5	11.5	6	36.00
32	96	102	10	14	−4	16.00

Sum of the squared differences $\rightarrow \Sigma d^2 = $ **1043.50**

We can now calculate the limit of our acceptance region:

$$\rho_{s_{H_0}} + 2.33\,\sigma_{r_s} = 0 + 2.33\left(\frac{1}{\sqrt{n-1}}\right)$$

$$= 0 + \frac{2.33}{\sqrt{31}}$$

$$= 0 + \frac{2.33}{5.568}$$

$$= 0.42 \leftarrow \text{upper limit of acceptance region}$$

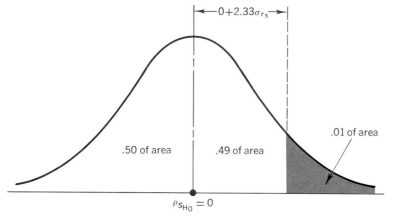

FIG. 12·8
One-tailed
hypothesis test at
the .01 level of
significance

$0+2.33\sigma_{r_s}$

.01 of area

.50 of area

.49 of area

$\rho_{s_{H_0}} = 0$

Figure 12·9 shows the limit of the acceptance region and the rank correlation coefficient we calculated from the IQ data. In Fig. 12·9, we can see that the rank correlation coefficient of .809 lies far outside the acceptance region. Therefore, we would reject the null hypothesis of no correlation and conclude that bright people tend to choose bright spouses.

Interpreting the results

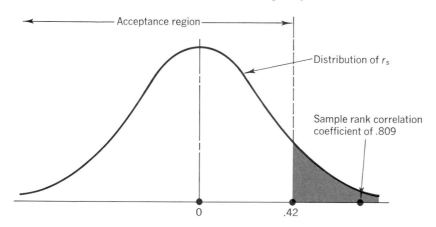

Acceptance region

Distribution of r_s

Sample rank correlation coefficient of .809

0

.42

FIG. 12·9
One-tailed
hypothesis test of
r_s at the .01 level
of significance
(with $n = 32$),
showing the
acceptance region
and the sample
rank correlation
coefficient

A special property of rank correlation

Rank correlation has a useful advantage over the correlation method we discussed in Chapter 11. Suppose we have cases in which one or several very extreme observations exist in the original data. Using numerical values as we did in Chapter 11, the correlation coefficient may not be a good description of the association that exists between two variables. Yet extreme observations in a *rank* correlation test will never produce a large rank difference.

Advantage of rank correlation

Consider the following data array of two variables X and Y:

X	10	13	16	19	25
Y	34	40	45	51	117

Because of the large value of the fifth Y term, we would get two significantly different answers for r using the conventional and the rank correlation methods. In this case, the rank correlation method would be less sensitive to the extreme value. We would assign a rank order of 5 to the numerical value of 117 and avoid the unduly large effect on the value of the correlation coefficient.

EXERCISES

28. Below are ratings of aggressiveness (X) and amount of sales in the last year (Y) for 8 salespeople. Is there a significant correlation between the two measures? Use the .05 significance level.

X	30	17	35	28	42	25	19	34
Y	35	31	40	46	50	32	33	42

29. A plant supervisor ranked a sample of 8 workers on the number of hours worked overtime and length of employment. Is the correlation between the two measures significant at the .01 level?

Amt. of overtime	5.0	8.0	2.0	4.0	3.0	7.0	1.0	6.0
Years employed	1.0	6.0	4.5	2.0	7.0	8.0	4.5	3.0

30. Because of the value 18 lying far from the other values, this data would not be suitable for usual correlation analysis, but a rank correlation might still tell us something. Test the significance of the rank correlation for these data, using the .05 significance level.

X	70	86	93	82	56	84
Y	41	52	49	18	51	54

31. The Occupational Safety and Health Administration (OSHA) was conducting a study of the relationship of expenditures for plant safety and the accident rate in the plants. OSHA had confined its studies to the synthetic chemical industry. To adjust for the size differential that existed between some of the plants, OSHA had converted its data into expenditures per production employee. The results of the data are listed below.

Expenditure by chemical companies per production employee in relation to accidents per year

Company	A	B	C	D	E	F	G	H	I	J	K
Expenditure	$60	$45	$30	$20	$28	$42	$39	$54	$48	$58	$26
Accidents	3	7	6	9	7	4	8	2	4	3	8

Is there a significant correlation between expenditures and accidents in the chemical company plants? Use a rank correlation (with 1 representing highest expenditure and accident rate) to support your conclusion. Test at 5 percent significance.

32. As part of a study in psychophysics, subjects are asked to perform fine weight discriminations, so that experimenters can see how well humans can detect weight differences. They are not asked to estimate the weights of objects, only to order them in terms of increasing weight. Below are the results of one subject's rankings and the actual ranks as determined by a scale.

Subject's	9	1	10	3	5	2	4	6	7	8
Actual	7	2	6	4	3	5	1	10	8	9

Use the rank correlation test at 5 percent significance to decide if the subject is able to detect weight differences in objects.

33. At the Lakeview Sanitarium, incoming patients are diagnosed and evaluated by one of two psychologists. Though they were trained at different schools and often use different techniques in their analysis of a patient, both psychologists are thought to be good diagnosticians. An administrator at the sanitarium wondered how closely their diagnoses would agree, so he had both psychologists independently evaluate 14 new patients. They ranked the patients in terms of their degree of pathology. The results are given below. Use a rank correlation at the 1 percent significance level and determine if there is significant positive correlation between the two psychologists' rankings.

Patient	1	2	3	4	5	6	7	8	9	10	11	12	13	14
Psychologist 1	1	11	12	2	13	10	3	4	14	5	6	9	7	8
Psychologist 2	4	12	11	2	13	10	1	3	14	8	6	5	9	7

34. Nancy Holsner, a social psychologist running a study of game playing, feels it is very important that subjects be unaware of the manipulation involved in her study; therefore, she asks subjects during their debriefing not to discuss it. However, she is afraid that they do not recognize the importance of the secrecy and that her results may be contaminated. Fearing that the true purpose of the study may have leaked out, she decides to see if the dependent variable scores of subjects show any relationship to the day on which they participated. The results are presented here.

Subject number	Game score	Day of experiment	Subject number	Game score	Day of experiment
1	6.0	10	6	2.0	5
2	7.0	4	7	6.0	2
3	8.0	6	8	0.5	8
4	4.0	1	9	9.0	9
5	5.0	3	10	3.0	7

Can Nancy conclude at a 1 percent significance level that there is no positive correlation between the subjects' scores and the day they participated in the study?

35. A scholarship selection committee for a private science foundation has used scores on a standardized test in the sciences and scores given by committee members after interviewing applicants. Until recently, only students from a particular region were allowed to apply for the scholarship, so the cost of traveling to various locales for interviews was not unreasonable. But now that the competition has been opened to the entire United States, it would be extremely costly to conduct individual interviews with all applicants, and the committee is not sure the interviews are valuable. If the committee can be confident (using 1 percent significance level) that the population rank correlation between the applicants' test scores and their interview scores is positive, it will feel justified in relying on the tests to make the selection and in discontinuing the interviews. They have drawn a sample of 35 applicants from the competitions in the last two years. On the basis of this sample (shown on the next page) what should the committee recommend?

Individual	Interview score	Test score	Individual	Interview score	Test score
1	81	113	19	81	111
2	88	88	20	84	121
3	55	76	21	82	83
4	83	129	22	90	79
5	78	99	23	63	71
6	93	142	24	78	108
7	65	93	25	73	68
8	87	136	26	79	121
9	95	82	27	72	109
10	76	91	28	95	121
11	60	83	29	81	140
12	85	96	30	87	132
13	93	126	31	93	135
14	66	108	32	85	143
15	90	95	33	91	118
16	69	65	34	94	147
17	87	96	35	94	138
18	68	101			

6 TERMS INTRODUCED IN CHAPTER 12

- **MANN-WHITNEY U TEST** A nonparametric method used to determine whether two independent samples have been drawn from populations with the same distribution.

- **NONPARAMETRIC TESTS** Statistical techniques that do not make restrictive assumptions about the shape of a population distribution when performing a hypothesis test.

- **ONE-SAMPLE RUNS TEST** A nonparametric method for determining the randomness with which sampled items have been selected.

- **RANK CORRELATION** A method for doing correlational analysis when the data are not available to use in numerical form, but when information is sufficient to rank the data.

- **RANK CORRELATION COEFFICIENT** A measure of the degree of association between two variables that is based on the ranks of observations, not their numerical values.

- **RANK SUM TESTS** A family of nonparametric tests that make use of the order information in a set of data.

- **RUN** A sequence of identical occurrences preceded and followed by different occurrences or by none at all.

- **SIGN TEST** A test for the difference between paired observations where + and − signs are substituted for quantitative values.

- **THEORY OF RUNS** A theory developed to allow us to test samples for the randomness of their order.

7 EQUATIONS INTRODUCED IN CHAPTER 12

p. 486:

$$U = n_1 n_2 + \frac{n_1(n_1 + 1)}{2} - R_1$$

[12·1]

To apply the Mann-Whitney U test, you need this formula to derive the U statistic, a measurement of the difference between the ranked observations of the two variables. R_1 is the sum of the ranked observations of variable 1; n_1 and n_2 the number of items in samples 1 and 2, respectively. Both samples need not be of the same size.

p. 486:
$$\mu_U = \frac{n_1 n_2}{2}$$
[12·2]

If the null hypothesis of a Mann-Whitney U test is that $n_1 + n_2$ observations came from identical populations, then the U statistic has a sampling distribution with a mean equal to the product of n_1 and n_2 divided by 2.

p. 486:
$$\sigma_U = \sqrt{\frac{n_1 n_2 (n_1 + n_2 + 1)}{12}}$$
[12·3]

This formula enables us to derive the *standard error of the U statistic* of a Mann-Whitney U test.

p. 488:
$$U = n_1 n_2 + \frac{n_2 (n_2 + 1)}{2} - R_2$$
[12·4]

This formula and Equation 12·1 can be used interchangeably to derive the U statistic in a Mann-Whitney U test. Use this formula if the number of observations of variable 2 is significantly smaller than the number of observations of variable 1.

p. 492:
$$\mu_r = \frac{2 n_1 n_2}{n_1 + n_2}$$
[12·5]

When doing a one-sample runs test, use this formula to derive the mean of the sampling distribution of the r statistic. This r statistic is equal to the *number of runs* in the sample being tested.

p. 493:
$$\sigma_r = \sqrt{\frac{2 n_1 n_2 (2 n_1 n_2 - n_1 - n_2)}{(n_1 + n_2)^2 (n_1 + n_2 - 1)}}$$
[12·6]

This formula enables us to derive the *standard error of the r statistic* in a one-sample runs test.

p. 497:
$$r_s = 1 - \frac{6 \Sigma d^2}{n(n^2 - 1)}$$
[12·7]

The *coefficient of rank correlation*, r_s, is a measure of the closeness of association between two ranked variables.

p. 501:
$$\sigma_{r_s} = \frac{1}{\sqrt{n - 1}}$$
[12·8]

This formula enables us to calculate the *standard error of* r_s in a hypothesis test on the coefficient of rank correlation.

8 CHAPTER REVIEW EXERCISES

36. A college football coach has a theory that in athletics, success feeds on itself. In other words, he feels that winning a championship one year increases the team's motivation to win it the next year. He expressed his theory to a student of statistics, who asked him for the records of the team's wins and losses over the last several years. The coach gave him a list, specifying whether the team had won (W) or lost (L) the championship that year. The results of this tally are presented below.

W, W, W, W, W, W, L, W, W, W, W, W, L, W, W, W, W, L, L, W, W, W, W, W, W

 a) At a 10 percent significance level, is the occurrence of wins and losses a random one?

 b) Does your answer to question a, combined with a sight inspection of the data, tell you anything about the one-sample runs test?

37. A small metropolitan airport recently opened a new runway, creating a new flight path over an upper-income residential area. Complaints of excessive noise had deluged the airport authority to the point that the two major airlines servicing the city had installed special engine baffles on the turbines of the jets, to reduce noise and help ease the pressure on the authority. Both airlines wanted to see if the baffles had helped to reduce the number of complaints that had been brought against the airport. If they had not, the baffles would be removed, because they increased fuel consumption. Based on the following data, can it be said at the .045 level of significance that installing the baffles has reduced the number of complaints?

Complaints per day before and after baffles were installed

Before	15	20	24	18	30	46	15	29	17	21	18
After	23	19	12	9	16	12	28	20	16	14	11

38. The American Broadcasting System (ABS) had invested a sizeable amount of money into a new program for television, *High Times*. *High Times* was ABS's entry into the situation comedy market and featured the happy-go-lucky life in a college dormitory. Unfortunately, the program had not done as well as expected, and the sponsor was considering cancelling. To beef up the ratings, ABS introduced co-ed dormitories into the series. Presented below are the results of telephone surveys before and after the change in the series. Surveys were conducted in several major metropolitan areas, so the results are a composite from the cities.

 a) Using a *U* test, can you infer at the .05 significance level that the change in the series format helped the ratings?

 b) Do the results of your test say anything about the effect of sex on TV program ratings?

Share of audience before and after change to co-ed dormitories

Before	22	18	19	20	27	22	25	19	22	24	18	16	14	28	30	15	16
After	25	28	18	30	33	25	29	29	19	16	30	33	13	25			

39. The dietician at a health club decided to try out two diets on the club members. After pairing them on body type and amount of weight loss desired, they were placed on the diet for one month. Below are the weight losses for members of each group.

Diet 1	20	22	18	15	17	14	8	10	9	12	19	20
Diet 2	16	17	20	15	14	13	10	11	7	10	15	16

Test the hypothesis of no difference between the success of the two diets at the .10 significance level.

40. The Ways and Means Committee of the U.S. House of Representatives was attempting to evaluate the results of a tax cut given to individuals during the preceding year. The intended purpose had been to stimulate the economy, the theory being that with a tax reduction, the consumer would spend the tax savings. The committee had employed an independent consumer research group to select a sample of households and maintain records of consumer spending both before and after the legislation was put into effect. A portion of the data from the research group is listed below.

Schedule of consumer spending

Household	Before legislation	After legislation	Household	Before legislation	After legislation
1	$ 3,578	$ 4,296	17	$11,597	$12,093
2	10,856	9,000	18	9,612	8,375
3	7,450	8,200	19	3,461	3,740
4	9,200	9,200	20	4,500	4,500
5	8,760	8,840	21	8,341	8,500
6	4,500	4,620	22	7,589	7,609
7	15,000	14,500	23	25,750	24,321
8	22,350	22,500	24	14,673	13,500
9	7,346	7,250	25	5,003	6,072
10	10,345	10,673	26	10,940	11,398
11	5,298	5,349	27	8,000	9,007
12	6,950	7,000	28	14,256	14,500
13	34,782	33,892	29	4,322	4,258
14	12,837	12,650	30	6,828	7,204
15	7,926	8,437	31	7,549	7,678
16	5,789	6,006	32	8,129	8,125

At a significance level of 4 percent, use a sign test to determine if the tax reduction policy has achieved its desired goals.

41. Carolyn Ezzell, candidate from California for the U.S. Senate, wants to bring her campaign to the people. Her campaign manager thinks that going into the inner city neighborhoods and small communities will win a lot of supporters. So after the primary in March, which Ezzell is expected to win, she is planning a number of outdoor appearances in small towns in the state and in city neighborhoods. In planning the trip, the campaign manager is concerned about the weather, since the crowds will have to stand outside for many of the appearances. When establishing the tour route, should she consider inclement weather (rain, snow, too hot, too cold) as a random event?

42. Two television weathermen got into a discussion one day about whether years with heavy rainfall tended to occur in spurts. One of them said he thought that there were patterns of annual rainfall amounts, and that several wet years were often followed by a number of drier than average years. The other weatherman was skeptical and said he thought that the amount of rainfall for consecutive years was fairly random. To investigate the question, they decided to look at the annual rainfall for several years back. They found the median amount and classified the rainfall as below (B) or above (A) the median annual rainfall. A summary of their results is presented at the top of the next page.

A, A, B, B, B, B, A, B, A, A, A, A, B, A, B, A, B, A, A, B, B, B,
A, A, B, A, B, A, A, B, B, B, B, A, B, B, B, A, B, A, A, A, A, B, A,
A, A, B, A, B, B, A

If the weathermen test at a 5 percent significance level, will they conclude that the annual rainfall amounts do not occur in patterns?

43. The Metro City Police Department recently invested in a programmed learning course for members of the force. The instructional package was designed to teach patrolmen how to observe people and events more carefully, how to spot unusual activity, and how to retain details of description. The distributors of the course were quite positive about its ability to produce results. The Metro City Police Chief, on the other hand, had his doubts. He gathered the following data from police records over two 3-week periods.

Usable descriptions of people/events made by patrolmen before and after course

Patrolman	Alioto	Kelly	O'Casey	Hunt	Krebs	Lee	Doan	Hudson	Hall	Chen	Kent
Before	17	26	19	22	11	32	13	15	9	24	26
After	19	30	14	25	15	29	19	16	6	26	22

The chief has let his skepticism be known around City Hall, and before he approves any further use of the programmed course, he wants to be 95 percent sure that there is a significant improvement in his patrolmen's observation ability. Can he be that sure?

44. The National Association of Better Advertising for Children (NABAC), a consumer group for improving children's television, was conducting a study on the impact of Saturday morning advertising. Specifically, the group wanted to know if a significant degree of purchasing was stimulated by advertising directed at children and if there was a positive correlation between Saturday morning TV advertising time and product sales.

NABAC chose the children's breakfast cereal market as a sample group. They selected products whose advertising message was aimed entirely at children. The results of the study are presented below. (The highest selling cereal has sales rank 1).

Comparison of TV advertising time and product sales

Product	Advertising time in minutes	Sales rank
Captain Grumbles	0.50	10
Obnoxious Berries	3.00	1
Fruity Hoops	1.25	9
OO La Granola	2.00	5
Sweet Tweets	3.50	2
Chocolate Chumps	1.00	11
Sugar Spots	4.00	3
Count Cavity	2.50	8
Crunchy Munchies	1.75	6
Karamel Kooks	2.25	4
Flakey Flakes	1.50	7

Can the group conclude that there is a *positive* rank correlation between the amount of Saturday morning advertising time and sales volume of breakfast cereals? Test at a 5 percent significance level.

45. The following ratings were made by people who used two detergents for three weeks. Test the hypothesis that the users found no difference in the two products. Use the .05 level of significance.

| Product 1 | 4 | 4 | 5 | 5 | 3 | 2 | 5 | 3 | 1 | 2 | 5 | 3 | 4 | 2 | 5 | 5 |
| Product 2 | 2 | 3 | 3 | 3 | 3 | 3 | 3 | 4 | 3 | 2 | 3 | 2 | 2 | 3 | 3 | 4 |

46. As part of a survey on political attitudes, Chuck Dangler, a political science researcher, asked voters to rate two gubernatorial candidates. On a scale of 1 to 10, voters were to rate characteristics such as the candidates' ability to get things accomplished, their knowledge of important issues, and their sensitivity to the needs of the people.

After the data were collected, a campaign worker for one of the candidates proposed that various statistical tests be performed. He specifically mentioned that he would like to see a mean and standard deviation for the responses to each question about each candidate, in order to see who had scored better. Several of Dangler's colleagues argued against the campaign worker's suggestion, noting that the quality of the input data would not justify a detailed statistical analysis. The other researchers argued that what was important was the voters' rankings of the two candidates. Evaluate the arguments presented by the campaign worker and the researchers.

47. Candidates for a private pilot's license were cautioned not to talk about the test after they took it. The examiner, however, suspected that the later ones knew something about what was going on. Were her suspicions correct? To find out, rank the scores received by subjects. Then test the significance of the rank correlation coefficient between scores and day number. Use the .05 significance level.

Day	Score	Day	Score
1	25	11	37
2	29	12	41
3	28	13	38
4	33	14	24
5	30	15	45
6	32	16	43
7	39	17	44
8	34	18	66
9	35	19	47
10	42	20	50

48. A traffic signal had been in place at Kenmore and Pine streets for more than 3 years, and the Public Safety Division wanted to know if it had reduced accidents at this hazardous intersection. The division had collected these data:

Accidents at Pine and Kenmore

	Jan.	Feb.	Mar.	Apr.	May	June	July	Aug.	Sept.	Oct.	Nov.	Dec.
1976	5	3	4	2	6	4	3	3	2	4	5	3
1977	4	4	3	3	3	4	0	5	4	2	0	1
1978	3	2	1	1	0	2	4	3	2	1	1	2
1979	2	1	0	0	1	2						

a) Determine the median number of accidents per month. If the light has been effective, we should find early months falling above the median and later months below the median. Accordingly, there will be a small number of runs above and below the median. Determine the points above and below the median. Conduct a test at a .03 level of significance to see if the accidents are randomly distributed.

b) What can you conclude about the effectiveness of the traffic light?

Answer true *or* false. *Answers are in the back of the book.*

1. One advantage of nonparametric methods is that some of the tests do not require us even to rank the observations.

2. The Mann-Whitney U test is one of a family of tests known as rank difference tests.

3. A sign test for paired data is based upon the binomial distribution, but can often be approximated by the normal distribution.

4. One disadvantage of nonparametric methods is that they tend to ignore a certain amount of information.

5. In the Mann-Whitney U test, two samples of size n_1 and n_2 are taken to determine the U statistic. The sampling distribution of the U statistic can be approximated by the normal distribution when either n_1 or n_2 is greater than 10.

6. The Mann-Whitney U test tends to waste less data than the sign test.

7. Assume that in a rank test, two elements are tied for the 10th rank position. We assign each of them a rank of 10.5 and the next element after these two receives a rank of 11.

8. In contrast to regression analysis where one may compute a coefficient of correlation, an equivalent measure may be determined in a ranking of two variables in nonparametric testing. This equivalent measure is called a rank correlation coefficient.

9. In a one-sample runs test, the number of runs is a statistic having its own sampling distribution.

10. One disadvantage in using the rank correlation coefficient is that it is very sensitive to extreme observations in the data set.

Afterword

We've come a long way and looked at many techniques that statisticians use to aid decision makers. We always concentrated on *appropriate* uses of the methods, but you will surely encounter *inappropriate* uses. Sometimes, ignorance is to blame, but too often there is an unscrupulous attempt to mislead us. The first section below will make you aware of potential abuses.

Many statistical techniques are beyond the technical scope or space limitations of this book. Often, an entire course is devoted to one technique. To give you some idea of the kinds of issues these more advanced techniques address, sections 2, 3, and 4 of the Afterword contain brief descriptions of design of experiments, time series, and index numbers.

1 FALLING OFF THE TRUE PATH

1 FALLING OFF THE TRUE PATH   Benjamin Disraeli once made the statement, "there are three kinds of lies: lies, damned lies, and statistics." This rather severe castigation of statistics, made so many years ago, has come to be a rather apt description of many of the statistical deceptions we encounter in our everyday lives. Darrell Huff, in an enjoyable little book, *How to Lie with Statistics,* noted that "The crooks already know these tricks; honest men must learn them in self-defense." The purpose of this next section is to review some of the common ways statistics are used incorrectly, whether out of honest lack of knowledge or in an attempt to deceive the user. In either case, users of statistics who do not know how to cope with such deception cannot derive much real value from this discipline.

How to lie with statistics

Biased samples

Statistics professors often use classroom demonstrations to prove one point or another. One of the most common ones involves tossing a coin to show that the long-run tendency is for the coin (if it's a fair one) to come up heads half of the time and tails the other half of the time. Suppose our professor tosses a fair coin 10 times and it comes up heads on 8 of these tosses. What should he do? One explanation for the class is that this coin is biased (not too likely an explanation, since the work involved in biasing a standard coin so that it will behave this way is rather substantial). Another explanation is that he has not tossed the coin a sufficient number of times. The second explanation is more likely to be the one used by the professor. He will more than likely continue to toss the coin until the proportion of heads and tails that appear becomes more even.

Bias

But suppose the purpose of such an experiment was to provide "statistical evidence" that was to be used to convince people to change their minds about things other than coins. If you and I interview 10 people concerning their political views, we may find that all ten are staunch Democrats. Does this give us the evidence we need to assert publicly, for political purposes that "all those interviewed supported the Democratic platform"? Of course not. But unless the user of this information understands the sampling issue involved and unless we are given complete information about the sampling process, how are we to react? How can we be sure that the pollster didn't "start out to find a biased coin" and then stop the polling process when an insufficient sample size "uncovered one for him," instead of making sure the sampling procedure was adequate? The answer is that without more complete information or a previous reputation for statistically accurate polling, we cannot be sure. We can, however, be alert to the risks we take when we do not ask for additional information.

Averages and comparisons

One of the oldest stories about improper use of averages is about a recruiter whose job it was to select pilots for the air force. His mandate was to ensure that the average pilot was 6 feet tall. As the story goes, he went out and found two unusual gentlemen, one only 4 feet tall and the other 8 feet tall. Of course, neither qualified, since the smaller one couldn't see over the cockpit dashboard, and the taller one couldn't even fit in the cockpit. In fact, however, they did average 6 feet. In our previous study of variation, we have learned to be sensitive to variation whenever we are using a measure of central tendency to describe data. In many cases, however, the statistical deception employed is much less obvious than that in our pilot example.

One school, which trains private pilots for their instrument examination, advertised that "our graduates score higher on the instrument written examination than graduates of other schools." To the unsuspecting reader, this seems perfectly clear. If you want to score higher on your instrument written examination, then this school is your best bet.

In fact, however, whenever we are using tests, we have to deal with standard error. Specifically, we need some measure of the precision of the test instrument, usually represented by standard error. This would tell us how large a difference in one school's grades would have to be for it to be statistically significant. Unfortunately, the advertisement did not offer such data; it merely asserted that "our graduates do better."

People who are sensitive to how tax monies are used often get upset over the seemingly low classroom utilization in public colleges and universities. One sometimes sees statements like, "at Eastern University, the average classroom utilization is only 34 percent." How are we to react to such assertions? The figure that is being quoted is a mean and includes potential scheduling hours that may have little or no value to the educational process.

For example, classes held late at night (say after 9:00) may be unproductive in terms of learning because of fatigue. And then, classes scheduled before 8:00 a.m. may not be feasible because of interruption of the normal sleep period and lack of adequate transportation facilities in these very early hours. Finally, classes scheduled on Saturdays and Sundays may encounter resistance because of religious traditions. When all these factors are considered, classroom utilization may be significantly higher than 34 percent, but until we know the basis for computing the average, we cannot react intelligently to such a published figure.

Making big jumps

College students often see ads for learning aids. One very popular such aid is a combination outline, study guide, and question set for various courses. Advertisements about such items often claim better examination scores with less studying time. Suppose a "study guide" for a basic statistics course is available through an organization that produces such guides for 50 different courses. If their study guide for basic statistics has been tested (and let us assume properly), they may advertise that "our study guides have been statistically proven to raise grades and lower study time." Of course their assertion is quite true, but only as it applies to their basic statistics experience. There may be no evidence of statistical significance that establishes the same kind of results for the other 49 guides.

Another product may be advertised as being beneficial in removing crab grass from your lawn and may assert that the product has been "thoroughly tested" on real lawns. Even if we assume that the proper statistical procedures were, in fact, used during the tests, such claims still involve "big jumps." Suppose that the test plot was in Florida, and your lawn problems are in Utah. Differences in rainfall, soil fertility, airborne pollutants, temperature, dormancy hours, and germination conditions may vary widely between these two locations. Claiming results for a statistically valid test under a completely different set of test conditions is invalid. One such test cannot measure effectiveness under a wide variety of environmental conditions.

A national association of trucklines claimed in an advertisement that "75 percent of everything you use travels by truck." This might lead us to believe that cars, railroads, airplanes, ships, and other forms of transportation carried only 25 percent of what we use. Reaching such a conclusion is easy but not enlightening. Missing from the trucking assertion is the question of "double counting." What did they do when something was carried to your city by rail and delivered to your house by truck? Or how were packages treated if they went by air mail and then by scooter? When the double-counting issue (a very complex one to treat) is resolved, it turns out that trucks carry a much lower proportion of the goods you use than truckers claimed. Although trucks are involved in *delivering* a relatively high proportion of what you use, railroads and ships still carry more goods for more total miles.

And then, of course, there is always some study purporting to show that one variable causes some change in another variable. Many years ago, Professor Stanley Jevons demonstrated that changes in sun spots caused business cycles. Even if this assertion were supported by statistically sound regressions, we must still remember that people who apply regression analysis sometimes find a relationship between two variables that have no real common bond. And when they find such a statistical relationship, they may

Projecting too far

Results in different places

Double counting

What causes what?

incorrectly assert that one variable "causes" a change in the other. In this regard, if one were to run a large number of regressions between many pairs of variables, it would probably be possible to get some rather interesting suggested "relationships." It might be possible, for example, to find a high statistical relationship between your income and the amount of beer that is consumed in the United States, or even between the length of a freight train (in cars) and the weather. But in neither case is there a factor common to both variables; hence, such "relationships" are meaningless. As in most other statistical "situations," it takes *both* knowledge of the inherent limitations of the technique that is used *and* a large dose of common sense, to avoid coming to unwarranted conclusions.

Finding things that do not exist

Graphs and pictures

Pictures lie, too

How many times have we heard the old aphorism "one picture is worth a thousand words"? And if that picture (or graph in the case of statistical presentations) leads us to an invalid conclusion, what is it worth then? Granted, the impact of pictorial displays is high. Glancing at one graph is easier than reading 5 pages of summary description, but the opportunity to mislead through the use of charts and graphs is perhaps the greatest challenge of all to someone practicing "statistical deception."

Suppose a California community is trying to attract retirees from colder climates. Further suppose that the community competes with Florida communities in this regard. Now, in the minds of potential retirees, the average winter temperature ranks high on the list of criteria affecting their choice of a retirement community. The same graphed data can possibly create two different impressions among those persons who see it.

First, assume that the average winter temperature in the California community is two degrees higher than that of competing Florida communities. How should we illustrate this situation graphically, so that it will be interpreted in our favor? In Fig. A·1, we illustrate average winter temperature graphically, without supplying the numbers for the vertical scale. A quick glance at Fig. A·1 suggests that winter temperatures in the California community are *much* higher than those in Florida communities. That is simply because we have neglected to indicate that each mark on the vertical scale

When the scales are left off

FIG. A·1
Average winter temperature in California and Florida communities

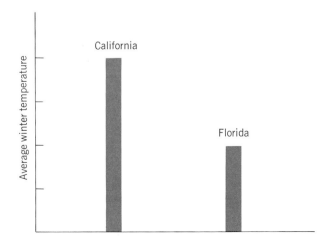

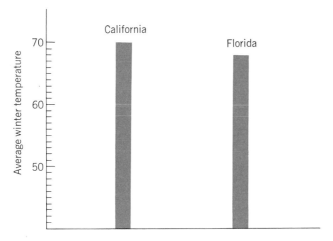

FIG. A·2
**Average winter
temperatures in
California and
Florida
communities**

is only 1 degree; hence, readers will tend to use relative heights on the two graphs in Fig. A·1 to form an opinion (which will likely be wrong).

Now suppose that you are employed by the Chamber of Commerce in the Florida community. You realize that your community has a two-degree disadvantage over your California competition, and you want to minimize that difference in graphical presentations. How to do it? The graph in Fig. A·2, using the same data, minimizes the adverse difference for the Florida community. It does this in two ways: by supplying numbers for the vertical scale and by changing the dimensions of the vertical scale to minimize the visual difference between the two bars. Viewers of the graph in Fig. A·2 would get the impression that winter temperatures are just about the same in both locations, an impression highly desired by the Florida community.

Turning things around for your benefit

We have included several exercises involving statistical deception, so that you can decide how much to believe of what you read and see.

EXERCISES

1. Recently, a student group surveyed attitudes toward a number of economic, social, and political issues. To obtain their data, they went in groups of two or three to several shopping centers around the city in the afternoons after school. Here they surveyed people for a couple of hours every day, asking them several questions on current topics of interest. Is their sample representative of the population of the city? Why or why not?

2. A private school has been suffering from decreased enrollment in the last two years and has launched a recruiting campaign in newspaper ads. One ad claims "our high school seniors score 25 points higher on the average on college entrance exams than do seniors in the public schools." Should parents rush to enroll their children in the private school, so that they will do well on these exams?

3. Research psychologists often remark how much we know about the typical college sophomore, because they are the most common subjects in university research. What are some of the shortcomings of using college sophomores in so much of our research?

4. An advertisement for a headache and pain remedy claims "laboratory tests showed that" their product "relieves pain twice as fast" as their leading competitor's product. What questions should occur to us in hearing this statement?

5. After a recent newspaper article criticizing state legislators for poor attendance during legislative sessions, one member of the legislature told a press conference, "Statistics show that the average attendance for days when the legislature is in session is 180, or 75 percent, of the total of 240 legislators." Should we conclude that the newspaper article was based on faulty information?

6. A commercial for sugarless chewing gum advertises "Four out of five dentists surveyed recommend sugarless gum for their patients who chew gum." Why should we be cautious of statements like this?

7. Recently, one the major automakers came under fire from consumer groups because of the unsafe location of the gas tank in the car, which could cause a fire or explosion if hit from the rear. In one television news report, attention had been focused on the small subcompact model put out by this company. One person in a consumer organization was quick to point out that the small car was not the only model about which to be concerned. She said that while 19 fatalities had resulted from fires in the small model cars, 35 deaths attributable to auto fires had occurred in a larger model car made by the same manufacturer. Why are the numbers of fatalities resulting from fire in the two car models inappropriate values for comparison?

8. An advertisement in a popular magazine reads as follows: "In a major hospital, _____ relieved pain better than regular aspirin. Here's why that could be important when you have a headache." The ad goes on to say in small print that because headache pain is not readily measured, the product was tested against aspirin on different kinds of pain. Still the ad urges use of the product for headache pain, based on this evidence. What is wrong with this type of reasoning?

9. A recent survey of working people showed a high correlation between the income level of an individual and the age at which he first married. Should we conclude from this that the later we get married the more income we're apt to make?

10. As you examine these graphs, what problems do you see with their depiction and interpretation of the data?

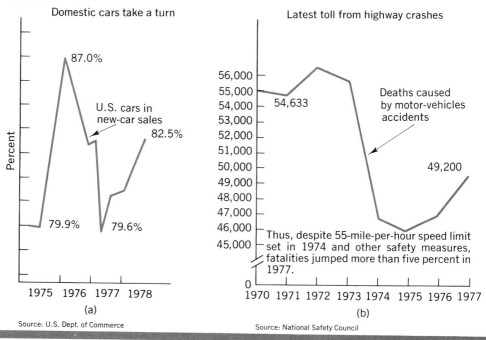

(a) Domestic cars take a turn — Source: U.S. Dept. of Commerce

(b) Latest toll from highway crashes — Source: National Safety Council

2 **DESIGN OF EXPERIMENTS** We have encountered the term *experiment* in Chapter 5, "Probability I." There we defined an *event* as one or more of the possible outcomes of doing something, and an *experiment* as an activity that would produce such events. In a coin toss experiment, the possible events would be heads and tails.

Events and experiments revisited

Planning experiments

If we are to conduct experiments that produce meaningful results in the form of usable conclusions, the way in which these experiments are designed is of the utmost importance. A good part of Chapter 7, "Sampling and sampling distributions," was taken up with ways of ensuring that random sampling was indeed being done. The way in which sampling is conducted is only a *part* of the total design of an experiment. In fact, the design of experiments is, itself, the subject of quite a number of books, some of them rather formidable, both in scope and volume.

Sampling is only one part

Phases of experimental design

To get a better feel for the complexity of experimental design without actually getting involved with the complex details, take an example from the many that confront us every day, and follow that example through from beginning to end.

The statement is made that a Crankmaster Battery will start your car's engine better than Battery X. Crankmaster might design its experiment this way:

A claim is made

Objective. This is our beginning point. Crankmaster wants to test its battery against the leading competitor. Although it is possible to design an experiment that would test the two batteries on several characteristics (life, size, cranking power, weight, and cost, to name but a few), Crankmaster has decided to limit this experiment to cranking power.

Objectives are set

What is to be measured. This is often referred to as the response variable. If Crankmaster is to design an experiment that measures cranking power of its battery against that of another, it must define how cranking power is to be measured. Again, there are quite a few ways in which this can be done. For example, Crankmaster could measure (1) the time it took for the batteries to run down completely while cranking engines, (2) the total number of engine starts it took to run down the batteries, or (3) the number of months in use that the two batteries could be expected to last. Crankmaster decides that the response variable in its experiment will be (1) the time it takes for batteries to run down completely while cranking engines.

The response variable is selected

How large a sample size. Crankmaster wants to be sure that it chooses a sample size large enough to support claims it makes for its battery, without fear of being challenged; however, it knows that the more batteries it tests, the higher the cost of conducting the experiment. As we pointed out in Chapter 7, "Sampling and sampling distributions," there is a

How many to test

diminishing return in sampling; and although sampling more items does, in fact, decrease the standard error, the benefit may not be worth the cost (a statistician would say that "the increased precision is not worth the additional cost"). Not wishing to choose a sample size that is too expensive to contend with, Crankmaster decides that comparing 10 batteries from each of the two companies will suffice.

Type I and Type II error. Before Crankmaster can conduct an experiment, it has to decide how serious these two types of errors are in relation to each other. You will recall from Chapter 9, "Testing hypotheses," that a Type I error is rejecting a null hypothesis when it is true; a Type II error is accepting the null hypothesis when it is false.

Setting a significance level

Crankmaster can reduce the probability of making one of those types of errors only if it is willing to increase the probability of making the other type error. To decide on an appropriate level of significance, Crankmaster needs to consider the seriousness of the two different types of error. In this case, making a Type I error (rejecting the null hypothesis when it's true) would run the risk of serious intervention by consumer groups, competitors, and govenment agencies; therefore, Crankmaster wisely chooses a rather low significance level, .01 in this instance.

Conducting the experiment. Crankmaster must be careful to conduct its experiment under controlled conditions; that is, it has to be sure that it is measuring *cranking power,* and that the other variables (such as temperature, age of engine, and condition of battery cables, to name only a few) are held as nearly constant as practicable. In an effort to accomplish just this, Crankmaster's statistical group uses new cars of the same make and model, conducts the tests at the same outside air temperature, and is careful to be quite precise in measuring the time variable. Crankmaster gathers experimental data on the performance of the 10 batteries from each manufacturer in this manner.

Experimental conditions

Analyzing the data. Data on the 20 individual battery tests are subjected to hypothesis testing in the same way that we introduced it in Chapter 9, "Testing hypotheses." Crankmaster is interested in whether there is a significant difference between the cranking power of its battery and that of its competitor. The proper test for Crankmaster to use here would be a right-tailed test of the difference between two means. It turns out that the difference between the mean cranking life of Crankmaster's battery and that of its competitor *is* significant. Crankmaster incorporates the result of this experiment into its advertising campaign.

Data are analyzed

Reacting to experimental claims

How should the consumer react?

How should we, as consumers, react to Crankmaster's new battery life claims in its latest advertising? Should we conclude from the tests it has run that the Crankmaster battery *is* superior to the competitive battery? If we stop for a moment to consider the nature of the experiment and the hypothesis test that accompanied it, we may not be too quick to come to such a conclusion.

How do we know that the ages and conditions of the cars' engines in the experiment *were* identical? And are we absolutely sure that the battery cables were identical in size and resistance to current? And what about the air temperature during the tests: was it the same? These are the normal kinds of questions that we should ask.

Are we sure?

How should we react to the statement, if it is made, that "we subjected the experimental results to extensive statistical testing"? The answer to that is to recall that in Chapter 9 we pointed out that hypothesis testing, while able to point out (as it did in Crankmaster's case) that an observed difference between cranking power was too large to be attributed to chance, *cannot* tell us exactly why such a difference occurred.

Hypothesis testing revisited

Other options open

Of course, Crankmaster would have had the same concerns we did, and in all likelihood would *not* have made significant advertising claims solely on the basis of the experimental design we have just described. One possible course of action to avoid criticism is to *ensure* that all variables except the one being measured have indeed been controlled. Despite the care taken to produce such controlled conditions, it turns out that these overcontrolled experiments do not really solve our problem. Normally, instead of investing resources in attempts to *eliminate* experimental variations, we choose a *completely different route*. The next section shows how we can accomplish this.

Another route for Crankmaster

Factorial experiments

In the Crankmaster situation, we had two competing batteries (let's refer to them now as A and B) and three test conditions that were of some concern to us: (1) temperature (2) age of the engine and (3) condition of the battery cable. Let's introduce the notion of factorial experiments by using this notation:

Handling all the test conditions at the same time

H = hot temperature   N = new engine   G = good cable
C = cold temperature   O = old engine   W = worn cable

Of course, in most experiments, we could find more than two temperature conditions and, for that matter, more than two conditions for engine condition and battery cable condition. But it's better to introduce the idea of factorial experiments using a somewhat simplified example.

Now, since there are two batteries, two temperature possibilities, two engine condition possibilities, and two battery cable condition possibilities, there are $2 \times 2 \times 2 \times 2$ (or 16) possible combinations of factors. If we wanted to write these sixteen possibilities down, they would look like this:

How many combinations?

Test	Battery	Temperature	Engine condition	Cable condition
1	A	H	N	G
2	A	H	N	W
3	A	H	O	G
4	A	H	O	W
5	A	C	N	G
6	A	C	N	W
7	A	C	O	G
8	A	C	O	W
9	B	H	N	G
10	B	H	N	W
11	B	H	O	G
12	B	H	O	W
13	B	C	N	G
14	B	C	N	W
15	B	C	O	G
16	B	C	O	W

Having set up all of the possible combinations of factors involved in this experiment, we could now conduct the sixteen tests in the table. If we did this, we would have conducted a complete factorial experiment, because each of the two *levels* of each of the four *factors* would have been used once with each possible combination of other levels of other factors. Designing the experiment this way would permit us to use techniques from Chapter 10, "Chi-square test and analysis of variance," to test the effect of each of the factors.

We need to point out, before we leave this section, that in an actual experiment we would hardly conduct the tests in the order in which they appear in the table. They were arranged in that order to facilitate your counting the combinations and determining that all possible combinations were indeed represented. In actual practice, we would randomize the order of the tests, perhaps by putting sixteen numbers in a hat and drawing out the order of the experiment in that simple manner.

Being more efficient in experimental design

As you saw from our 4-factor experiment, 16 tests were required to compare all levels with all factors. If we were to compare the same 2 batteries, but this time with 5 levels of temperature, 4 measures of engine condition, and 3 measures of battery cable condition, it would take $2 \times 5 \times 4 \times 3 = 120$ tests for a complete factorial experiment.

Fortunately, statisticians have been able to help us reduce the number of tests in cases like this. To illustrate how this works, look at the consumer products company that wants to test market a new toothpaste in 4 different cities with 4 different kinds of package and with 4 different advertising pro-

Levels and factors to be handled

Randomizing

A bit of efficiency

grams. In such a case, a complete factorial experiment would take 4 × 4 × 4 = 64 tests. However, if we do some clever planning, we can actually do it with far fewer tests, 16 to be precise.

Let's use the notation:

A = city 1	I = package 1	1 = ad program 1
B = city 2	II = package 2	2 = ad program 2
C = city 3	III = package 3	3 = ad program 3
D = city 4	IV = package 4	4 = ad program 4

Now we arrange the cities, packages, and advertising programs in a design called a Latin square (Fig. A·3).

In the experimental design represented by the Latin square, we would need only 16 tests instead of 64 as originally calculated. Each combination of city, package, and advertising program would be represented in the 16 tests. The actual statistical analysis of the data obtained from such a Latin square experimental design would require a form of analysis of variance a bit beyond the scope of this book.

The statistical analysis

FIG. A·3
A Latin square.

3 TIME SERIES

Forecasting, or predicting, is an essential tool in any decision-making process. Its uses vary from determining teacher requirements for a local school system to estimating the annual GNP of the country. The quality of the forecasts made by decision makers is strongly related to the information that can be extracted and used from past data. *Time series analysis* is one quantitative method we use to determine patterns in data collected over time. Table A·1 is an example of time series data.

Definition of time series

TABLE A·1
Time series for gasoline consumption in U.S. (billions of gallons)

Year	1971	1972	1973	1974	1975	1976	1977	1978
Quantity	5.53	5.69	5.58	6.08	6.76	7.78	8.32	8.86

Time series analysis is used to detect patterns of change in statistical information over regular intervals of time. We *project* these patterns to arrive at an estimate for the future. Thus, time series analysis helps us cope with uncertainty about the future.

Use of time series

Variations in time series

We use the term *time series* to refer to any group of statistical information accumulated at regular intervals. There are four kinds of change, or variation, involved in time series analysis. They are:

1. secular trend
2. cyclical fluctuation
3. seasonal variation
4. irregular variation.

With the first type of change, *secular trend*, the value of the variable tends to increase or decrease over a long period of time. The steady increase in the cost of living recorded by the Consumer Price Index is an example of secular trend. From year to individual year, the cost of living varies a great deal; but if we examine a long-term period, we see that the trend is toward a steady increase. Figure A·4 (*a*) shows a secular trend in an increasing but fluctuating time series.

Secular trend

The second type of variation seen in a time series is *cyclical fluctuation*. The most common example of cyclical fluctuation is the economic cycle. Over time, there are years when the economic cycle hits a peak above the trend line. At other times, economic activity is likely to slump, hitting a low point below the trend line. The time between hitting peaks or falling to low points is at least one year, and it can be as many as 15 or 20 years. In Fig. A·4, chart *b* illustrates a typical pattern of cyclical fluctuation above and below a secular trend line. Note that the cyclical movements do not follow any definite trend but move in a somewhat unpredictable manner.

Cyclical fluctuation

The third kind of change in time series data is *seasonal variation*. As we might expect from the name, seasonal variation involves patterns of change within a year and tend to be repeated from year to year. For example, a physician can expect a substantial increase in the number of flu cases every winter and of poison ivy every summer. Since these are regular patterns, they are useful in forecasting the future. In *c,* Fig. A·4, we see a seasonal variation. Notice how it peaks in the fourth quarter of each time period.

Seasonal variation

Irregular variation is the fourth type of change discussed in time series analysis. In many situations, the value of a variable may be completely unpredictable, changing in a random manner. Irregular variations describe such movements. An example of irregular variation is the effect on gasoline sales in the U.S. when the Middle East conflict occurred in 1973 or the Iranian government changed in 1979. In Fig. A·4, *d* illustrates the characteristics of irregular variation.

Irregular variation

Thus far we have referred to a time series as exhibiting one or another of these four types of variation. In most instances, however, a time series will contain several of these components. Time series analysis describes the overall variation in a single time series in terms of these four different kinds of variation.

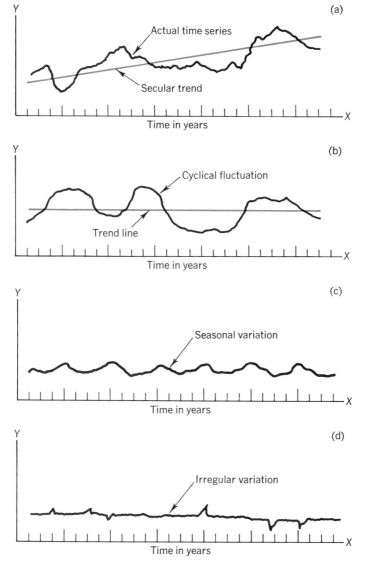

FIG. A·4
Time series variations

4 INDEX NUMBERS   At some time, everyone faces the question of how much something has changed over a period of time. A person may want to know how much the price of groceries has increased, so she can adjust her budget accordingly. A factory manager may wish to compare this month's per unit production cost with that of the past six months. Or a medical research team may wish to compare the number of flu cases reported this year with the number reported in previous years. In each of these situations, the degree of change needs to be determined and defined. Typically, we use *index numbers* to measure such differences.

Why use an index number?

An index number measures how much a variable changes over time. We calculate an index number by finding the ratio of the current value to a base value. Then we multiply the resulting number by 100 to express the index as a percentage. This final value is the *percentage relative.* Note that the index number for the base year is always 100.

Suppose EPA is interested in measuring the change in the number of cars in the U.S. over 14 years. The data EPA collects shows that 9.3 million cars were sold in 1965; 6.5 million in 1970; 9.6 million in 1973; 10.1 million in 1976; and 11.9 million in 1979. If 1965 is the base year, EPA calculates the index numbers reflecting sales changes using the process presented in Table A·2.

Calculating a simple index

Using these calculations, EPA finds that sales in 1970 had an index of 70 relative to 1965. Another way to state this is to say that sales in 1970 were 70 percent of the sales in 1965.

TABLE A·2
Calculation of index numbers (Base year = 1965)

Year (1)	Number of auto sales (millions) (2)	Ratio (3) = (2) ÷ (9.3)	Index or percentage relative (4) = (3) × 100
1965	9.3	$\frac{9.3}{9.3} = 1.00$	1.00 × 100 = 100
1970	6.5	$\frac{6.5}{9.3} = 0.70$	0.70 × 100 = 70
1973	9.6	$\frac{9.6}{9.3} = 1.03$	1.03 × 100 = 103
1976	10.1	$\frac{10.1}{9.3} = 1.09$	1.09 × 100 = 109
1979	11.9	$\frac{11.9}{9.3} = 1.28$	1.28 × 100 = 128

Types of index numbers

Price index

There are three principal types of indices: the price index, the quantity index, and the value index. A *price index* is the one most frequently used. It compares changes in price from one period to another. A price index familiar to everyone is the Consumer Price Index, tabulated by the Bureau of Labor Statistics. It measures overall price changes of a variety of consumer goods and services and is used to define the cost of living.

Quantity index

A *quantity index* measures how much the number, or quantity, of a variable changes over time. Our example using car sales determined a quantity index relating the volume of sales in 1970, 1973, 1976 and 1979 to that in 1965.

Value index

The last type of index, the *value index,* measures changes in total monetary worth. That is, it measures changes in the dollar value of a variable. In effect, the value index combines price and quantity changes to present a more informative index. In our auto example, we determined only a quantity index.

Usually, an index measures change in a variable over a period of time, such as in a time series. However, it can also be used to measure differences in a given variable in different locations. This is done by simultaneously collecting data in different locations and then comparing the data. The comparative cost of living index, for example, shows that in terms of the cost of goods and services, it is cheaper to live in Austin, Texas, than in New York City.

A single index may reflect a composite, or group, of changing variables. The Consumer Price Index measures the general price level for specific goods and services in the economy. It combines the individual prices of the goods and services to form a composite price index number.

Composite index numbers

Uses of index numbers

Index numbers can be used in several ways. It is most common to use them by themselves, as an end result. Index numbers such as the Consumer Price Index are often cited in news reports as general indicators of the nation's economic condition.

Decision makers use index numbers as part of an intermediate computation to understand other information better. In time series analysis, seasonal indices are used to modify and improve estimates of the future. The use of the Consumer Price Index to determine the "real" buying power of money is another example of how index numbers help increase knowledge of other factors.

One use of the Consumer Price Index

Problems related to index numbers

Several things can distort index numbers. The four common causes of distortion are discussed below:

1. Sometimes there is difficulty in finding suitable data to compute an index. Suppose the Federal Aviation Administration is computing an index describing seasonal variation in aircraft accidents. If accident data are reported only on an annual basis, it would be unable to determine the seasonal pattern.

Limited data

2. Incomparability of indices occurs when attempts are made to compare one index with another after there has been a basic change in what is being measured. If the managers of Phoenix Television Company compare price indices of TV sets from 1956 to 1979, they find that prices have increased substantially. However, this comparison does not take into consideration technological advances in the quality of TV sets made over the time period in consideration.

Incomparability

3. Inappropriate weighting of factors can also distort an index. In developing a composite index, such as the Consumer Price Index, we must

Inappropriate weighting

consider changes in some variables to be more important than changes in others. The effect on the economy of a 50¢ per gallon increase in the price of gasoline cannot be counterbalanced by a 50¢ decrease in the price of cars. It must be realized that the 50¢ per gallon increase in gas cost has a much greater effect on consumers. Thus, greater weight has to be assigned to the gas price increase than to increase in the cost of cars.

Use of an improper base

4. Distortion of index numbers also occurs when **selection of an improper base** occurs. Sometimes a firm selects a base that automatically leads to a result in its own interest and proves its initial assumption. If Consumers Against Oil Waste wants to portray oil companies in a bad light, it might measure this year's profits, with a recession year as its base for oil profits. This would produce an index that shows oil profits have increased substantially. On the other hand, if Consumers for Unlimited Oil Use wishes to show that this year's profits are minimal, it would select a year with high profits for its base year. Using high profit as a base would probably result in an index indicating a small increase or maybe even a decline in oil profits this year. Therefore, we must always consider how and why the base period was selected before accepting a claim based on the result of comparing index numbers.

Sources of index numbers

Sources of data for index numbers

When people apply index numbers to everyday problems, they use many sources to obtain the necessary information. The source depends on their information requirements. In dealing with broad areas of national economy and the general level of business activity, publications such as the *Federal Reserve Bulletin, Moody's, Monthly Labor Review,* and the *Consumer Price Index* provide a wealth of data. Many federal and state publications are listed in the U.S. Department of Commerce pamphlet, *Measuring Markets.* Almost all governmental agencies distribute data about their activities, from which index numbers can be computed. Many financial newspapers and magazines provide information from which index numbers can be computed. When you read these sources, you will find that many of them use index numbers themselves.

5 CONCLUSION   We've spent about 500 pages introducing you to statistics, all the way from "where you get them" through "how to use them" and finally to "how to guard against misuse of statistics." We realize that most of the readers of this book will not go on to become professional statisticians; however, as citizens who are more sophisticated about the uses of statistics in modern life, you will be better prepared to make the kinds of consequential decisions you will face. Being a part of your development process has been enjoyable for us. Good luck to you.

Answers to selected even-numbered exercises

Some answers or parts of answers have been omitted. When an exercise calls for simple rearrangement or display of data with no computation or interpretation, the resulting answer may not appear.

Chapter 2

2. The 30 students that make up the average are chosen as being representative of the entire collection of third graders from 5 states. They form a *sample* of the larger *population* of third graders. From the sample, an inference is drawn about the population as a whole; when the average goes down, all IQ's as a whole are considered to have gone down.

4. The sample may have contained a greater percentage of conservatives (or liberals) than the actual voting population, a greater percentage of Republicans (or Democrats) than the actual voting population, or maybe a greater percentage of farm persons (or city dwellers) than the actual voting population. The sample could also have been biased. Maybe the questions were slanted toward one candidate, or perhaps the pollsters did not ask the right questions. The sample could also have been too small. (*Note:* Experts have attributed the real cause of the turnabout to the fact that Dewey supporters became so confident of victory that they did not bother to vote, whereas Truman supporters did.)

6. We cannot draw any conclusions from these data as they exist. We would first need to do a certain amount of rearranging, such as listing the grades from highest to lowest or determining the most frequent grade pair.

8. No. In this case, the raw data would be a list of sample units indicating whether or not they were defective. The quality control section has already performed an analysis on these data to calculate the averages contained in the report.

10.
450–499.9	1	550–599.9	9	650–699.9	4
500–549.9	10	600–649.9	15	700–749.9	1

12. (a) 7 equal intervals

Class	Frequency
30–39	.02
40–49	.08
50–59	.20
60–69	.28
70–79	.22
80–89	.12
90–99	.08
	1.00

(b) 13 equal intervals

Class	Frequency
35–39	.02
40–44	.04
45–49	.04
50–54	.08
55–59	.12
60–64	.12
65–69	.16
70–74	.12
75–79	.10
80–84	.06
85–89	.06
90–94	.04
95–99	.04
	1.00

16. (a)

Spread	SAT differential	Spread	SAT differential
1.3	150	0.0	0
0.8	160	−0.1	5
0.6	75	−0.2	−10
0.4	45	−0.2	−10
0.3	60	−0.5	−40
0.3	50	−0.5	−80
0.2	10	−0.6	−110
0.1	5	−0.7	−110
0.1	5	−1.1	−130
0.1	0		
0.1	−5		

(b) From the array, we can see that the most common spread is 0.1, which occurs 4 times.

(c) For a spread of 0.1, the most common SAT differential is +5 which occurs twice out of the 4 data points.

18.

Classes (lbs./sq. in.)	Relative frequencies
2490.0–2493.9	.150
2494.0–2497.9	.175
2498.0–2501.9	.325
2502.0–2505.9	.225
2506.0–2509.9	.125

20. (a) 29.995–34.995
34.995–39.995
39.995–44.995
44.995–49.995
49.995–54.995
54.995–59.995

(b) 30.00–34.99
35.00–39.99
40.00–44.99
50.00–54.99
55.00–59.99

(c) 32.5, 37.5, 42.5, 47.5, 52.5, 57.5

22.

Class marks	Stated limits	Real limits
8.50	7.0– 9.99	6.995– 9.995
11.50	10.0–12.99	9.995–12.995
14.50	13.0–15.99	12.995–15.995
17.50	16.0–18.99	15.995–18.995
20.50	19.0–21.99	18.995–21.995
23.50	22.0–24.99	21.995–24.995
26.50	25.0–27.99	24.995–27.995
29.50	28.0–30.99	27.995–30.995

24.

Closed	Open
Single	Single
Married	Married
Divorced	Other
Separated	
Widowed	

26.

Class marks	Real limits	Stated limits
102.45	99.95–104.95	100.0–104.9
107.45	104.95–109.95	105.0–109.9
112.45	109.95–114.95	110.0–114.9
117.45	114.95–119.95	115.0–119.9
122.45	119.95–124.95	120.0–124.9
127.45	124.95–129.95	125.0–129.9
132.45	129.95–134.95	130.0–134.9
137.45	134.95–139.95	135.0–139.9

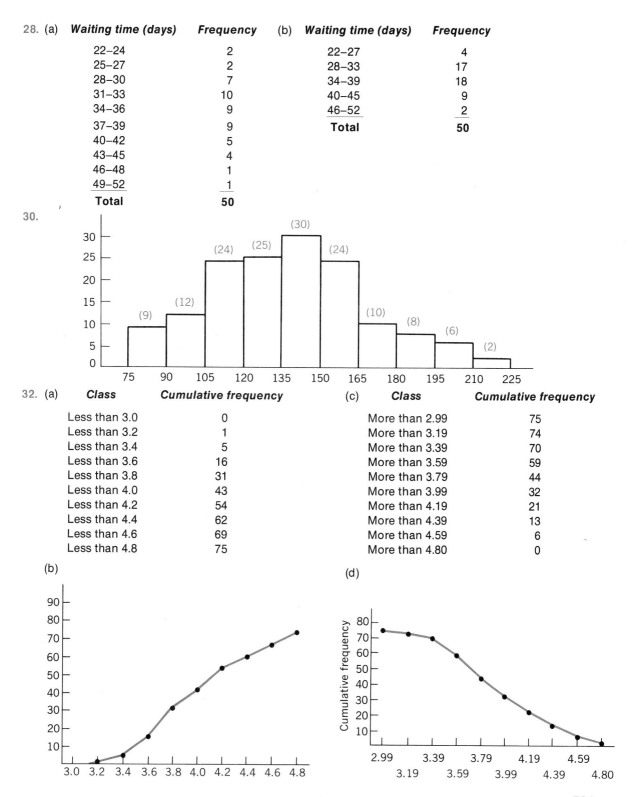

28. (a)

Waiting time (days)	Frequency
22–24	2
25–27	2
28–30	7
31–33	10
34–36	9
37–39	9
40–42	5
43–45	4
46–48	1
49–52	1
Total	**50**

(b)

Waiting time (days)	Frequency
22–27	4
28–33	17
34–39	18
40–45	9
46–52	2
Total	**50**

30.

32. (a)

Class	Cumulative frequency
Less than 3.0	0
Less than 3.2	1
Less than 3.4	5
Less than 3.6	16
Less than 3.8	31
Less than 4.0	43
Less than 4.2	54
Less than 4.4	62
Less than 4.6	69
Less than 4.8	75

(c)

Class	Cumulative frequency
More than 2.99	75
More than 3.19	74
More than 3.39	70
More than 3.59	59
More than 3.79	44
More than 3.99	32
More than 4.19	21
More than 4.39	13
More than 4.59	6
More than 4.80	0

(b)

(d)

34. (a)

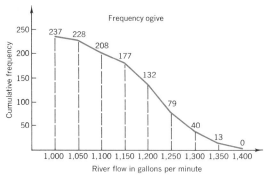

Frequency ogive

River flow "more than"	1,000	1,050	1,100	1,150	1,200	1,250	1,300	1,350	1,400
Cumulative frequency	237	228	208	177	132	79	40	13	0

(b)

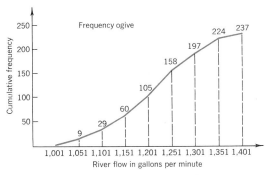

Frequency ogive

River flow "less than"	1001	1051	1101	1151	1201	1251	1301	1351	1401
Cumulative frequency	0	9	29	60	105	158	197	224	237

36. (b)

Minutes to set type	Frequency	Minutes to set type "less than"	Cumulative frequency
19.0–19.7	4	19.0	0
19.8–20.5	4	19.8	4
20.6–21.3	10	20.6	8
21.4–22.1	5	21.4	18
22.2–22.9	7	22.2	23
23.0–23.7	5	23.0	30
23.8–24.5	11	23.8	35
24.6–25.3	4	24.6	46
Total	**50**	25.4	50

(c)

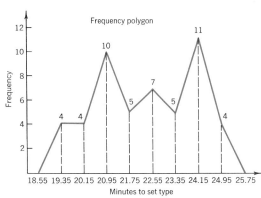

Frequency polygon

(d)

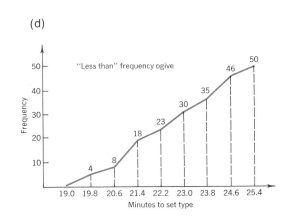

"Less than" frequency ogive

38. No, because it has been analyzed to some extent to get averages for each week's attendance and to get percentages of full attendance. Raw data would be the actual number of absences for each day or week of the time period.

40. To calculate the answer, construct a "less than" cumulative distribution.

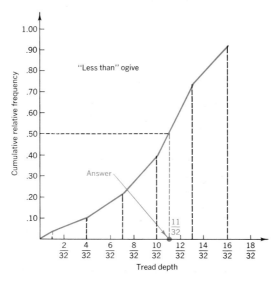

Tread depth (inches) "less than"	0/32	1/32	4/32	7/32	10/32	13/32	16/32	17/32
Cumulative relative frequency	0	.03	.10	.22	.42	.75	.92	1.00

42. **(b)** Here we could use several different interval classifications. The following distribution is based on 5-minute intervals.

Classes (mins.)	Frequency
1–5	2
6–10	3
11–15	5
16–20	4
21–25	3
26–30	3
Total	**20**

(c) The following distribution is based on 3-minute intervals.

Classes (mins.)	Frequency
1–3	1
4–6	1
7–9	3
10–12	1
13–15	4
16–18	1
19–21	3
22–24	3
25–27	1
28–30	2
Total	**20**

533

(a)

	1st ten		**Largest ten**
226	198	267	264
210	233	259	258
222	175	257	252
215	191	248	245
201	175	244	243

(b) No. The sample of the first ten elements is more representative because the items in it more adequately represent the distribution of items in the whole population in terms of their values.

46.

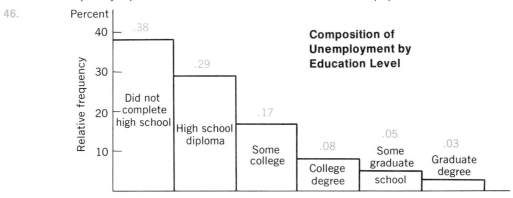

48.

50.

Real limits	**Stated limits**	**Class marks**
.05– .35	.10– .30	.20
.35– .65	.40– .60	.50
.65– .95	.70– .90	.80
.95–1.25	1.00–1.20	1.10
1.25–1.55	1.30–1.50	1.40
1.55–1.85	1.60–1.80	1.70
1.85–2.15	1.90–2.10	2.00
2.15–2.45	2.20–2.40	2.30
2.45–2.75	2.50–2.70	2.60
2.75–3.05	2.80–3.00	2.90
3.05–	3.10–	none

52. (a)

Journal number	Frequency	Relative frequency
1	1	.0417
2	2	.0833
3	2	.0833
4	0	.0000
5	2	.0833
6	2	.0833
7	3	.1250
8	1	.0417
9	2	.0833
10	1	.0417
11	2	.0833
12	2	.0833
13	0	.0000
14	2	.0833
15	2	.0833
Total	**24**	**.9998**

(b)

Branch	Frequency	Relative frequency
North	11	.4583
West	8	.3333
South	5	.2083
Total	**24**	**.9999**

(c)

No. of publications	Frequency	Relative frequency
1–3	6	.2500
4–6	5	.2083
7–9	4	.1667
10–12	4	.1667
13–15	2	.0833
16–18	2	.0833
19–21	1	.0417
Total	**24**	**1.000**

54. ① Age—This distribution would be quantitative and discrete. The distribution would be discrete because people report age at their last birthday. The distribution would also be most likely to be open on both ends, because the company would not be interested in very young or very old age classifications for marketing purposes.

② Income—This is again a quantitative, continuous distribution because of the numerical data. Here the company would most certainly use group, rather than individual, categories. Also, the distribution would most likely be open-ended for the same reasons as the age distribution.

③ Marital Status—Here we have a qualitative, discrete distribution because the categories of marital status would not be numerical. Depending on the choice of categories, the list might or might not be open-ended. The possible answers to the question are sufficiently limited that the company might wish to break down the data into all possible categories. However, the company might also choose to limit the categories and use "other" to make the distribution open-ended.

④ and ⑤ Where and Why—Both of these distributions would be qualitative and discrete. However, in contrast to "Marital Status," the answers could be so varied that it is unlikely that the company would choose to list all possibilities. Instead, the most frequent responses would be used and "other" included to cover all other possibilities. Therefore, these distributions would most likely be open-ended.

56. **Group 1**

None	Mild	Moderate	Severe
None	Mild	Moderate	Severe
None	Mild	Moderate	Severe
	Mild	Moderate	
	Mild	Moderate	
	Mild		
	Mild		

or:

None	3
Mild	7
Moderate	5
Severe	3

Group 2

None	Mild	Moderate	Severe
	Mild	Moderate	Severe
	Mild	Moderate	Severe
	Mild	Moderate	Severe
		Moderate	Severe
		Moderate	
		Moderate	
		Moderate	

or:

None	1
Mild	4
Moderate	8
Severe	5

It is better because one can more easily compare the two groups' responses to see if they are experiencing different degrees of discomfort as a whole. In this case, group 2 appears to be more uncomfortable.

58. The problem asks for a distribution by specialty, not combinations of specialties. Therefore, the only categories would be social psychology, clinical psychology, experimental psychology, developmental psychology, and no publications. Faculty members with more than one specialty would be double counted in this particular distribution.

Type of specialty	Frequency	Relative frequency
Social	17	.140
Clinical	41	.339
Experimental	40	.331
Developmental	22	.182
No publications	1	.008
	121	**1.000**

60.

Classes	Frequency	Relative frequency
0–10	1	.05
11–20	0	.00
21–30	1	.05
31–40	6	.30
41–50	3	.15
51–60	4	.20
61–70	2	.10
71–80	0	.00
81–90	3	.15
91–100	0	.00
	20	**1.00**

Chapter 3

2.

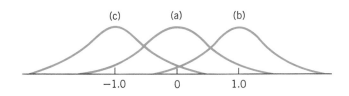

4. (a) *B*   (b) *A*   (c) *B*   (d) *A*   (e) neither

6. (a) 13.75   (b) 1.399   (c) 412.33   (d) 48

8. (a)

Interval	5–44	45–54	55–64	65–74	75–84	85–94	95–104
Class mark	40	50	60	70	80	90	100
Frequency	1	6	4	10	9	7	3

   (b) 72.225   (c) 73.25   (d) Very close to each other

10. 43,096 students   12. 23.195 seconds

14. 88.4   86.2   90.4   85.75   89.7

16. 1.976 children   18. $.117/ounce   20. 1.097

22. 1.32   24. $122.22   26. 673

28. (a) 330–349.5   (b) ave. of 150th and 151st   (c) .694   (d) 326.73

30. 29.5 min.   32. Approximately $402   34. (a) 65.9   (b) 65.85

36.

   (a) 16   (b)

Class	Frequency
10–14.9	12
15–19.9	18
20–24.9	7
25–29.9	3
	40

   (c) 16.76   (d) Fairly close to each other

38. 1.5 days and 4.4 days

40. (a) negatively skewed   (b) positively skewed   (c) negatively skewed

42. (a) Saturday   (b) 24.00–24.99 kilotons is the modal class; 24.6 kilotons is the modal value.

44. 440 children   46. $.887

48. (a) bimodal: 6″ and 5⅞″   (b) 9.0–9.9   (c) 9.75

50. There are 3 possible answers to this problem, depending on how risk averse you are. Since both curves are symmetrical and centered on zero winnings, the choice depends on the dispersion of possible outcomes. Students who are risk neutral would be indifferent between the two curves. They would attach no significance to the greater dispersion of distribution *A* relative to distribution *B*. However, if the student is risk averse, he should choose distribution *B*, because the most likely outcome is the same for both distributions, but the dispersion of distribution *B* is less. A student who is actually risk favorable might choose distribution *A* over *B*. This would indicate the willingness to accept additional risk without any expectation of greater winnings.

52. 7,110 tons   54. 20.35 mpg   56. 5.3%   58. 11.55%

60. 8.68%   62. 14.6 lbs   64. 7.47 secs

Chapter 4

2. *a*, because the values tend to cluster more around the mean.   4. *c*

6. There are many ways that the concept may be involved. Certainly, the FTC would examine the price variability for the industry and compare the result to that of the suspect companies. The agency might examine price distributions for similar products, for the same products in a city, or for the same products in different cities. If the variability was significantly different in any of these cases, this result might constitute evidence of a conspiracy to set prices at the same levels.

8. 4.73 4.90 5.02 5.10 5.24
   4.81 4.96 5.03 5.13 5.25
   4.85 4.97 5.07 5.17 5.31
   4.88 5.00 5.09 5.18 5.43
   (a) .18   (b) .09   (c) .70

10. (a) 1260   (b) 407   (c) 281   (d) 140.5   (e) 169
    (f) 140.5 measures the average range of ¼ of the data, whereas 169 measures the range of a particular ¼ of the data.

12. No

14.
Percentile	20	40	60	80
Interfractile range	4	2	2	

16. (a) 3.5
    (b) 22.456
    (c) 4.74

18. 56,566.67

20. (a) $\mu=7.1$ days; $\sigma=3.87$ days
    (b) 195 should be; 249 are
    (c) 247

22. 5 games below; 1 game above

24. The first paragraph

26. The first group

28. City 3

30. The second machine

32. $9116

34. A person making this statement has missed the point of variability. By definition, you *always* have an equal chance of falling above or below the median, regardless of the variability. We are not considering the average result when we consider variability, however. Instead, we are looking at what a single outcome might be. Even though the outcome might be the same on average, you wouldn't want to be the "one guy in a thousand" that happens to get the lowest possible result. The variability of a distribution describes this risk of having an outcome other than the mean value. You might say that variability tells us how bad or how good the outcome might be and gives us a relative measure of how likely these extreme values are.

36. California

38. (a) 4.3 tons
    (b) .81 tons
    (c) .41 tons
    (d) .62 tons

40. Brackson, if we consider the relative deviation

42. Missile 2   44. Var. 37778.57   Std. Dev. 194.367

48. (a) Day 1 = 1.9
       Day 2 = 1.0
    (b) In this case it is not an accurate measure, since the Day 1 range is distorted by one large value.

Chapter 5

2. The Surgeon General's office undoubtedly studied the incidence of sickness and death among smokers and nonsmokers. Evidence indicated that smokers were more likely to have poor health (or earlier deaths) than nonsmokers. From his samples, the Surgeon General has assigned a higher probability of poor health occurring in the smoking population than in the nonsmoking population.

4. This decision involves estimates of enrollment, costs, physical plant additions required, and new staff necessary. Each of these estimates involves a degree of uncertainty and therefore a probability estimate.

6. (b) and (d) are mutually exclusive.

8. (a) 0　　(b) $\frac{1}{36}$　　(c) $\frac{5}{36}$　　(d) $\frac{6}{36}$　　(e) $\frac{4}{36}$　　(f) $\frac{3}{36}$　　(g) $\frac{2}{36}$

10. (a) They are collectively exhaustive; they are not mutually exclusive.

(b)	M only	P only	M, P only	B, P only	P, L only
	B only	L only	M, L only	W, P only	M, W, P only
	W only	M, W only	B, W only	W, L only	W, P, L only

(c) M, L　　M, W, P　　B, W　　W, P, L　　B, P

12. (a) $\frac{1}{13}$　　(b) $\frac{1}{4}$　　(c) $\frac{1}{26}$　　(d) $\frac{1}{2}$　　(e) $\frac{3}{13}$ (These are classical probabilities.)

14. (a) $\frac{1}{13}, \frac{2}{13}, \frac{4}{13}, \frac{4}{13}, \frac{2}{13}$
    (b) Probability of greater than 400,000 or less than 200,000 is zero.

16. (a) subjective　　　　(d) relative frequency
    (b) relative frequency　　(e) classical
    (c) subjective

18. .30, .40, .62　　　　20. .0001768　　　　22. .145

24. (a) $\frac{1}{2}$　(b) $\frac{1}{2}$　　　26. (a) $\frac{1}{4}$　(b) $\frac{1}{2}$　(c) $\frac{1}{13}$

28. (a) .05　(b) .02　(c) .001　　　　30. .75

32. (a) .57　(b) .15　(c) .60　(d) .45　　　34. (a) .60　(b) .30

36. (a) .18　(b) .54, .12, .42　(c) .28　(d) .554　　38. (a) .48　(b) .39　(c) .13

40. Indianapolis　　　　42. (a) .005, .0025, .0140　(b) .23, .45, .32　(c) .0022

44. (a) $\frac{3}{8}$　(b) $\frac{1}{8}$　(c) $\frac{5}{16}$　(d) $\frac{3}{16}$　　　46. (c)

48. The chances of a player continually getting very good hands or a long string of successful rolls are slim.

50. (a) .36　(b) .64　　　　52. (c) and (e)

54. (a) .333　(b) relative frequency　(c) The number of observations is small, and they were made for a plant not exactly like the proposed one; therefore, this estimate takes on the characteristics of a subjective estimate.

56. The increase of movement is purely a subjective estimate and cannot be answered from the data given.

58. (a) should not divert　(b) should divert　　　60. (e)

62. (a) .85　(b) .85

64. (a) .60　(b) .50

66. P (both over 120) = .144. The probability of scoring over 120 if an older sibling scored under 120 is needed to answer the second part.

Chapter 6

2.

Value	Probability
2	.05
3	.15
4	.30
5	.10
6	.20
7	.15
8	.05
	1.00

4. (a), (b), and (e)

6.

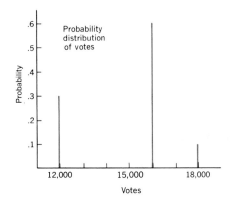

Votes	Probability
12,000	.3
16,000	.6
18,000	.1
	1.0

8. (a)

Outcome	Probability
$ 8,000	.05
9,000	.15
10,000	.25
11,000	.30
12,000	.20
13,000	.05
	1.00

(b) $10,600

10. (a)

Outcome	Probability
0	.03
1	.08
2	.30
3	.42
4	.12
5	.05
	1.00

(b) 2.67

12. 14 months

14. 30 bags

16. (a) .0102 (b) .0109 (c) .4202 (d) .2557

18. (a) $\mu=3$; $\sigma=1.5$ (b) $\mu=15$; $\sigma=2.45$ (c) $\mu=50$; $\sigma=6.71$ (d) $\mu=2$; $\sigma=1.38$
(e) $\mu=2137.5$; $\sigma=10.34$ **20.** (a) .3456 (b) .1296

22. (a)

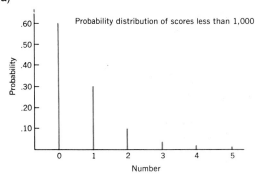

Number of scores less than 1,000	Probability
0	.5905
1	.3280
2	.0729
3	.0081
4	.0005
5	.0000
	1.0000

(b) .1319 decrease

24. (a) .3209 (b) .4633 (c) .0771   26. (a) .1913 (b) .9985 (c) .4493

28. .5859; replace the bird   30. (a) .8647 (b) .1353

32. (a) .2805 (b) .1587 (c) .1949 (d) .3425   34. (a) 23.76 (b) 18.64

36. (a) .7644 (b) .9162 (c) .3446 (d) .3202

38. 112.8 points   40. (a) .1335 (b) .5160   42. .8585   44. 84

46. A fixed number of trials of relatively small size with two possible outcomes implies binomial. An infinite number of trials implies Poisson; both are for discrete variables. The normal distribution is for continuous variables or the case where trials are so numerous that the distribution may be approximated by a continuous function.

48. One solution is to use the services of a professional statistician who is trained to make these kinds of decisions. As an alternative, it is reasonable to expect that someone who completes this book will be able to employ the tests described to determine whether an observed distribution can be described by one of the probability distributions we will study.

50. It is probably not Bernoulli. We don't know if the production of these timers is independent.

52. (a) discrete (b) continuous (c) continuous   54. 206

56. (a) .1680 (b) .1852   58. (a) .8491 (b) .5000 (c) .0592

60. (a)

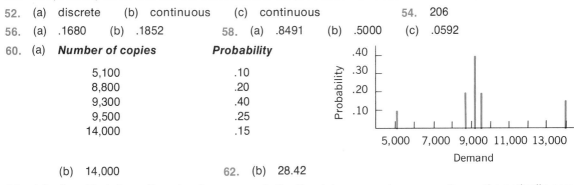

Number of copies	Probability
5,100	.10
8,800	.20
9,300	.40
9,500	.25
14,000	.15

(b) 14,000   62. (b) 28.42

64. (a) Specific information about an unusual situation takes precedence over the mathematically computed value.
    (b) Data obtained from the statistical model are dominated by specific knowledge of an unusual situation.

66. (a) .2708 (b) .4206   68. (a) .2676 (b) .8238

70. (a) .8485 (b) .5000 (c) .0606 The worst error is only about 2 percent off.

Chapter 7

2. If we are only generalizing to the same group of observations in making our conclusions, then the set of elements is the population under study. If, however, we wish to draw conclusions about some larger group of which these elements are only a part, then the set of observations would be considered a sample.

4. Probability samples involve more statistical analysis and planning at the beginning of a study and usually take more time and money than judgment samples.

6. From what we've been told in the problem, Jean's position is apparently quite defensible. Perhaps what makes statistical sampling unique is that it permits statistical inference to be made about a population and its parameters. This is apparently what Jean has done. There are no hard and fast rules as to the size of the sample that must be drawn before inferences can be made. Specifically, there is nothing magic about the 50 percent mark. Common sense would seem to point out that gathering data from 50 percent of some populations might tend to be just as difficult as gathering data from the entire

population—for instance, the population of the United States or the world. The defense for Jean's position lies in empirical evidence and some explanation and reasoning with the project leaders, educating them about the abilities of statistical inference.

8. Situation *b*. The distributions have greater between-group variance and less within-group variance than in *a*.

10. Jan. 3, 17, 31; Feb. 14, 28; Mar. 14, 28; Apr. 11, 25; May 9, 23; June 6, 20; July 4,18;
Aug. 1, 15, 29; Sept. 12, 26; Oct. 10, 24; Nov. 7, 21; Dec. 5, 19.

12. No. They have restricted themselves to a subset of the population which has already shown a preference by virtue of the fact that its members (department chairpersons) have done quite well academically and probably will respond with a bias in favor of the better schools.

14. One chance in ten for 3, 6, or 8. We would expect to see each number 12.5 times. In actuality, 3 appears 6 times, 6 appears 20 times, and 8 appears 19 times.

16. Yes. The accident times during the year are probably randomly distributed, making that a good candidate for systematic sampling.

18. The older nurse is correct. It is strictly chance that the pressures average 120. The probability of their not averaging 120 is greater than the probability that they will.

20. No. the mean of a sample does not exactly equal the mean of the population, because of sampling error. If it differed greatly, there might be doubts, but that is not the case here.

22. No. This is not a sampling distribution of the mean because a sampling distribution refers to a theoretical distribution based on all samples of a given size from a population. These samples by region are of varying size.

24. No. The standard error of the mean is a measure of the dispersion of the theoretical sampling distribution and not an allowable deviation from the population mean as the secretary believes.

26. (a) .8480 (b) .9689   28. about 16

30. (a) .2776 (b) .2033; less

32. (a) .4681 (b) .4168   34. No; the statement can be made only with a .9648 probability.

36. .7242   38. (a) .483   (b) .268   (c) .166

40. (a) .9544 (b) 22   42. .9451

44. Proponents of random sampling often argue that it is better than judgmental sampling because more statistical inferences can be made and the information obtained is more reliable. In this case, the opposite may be true. Judgmental sampling seems to have worked quite well; and given the cost of the proposed alternative, it may be better to leave well enough alone.

46. No, cluster sampling will not give representative results. Problems in one department may be quite different from those in another.

48. In this case the fire department is not constrained by cost, time, or destruction of the population; but given the effort it would require to poll all households, sampling is probably a better alternative.

50. A sampling distribution of means is a frequency distribution of the means of all possible samples. It is not in any sense a graph of the individual observations in sample combinations.

52. .0475   54. .0640   56. .0359

Chapter 8

2. Measuring an entire population may not be feasible because of time and cost considerations. A sample yields only an estimate and is subject to sampling errors.

4. An estimator is a sample statistic used to estimate a population parameter. An estimate is a specific numerical value for an estimator.

6. It assures us that the estimator becomes more reliable with larger samples.

8. 17.45   33.08   10.   .568

12. (a) .4   (b) $\bar{x} \pm 2\sigma_{\bar{x}}$ 14. (a) 67.15–68.85   (b) 65.45–70.55

16. (a) 34,000–38,000   (b) 32,000–40,000

18. The confidence interval is the range of an estimate; i.e., the interval between and including the upper confidence limit and the lower confidence limit.

20. (a) High confidence levels produce large confidence intervals; thus the more confident we are, the less we can be confident of.
    (b) Narrow confidence intervals produce low confidence levels; thus the more definite our estimate, the less confidence we can have in it.

22. No.   24. (a) 1.5   (b) 117.54–122.46   26. (a) .573   (b) 24.06–25.94

28. (a) 1.8   (b) .2   (c) 26 ± .33   30. (a) 150   (b) 25   (c) 1800 ± 58.25

32. (a) .056   (b) .72 ± .092   34. (a) .0215   (b) .3247–.3953

36. (a) .6396–.8004   (b) 1599–2001   38. 80% ± 11.1%

40. (a) 90%   (b) 95%   (c) 99%   42. 49.35–62.65

44. 34.6 ± 2.939   46. 1068   48. 316   50. 78   52. 2.45 ± .3

54. To indicate the statistic's reliability as an estimator of a population parameter. The smaller the standard error, the more likely it is that the statistic falls close to the true value of the population parameter.

56. The 95 percent interval has a higher probability of including the true mean, but it is also over 12 points wide, which makes it more approximate. The 75 percent interval narrows down the range of the estimate but does not have as great a probability of including the true mean value.

58. (a) 95.5%   (b) 98.76%   (c) 59.9%

60. (a) estimate p by .045   (b) estimate σ_p by .012

62. 93   64. 58   66. (a) .069   (b) 48% ± 17.8%

68. (a) .2   (b) 3.5 ± .392   70. (a) 28,640 ± 166.6   (b) 28,640 ± 198.05

72. (a) 2.120 and .415   (b) .415   (c) 2.120 ± .695   74. .4 ± .065   76. $10.50 ± $2.09

Chapter 9

2. Theoretically, one could toss a coin a large number of times to see if the proportion of heads was very different from .5. Similarly, by recording the outcomes of many dice rolls, one could see if the proportion of any side was very different from ⅙. A large number of trials would be needed for each of these examples.

4. (a) Assume a hypothesis about a population.   (b) Collect sample data.   (c) Calculate a sample statistic.   (d) Use the sample statistic to evaluate the hypothesis.

6. We mean that we would not have reasonably expected to find that particular sample if in fact the hypothesis had been true.

8. There is a 31.74 percent probability of mistakenly rejecting the hypothesis.

10. No, we cannot.   12. We should accept the claim.

14. A null hypothesis represents the hypothesis you are trying to reject; the alternative hypothesis represents all other possibilities.

16. Type I: the probability that we will reject the null hypothesis when in fact it is true.
    Type II: the probability that we will accept the null hypothesis when in fact it is false.

18. The significance level of a test indicates the probability of a type I error.

20. (a) normal   (b) normal   (c) t with 15 d.f.   (d) normal   (e) t with 24 d.f.

22. A one-tailed test would be used when we are testing whether the population mean is lower or higher than some hypothesized value. A two-tailed test will reject the null hypothesis if the sample mean is significantly higher or lower than the hypothesized population mean.

24. $H_0 : \mu = 10$ tons $H_1 : \mu > 10$ tons

26. Reject H_0 28. (a) $H_0 : \mu = 76$; $H_1 : \mu \neq 76$ (b) Reject H_0 30. Accept H_0

32. $9.5 : 1 - \beta = .0202$; $10.0 : 1 - \beta = .0968$; $10.5 : 1 - \beta = .2912$

34. $9.5 : 1 - \beta = .0099$; $10.0 : 1 - \beta = .0582$; $10.5 : 1 - \beta = .2061$

36. Accept H_0 38. (a) Reject H_0 (b) Accept H_0 40. Reject H_0

42. Accept H_0 44. Accept H_0 46. Reject H_0

48. (a) 2.736   (b) Reject H_0 50. (a) 8.75   (b) $s = 8.74$; $\hat{\sigma}_{\bar{x}} = 3.09$ (c) Reject H_0

52. Accept H_0 54. Accept H_0 56. Reject H_0

58. Reject H_0 60. Reject H_0 62. Accept H_0

64. A Type I error would not be serious for the patient. A Type II error would be serious because the battery might run down before the next scheduled operation.

66. Reject H_0 68. Reject H_0 70. Accept H_0

72. 1.28 std. errors; 1.64 std. errors   74. Accept H_0

76. Accept H_0 78. Accept H_0 80. Accept H_0

82. Reject H_0 84. Reject H_0 86. Reject H_0

88. (a) .1401   (b) .4013   (c) .5987   90. (a) .2611   (b) .6406   (c) .9131

92. (a) .0485   (b) .2033   (c) .3707   94. Accept H_0

Chapter 10

2. To determine whether or not three or more populations means can be considered equal.

4. (a) chi-square   (b) analysis of variance   (c) normal or t distribution   (d) F distribution

6. (a) 4   (b) 6   (c) 15   (d) 16   (e) 10

8. (a) 1.723   (b) $H_0 : p_A = p_B = p_C = p_D$ (c) Accept H_0
   $H_1 : p_A, p_B, p_C, p_D$ are not equal

10. (a) $H_0 : p_D = p_L = p_B = p_S$ (b) 17.475   (c) Reject H_0
    $H_1 : p_D, p_L, p_B, p_S$ are not equal

12. Reject H_0 14. Accept H_0

16. (a) The probabilities for the 5 classes are: .0548; .2195; .3811; .2638; .0808   (b) The expected frequencies are: 4, 18, 31, 21, 6.   (c) .653   (d) Accept H_0 18. Accept H_0

20. Accept H_0 22. Accept H_0 24. Reject H_0 26. Accept H_0

28. (a) 69, 62.33, 58, 63.11   (b) 30.706   (c) 184.236   (d) 68.889   (e) 3.68 Accept H_0

30. Reject H_0 32. Accept H_0

34. 9.22–40.48 is a 95% interval   36. Accept H_0

38. Accept H_0 40. 32.92 – 104.51 is a 95% interval

42. Accept H_0 44. Accept H_0 46. Accept H_0 48. Reject H_0

50. 40.34–120.19 is a 90% interval.

52. (a) $H_0 : p_L = p_M = p_U$ (b) 1.587   (c) Accept H_0
    $H_1 : p_L, p_M, p_U$ are not equal

54. (a) *t* test   (b) *F* distribution   (c) normal   (d) chi-square   56. Reject H_0

58. (b) 8.4398   (c) H_0 : attendance is independent of income level
         H_1 : attendance and income level are not independent
    (d) Accept H_0

60. Accept H_0 62. Reject H_0 64. Accept H_0

66. Accept H_0

Chapter 11

2. An estimating equation is the mathematical relationship describing the association between a dependent variable and one or more independent variables.

4. The term *direct relationship* applies to the situation in which the dependent variable increases as the independent variable(s) increases. The term *indirect relationship* describes the situation in which the dependent variable decreases as the independent variable(s) increases.

6. A linear relationship describes the situation in which the dependent variable changes a constant amount for equal incremental changes in the independent variable(s). A curvilinear relationship describes the situation in which the dependent variable changes at an increasing (or decreasing) rate with equal incremental changes in the independent variable(s).

8. It is the process that determines the relationship between a dependent variable and more than one independent variable.

10. (a) Yes, direct and linear.   (b) Yes, inverse and curvilinear.   (c) Yes, inverse and curvilinear.

12. (b) $\hat{Y} = .384 + .277 X$ (c) $X = 4: \hat{Y} = 1.492$
         $X = 9: \hat{Y} = 2.877$
         $X = 12: \hat{Y} = 3.708$

14. (a) $\hat{Y} = -.4869 + .1754X$ (b) .3474   (c) $6.338 - 8.124$

16. (a) $\hat{Y} = -20.47 + 5.21 X$ (b) 2.45   (c) $5.33 - 16.25$

18. (b) $\hat{Y} = 70.5 - 2.8 X$ (c) 20.1   20. (b) $\hat{Y} = 11.81 - .176 X$ (c) 6.53

22. $r^2 = .6244; r = -.7902$ 24. $r^2 = .8484; r = -.9211$

26. (a) positive   (b) positive   (c) positive   (d) zero

28. (a) .062   (b) $.6178 - .8822$ 30. Accept H_0

32. (a) $.095 - .185$ (the slope has changed)   (b) $.077 - .203$ (the slope has not changed)

34. Because it is *determined* by the relationship of the values over a certain range and therefore may be different for other values.

36. The relationship may no longer be the same as it was in earlier studies made under different conditions. New variables may be present, which will affect the relationship between variables studied earlier.

38. (a) $\hat{Y} = 14.52 + 1.12 X_1 + 2.2 X_2$ (b) 28.696   40. .7083

42. (a) $\hat{Y} = -6.3666 + .1553 X_1 + .7262 X_2$ (b) 7.6236   (c) $1.3017; 5.0007 - 10.2465$

44. It is the percentage of the variation of the dependent variable which is explained by the independent variables.

46. $r^2 = .7734; r = .8794$ 48. $r^2 = .8810; r = -.9386$

50. (a) 1, +   (b) 2, +   (c) 2, −   (d) 2, −

52. $r^2 = .9088; r = .9533$ 54. Reject H_0

56. (b) $\hat{Y} = 41.95 + 1.35 X$

58. (a) $\hat{Y} = -17.7930 + .9121 X_1 + .1589 X_2$ (b) 11.9535

60. (a) $\hat{Y} = 5.94 + .81 X$ (b) 6.99   (c) $12.27 - 48.21$ 62. Accept H_0

2. *b*

4. They do not use all the information in the data, since they usually rely on ranks or counts.

6. Yes, a great deal of information is sacrificed by using a ranking test. If the data were examined, it could be seen that there is a very distinct bi-modal distribution. In this instance, choice of two colors might well be the better alternative.

8. Accept H_0

10. (a) The meteorologist has a bit of a point, but it is not strong. A nonparametric test does not address itself to his concern.
    (b) Accept H_0

12. (a) Accept H_0 (b) Accept H_0 (c) Reject H_0 (d) The only change in the computations of *b* and *c* is in the standard error of the mean. With only 13 responses, the reliability of our sample was lessened by the small size of the sample.

14. Accept H_0 Accept H_0 Reject H_0 Accept H_0; $n = 28$;
    $n_2 = 30$; $R_2 = 881$

22. Reject the hypothesis of randomness.   24.   Reject the hypothesis of randomness.

26. (a) Accept Kathy's hypothesis   (b) No.

28. Yes, there is a significant correlation in the ranked data.

30. Accept H_0; there is no significant correlation in the ranked data.

32. Yes, the subject does significantly well in detecting weight differences.

34. There is no significant positive correlation. In fact, r_s is negative.

36. (a) Yes. (b) Even though there are many W's, our test still shows the data to be randomly distributed. The test then confines itself only to the occurrences listed and does not make any inferences about the underlying population distribution. In other words, our procedure tests only how items are distributed in the sample and ignores frequency of occurrences.

38. (a) The ratings have improved.   (b) Possibly, but before a definitive answer can be given, we would have to analyze the other variables that might have had an effect on the ratings: quality of acting, competitive shows, weather, and others. A change in any one of these could have caused the ratings to change.

40. Yes, it has achieved its goals.

42. Yes, they will conclude the annual rainfall does not occur in patterns.

44. There is a positive correlation.

46. Two reasons that might be advanced for favoring the campaign worker's proposal are (1) statistical tests on real data are sharper than nonparametric tests, and (2) no information is lost as it perhaps would be with ranking.

    Although there is some validity in the first reason, the second one falls short. The quality of information received must first be evaluated before determining what will be lost by using "distribution-free" tests.

    In this instance, given a scale of such a broad range and without established norms as to how the form will be filled out, the actual numbers supplied take on little significance. What is really important is the ranking of the data.

48. (a) The number of accidents is not randomly distributed.   (b) Whereas the number of accidents seems to have declined since the light was installed, we can only infer *relationship* not *causality*. We don't know what other factors may also be involved. The runs test does not permit us to conclude that accidents have declined, only that the number is not random. A linear regression of number of accidents on time would give a better test.

ANSWERS TO CHAPTER CONCEPTS TESTS

Chap. 2	Chap. 4	Chap. 6	Chap. 8	Chap. 10	Chap. 12
1. T	1. T	1. F	1. F	1. T	1. T
2. T	2. T	2. F	2. F	2. T	2. F
3. F	3. F	3. T	3. T	3. F	3. T
4. F	4. T	4. F	4. T	4. T	4. T
5. T	5. T	5. T	5. F	5. T	5. F
6. T	6. F	6. T	6. F	6. F	6. T
7. T	7. T	7. F	7. T	7. T	7. F
8. T	8. F	8. F	8. T	8. T	8. T
9. F	9. T	9. T	9. T	9. T	9. T
10. F	10. F	10. T	10. T	10. F	10. F

Chap. 3	Chap. 5	Chap. 7	Chap. 9	Chap. 11
1. F	1. F	1. T	1. F	1. F
2. T	2. F	2. F	2. T	2. F
3. F	3. T	3. F	3. F	3. F
4. F	4. T	4. F	4. T	4. F
5. F	5. T	5. T	5. T	5. T
6. T	6. F	6. T	6. F	6. T
7. T	7. F	7. T	7. T	7. T
8. F	8. T	8. T	8. T	8. T
9. T	9. T	9. T	9. T	9. F
10. F	10. T	10. F	10. F	10. T

Appendix tables

Areas under the Standard Normal Probability Distribution
between the Mean and Successive Value of *z*.*

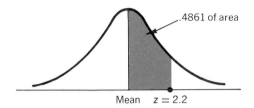

.4861 of area

Mean $z = 2.2$

EXAMPLE: To find the area under the curve between the mean and a point 2.2 standard deviations to the right of the mean, look up the value opposite 2.2 in the table; .4861 of the area under the curve lies between the mean and a *z* value of 2.2.

z	.00	.01	.02	.03	.04	.05	.06	.07	.08	.09
0.0	.0000	.0040	.0080	.0120	.0160	.0199	.0239	.0279	.0319	.0359
0.1	.0398	.0438	.0478	.0517	.0557	.0596	.0636	.0675	.0714	.0753
0.2	.0793	.0832	.0871	.0910	.0948	.0987	.1026	.1064	.1103	.1141
0.3	.1179	.1217	.1255	.1293	.1331	.1368	.1406	.1443	.1480	.1517
0.4	.1554	.1591	.1628	.1664	.1700	.1736	.1772	.1808	.1844	.1879
0.5	.1915	.1950	.1985	.2019	.2054	.2088	.2123	.2157	.2190	.2224
0.6	.2257	.2291	.2324	.2357	.2389	.2422	.2454	.2486	.2517	.2549
0.7	.2580	.2611	.2642	.2673	.2704	.2734	.2764	.2794	.2823	.2852
0.8	.2881	.2910	.2939	.2967	.2995	.3023	.3051	.3078	.3106	.3133
0.9	.3159	.3186	.3212	.3238	.3264	.3289	.3315	.3340	.3365	.3389
1.0	.3413	.3438	.3461	.3485	.3508	.3531	.3554	.3577	.3599	.3621
1.1	.3643	.3665	.3686	.3708	.3729	.3749	.3770	.3790	.3810	.3830
1.2	.3849	.3869	.3888	.3907	.3925	.3944	.3962	.3980	.3997	.4015
1.3	.4032	.4049	.4066	.4082	.4099	.4115	.4131	.4147	.4162	.4177
1.4	.4192	.4207	.4222	.4236	.4251	.4265	.4279	.4292	.4306	.4319
1.5	.4332	.4345	.4357	.4370	.4382	.4394	.4406	.4418	.4429	.4441
1.6	.4452	.4463	.4474	.4484	.4495	.4505	.4515	.4525	.4535	.4545
1.7	.4554	.4564	.4573	.4582	.4591	.4599	.4608	.4616	.4625	.4633
1.8	.4641	.4649	.4656	.4664	.4671	.4678	.4686	.4693	.4699	.4706
1.9	.4713	.4719	.4726	.4732	.4738	.4744	.4750	.4756	.4761	.4767
2.0	.4772	.4778	.4783	.4788	.4793	.4798	.4803	.4808	.4812	.4817
2.1	.4821	.4826	.4830	.4834	.4838	.4842	.4846	.4850	.4854	.4857
2.2	.4861	.4864	.4868	.4871	.4875	.4878	.4881	.4884	.4887	.4890
2.3	.4893	.4896	.4898	.4901	.4904	.4906	.4909	.4911	.4913	.4916
2.4	.4918	.4920	.4922	.4925	.4927	.4929	.4931	.4932	.4934	.4936
2.5	.4938	.4940	.4941	.4943	.4945	.4946	.4948	.4949	.4951	.4952
2.6	.4953	.4955	.4956	.4957	.4959	.4960	.4961	.4962	.4963	.4946
2.7	.4965	.4966	.4967	.4968	.4969	.4970	.4971	.4972	.4973	.4974
2.8	.4974	.4975	.4976	.4977	.4977	.4978	.4979	.4979	.4980	.4981
2.9	.4981	.4982	.4982	.4983	.4984	.4984	.4985	.4985	.4986	.4986
3.0	.4987	.4987	.4987	.4988	.4988	.4989	.4989	.4989	.4990	.4990

* From Robert D. Mason, *Essentials of Statistics*, Prentice-Hall, Inc., 1976.

APPENDIX TABLE 2

Areas in Both Tails Combined for Student's *t* Distribution.*

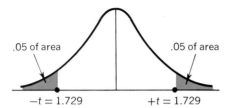

.05 of area   .05 of area

$-t = 1.729$ $+t = 1.729$

EXAMPLE: To find the value of *t* which corresponds to an area of .10 in both tails of the distribution combined, when there are 19 degrees of freedom, look under the .10 column, and proceed down to the 19 degrees of freedom row; the appropriate *t* value there is 1.729.

Degrees of freedom	Area in both tails combined			
	.10	.05	.02	.01
1	6.314	12.706	31.821	63.657
2	2.920	4.303	6.965	9.925
3	2.353	3.182	4.541	5.841
4	2.132	2.776	3.747	4.604
5	2.015	2.571	3.365	4.032
6	1.943	2.447	3.143	3.707
7	1.895	2.365	2.998	3.499
8	1.860	2.306	2.896	3.355
9	1.833	2.262	2.821	3.250
10	1.812	2.228	2.764	3.169
11	1.796	2.201	2.718	3.106
12	1.782	2.179	2.681	3.055
13	1.771	2.160	2.650	3.012
14	1.761	2.145	2.624	2.977
15	1.753	2.131	2.602	2.947
16	1.746	2.120	2.583	2.921
17	1.740	2.110	2.567	2.898
18	1.734	2.101	2.552	2.878
19	1.729	2.093	2.539	2.861
20	1.725	2.086	2.528	2.845
21	1.721	2.080	2.518	2.831
22	1.717	2.074	2.508	2.819
23	1.714	2.069	2.500	2.807
24	1.711	2.064	2.492	2.797
25	1.708	2.060	2.485	2.787
26	1.706	2.056	2.479	2.779
27	1.703	2.052	2.473	2.771
28	1.701	2.048	2.467	2.763
29	1.699	2.045	2.462	2.756
30	1.697	2.042	2.457	2.750
40	1.684	2.021	2.423	2.704
60	1.671	2.000	2.390	2.660
120	1.658	1.980	2.358	2.617
Normal Distribution	1.645	1.960	2.326	2.576

*Taken from Table III of Fisher and Yates, *Statistical Tables for Biological, Agricultural and Medical Research,* published by Longman Group Ltd., London (previously published by Oliver & Boyd, Edinburgh) and by permission of the authors and publishers.

The Cumulative Binomial Distribution*

EXAMPLE: These tables describe the cumulative binomial distribution; a sample problem will illustrate how they are used. Suppose that we are grading bar examinations and wish to find the probability of finding 7 or more failures in a batch of 15, when the probability that any one exam is a failure is .20.

In binomial notation, the elements in this example can be represented:

$$n = 15 \text{ (number of exams to be graded)}$$
$$p = .20 \text{ (probability that any one exam will be a failure)}$$
$$r = 7 \text{ (number of failures in question)}$$

Steps for solution:

1. Since the problem involves 15 trials or inspections, first find the table for $n = 15$.
2. The probability of a failing examination is .20; therefore, we look through the $n = 15$ table until we find the column where $p = 20$.
3. We then move down the $p = 20$ column until we are opposite the $r = 7$ row.
4. The answer there is found to be 0181; this is interpreted to be a probability of .0181.

This problem asked for the probability of *7 or more* failures. Had it asked for the probability of *more than 7* failures, we would have looked up the probability of *8 or more*.

Note that this table only goes up to $p = .50$. When p is *larger* than .50, q $(1 - p)$ is *less* than .50. Therefore the problem is worked in terms of q and the number of passing exams $(n - r)$ rather than in terms of p and r (the number of failures). For example, suppose $p = .60$ and $n = 15$. What is the probability of more than 12 failures? More than 12 failures (13, 14, or 15 failures) is the same as 2 or fewer successes. The probability of 2 or fewer successes is 1 − the probability of 3 or more successes. We look in the $n = 15$ table for the $p = 40$ column and the $r = 3$ row. There we see the number 9729, which we interpret as a probability of .9729; so the answer is $1 - .9729$, or .0271.

					n = 1					
P	01	02	03	04	05	06	07	08	09	10
R										
1	0100	0200	0300	0400	0500	0600	0700	0800	0900	1000
P	11	12	13	14	15	16	17	18	19	20
R										
1	1100	1200	1300	1400	1500	1600	1700	1800	1900	2000
P	21	22	23	24	25	26	27	28	29	30
R										
1	2100	2200	2300	2400	2500	2600	2700	2800	2900	3000
P	31	32	33	34	35	36	37	38	39	40
R										
1	3100	3200	3300	3400	3500	3600	3700	3800	3900	4000
P	41	42	43	44	45	46	47	48	49	50
R										
1	4100	4200	4300	4400	4500	4600	4700	4800	4900	5000

P R	01	02	03	04	05	06	07	08	09	10
1	0199	0396	0591	0784	0975	1164	1351	1536	1719	1900
2	0001	0004	0009	0016	0025	0036	0049	0064	0081	0100

P R	11	12	13	14	15	16	17	18	19	20
1	2079	2256	2431	2604	2775	2944	3111	3276	3439	3600
2	0121	0144	0169	0196	0225	0256	0289	0324	0361	0400

P R	21	22	23	24	25	26	27	28	29	30
1	3759	3916	4071	4224	4375	4524	4671	4816	4959	5100
2	0441	0484	0529	0576	0625	0676	0729	0784	0841	0900

P R	31	32	33	34	35	36	37	38	39	40
1	5239	5376	5511	5644	5775	5904	6031	6156	6279	6400
2	0961	1024	1089	1156	1225	1296	1369	1444	1521	1600

P R	41	42	43	44	45	46	47	48	49	50
1	6519	6636	6751	6864	6975	7084	7191	7296	7399	7500
2	1681	1764	1849	1936	2025	2116	2209	2304	2401	2500

P R	01	02	03	04	05	06	07	08	09	10
1	0297	0588	0873	1153	1426	1694	1956	2213	2464	2710
2	0003	0012	0026	0047	0073	0104	0140	0182	0228	0280
3				0001	0001	0002	0003	0005	0007	0010

P R	11	12	13	14	15	16	17	18	19	20
1	2950	3185	3415	3639	3859	4073	4282	4486	4686	4880
2	0336	0397	0463	0533	0608	0686	0769	0855	0946	1040
3	0013	0017	0022	0027	0034	0041	0049	0058	0069	0080

P R	21	22	23	24	25	26	27	28	29	30
1	5070	5254	5435	5610	5781	5948	6110	6268	6421	6570
2	1138	1239	1344	1452	1563	1676	1793	1913	2035	2160
3	0093	0106	0122	0138	0156	0176	0197	0220	0244	0270

P R	31	32	33	34	35	36	37	38	39	40
1	6715	6856	6992	7125	7254	7379	7500	7617	7730	7840
2	2287	2417	2548	2682	2818	2955	3094	3235	3377	3520
3	0298	0328	0359	0393	0429	0467	0507	0549	0593	0640

P R	41	42	43	44	45	46	47	48	49	50
1	7946	8049	8148	8244	8336	8425	8511	8594	8673	8750
2	3665	3810	3957	4104	4253	4401	4551	4700	4850	5000
3	0689	0741	0795	0852	0911	0973	1038	1106	1176	1250

P R	01	02	03	04	05	06	07	08	09	10
1	0394	0776	1147	1507	1855	2193	2519	2836	3143	3439
2	0006	0023	0052	0091	0140	0199	0267	0344	0430	0523
3			0001	0002	0005	0008	0013	0019	0027	0037
4									0001	0001

P R	11	12	13	14	15	16	17	18	19	20
1	3726	4003	4271	4530	4780	5021	5254	5479	5695	5904
2	0624	0732	0847	0968	1095	1228	1366	1509	1656	1808
3	0049	0063	0079	0098	0120	0144	0171	0202	0235	0272
4	0001	0002	0003	0004	0005	0007	0008	0010	0013	0016

P	21	22	23	24	25	26	27	28	29	30
R
1	6105	6298	6485	6664	6836	7001	7160	7313	7459	7599
2	1963	2122	2285	2450	2617	2787	2959	3132	3307	3483
3	0312	0356	0403	0453	0508	0566	0628	0694	0763	0837
4	0019	0023	0028	0033	0039	0046	0053	0061	0071	0081
P	31	32	33	34	35	36	37	38	39	40
R
1	7733	7862	7985	8103	8215	8322	8425	8522	8615	8704
2	3660	3837	4015	4193	4370	4547	4724	4900	5075	5248
3	0915	0996	1082	1171	1265	1362	1464	1569	1679	1792
4	0092	0105	0119	0134	0150	0168	0187	0209	0231	0256
P	41	42	43	44	45	46	47	48	49	50
R
1	8788	8868	8944	9017	9085	9150	9211	9269	9323	9375
2	5420	5590	5759	5926	6090	6252	6412	6569	6724	6875
3	1909	2030	2155	2283	2415	2550	2689	2834	2977	3125
4	0283	0311	0342	0375	0410	0448	0488	0531	0576	0625

P	01	02	03	04	05	06	07	08	09	10
R
1	0490	0961	1413	1846	2262	2661	3043	3409	3760	4095
2	0010	0038	0085	0148	0226	0319	0425	0544	0674	0815
3		0001	0003	0006	0012	0020	0031	0045	0063	0086
4						0001	0001	0002	0003	0005
P	11	12	13	14	15	16	17	18	19	20
R
1	4416	4723	5016	5296	5563	5818	6061	6293	6513	6723
2	0965	1125	1292	1467	1648	1835	2027	2224	2424	2627
3	0112	0143	0179	0220	0266	0318	0375	0437	0505	0579
4	0007	0009	0013	0017	0022	0029	0036	0045	0055	0067
5				0001	0001	0001	0001	0002	0002	0003
P	21	22	23	24	25	26	27	28	29	30
R
1	6923	7113	7293	7464	7627	7781	7927	8065	8196	8319
2	2833	3041	3251	3461	3672	3883	4093	4303	4511	4718
3	0659	0744	0836	0933	1035	1143	1257	1376	1501	1631
4	0081	0097	0114	0134	0156	0181	0208	0238	0272	0308
5	0004	0005	0006	0008	0010	0012	0014	0017	0021	0024
P	31	32	33	34	35	36	37	38	39	40
R
1	8436	8546	8650	8748	8840	8926	9008	9084	9155	9222
2	4923	5125	5325	5522	5716	5906	6093	6276	6455	6630
3	1766	1905	2050	2199	2352	2509	2670	2835	3003	3174
4	0347	0390	0436	0486	0540	0598	0660	0726	0796	0870
5	0029	0034	0039	0045	0053	0060	0069	0079	0090	0102
P	41	42	43	44	45	46	47	48	49	50
R
1	9285	9344	9398	9449	9497	9541	9582	9620	9655	9688
2	6801	6967	7129	7286	7438	7585	7728	7865	7998	8125
3	3349	3525	3705	3886	4069	4253	4439	4625	4813	5000
4	0949	1033	1121	1214	1312	1415	1522	1635	1753	1875
5	0116	0131	0147	0165	0185	0206	0229	0255	0282	0313

P	01	02	03	04	05	06	07	08	09	10
R										
1	0585	1142	1670	2172	2649	3101	3530	3936	4321	4686
2	0015	0057	0125	0216	0328	0459	0608	0773	0952	1143
3		0002	0005	0012	0022	0038	0058	0085	0118	0159
4					0001	0002	0003	0005	0008	0013
5										0001

P	11	12	13	14	15	16	17	18	19	20
R										
1	5030	5356	5664	5954	6229	6487	6731	6960	7176	7379
2	1345	1556	1776	2003	2235	2472	2713	2956	3201	3446
3	0206	0261	0324	0395	0473	0560	0655	0759	0870	0989
4	0018	0025	0034	0045	0059	0075	0094	0116	0141	0170
5	0001	0001	0002	0003	0004	0005	0007	0010	0013	0016
6										0001

P	21	22	23	24	25	26	27	28	29	30
R										
1	7569	7748	7916	8073	8220	8358	8487	8607	8719	8824
2	3692	3937	4180	4422	4661	4896	5128	5356	5580	5798
3	1115	1250	1391	1539	1694	1856	2023	2196	2374	2557
4	0202	0239	0280	0326	0376	0431	0492	0557	0628	0705
5	0020	0025	0031	0038	0046	0056	0067	0079	0093	0109
6	0001	0001	0001	0002	0002	0003	0004	0005	0006	0007

P	31	32	33	34	35	36	37	38	39	40
R										
1	8921	9011	9095	9173	9246	9313	9375	9432	9485	9533
2	6012	6220	6422	6619	6809	6994	7172	7343	7508	7667
3	2744	2936	3130	3328	3529	3732	3937	4143	4350	4557
4	0787	0875	0969	1069	1174	1286	1404	1527	1657	1792
5	0127	0148	0170	0195	0223	0254	0288	0325	0365	0410
6	0009	0011	0013	0015	0018	0022	0026	0030	0035	0041

P	41	42	43	44	45	46	47	48	49	50
R										
1	9578	9619	9657	9692	9723	9752	9778	9802	9824	9844
2	7819	7965	8105	8238	8364	8485	8599	8707	8810	8906
3	4764	4971	5177	5382	5585	5786	5985	6180	6373	6563
4	1933	2080	2232	2390	2553	2721	2893	3070	3252	3438
5	0458	0510	0566	0627	0692	0762	0837	0917	1003	1094
6	0048	0055	0063	0073	0083	0095	0108	0122	0138	0156

P	01	02	03	04	05	06	07	08	09	10
R										
1	0679	1319	1920	2486	3017	3515	3983	4422	4832	5217
2	0020	0079	0171	0294	0444	0618	0813	1026	1255	1497
3		0003	0009	0020	0038	0063	0097	0140	0193	0257
4				0001	0002	0004	0007	0012	0018	0027
5								0001	0001	0002

P	11	12	13	14	15	16	17	18	19	20
R										
1	5577	5913	6227	6521	6794	7049	7286	7507	7712	7903
2	1750	2012	2281	2556	2834	3115	3396	3677	3956	4233
3	0331	0416	0513	0620	0738	0866	1005	1154	1313	1480
4	0039	0054	0072	0094	0121	0153	0189	0231	0279	0333
5	0003	0004	0006	0009	0012	0017	0022	0029	0037	0047
6					0001	0001	0001	0002	0003	0004

P	21	22	23	24	25	26	27	28	29	30
R
1	8080	8243	8395	8535	8665	8785	8895	8997	9090	9176
2	4506	4775	5040	5298	5551	5796	6035	6266	6490	6706
3	1657	1841	2033	2231	2436	2646	2861	3081	3304	3529
4	0394	0461	0536	0617	0706	0802	0905	1016	1134	1260
5	0058	0072	0088	0107	0129	0153	0181	0213	0248	0288
6	0005	0006	0008	0011	0013	0017	0021	0026	0031	0038
7					0001	0001	0001	0001	0002	0002

P	31	32	33	34	35	36	37	38	39	40
R
1	9255	9328	9394	9454	9510	9560	9606	9648	9686	9720
2	6914	7113	7304	7487	7662	7828	7987	8137	8279	8414
3	3757	3987	4217	4447	4677	4906	5134	5359	5581	5801
4	1394	1534	1682	1837	1998	2167	2341	2521	2707	2898
5	0332	0380	0434	0492	0556	0625	0701	0782	0869	0963
6	0046	0055	0065	0077	0090	0105	0123	0142	0164	0188
7	0003	0003	0004	0005	0006	0008	0009	0011	0014	0016

P	41	42	43	44	45	46	47	48	49	50
R
1	9751	9779	9805	9827	9848	9866	9883	9897	9910	9922
2	8541	8660	8772	8877	8976	9068	9153	9233	9307	9375
3	6017	6229	6436	6638	6836	7027	7213	7393	7567	7734
4	3094	3294	3498	3706	3917	4131	4346	4563	4781	5000
5	1063	1169	1282	1402	1529	1663	1803	1951	2105	2266
6	0216	0246	0279	0316	0357	0402	0451	0504	0562	0625
7	0019	0023	0027	0032	0037	0044	0051	0059	0068	0078

P	01	02	03	04	05	06	07	08	09	10
R
1	0773	1492	2163	2786	3366	3904	4404	4868	5297	5695
2	0027	0103	0223	0381	0572	0792	1035	1298	1577	1869
3	0001	0004	0013	0031	0058	0096	0147	0211	0289	0381
4			0001	0002	0004	0007	0013	0022	0034	0050
5							0001	0001	0003	0004

P	11	12	13	14	15	16	17	18	19	20
R
1	6063	6404	6718	7008	7275	7521	7748	7956	8147	8322
2	2171	2480	2794	3111	3428	3744	4057	4366	4670	4967
3	0487	0608	0743	0891	1052	1226	1412	1608	1815	2031
4	0071	0097	0129	0168	0214	0267	0328	0397	0476	0563
5	0007	0010	0015	0021	0029	0038	0050	0065	0083	0104
6		0001	0001	0002	0002	0003	0005	0007	0009	0012
7									0001	0001

P	21	22	23	24	25	26	27	28	29	30
R
1	8483	8630	8764	8887	8999	9101	9194	9278	9354	9424
2	5257	5538	5811	6075	6329	6573	6807	7031	7244	7447
3	2255	2486	2724	2967	3215	3465	3718	3973	4228	4482
4	0659	0765	0880	1004	1138	1281	1433	1594	1763	1941
5	0129	0158	0191	0230	0273	0322	0377	0438	0505	0580
6	0016	0021	0027	0034	0042	0052	0064	0078	0094	0113
7	0001	0002	0002	0003	0004	0005	0006	0008	0010	0013
8									0001	0001

P R	31	32	33	34	35	36	37	38	39	40
1	9486	9543	9594	9640	9681	9719	9752	9782	9808	9832
2	7640	7822	7994	8156	8309	8452	8586	8711	8828	8936
3	4736	4987	5236	5481	5722	5958	6189	6415	6634	6846
4	2126	2319	2519	2724	2936	3153	3374	3599	3828	4059
5	0661	0750	0846	0949	1061	1180	1307	1443	1586	1737
6	0134	0159	0187	0218	0253	0293	0336	0385	0439	0498
7	0016	0020	0024	0030	0036	0043	0051	0061	0072	0085
8	0001	0001	0001	0002	0002	0003	0004	0004	0005	0007

P R	41	42	43	44	45	46	47	48	49	50
1	9853	9872	9889	9903	9916	9928	9938	9947	9954	9961
2	9037	9130	9216	9295	9368	9435	9496	9552	9602	9648
3	7052	7250	7440	7624	7799	7966	8125	8276	8419	8555
4	4292	4527	4762	4996	5230	5463	5694	5922	6146	6367
5	1895	2062	2235	2416	2604	2798	2999	3205	3416	3633
6	0563	0634	0711	0794	0885	0982	1086	1198	1318	1445
7	0100	0117	0136	0157	0181	0208	0239	0272	0310	0352
8	0008	0010	0012	0014	0017	0020	0024	0028	0033	0039

P R	01	02	03	04	05	06	07	08	09	10
1	0865	1663	2398	3075	3698	4270	4796	5278	5721	6126
2	0034	0131	0282	0478	0712	0978	1271	1583	1912	2252
3	0001	0006	0020	0045	0084	0138	0209	0298	0405	0530
4			0001	0003	0006	0013	0023	0037	0057	0083
5						0001	0002	0003	0005	0009
6										0001

P R	11	12	13	14	15	16	17	18	19	20
1	6496	6835	7145	7427	7684	7918	8131	8324	8499	8658
2	2599	2951	3304	3657	4005	4348	4685	5012	5330	5638
3	0672	0833	1009	1202	1409	1629	1861	2105	2357	2618
4	0117	0158	0209	0269	0339	0420	0512	0615	0730	0856
5	0014	0021	0030	0041	0056	0075	0098	0125	0158	0196
6	0001	0002	0003	0004	0006	0009	0013	0017	0023	0031
7						0001	0001	0002	0002	0003

P R	21	22	23	24	25	26	27	28	29	30
1	8801	8931	9048	9154	9249	9335	9411	9480	9542	9596
2	5934	6218	6491	6750	6997	7230	7452	7660	7856	8040
3	2885	3158	3434	3713	3993	4273	4552	4829	5102	5372
4	0994	1144	1304	1475	1657	1849	2050	2260	2478	2703
5	0240	0291	0350	0416	0489	0571	0662	0762	0870	0988
6	0040	0051	0065	0081	0100	0122	0149	0179	0213	0253
7	0004	0006	0008	0010	0013	0017	0022	0028	0035	0043
8			0001	0001	0001	0001	0002	0003	0003	0004

P R	31	32	33	34	35	36	37	38	39	40
1	9645	9689	9728	9762	9793	9820	9844	9865	9883	9899
2	8212	8372	8522	8661	8789	8908	9017	9118	9210	9295
3	5636	5894	6146	6390	6627	6856	7076	7287	7489	7682
4	2935	3173	3415	3662	3911	4163	4416	4669	4922	5174
5	1115	1252	1398	1553	1717	1890	2072	2262	2460	2666
6	0298	0348	0404	0467	0536	0612	0696	0787	0886	0994
7	0053	0064	0078	0094	0112	0133	0157	0184	0215	0250
8	0006	0007	0009	0011	0014	0017	0021	0026	0031	0038
9				0001	0001	0001	0001	0002	0002	0003

n = 9

P R	41	42	43	44	45	46	47	48	49	50
1	9913	9926	9936	9946	9954	9961	9967	9972	9977	9980
2	9372	9442	9505	9563	9615	9662	9704	9741	9775	9805
3	7866	8039	8204	8359	8505	8642	8769	8889	8999	9102
4	5424	5670	5913	6152	6386	6614	6836	7052	7260	7461
5	2878	3097	3322	3551	3786	4024	4265	4509	4754	5000
6	1109	1233	1366	1508	1658	1817	1985	2161	2346	2539
7	0290	0334	0383	0437	0498	0564	0637	0717	0804	0898
8	0046	0055	0065	0077	0091	0107	0125	0145	0169	0195
9	0003	0004	0005	0006	0008	0009	0011	0014	0016	0020

n = 10

P R	01	02	03	04	05	06	07	08	09	10
1	0956	1829	2626	3352	4013	4614	5160	5656	6106	6513
2	0043	0162	0345	0582	0861	1176	1517	1879	2254	2639
3	0001	0009	0028	0062	0115	0188	0283	0401	0540	0702
4			0001	0004	0010	0020	0036	0058	0088	0128
5					0001	0002	0003	0006	0010	0016
6									0001	0001

P R	11	12	13	14	15	16	17	18	19	20
1	6882	7215	7516	7787	8031	8251	8448	8626	8784	8926
2	3028	3417	3804	4184	4557	4920	5270	5608	5932	6242
3	0884	1087	1308	1545	1798	2064	2341	2628	2922	3222
4	0178	0239	0313	0400	0500	0614	0741	0883	1039	1209
5	0025	0037	0053	0073	0099	0130	0168	0213	0266	0328
6	0003	0004	0006	0010	0014	0020	0027	0037	0049	0064
7			0001	0001	0001	0002	0003	0004	0006	0009
8									0001	0001

P R	21	22	23	24	25	26	27	28	29	30
1	9053	9166	9267	9357	9437	9508	9570	9626	9674	9718
2	6536	6815	7079	7327	7560	7778	7981	8170	8345	8507
3	3526	3831	4137	4442	4744	5042	5335	5622	5901	6172
4	1391	1587	1794	2012	2241	2479	2726	2979	3239	3504
5	0399	0479	0569	0670	0781	0904	1037	1181	1337	1503
6	0082	0104	0130	0161	0197	0239	0287	0342	0404	0473
7	0012	0016	0021	0027	0035	0045	0056	0070	0087	0106
8	0001	0002	0002	0003	0004	0006	0007	0010	0012	0016
9							0001	0001	0001	0001

P R	31	32	33	34	35	36	37	38	39	40
1	9755	9789	9818	9843	9865	9885	9902	9916	9929	9940
2	8656	8794	8920	9035	9140	9236	9323	9402	9473	9536
3	6434	6687	6930	7162	7384	7595	7794	7983	8160	8327
4	3772	4044	4316	4589	4862	5132	5400	5664	5923	6177
5	1679	1867	2064	2270	2485	2708	2939	3177	3420	3669
6	0551	0637	0732	0836	0949	1072	1205	1348	1500	1662
7	0129	0155	0185	0220	0260	0305	0356	0413	0477	0548
8	0020	0025	0032	0039	0048	0059	0071	0086	0103	0123
9	0002	0003	0003	0004	0005	0007	0009	0011	0014	0017
10								0001	0001	0001

P R	41	42	43	44	45	46	47	48	49	50
1	9949	9957	9964	9970	9975	9979	9983	9986	9988	9990
2	9594	9645	9691	9731	9767	9799	9827	9852	9874	9893
3	8483	8628	8764	8889	9004	9111	9209	9298	9379	9453
4	6425	6665	6898	7123	7340	7547	7745	7933	8112	8281
5	3922	4178	4436	4696	4956	5216	5474	5730	5982	6230
6	1834	2016	2207	2407	2616	2832	3057	3288	3526	3770
7	0626	0712	0806	0908	1020	1141	1271	1410	1560	1719
8	0146	0172	0202	0236	0274	0317	0366	0420	0480	0547
9	0021	0025	0031	0037	0045	0054	0065	0077	0091	0107
10	0001	0002	0002	0003	0003	0004	0005	0006	0008	0010

P R	01	02	03	04	05	06	07	08	09	10
1	1047	1993	2847	3618	4312	4937	5499	6004	6456	6862
2	0052	0195	0413	0692	1019	1382	1772	2181	2601	3026
3	0002	0012	0037	0083	0152	0248	0370	0519	0695	0896
4			0002	0007	0016	0030	0053	0085	0129	0185
5					0001	0003	0005	0010	0017	0028
6								0001	0002	0003

P R	11	12	13	14	15	16	17	18	19	20
1	7225	7549	7839	8097	8327	8531	8712	8873	9015	9141
2	3452	3873	4286	4689	5078	5453	5811	6151	6474	6779
3	1120	1366	1632	1915	2212	2521	2839	3164	3494	3826
4	0256	0341	0442	0560	0694	0846	1013	1197	1397	1611
5	0042	0061	0087	0119	0159	0207	0266	0334	0413	0504
6	0005	0008	0012	0018	0027	0037	0051	0068	0090	0117
7		0001	0001	0002	0003	0005	0007	0010	0014	0020
8							0001	0001	0002	0002

P R	21	22	23	24	25	26	27	28	29	30
1	9252	9350	9436	9511	9578	9636	9686	9730	9769	9802
2	7065	7333	7582	7814	8029	8227	8410	8577	8730	8870
3	4158	4488	4814	5134	5448	5753	6049	6335	6610	6873
4	1840	2081	2333	2596	2867	3146	3430	3719	4011	4304
5	0607	0723	0851	0992	1146	1313	1493	1685	1888	2103
6	0148	0186	0231	0283	0343	0412	0490	0577	0674	0782
7	0027	0035	0046	0059	0076	0095	0119	0146	0179	0216
8	0003	0005	0007	0009	0012	0016	0021	0027	0034	0043
9			0001	0001	0001	0002	0002	0003	0004	0006

P R	31	32	33	34	35	36	37	38	39	40
1	9831	9856	9878	9896	9912	9926	9938	9948	9956	9964
2	8997	9112	9216	9310	9394	9470	9537	9597	9650	9698
3	7123	7361	7587	7799	7999	8186	8360	8522	8672	8811
4	4598	4890	5179	5464	5744	6019	6286	6545	6796	7037
5	2328	2563	2807	3059	3317	3581	3850	4122	4397	4672
6	0901	1031	1171	1324	1487	1661	1847	2043	2249	2465
7	0260	0309	0366	0430	0501	0581	0670	0768	0876	0994
8	0054	0067	0082	0101	0122	0148	0177	0210	0249	0293
9	0008	0010	0013	0016	0020	0026	0032	0039	0048	0059
10	0001	0001	0001	0002	0002	0003	0004	0005	0006	0007

P R	41	42	43	44	45	46	47	48	49	50
1	9970	9975	9979	9983	9986	9989	9991	9992	9994	9995
2	9739	9776	9808	9836	9861	9882	9900	9916	9930	9941
3	8938	9055	9162	9260	9348	9428	9499	9564	9622	9673
4	7269	7490	7700	7900	8089	8266	8433	8588	8733	8867
5	4948	5223	5495	5764	6029	6288	6541	6787	7026	7256
6	2690	2924	3166	3414	3669	3929	4193	4460	4729	5000
7	1121	1260	1408	1568	1738	1919	2110	2312	2523	2744
8	0343	0399	0461	0532	0610	0696	0791	0895	1009	1133
9	0072	0087	0104	0125	0148	0175	0206	0241	0282	0327
10	0009	0012	0014	0018	0022	0027	0033	0040	0049	0059
11	0001	0001	0001	0001	0002	0002	0002	0003	0004	0005

P R	01	02	03	04	05	06	07	08	09	10
1	1136	2153	3062	3873	4596	5241	5814	6323	6775	7176
2	0062	0231	0486	0809	1184	1595	2033	2487	2948	3410
3	0002	0015	0048	0107	0196	0316	0468	0652	0866	1109
4		0001	0003	0010	0022	0043	0075	0120	0180	0256
5				0001	0002	0004	0009	0016	0027	0043
6							0001	0002	0003	0005
7										0001

P R	11	12	13	14	15	16	17	18	19	20
1	7530	7843	8120	8363	8578	8766	8931	9076	9202	9313
2	3867	4314	4748	5166	5565	5945	6304	6641	6957	7251
3	1377	1667	1977	2303	2642	2990	3344	3702	4060	4417
4	0351	0464	0597	0750	0922	1114	1324	1552	1795	2054
5	0065	0095	0133	0181	0239	0310	0393	0489	0600	0726
6	0009	0014	0022	0033	0046	0065	0088	0116	0151	0194
7	0001	0002	0003	0004	0007	0010	0015	0021	0029	0039
8					0001	0001	0002	0003	0004	0006
9										0001

P R	21	22	23	24	25	26	27	28	29	30
1	9409	9493	9566	9629	9683	9730	9771	9806	9836	9862
2	7524	7776	8009	8222	8416	8594	8755	8900	9032	9150
3	4768	5114	5450	5778	6093	6397	6687	6963	7225	7472
4	2326	2610	2904	3205	3512	3824	4137	4452	4765	5075
5	0866	1021	1192	1377	1576	1790	2016	2254	2504	2763
6	0245	0304	0374	0453	0544	0646	0760	0887	1026	1178
7	0052	0068	0089	0113	0143	0178	0219	0267	0322	0386
8	0008	0011	0016	0021	0028	0036	0047	0060	0076	0095
9	0001	0001	0002	0003	0004	0005	0007	0010	0013	0017
10						0001	0001	0001	0002	0002

P R	31	32	33	34	35	36	37	38	39	40
1	9884	9902	9918	9932	9943	9953	9961	9968	9973	9978
2	9256	9350	9435	9509	9576	9634	9685	9730	9770	9804
3	7704	7922	8124	8313	8487	8648	8795	8931	9054	9166
4	5381	5681	5973	6258	6533	6799	7053	7296	7528	7747
5	3032	3308	3590	3876	4167	4459	4751	5043	5332	5618
6	1343	1521	1711	1913	2127	2352	2588	2833	3087	3348
7	0458	0540	0632	0734	0846	0970	1106	1253	1411	1582
8	0118	0144	0176	0213	0255	0304	0359	0422	0493	0573
9	0022	0028	0036	0045	0056	0070	0086	0104	0127	0153
10	0003	0004	0005	0007	0008	0011	0014	0018	0022	0028
11				0001	0001	0001	0001	0002	0002	0003

R\P	41	42	43	44	45	46	47	48	49	50
1	9982	9986	9988	9990	9992	9994	9995	9996	9997	9998
2	9834	9860	9882	9901	9917	9931	9943	9953	9961	9968
3	9267	9358	9440	9513	9579	9637	9688	9733	9773	9807
4	7953	8147	8329	8498	8655	8801	8934	9057	9168	9270
5	5899	6175	6443	6704	6956	7198	7430	7652	7862	8062
6	3616	3889	4167	4448	4731	5014	5297	5577	5855	6128
7	1765	1959	2164	2380	2607	2843	3089	3343	3604	3872
8	0662	0760	0869	0988	1117	1258	1411	1575	1751	1938
9	0183	0218	0258	0304	0356	0415	0481	0555	0638	0730
10	0035	0043	0053	0065	0079	0095	0114	0137	0163	0193
11	0004	0005	0007	0009	0011	0014	0017	0021	0026	0032
12				0001	0001	0001	0001	0001	0002	0002

R\P	01	02	03	04	05	06	07	08	09	10
1	1225	2310	3270	4118	4867	5526	6107	6617	7065	7458
2	0072	0270	0564	0932	1354	1814	2298	2794	3293	3787
3	0003	0020	0062	0135	0245	0392	0578	0799	1054	1339
4		0001	0005	0014	0031	0060	0103	0163	0242	0342
5				0001	0003	0007	0013	0024	0041	0065
6						0001	0001	0003	0005	0009
7									0001	0001

R\P	11	12	13	14	15	16	17	18	19	20
1	7802	8102	8364	8592	8791	8963	9113	9242	9354	9450
2	4270	4738	5186	5614	6017	6396	6751	7080	7384	7664
3	1651	1985	2337	2704	3080	3463	3848	4231	4611	4983
4	0464	0609	0776	0967	1180	1414	1667	1939	2226	2527
5	0097	0139	0193	0260	0342	0438	0551	0681	0827	0991
6	0015	0024	0036	0053	0075	0104	0139	0183	0237	0300
7	0002	0003	0005	0008	0013	0019	0027	0038	0052	0070
8			0001	0001	0002	0003	0004	0006	0009	0012
9								0001	0001	0002

R\P	21	22	23	24	25	26	27	28	29	30
1	9533	9604	9666	9718	9762	9800	9833	9860	9883	9903
2	7920	8154	8367	8559	8733	8889	9029	9154	9265	9363
3	5347	5699	6039	6364	6674	6968	7245	7505	7749	7975
4	2839	3161	3489	3822	4157	4493	4826	5155	5478	5794
5	1173	1371	1585	1816	2060	2319	2589	2870	3160	3457
6	0375	0462	0562	0675	0802	0944	1099	1270	1455	1654
7	0093	0120	0154	0195	0243	0299	0365	0440	0527	0624
8	0017	0024	0032	0043	0056	0073	0093	0118	0147	0182
9	0002	0004	0005	0007	0010	0013	0018	0024	0031	0040
10			0001	0001	0001	0002	0003	0004	0005	0007
11									0001	0001

R\P	31	32	33	34	35	36	37	38	39	40
1	9920	9934	9945	9955	9963	9970	9975	9980	9984	9987
2	9450	9527	9594	9653	9704	9749	9787	9821	9849	9874
3	8185	8379	8557	8720	8868	9003	9125	9235	9333	9421
4	6101	6398	6683	6957	7217	7464	7698	7917	8123	8314
5	3760	4067	4376	4686	4995	5301	5603	5899	6188	6470
6	1867	2093	2331	2581	2841	3111	3388	3673	3962	4256
7	0733	0854	0988	1135	1295	1468	1654	1853	2065	2288
8	0223	0271	0326	0390	0462	0544	0635	0738	0851	0977
9	0052	0065	0082	0102	0126	0154	0187	0225	0270	0321
10	0009	0012	0015	0020	0025	0032	0040	0051	0063	0078
11	0001	0001	0002	0003	0003	0005	0006	0008	0010	0013
12							0001	0001	0001	0001

$$n = 13$$

P R	41	42	43	44	45	46	47	48	49	50
1	9990	9992	9993	9995	9996	9997	9997	9998	9998	9999
2	9895	9912	9928	9940	9951	9960	9967	9974	9979	9983
3	9499	9569	9630	9684	9731	9772	9808	9838	9865	9888
4	8492	8656	8807	8945	9071	9185	9288	9381	9464	9539
5	6742	7003	7254	7493	7721	7935	8137	8326	8502	8666
6	4552	4849	5146	5441	5732	6019	6299	6573	6838	7095
7	2524	2770	3025	3290	3563	3842	4127	4415	4707	5000
8	1114	1264	1426	1600	1788	1988	2200	2424	2659	2905
9	0379	0446	0520	0605	0698	0803	0918	1045	1183	1334
10	0096	0117	0141	0170	0203	0242	0287	0338	0396	0461
11	0017	0021	0027	0033	0041	0051	0063	0077	0093	0112
12	0002	0002	0003	0004	0005	0007	0009	0011	0014	0017
13							0001	0001	0001	0001

$$n = 14$$

P R	01	02	03	04	05	06	07	08	09	10
1	1313	2464	3472	4353	5123	5795	6380	6888	7330	7712
2	0084	0310	0645	1059	1530	2037	2564	3100	3632	4154
3	0003	0025	0077	0167	0301	0478	0698	0958	1255	1584
4		0001	0006	0019	0042	0080	0136	0214	0315	0441
5				0002	0004	0010	0020	0035	0059	0092
6						0001	0002	0004	0008	0015
7									0001	0002

P R	11	12	13	14	15	16	17	18	19	20
1	8044	8330	8577	8789	8972	9129	9264	9379	9477	9560
2	4658	5141	5599	6031	6433	6807	7152	7469	7758	8021
3	1939	2315	2708	3111	3521	3932	4341	4744	5138	5519
4	0594	0774	0979	1210	1465	1742	2038	2351	2679	3018
5	0137	0196	0269	0359	0467	0594	0741	0907	1093	1298
6	0024	0038	0057	0082	0115	0157	0209	0273	0349	0439
7	0003	0006	0009	0015	0022	0032	0046	0064	0087	0116
8		0001	0001	0002	0003	0005	0008	0012	0017	0024
9						0001	0001	0002	0003	0004

P R	21	22	23	24	25	26	27	28	29	30
1	9631	9691	9742	9786	9822	9852	9878	9899	9917	9932
2	8259	8473	8665	8837	8990	9126	9246	9352	9444	9525
3	5887	6239	6574	6891	7189	7467	7727	7967	8188	8392
4	3366	3719	4076	4432	4787	5136	5479	5813	6137	6448
5	1523	1765	2023	2297	2585	2884	3193	3509	3832	4158
6	0543	0662	0797	0949	1117	1301	1502	1718	1949	2195
7	0152	0196	0248	0310	0383	0467	0563	0673	0796	0933
8	0033	0045	0060	0079	0103	0132	0167	0208	0257	0315
9	0006	0008	0011	0016	0022	0029	0038	0050	0065	0083
10	0001	0001	0002	0002	0003	0005	0007	0009	0012	0017
11						0001	0001	0001	0002	0002

P R	31	32	33	34	35	36	37	38	39	40
1	9945	9955	9963	9970	9976	9981	9984	9988	9990	9992
2	9596	9657	9710	9756	9795	9828	9857	9881	9902	9919
3	8577	8746	8899	9037	9161	9271	9370	9457	9534	9602
4	6747	7032	7301	7556	7795	8018	8226	8418	8595	8757
5	4486	4813	5138	5458	5773	6080	6378	6666	6943	7207
6	2454	2724	3006	3297	3595	3899	4208	4519	4831	5141
7	1084	1250	1431	1626	1836	2059	2296	2545	2805	3075
8	0381	0458	0545	0643	0753	0876	1012	1162	1325	1501
9	0105	0131	0163	0200	0243	0294	0353	0420	0497	0583
10	0022	0029	0037	0048	0060	0076	0095	0117	0144	0175
11	0003	0005	0006	0008	0011	0014	0019	0024	0031	0039
12		0001	0001	0001	0001	0002	0003	0003	0005	0006
13										0001

P R	41	42	43	44	45	46	47	48	49	50
1	9994	9995	9996	9997	9998	9998	9999	9999	9999	9999
2	9934	9946	9956	9964	9971	9977	9981	9985	9988	9991
3	9661	9713	9758	9797	9830	9858	9883	9903	9921	9935
4	8905	9039	9161	9270	9368	9455	9532	9601	9661	9713
5	7459	7697	7922	8132	8328	8510	8678	8833	8974	9102
6	5450	5754	6052	6344	6627	6900	7163	7415	7654	7880
7	3355	3643	3937	4236	4539	4843	5148	5451	5751	6047
8	1692	1896	2113	2344	2586	2840	3105	3380	3663	3953
9	0680	0789	0910	1043	1189	1348	1520	1707	1906	2120
10	0212	0255	0304	0361	0426	0500	0583	0677	0782	0898
11	0049	0061	0076	0093	0114	0139	0168	0202	0241	0287
12	0008	0010	0013	0017	0022	0027	0034	0042	0053	0065
13	0001	0001	0001	0002	0003	0003	0004	0006	0007	0009
14										0001

P R	01	02	03	04	05	06	07	08	09	10
1	1399	2614	3667	4579	5367	6047	6633	7137	7570	7941
2	0096	0353	0730	1191	1710	2262	2832	3403	3965	4510
3	0004	0030	0094	0203	0362	0571	0829	1130	1469	1841
4		0002	0008	0024	0055	0104	0175	0273	0399	0556
5			0001	0002	0006	0014	0028	0050	0082	0127
6					0001	0001	0003	0007	0013	0022
7								0001	0002	0003

P R	11	12	13	14	15	16	17	18	19	20
1	8259	8530	8762	8959	9126	9269	9389	9490	9576	9648
2	5031	5524	5987	6417	6814	7179	7511	7813	8085	8329
3	2238	2654	3084	3520	3958	4392	4819	5234	5635	6020
4	0742	0959	1204	1476	1773	2092	2429	2782	3146	3518
5	0187	0265	0361	0478	0617	0778	0961	1167	1394	1642
6	0037	0057	0084	0121	0168	0227	0300	0387	0490	0611
7	0006	0010	0015	0024	0036	0052	0074	0102	0137	0181
8	0001	0001	0002	0004	0006	0010	0014	0021	0030	0042
9					0001	0001	0002	0003	0005	0008
10									0001	0001

P	21	22	23	24	25	26	27	28	29	30
R										
1	9709	9759	9802	9837	9866	9891	9911	9928	9941	9953
2	8547	8741	8913	9065	9198	9315	9417	9505	9581	9647
3	6385	6731	7055	7358	7639	7899	8137	8355	8553	8732
4	3895	4274	4650	5022	5387	5742	6086	6416	6732	7031
5	1910	2195	2495	2810	3135	3469	3810	4154	4500	4845
6	0748	0905	1079	1272	1484	1713	1958	2220	2495	2784
7	0234	0298	0374	0463	0566	0684	0817	0965	1130	1311
8	0058	0078	0104	0135	0173	0219	0274	0338	0413	0500
9	0011	0016	0023	0031	0042	0056	0073	0094	0121	0152
10	0002	0003	0004	0006	0008	0011	0015	0021	0028	0037
11			0001	0001	0001	0002	0002	0003	0005	0007
12									0001	0001

P	31	32	33	34	35	36	37	38	39	40
R										
1	9962	9969	9975	9980	9984	9988	9990	9992	9994	9995
2	9704	9752	9794	9829	9858	9883	9904	9922	9936	9948
3	8893	9038	9167	9281	9383	9472	9550	9618	9678	9729
4	7314	7580	7829	8060	8273	8469	8649	8813	8961	9095
5	5187	5523	5852	6171	6481	6778	7062	7332	7587	7827
6	3084	3393	3709	4032	4357	4684	5011	5335	5654	5968
7	1509	1722	1951	2194	2452	2722	3003	3295	3595	3902
8	0599	0711	0837	0977	1132	1302	1487	1687	1902	2131
9	0190	0236	0289	0351	0422	0504	0597	0702	0820	0950
10	0048	0062	0079	0099	0124	0154	0190	0232	0281	0338
11	0009	0012	0016	0022	0028	0037	0047	0059	0075	0093
12	0001	0002	0003	0004	0005	0006	0009	0011	0015	0019
13					0001	0001	0001	0002	0002	0003

P	41	42	43	44	45	46	47	48	49	50
R										
1	9996	9997	9998	9998	9999	9999	9999	9999	10000	10000
2	9958	9966	9973	9979	9983	9987	9990	9992	9994	9995
3	9773	9811	9843	9870	9893	9913	9929	9943	9954	9963
4	9215	9322	9417	9502	9576	9641	9697	9746	9788	9824
5	8052	8261	8454	8633	8796	8945	9080	9201	9310	9408
6	6274	6570	6856	7131	7392	7641	7875	8095	8301	8491
7	4214	4530	4847	5164	5478	5789	6095	6394	6684	6964
8	2374	2630	2898	3176	3465	3762	4065	4374	4686	5000
9	1095	1254	1427	1615	1818	2034	2265	2510	2767	3036
10	0404	0479	0565	0661	0769	0890	1024	1171	1333	1509
11	0116	0143	0174	0211	0255	0305	0363	0430	0506	0592
12	0025	0032	0040	0051	0063	0079	0097	0119	0145	0176
13	0004	0005	0007	0009	0011	0014	0018	0023	0029	0037
14			0001	0001	0001	0002	0002	0003	0004	0005

APPENDIX TABLE 4

Values of $e^{-\lambda}$ (for computing Poisson probabilities)

λ	$e^{-\lambda}$	λ	$e^{-\lambda}$	λ	$e^{-\lambda}$	λ	$e^{-\lambda}$
0.1	0.90484	2.6	0.07427	5.1	0.00610	7.6	0.00050
0.2	0.81873	2.7	0.06721	5.2	0.00552	7.7	0.00045
0.3	0.74082	2.8	0.06081	5.3	0.00499	7.8	0.00041
0.4	0.67032	2.9	0.05502	5.4	0.00452	7.9	0.00037
0.5	0.60653	3.0	0.04979	5.5	0.00409	8.0	0.00034
0.6	0.54881	3.1	0.04505	5.6	0.00370	8.1	0.00030
0.7	0.49659	3.2	0.04076	5.7	0.00335	8.2	0.00027
0.8	0.44933	3.3	0.03688	5.8	0.00303	8.3	0.00025
0.9	0.40657	3.4	0.03337	5.9	0.00274	8.4	0.00022
1.0	0.36788	3.5	0.03020	6.0	0.00248	8.5	0.00020
1.1	0.33287	3.6	0.02732	6.1	0.00224	8.6	0.00018
1.2	0.30119	3.7	0.02472	6.2	0.00203	8.7	0.00017
1.3	0.27253	3.8	0.02237	6.3	0.00184	8.8	0.00015
1.4	0.24660	3.9	0.02024	6.4	0.00166	8.9	0.00014
1.5	0.22313	4.0	0.01832	6.5	0.00150	9.0	0.00012
1.6	0.20190	4.1	0.01657	6.6	0.00136	9.1	0.00011
1.7	0.18268	4.2	0.01500	6.7	0.00123	9.2	0.00010
1.8	0.16530	4.3	0.01357	6.8	0.00111	9.3	0.00009
1.9	0.14957	4.4	0.01228	6.9	0.00101	9.4	0.00008
2.0	0.13534	4.5	0.01111	7.0	0.00091	9.5	0.00007
2.1	0.12246	4.6	0.01005	7.1	0.00083	9.6	0.00007
2.2	0.11080	4.7	0.00910	7.2	0.00075	9.7	0.00006
2.3	0.10026	4.8	0.00823	7.3	0.00068	9.8	0.00006
2.4	0.09072	4.9	0.00745	7.4	0.00061	9.9	0.00005
2.5	0.08208	5.0	0.00674	7.5	0.00055	10.0	0.00005

APPENDIX TABLE 5

Area in the Right Tail of a Chi-square (χ^2) Distribution.*

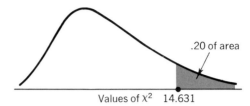

Values of χ^2 14.631

EXAMPLE: In a chi-square distribution with 11 degrees of freedom, if we want to find the appropriate chi-square value for .20 of the area under the curve (the colored area in the right tail) we look under the .20 column in the table and proceed down to the 11 degrees of freedom row; the appropriate chi-square value there is 14.631

Degrees of freedom	Area in right tail				
	.99	.975	.95	.90	.800
1	.00016	.00098	.00398	.0158	.0642
2	.0201	.0506	.103	.211	.446
3	.115	.216	.352	.584	1.005
4	.297	.484	.711	1.064	1.649
5	.554	.831	1.145	1.610	2.343
6	.872	1.237	1.635	2.204	3.070
7	1.239	1.690	2.167	2.833	3.822
8	1.646	2.180	2.733	3.490	4.594
9	2.088	2.700	3.325	4.168	5.380
10	2.558	3.247	3.940	4.865	6.179
11	3.053	3.816	4.575	5.578	6.989
12	3.571	4.404	5.226	6.304	7.807
13	4.107	5.009	5.892	7.042	8.634
14	4.660	5.629	6.571	7.790	9.467
15	5.229	6.262	7.261	8.547	10.307
16	5.812	6.908	7.962	9.312	11.152
17	6.408	7.564	8.672	10.085	12.002
18	7.015	8.231	9.390	10.865	12.857
19	7.633	8.907	10.117	11.651	13.716
20	8.260	9.591	10.851	12.443	14.578
21	8.897	10.283	11.591	13.240	15.445
22	9.542	10.982	12.338	14.041	16.314
23	10.196	11.689	13.091	14.848	17.187
24	10.856	12.401	13.848	15.658	18.062
25	11.524	13.120	14.611	16.473	18.940
26	12.198	13.844	15.379	17.292	19.820
27	12.879	14.573	16.151	18.114	20.703
28	13.565	15.308	16.928	18.939	21.588
29	14.256	16.047	17.708	19.768	22.475
30	14.953	16.791	18.493	20.599	23.364

*Taken from Table IV of Fisher and Yates, *Statistical Tables for Biological, Agricultural and Medical Research,* published by Longman Group Ltd., London (previously published by Oliver & Boyd, Edinburgh) and by permission of the authors and publishers.

		Area in right tail			Degrees of
.20	.10	.05	.025	.01	freedom
1.642	2.706	3.841	5.024	6.635	1
3.219	4.605	5.991	7.378	9.210	2
4.642	6.251	7.815	9.348	11.345	3
5.989	7.779	9.488	11.143	13.277	4
7.289	9.236	11.070	12.833	15.086	5
8.558	10.645	12.592	14.449	16.812	6
9.803	12.017	14.067	16.013	18.475	7
11.030	13.362	15.507	17.535	20.090	8
12.242	14.684	16.919	19.023	21.666	9
13.442	15.987	18.307	20.483	23.209	10
14.631	17.275	19.675	21.920	24.725	11
15.812	18.549	21.026	23.337	26.217	12
16.985	19.812	22.362	24.736	27.688	13
18.151	21.064	23.685	26.119	29.141	14
19.311	22.307	24.996	27.488	30.578	15
20.465	23.542	26.296	28.845	32.000	16
21.615	24.769	27.587	30.191	33.409	17
22.760	25.989	28.869	31.526	34.805	18
23.900	27.204	30.144	32.852	36.191	19
25.038	28.412	31.410	34.170	37.566	20
26.171	29.615	32.671	35.479	38.932	21
27.301	30.813	33.924	36.781	40.289	22
28.429	32.007	35.172	38.076	41.638	23
29.553	33.196	36.415	39.364	42.980	24
30.675	34.382	37.652	40.647	44.314	25
31.795	35.563	38.885	41.923	45.642	26
32.912	36.741	40.113	43.194	46.963	27
34.027	37.916	41.337	44.461	48.278	28
35.139	39.087	42.557	45.722	49.588	29
36.250	40.256	43.773	46.979	50.892	30

Values of F for F Distributions with .05 of the Area in the Right Tail.*

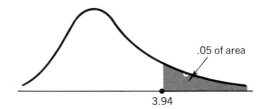

.05 of area

3.94

EXAMPLE: For a test at a significance level of .05 where we have 15 degrees of freedom for the numerator and 6 degrees of freedom for the denominator, the appropriate F value is found by looking under the 15 degrees of freedom column and proceeding down to the 6 degrees of freedom row; there we find the appropriate F value to be 3.94.

Degrees of freedom for numerator

	1	2	3	4	5	6	7	8	9	10	12	15	20	24	30	40	60	120	∞
1	161	200	216	225	230	234	237	239	241	242	244	246	248	249	250	251	252	253	254
2	18.5	19.0	19.2	19.2	19.3	19.3	19.4	19.4	19.4	19.4	19.4	19.4	19.4	19.5	19.5	19.5	19.5	19.5	19.5
3	10.1	9.55	9.28	9.12	9.01	8.94	8.89	8.85	8.81	8.79	8.74	8.70	8.66	8.64	8.62	8.59	8.57	8.55	8.53
4	7.71	6.94	6.59	6.39	6.26	6.16	6.09	6.04	6.00	5.96	5.91	5.86	5.80	5.77	5.75	5.72	5.69	5.66	5.63
5	6.61	5.79	5.41	5.19	5.05	4.95	4.88	4.82	4.77	4.74	4.68	4.62	4.56	4.53	4.50	4.46	4.43	4.40	4.37
6	5.99	5.14	4.76	4.53	4.39	4.28	4.21	4.15	4.10	4.06	4.00	3.94	3.87	3.84	3.81	3.77	3.74	3.70	3.67
7	5.59	4.74	4.35	4.12	3.97	3.87	3.79	3.73	3.68	3.64	3.57	3.51	3.44	3.41	3.38	3.34	3.30	3.27	3.23
8	5.32	4.46	4.07	3.84	3.69	3.58	3.50	3.44	3.39	3.35	3.28	3.22	3.15	3.12	3.08	3.04	3.01	2.97	2.93
9	5.12	4.26	3.86	3.63	3.48	3.37	3.29	3.23	3.18	3.14	3.07	3.01	2.94	2.90	2.86	2.83	2.79	2.75	2.71
10	4.96	4.10	3.71	3.48	3.33	3.22	3.14	3.07	3.02	2.98	2.91	2.85	2.77	2.74	2.70	2.66	2.62	2.58	2.54
11	4.84	3.98	3.59	3.36	3.20	3.09	3.01	2.95	2.90	2.85	2.79	2.72	2.65	2.61	2.57	2.53	2.49	2.45	2.40
12	4.75	3.89	3.49	3.26	3.11	3.00	2.91	2.85	2.80	2.75	2.69	2.62	2.54	2.51	2.47	2.43	2.38	2.34	2.30
13	4.67	3.81	3.41	3.18	3.03	2.92	2.83	2.77	2.71	2.67	2.60	2.53	2.46	2.42	2.38	2.34	2.30	2.25	2.21
14	4.60	3.74	3.34	3.11	2.96	2.85	2.76	2.70	2.65	2.60	2.53	2.46	2.39	2.35	2.31	2.27	2.22	2.18	2.13
15	4.54	3.68	3.29	3.06	2.90	2.79	2.71	2.64	2.59	2.54	2.48	2.40	2.33	2.29	2.25	2.20	2.16	2.11	2.07
16	4.49	3.63	3.24	3.01	2.85	2.74	2.66	2.59	2.54	2.49	2.42	2.35	2.28	2.24	2.19	2.15	2.11	2.06	2.01
17	4.45	3.59	3.20	2.96	2.81	2.70	2.61	2.55	2.49	2.45	2.38	2.31	2.23	2.19	2.15	2.10	2.06	2.01	1.96
18	4.41	3.55	3.16	2.93	2.77	2.66	2.58	2.51	2.46	2.41	2.34	2.27	2.19	2.15	2.11	2.06	2.02	1.97	1.92
19	4.38	3.52	3.13	2.90	2.74	2.63	2.54	2.48	2.42	2.38	2.31	2.23	2.16	2.11	2.07	2.03	1.98	1.93	1.88
20	4.35	3.49	3.10	2.87	2.71	2.60	2.51	2.45	2.39	2.35	2.28	2.20	2.12	2.08	2.04	1.99	1.95	1.90	1.84
21	4.32	3.47	3.07	2.84	2.68	2.57	2.49	2.42	2.37	2.32	2.25	2.18	2.10	2.05	2.01	1.96	1.92	1.87	1.81
22	4.30	3.44	3.05	2.82	2.66	2.55	2.46	2.40	2.34	2.30	2.23	2.15	2.07	2.03	1.98	1.94	1.89	1.84	1.78
23	4.28	3.42	3.03	2.80	2.64	2.53	2.44	2.37	2.32	2.27	2.20	2.13	2.05	2.01	1.96	1.91	1.86	1.81	1.76
24	4.26	3.40	3.01	2.78	2.62	2.51	2.42	2.36	2.30	2.25	2.18	2.11	2.03	1.98	1.94	1.89	1.84	1.79	1.73
25	4.24	3.39	2.99	2.76	2.60	2.49	2.40	2.34	2.28	2.24	2.16	2.09	2.01	1.96	1.92	1.87	1.82	1.77	1.71
30	4.17	3.32	2.92	2.69	2.53	2.42	2.33	2.27	2.21	2.16	2.09	2.01	1.93	1.89	1.84	1.79	1.74	1.68	1.62
40	4.08	3.23	2.84	2.61	2.45	2.34	2.25	2.18	2.12	2.08	2.00	1.92	1.84	1.79	1.74	1.69	1.64	1.58	1.51
60	4.00	3.15	2.76	2.53	2.37	2.25	2.17	2.10	2.04	1.99	1.92	1.84	1.75	1.70	1.65	1.59	1.53	1.47	1.39
120	3.92	3.07	2.68	2.45	2.29	2.18	2.09	2.02	1.96	1.91	1.83	1.75	1.66	1.61	1.55	1.50	1.43	1.35	1.25
∞	3.84	3.00	2.60	2.37	2.21	2.10	2.01	1.94	1.88	1.83	1.75	1.67	1.57	1.52	1.46	1.39	1.32	1.22	1.00

Degrees of freedom for denominator

*Source: M. Merrington and C. M. Thompson, *Biometrika*, vol. 33(1943).

Values of F for F Distributions with .01 of the Area in the Right Tail.

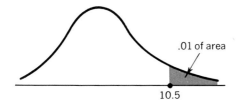

.01 of area

10.5

EXAMPLE: For a test at a significance level of .01 where we have 7 degrees of free-dom for the numerator and 5 degrees of freedom for the denominator, the appropri-ate F value is found by looking under the 7 degrees of freedom column and pro-ceeding down to the 5 degrees of freedom row; there we find the appropriate F value to be 10.5.

Degrees of freedom for numerator

	1	2	3	4	5	6	7	8	9	10	12	15	20	24	30	40	60	120	∞
1	4,052	5,000	5,403	5,625	5,764	5,859	5,928	5,982	6,023	6,056	6,106	6,157	6,209	6,235	6,261	6,287	6,313	6,339	6,366
2	98.5	99.0	99.2	99.2	99.3	99.3	99.4	99.4	99.4	99.4	99.4	99.4	99.4	99.5	99.5	99.5	99.5	99.5	99.5
3	34.1	30.8	29.5	28.7	28.2	27.9	27.7	27.5	27.3	27.2	27.1	26.9	26.7	26.6	26.5	26.4	26.3	26.2	26.1
4	21.2	18.0	16.7	16.0	15.5	15.2	15.0	14.8	14.7	14.5	14.4	14.2	14.0	13.9	13.8	13.7	13.7	13.6	13.5
5	16.3	13.3	12.1	11.4	11.0	10.7	10.5	10.3	10.2	10.1	9.89	9.72	9.55	9.47	9.38	9.29	9.20	9.11	9.02
6	13.7	10.9	9.78	9.15	8.75	8.47	8.26	8.10	7.98	7.87	7.72	7.56	7.40	7.31	7.23	7.14	7.06	6.97	6.88
7	12.2	9.55	8.45	7.85	7.46	7.19	6.99	6.84	6.72	6.62	6.47	6.31	6.16	6.07	5.99	5.91	5.82	5.74	5.65
8	11.3	8.65	7.59	7.01	6.63	6.37	6.18	6.03	5.91	5.81	5.67	5.52	5.36	5.28	5.20	5.12	5.03	4.95	4.86
9	10.6	8.02	6.99	6.42	6.06	5.80	5.61	5.47	5.35	5.26	5.11	4.96	4.81	4.73	4.65	4.57	4.48	4.40	4.31
10	10.0	7.56	6.55	5.99	5.64	5.39	5.20	5.06	4.94	4.85	4.71	4.56	4.41	4.33	4.25	4.17	4.08	4.00	3.91
11	9.65	7.21	6.22	5.67	5.32	5.07	4.89	4.74	4.63	4.54	4.40	4.25	4.10	4.02	3.94	3.86	3.78	3.69	3.60
12	9.33	6.93	5.95	5.41	5.06	4.82	4.64	4.50	4.39	4.30	4.16	4.01	3.86	3.78	3.70	3.62	3.54	3.45	3.36
13	9.07	6.70	5.74	5.21	4.86	4.62	4.44	4.30	4.19	4.10	3.96	3.82	3.66	3.59	3.51	3.43	3.34	3.25	3.17
14	8.86	6.51	5.56	5.04	4.70	4.46	4.28	4.14	4.03	3.94	3.80	3.66	3.51	3.43	3.35	3.27	3.18	3.09	3.00
15	8.68	6.36	5.42	4.89	4.56	4.32	4.14	4.00	3.89	3.80	3.67	3.52	3.37	3.29	3.21	3.13	3.05	2.96	2.87
16	8.53	6.23	5.29	4.77	4.44	4.20	4.03	3.89	3.78	3.69	3.55	3.41	3.26	3.18	3.10	3.02	2.93	2.84	2.75
17	8.40	6.11	5.19	4.67	4.34	4.10	3.93	3.79	3.68	3.59	3.46	3.31	3.16	3.08	3.00	2.92	2.83	2.75	2.65
18	8.29	6.01	5.09	4.58	4.25	4.01	3.84	3.71	3.60	3.51	3.37	3.23	3.08	3.00	2.92	2.84	2.75	2.66	2.57
19	8.19	5.93	5.01	4.50	4.17	3.94	3.77	3.63	3.52	3.43	3.30	3.15	3.00	2.92	2.84	2.76	2.67	2.58	2.49
20	8.10	5.85	4.94	4.43	4.10	3.87	3.70	3.56	3.46	3.37	3.23	3.09	2.94	2.86	2.78	2.69	2.61	2.52	2.42
21	8.02	5.78	4.87	4.37	4.04	3.81	3.64	3.51	3.40	3.31	3.17	3.03	2.88	2.80	2.72	2.64	2.55	2.46	2.36
22	7.95	5.72	4.82	4.31	3.99	3.76	3.59	3.45	3.35	3.26	3.12	2.98	2.83	2.75	2.67	2.58	2.50	2.40	2.31
23	7.88	5.66	4.76	4.26	3.94	3.71	3.54	3.41	3.30	3.21	3.07	2.93	2.78	2.70	2.62	2.54	2.45	2.35	2.26
24	7.82	5.61	4.72	4.22	3.90	3.67	3.50	3.36	3.26	3.17	3.03	2.89	2.74	2.66	2.58	2.49	2.40	2.31	2.21
25	7.77	5.57	4.68	4.18	3.86	3.63	3.46	3.32	3.22	3.13	2.99	2.85	2.70	2.62	2.53	2.45	2.36	2.27	2.17
30	7.56	5.39	4.51	4.02	3.70	3.47	3.30	3.17	3.07	2.98	2.84	2.70	2.55	2.47	2.39	2.30	2.21	2.11	2.01
40	7.31	5.18	4.31	3.83	3.51	3.29	3.12	2.99	2.89	2.80	2.66	2.52	2.37	2.29	2.20	2.11	2.02	1.92	1.80
60	7.08	4.98	4.13	3.65	3.34	3.12	2.95	2.82	2.72	2.63	2.50	2.35	2.20	2.12	2.03	1.94	1.84	1.73	1.60
120	6.85	4.79	3.95	3.48	3.17	2.96	2.79	2.66	2.56	2.47	2.34	2.19	2.03	1.95	1.86	1.76	1.66	1.53	1.38
∞	6.63	4.61	3.78	3.32	3.02	2.80	2.64	2.51	2.41	2.32	2.18	2.04	1.88	1.79	1.70	1.59	1.47	1.32	1.00

Degrees of freedom for denominator

APPENDIX TABLE 7

Values for Spearman's Rank Correlation (r_s) for
Combined Areas in Both Tails.*

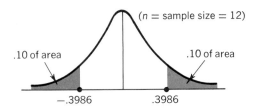

(n = sample size = 12)

.10 of area   .10 of area

$-.3986$ $.3986$

EXAMPLE: For a two-tailed test of significance at the .20 level, with $n = 12$, the appropriate value for r_s can be found by looking under the .20 column and proceeding down to the 12 row; there we find the appropriate r_s value to be .3986.

n	.20	.10	.05	.02	.01	.002
4	.8000	.8000				
5	.7000	.8000	.9000	.9000		
6	.6000	.7714	.8286	.8857	.9429	
7	.5357	.6786	.7450	.8571	.8929	.9643
8	.5000	.6190	.7143	.8095	.8571	.9286
9	.4667	.5833	.6833	.7667	.8167	.9000
10	.4424	.5515	.6364	.7333	.7818	.8667
11	.4182	.5273	.6091	.7000	.7455	.8364
12	.3986	.4965	.5804	.6713	.7273	.8182
13	.3791	.4780	.5549	.6429	.6978	.7912
14	.3626	.4593	.5341	.6220	.6747	.7670
15	.3500	.4429	.5179	.6000	.6536	.7464
16	.3382	.4265	.5000	.5824	.6324	.7265
17	.3260	.4118	.4853	.5637	.6152	.7083
18	.3148	.3994	.4716	.5480	.5975	.6904
19	.3070	.3895	.4579	.5333	.5825	.6737
20	.2977	.3789	.4451	.5203	.5684	.6586
21	.2909	.3688	.4351	.5078	.5545	.6455
22	.2829	.3597	.4241	.4963	.5426	.6318
23	.2767	.3518	.4150	.4852	.5306	.6186
24	.2704	.3435	.4061	.4748	.5200	.6070
25	.2646	.3362	.3977	.4654	.5100	.5962
26	.2588	.3299	.3894	.4564	.5002	.5856
27	.2540	.3236	.3822	.4481	.4915	.5757
28	.2490	.3175	.3749	.4401	.4828	.5660
29	.2443	.3113	.3685	.4320	.4744	.5567
30	.2400	.3059	.3620	.4251	.4665	.5479

*Source: W.J. Conover, *Practical Nonparametric Statistics*, John Wiley & Sons, Inc., New York, 1971.

APPENDIX TABLE 8

Square Roots for Numbers from 1 to 400.*

1	1.00	41	6.40	81	9.00	121	11.00	161	12.69
2	1.41	42	6.48	82	9.06	122	11.05	162	12.73
3	1.73	43	6.56	83	9.11	123	11.09	163	12.77
4	2.00	44	6.63	84	9.17	124	11.14	164	12.81
5	2.24	45	6.71	85	9.22	125	11.18	165	12.85
6	2.45	46	6.78	86	9.27	126	11.23	166	12.88
7	2.65	47	6.86	87	9.33	127	11.27	167	12.92
8	2.83	48	6.93	88	9.38	128	11.31	168	12.96
9	3.00	49	7.00	89	9.43	129	11.36	169	13.00
10	3.16	50	7.07	90	9.49	130	11.40	170	13.04
11	3.32	51	7.14	91	9.54	131	11.45	171	13.08
12	3.46	52	7.21	92	9.59	132	11.49	172	13.11
13	3.61	53	7.28	93	9.64	133	11.53	173	13.15
14	3.74	54	7.35	94	9.70	134	11.58	174	13.19
15	3.87	55	7.42	95	9.75	135	11.62	175	13.23
16	4.00	56	7.48	96	9.80	136	11.66	176	13.27
17	4.12	57	7.55	97	9.85	137	11.70	177	13.30
18	4.24	58	7.62	98	9.90	138	11.74	178	13.34
19	4.36	59	7.68	99	9.95	139	11.79	179	13.38
20	4.47	60	7.75	100	10.00	140	11.83	180	13.42
21	4.58	61	7.81	101	10.05	141	11.87	181	13.45
22	4.69	62	7.87	102	10.10	142	11.92	182	13.49
23	4.80	63	7.94	103	10.15	143	11.96	183	13.53
24	4.90	64	8.00	104	10.20	144	12.00	184	13.56
25	5.00	65	8.06	105	10.25	145	12.04	185	13.60
26	5.10	66	8.12	106	10.30	146	12.08	186	13.64
27	5.20	67	8.19	107	10.34	147	12.12	187	13.67
28	5.29	68	8.25	108	10.39	148	12.17	188	13.71
29	5.39	69	8.31	109	10.44	149	12.21	189	13.75
30	5.48	70	8.37	110	10.49	150	12.25	190	13.78
31	5.57	71	8.43	111	10.54	151	12.29	191	13.82
32	5.66	72	8.49	112	10.58	152	12.33	192	13.86
33	5.74	73	8.54	113	10.63	153	12.37	193	13.89
34	5.83	74	8.60	114	10.68	154	12.41	194	13.93
35	5.92	75	8.66	115	10.72	155	12.45	195	13.96
36	6.00	76	8.72	116	10.77	156	12.49	196	14.00
37	6.08	77	8.77	117	10.82	157	12.53	197	14.04
38	6.16	78	8.83	118	10.86	158	12.57	198	14.07
39	6.25	79	8.89	119	10.91	159	12.61	199	14.11
40	6.32	80	8.94	120	10.95	160	12.65	200	14.14

* Source: Levin and Kirkpatrick, *Quantitative Approaches to Management,* 4th ed., New York: McGraw-Hill Book Co., Inc., 1978.

201	14.18	241	15.52	281	16.76	321	17.92	361	19.00
202	14.21	242	15.56	282	16.79	322	17.94	362	19.03
203	14.25	243	15.59	283	16.82	323	17.97	363	19.05
204	14.28	244	15.62	284	16.85	324	18.00	364	19.08
205	14.32	245	15.65	285	16.88	325	18.03	365	19.11
206	14.35	246	15.68	286	16.91	326	18.06	366	19.13
207	14.39	247	15.72	287	16.94	327	18.08	367	19.16
208	14.42	248	15.75	288	16.97	328	18.11	368	19.18
209	14.46	249	15.78	289	17.00	329	18.14	369	19.21
210	14.49	250	15.81	290	17.03	330	18.17	370	19.24
211	14.53	251	15.84	291	17.06	331	18.19	371	19.26
212	14.56	252	15.87	292	17.09	332	18.22	372	19.29
213	14.59	253	15.91	293	17.12	333	18.25	373	19.31
214	14.63	254	15.94	294	17.15	334	18.28	374	19.34
215	14.66	255	15.97	295	17.18	335	18.30	375	19.36
216	14.70	256	16.00	296	17.20	336	18.33	376	19.39
217	14.73	257	16.03	297	17.23	337	18.36	377	19.42
218	14.76	258	16.06	298	17.26	338	18.38	378	19.44
219	14.80	259	16.09	299	17.29	339	18.41	379	19.47
220	14.83	260	16.12	300	17.32	340	18.44	380	19.49
221	14.87	261	16.16	301	17.35	341	18.47	381	19.52
222	14.90	262	16.19	302	17.38	342	18.49	382	19.54
223	14.93	263	16.22	303	17.41	343	18.52	383	19.57
224	14.97	264	16.25	304	17.44	344	18.55	384	19.60
225	15.00	265	16.28	305	17.46	345	18.57	385	19.62
226	15.03	266	16.31	306	17.49	346	18.60	386	19.65
227	15.07	267	16.34	307	17.52	347	18.63	387	19.67
228	15.10	268	16.37	308	17.55	348	18.65	388	19.70
229	15.13	269	16.40	309	17.58	349	18.68	389	19.72
230	15.17	270	16.43	310	17.61	350	18.71	390	19.75
231	15.20	271	16.46	311	17.64	351	18.74	391	19.77
232	15.23	272	16.49	312	17.66	352	18.76	392	19.80
233	15.26	273	16.52	313	17.69	353	18.79	393	19.82
234	15.30	274	16.55	314	17.72	354	18.81	394	19.85
235	15.33	275	16.58	315	17.75	355	18.84	395	19.87
236	15.36	276	16.61	316	17.78	356	18.87	396	19.90
237	15.39	277	16.64	317	17.80	357	18.89	397	19.92
238	15.43	278	16.67	318	17.83	358	18.92	398	19.95
239	15.46	279	16.70	319	17.86	359	18.95	399	19.98
240	15.49	280	16.73	320	17.89	360	18.97	400	20.00

Bibliography

Introductions to statistics for the layman

FEDERER, W. T. *Statistics and Society.* New York: Marcel Dekker, 1973.

HUFF, D. *How to Lie with Statistics.* New York: Norton, 1954.

LEVINSON, H. C. *Chance, Luck, and Statistics.* New York: Dover, 1963.

MORONEY, M. J. *Facts from Figures.* London: Pelican Books, 1956.

MOSTELLER, F., W. H. KRUSKAL, R. S. PIETERS, G. R. RISING, and R. F. LINK. *Statistics by Example.* Reading, Mass.: Addison-Wesley, 1973.

REICHMAN, W. J. *Use and Abuse of Statistics.* New York: Oxford Univ. Press, 1963.

TANUR, J. M., et al. *Statistics: A Guide to the Unknown.* San Francisco: Holden-Day, 1972.

WILLIAMS, J. D. *The Compleat Strategyst, revised.* New York: McGraw-Hill, 1965.

General statistics

CLELLAND, R. C., J. S. DeCANI, and F. E. BROWN. *Basic Statistics with Business Applications.* 2nd ed. New York: Wiley, 1973.

DIXON, W. and F. MASSEY, JR. *Introduction to Statistical Analysis.* 3rd ed. New York: McGraw-Hill, 1969.

FREUND, J. E. *Mathematical Statistics,* 2nd ed. Englewood Cliffs, N.J.: Prentice-Hall, 1971.

FREUND, J. E. *Modern Elementary Statistics,* 5th ed. Englewood Cliffs, N.J.: Prentice-Hall, 1979.

HOEL, P. *Introduction to Mathematical Statistics.* 4th ed. New York, Wiley, 1971.

PETERS, W. S. *Readings in Applied Statistics.* Englewood Cliffs, N.J.: Prentice-Hall, 1969.

Probability

FELLER, W. *An Introduction to Probability Theory and Its Applications.* Vol. 2, 2nd ed. New York: Wiley, 1971.

FREUND, J. E., *Introduction to Probability.* Encino, Calif.: Dickenson, 1973.

GOLDBERG, S. *Probability: An Introduction.* Englewood Cliffs, N.J.: Prentice-Hall, 1960.

HODGES, J. L., JR., and E. LEHMANN. *Basic Concepts of Probability and Statistics.* San Francisco: Holden-Day, 1964.

HODGES, J. L., and E. L. LEHMANN. *Elements of Finite Probability.* San Francisco: Holden-Day, 1965.

MOSTELLER, F., R. E. K. ROURKE, and G. B. THOMAS, JR. *Probability and Statistics.* Reading, Mass.: Addison-Wesley, 1961.

MOSTELLER, F., R. E. K. ROURKE, and G. B. THOMAS, JR. *Probability with Statistical Applications.* 2nd ed. Reading, Mass.: Addison-Wesley, 1970.

Sampling theory and techniques

COCHRAN, W. G. *Sampling Techniques,* 2nd ed. New York: Wiley, 1963.

DEMING, W. E. *Sample Designs in Business Research.* New York: Wiley, 1953.

HANSEN, M. H., W. N. HURWITZ, and W. G. MADOW. *Sample Survey Methods and Theory,* Vol. 1: *Methods and Applications;* Vol. II: *Theory.* New York: Wiley, 1963.

KISH, L. *Survey Sampling.* New York: Wiley, 1965.

NAMIAS, J. *Handbook of Selected Sample Surveys in the Federal Government,* New York: St. John's Univ. Press, 1969.

SLONIM, M. J. *Sampling in a Nutshell.* New York: Simon & Schuster, 1973.

STUART, A. *Basic Ideas of Scientific Sampling.* New York: Hafner, 1962.

Nonparametric statistics

BRADLEY, J. V. *Distribution-Free Statistical Tests.* Englewood Cliffs, N.J.: Prentice-Hall, 1968.

CONOVER, W. J. *Practical Non-Parametric Statistics.* New York: Wiley, 1971.

GIBBONS, J. D. *Nonparametric Statistical Inference.* New York: McGraw-Hill, 1971.

KRAFT, C. H. and C. VAN EEDEN. *A Nonparametric Introduction to Statistics.* New York: Macmillan, 1968.

NOETHER, G. E. *Elements of Nonparametric Statistics.* New York: Wiley, 1967.

SIEGEL, S. *Nonparametric Statistics for the Behavioral Sciences.* New York: McGraw-Hill, 1956.

Statistical decision theory

DeGROOT, M. H. *Optimal Statistical Decisions.* New York: McGraw-Hill, 1970.

LEVIN, R. I. and C. KIRKPATRICK. *Quantitative Approaches to Management.* 4th ed., New York: McGraw-Hill, 1978.

LINDLEY, D. V. *Introduction to Probability and Statistics from a Bayesian Viewpoint.* New York: Cambridge Univ. Press, 1965.

LUCE R. D. and H. RAIFFA. *Games and Decisions.* New York: Wiley, 1957.

MORRIS, W. T. *Management Science—A Bayesian Introduction.* Englewood Cliffs, N.J.: Prentice-Hall, 1968.

RAIFFA, H. *Decision Analysis, Introductory Lectures on Choices under Uncertainty.* Reading, Mass.: Addison-Wesley, 1968.

RAIFFA, H. and R. SCHLAIFER. *Applied Statistical Decision Theory.* Cambridge, Mass.: Division of Research, Graduate School of Business Administration, Harvard Univ., 1961.

SCHLAIFER, R. *Probability and Statistics for Business Decisions.* New York: McGraw-Hill, 1959.

WINKLER, R. L. *Introduction to Bayesian Inference Decision.* New York: Holt, 1972.

Special statistical topics

EZEKIEL, M., and K. FOX. *Methods of Correlation and Regression Analysis.* 3rd ed. New York: Wiley, 1959.

HARRIS, R. J. *A Primer of Multivariate Statistics.* New York: Academic Press, 1974.

MUDGETT, B. D. *Index Numbers.* New York: Wiley, 1951.

U.S. DEPT. OF LABOR. *The Consumer Price Index: History and Techniques.* Bureau of Labor Statistics Bulletin 1517, updated, Washington, D.C.

Sources of statistical data

COMAN, E. T. *Sources of Business Information.* rev. ed. Englewood Cliffs, N.J.: Prentice-Hall, 1964.

NATIONAL REFERRAL CENTER FOR SCIENCE AND TECHNOLOGY. *A Directory of Information Resources in the United States, Social Sciences.* Library of Congress, Washington, D.C.: U.S. Government Printing Office, October 1967.

SILK, L. S., and M. L. CURLEY. *A Primer on Business Forecasting with a Guide to Sources of Business Data.* New York: Random House, 1970.

WASSERMAN, P., E. ALLEN, A. KRUZAS, and C. GEORGI. *Statistics Sources.* 4th ed. Detroit: Gale Research Co., 1971.

Statistical tables

BURRINGTON, R. S., and D. C. MAY. *Handbook of Probability and Statistics with Tables.* 2nd ed. New York: McGraw-Hill, 1970.

HALD, A. *Statistical Tables and Formulas.* New York: Wiley, 1952.

NATIONAL BUREAU OF STANDARDS. *Tables of the Binomial Probability Distribution.* Washington, D.C.: Government Printing Office, 1950.

OWEN, D. *Handbook of Statistical Tables.* Reading, Mass.: Addison-Wesley, 1962.

PEARSON, E. S., and H. O. HARTLEY. *Biometrika Tables for Statisticians.* 2nd ed. Cambridge, Eng.: Cambridge Univ. Press, 1962.

RAND CORPORATION. *A Million Random Digits with 100,000 Normal Deviates.* New York: Free Press, 1955.

SHELBY, S. *Standard Mathematical Tables.* 17th ed. Cleveland, Ohio, Chemical Rubber Co., 1969.

Dictionaries and general reference works

FREUND, J. and F. WILLIAMS. *Dictionary/Outline of Basic Statistics.* New York: McGraw-Hill, 1966.

KENDALL, M. G., and W. R. BUCKLAND. *A Dictionary of Statistical Terms.* 3rd ed. New York: Hafner Press, 1971.

SNEDECOR, G. W., and W. G. COCHRAN. *Statistical Methods.* 6th ed. Ames, Iowa: The Iowa State Univ. Press, 1967.

Index

A

Absolute value, 93
Absolute value ($\|$), symbol, 420
Absolute value of error, using to
    measure best fit, 419–20
Acceptance region:
    determining limits of, 311, 312–13,
      372–73
    equations for limits of, 446, 468
    for null hypothesis, 303
Actual mean, compared to estimated
    mean, 52
Addition rule for mutually exclusive
    events, 126–27, 152
Addition rule for not mutually
    exclusive events, 127–29, 152–
    53
Alpha (α), the probability of a Type I
    error, 305, 314, 346
Alternative hypothesis, defined, 302,
    346
Analysis of variance:
    basic concepts, 376–77
    calculating variance among sample
      means, 377–79
    calculating variance within
      samples, 379–80
    defined, 6
    F hypothesis test, 380–85
    function of, 6, 356, 374–75, 399
    statement of hypothesis, 376
    statement of problem, 375–76
A priori probability, 120, 151
Arithmetic mean, 50–56
Average absolute deviation:
    calculating, 93–94
    defined, 92, 106
    equations, 93, 107–108
Average deviation measures, 92–
    102

B

Bayes' theorem, 145, 151
Bernoulli process:
    defined, 175–76, 214
    meeting conditions for using, 183
    use of, 175–77
Best-fitting line, finding, 421
Best-fitting multiple regression plane,
    equations for, 454, 468–69
Beta (β), the probability of a Type I
    error, 305, 314–15, 346
Between-column variance:
    defined, 377, 399
    equation, 378, 401
Bimodal distribution, defined, 73, 76

Binomial distribution, 175–83
    defined, 175, 214
    general appearance of, 180–81
    graphic illustrations of, 177–81
    and interval estimates, 277–78
    mean of, equation for, 182, 215
    measures of central tendency and
      dispersion for, 182–83
    normal distribution as an
      approximation of, 201–203,
      277
    Poisson distribution as an
      approximation of, 189, 213–
      14, 216
    review of, 277
    shortcomings of, 277
    standard deviation of, equation,
      182, 216
Binomial formula, 176–77, 215
Binomial tables, use of, 181–82

C

Causal relationship, 411–12
Cause and effect, as mistaken
    assumption in regression and
    correlation analysis, 450
Cells:
    calculating expected frequencies
      for, 365–66
    defined, 360, 364
    estimating proportions in, 365
Census, defined, 226, 253
Central limit theorem, 245–47, 253
Central tendency, measures of, 46, 76
Characteristic probability, 176
Chebyshev's theorem, 97–98, 106
Chi-square, as a test of
    independence, 357–67
Chi-square distribution, 361, 399
Chi-square goodness-of-fit test,
    using, 372–73
Chi-square statistic:
    calculating, 360–61, 371
    equation for, 360, 400
    interpreting, 360–61
Chi-square tests:
    defined, 5–6
    equation for, degrees of freedom,
      362, 400
    precautions about using, 367
    reasoning intuitively about, 360
    and regression analysis, 410
    as test of goodness-of-fit, 369–73
    using the, 356, 363–64
Class intervals:
    construction of, 25
    defined, 21
    equation for determining, 22, 37

Class limits, 24–25
    defined, 24, 36
Class mark:
    and coding, 54–55
    defined, 24, 26
    equation for determining, 24–25,
      37
Classes:
    all-inclusive, 17
    continuous, 18
    defined, 16
    discrete, 18
    mutually exclusive, 17
    open-ended, 18, 36
    sorting data points into, 22–23
Classical probability, 119–20, 151
    equation, 119, 152
Cluster, 25
Cluster sampling, defined, 233, 253
Coding, defined, 54, 76
Coefficient of correlation, 442–44
    defined, 443, 463
    misinterpretation of, 450
Coefficient of determination, 435–36,
    463
    misinterpretation of, 450
Coefficient of multiple correlation
    (R),
    defined, 459, 463
Coefficient of multiple
    determination(R^2), 459–60
    defined, 459, 463
    equation, 459, 469
Coefficient of rank correlation, 497–
    99
    equation, 497, 507
Coefficient of variation:
    defined, 104–105, 106
    equation, 105, 109
Collectively exhaustive events, 118,
    151
Columns, contingency table, 358
Complete enumeration, defined, 226
Composite index numbers, 527
Computer, and multiple regression,
    461–63, 464–65
Conditional probabilities:
    under statistical dependence, 138–
    41
    under statistical independence,
      135–36
Conditional probability:
    defined, 136, 151
    equations, 136, 140, 153
Confidence interval:
    compared to $\hat{Y}$ approximate
      prediction intervals, 430–31
    and confidence limits, 271
    constructing, for a variance, 389–
      90

Confidence interval (*cont.*)
  defined, 271, 291
  and interval estimates, 270–72
  multiple regression, 458–59
  and regression line, 448
Confidence level:
  defined, 271, 291
  finding a 90%, 274, 275–76
  finding a 95%, 273–74
  finding a 99%, 279
  relationship to confidence interval, 271–72
Confidence limits:
  defined, 271, 291
  equations for, under various conditions, 285
  two ways to express, 286–87
  using t table to compute, 283–84
Consistent estimator, defined, 263, 291
Consumer Price Index, 526, 527, 528
Contingency table:
  calculating degrees of freedom in, 362–63
  description of, 357–58, 399–400
  with more than two rows, 364–67
Continuity correction factor, 202, 214
Continuous data, defined, 18, 36
Continuous distribution:
  defined, 191
  expected value decision making with, 205–212
Continuous probability distribution, defined, 164, 214
Continuous random variable, 166, 214
Coordinates, 416
Correlation analysis, 435–44 (*see also* Rank correlation)
  defined, 6, 435, 463
  misuse of, 449–50
Cumulative frequency distribution, defined, 31, 36
Cumulative probabilities, derivation of, 208–209
Cumulative probability, 207, 208
Curvilinear regressions, 462
Curvilinear relationship, defined, 413–14, 463
Cyclical fluctuation, 524, 525

D

Data:
  arrangement of, 10–12, 14–18
  collection of, 10–11
  continuous, 18, 36
  defined, 10, 36
  discrete, 18, 36
  grouped, dealing with, 52
  historical, misuse of, 450
  illustration of, in chart form, 23
  raw, 13–14, 36
  tests for, 11
  ungrouped, dealing with, 51–52
Data array, 14–15
  defined, 14, 36
Data point, defined, 10, 36
Data set, defined, 10, 36
Deciles, defined, 89, 106

Decision makers, role in formulating hypotheses, 300
Decision making, use of expected value in, 172–74, 205–212
Decision theory, 3–4
Degrees of freedom:
  defined, 281, 291
  determining, in contingency tables, 362–63
  determining, in a goodness-of-fit test, 371–72
  equation for, a chi-square test, 362, 400
  equation for, between-column variance, 382, 402
  equation for, within-column variance, 382, 402
  function of, 282
  using F distribution, 382–83
  using t distribution, 282–85
Dependence, defined, 138
Dependent samples, defined, 333, 346
Dependent variable, defined, 410, 463
Descriptive statistics, defined, 3
Direct realtionship:
  defined, 411, 463
  graphing of, 411
  positive slope of, 417
Discrete data, defined, 18, 36
Discrete probability distribution, 164, 215
Discrete random variable, 166, 215
Dispersion:
  average deviation measures, 92–102
  defined, 86, 106
  distance measures, 88–90, 107
  measures of, 46, 76, 86
  need to measure, 86
  relative, 104–105
Distance measures, 88–90, 107
Distribution-free tests (*see* Non-parametric statistics)

E

Efficient estimator, defined, 262–63, 291
Equal classes, dividing range by, 21
Estimate, defined, 261, 291
Estimated mean, compared to actual mean, 52
Estimated regression coefficients, 456
Estimated standard error of the difference between two means, equations, 328, 329, 331, 347, 348
Estimated standard error of the difference between two proportions, equations, 338, 339, 348
Estimated standard error of the mean of a finite population, equation, 275, 292
Estimated standard error of the mean of an infinite population, equation, 284, 292
Estimated standard error of the proportion, equation, 278, 292

Estimates, 261
Estimating equation:
  checking, 426
  defined, 410, 463
  shortcoming of, 425
  using, for a straight line, 416–18
Estimating equation for multiple regression, 453, 468
Estimating line:
  equation, 418, 466
  fitting mathematically, 418
Estimating regression equation for three variables, equation, 453, 468
Estimation:
  determining sample size of, 286–90
  using the regression line, 415–33
Estimator:
  criteria of a good, 262–63
  defined, 261, 291
Event, 117, 151
Expected frequencies:
  calculating, 365–66, 370–71
  defined, 360, 400
  equation for, 365, 401
  finding, 364–65
Expected gain, maximizing, 174
Expected gain = expected loss, equation, 207, 216–17
Expected losses, calculating, 173–74
Expected marginal gain, defined, 206–207
Expected marginal loss, defined, 206–207
Expected value:
  defined, 168, 215
  of a random variable, 168–69, 215
  use of, in decision making, 172–74, 205–212
Experiment:
  defined, 117, 152
  design of, 519–23
  sampling, 519
Experimental claims, consumer reaction to, 520–21
Experimental design:
  efficiency in, 522–23
  factorial experiments, 521–22
  phases of, 519–20
Explained deviation, 440
Explained variation, 440–41
Extrapolation, as limited by range of data, 449–50
Extreme values, effect of, 56, 58, 74, 75

F

F distribution:
  described, 381–82, 400
  using, degrees of freedom, 382–83
F hypothesis test, 380–81
  calculating limit of acceptance region, 384
  precautions about using, 384–85
F ratio:
  defined, 381–400
  equation, 380, 401–402
F statistic:
  equations, 380, 395, 397, 403
  interpreting, 380–81

F table:
    using, 383
    using, with one-tailed test of two
      variances, 395–96
Factorial, 176
Factorial experiments, 521–22
Finance, use of dispersion measures
    in, 86
Finite population, defined, 229, 253
Finite population multiplier, 249–51,
    253
    equation, 251, 254
Fractile, defined, 88, 107
Frequency curve, defined, 30, 36
Frequency distribution:
    advantages and disadvantages, 15–
      16
    compared to probability
      distributions, 162, 164
    construction of, 21–25
    defined, 16, 36
    descriptive measures of, 46–48
    function of classes in, 16
    graphing, 28–34
Frequency polygons, 29–30
    defined, 29, 36
Frequency table (*see* Frequency
    distribution)

G

Gaussian distribution (*see* Normal
    distribution)
Geometric mean (GM), 61–63
    defined, 61, 76
    equation, 62, 77–78
Good fit, defined, 418
Goodness-of-fit test:
    determining degrees of freedom in,
      371–72
    function of, 369, 400
    using chi-square, 372–73
Grand mean, calculating, 375–76,
    400
Graphical solution, of mode of
    grouped data, 71–72
Graphs, function of, 28
Grouped data, 52
Growth rate (*see* Geometric mean)

H

Histograms, 28–29, 30
    defined, 28, 36
Horizontal axis, in graphing, 28
Hypothesis:
    defined, 298, 346
    formal statement of, 301–302
    formulating, 299
Hypothesis of no difference, testing a,
    480–81
Hypothesis testing:
    alternative hypothesis, 302
    concepts basic to procedure, 299–
      300
    defined, 5
    with dependent samples, 333–37
    for differences between means,
      325–37

    for differences between
      proportions, 338–43
    function of, 298
    of means, 309–313, 322–24
    measuring power of, 314–16
    null hypothesis, 301–302
    of proportions, 317–21
    and regression line, 445–48
    selecting the correct probability
      distribution, 305–306
    significance level in, 302–304

I

Improper base, use of, 528
Independence:
    defined, 130, 133
    test of, 357, 400
Independent variables, defined, 410,
    463
Index numbers, 525–28
    defined, 525–26
    problems related to, 527–28
    sources of, 528
    types of, 526–27
    uses of, 527
Indices, incomparability of, 527
Inferential statistics, defined, 3
Infinite population, defined, 230, 253
Interfractile range:
    calculating, 88–89
    definition, 88, 107
Interquartile range:
    computing, 89–90
    defined, 89, 107
    equation, 89, 107
Interval estimates:
    basic concepts, 267–69
    calculating, of the mean from large
      samples, 273–76
    calculating, of the proportion from
      large samples, 277–79
    and confidence intervals, 270–72
    defined, 261, 267, 291
    using the *t* distribution, 280–85
Inverse relationship:
    defined, 411, 463
    negative slope of, 417
Irregular variation, 524, 525

J

Joint probability, 131, 152
    equations, 131, 142, 153
    under statistical dependence, 141–
      42
    under statistical independence,
      131–35
Judgment sampling, defined, 227,
    253

K

k (symbol for classes), 371
k − 1 rule, 372
Kurtosis:
    defined, 47, 76
    measures of, 47–48

L

Latin square, 523
Least squares method, 418–22
    defined, 420–21, 463
    examples using, 422–25
    using to fit a regression plane,
      454
Left-tailed test of hypothesis, 307
Leptokurtic:
    curve, 48
    defined, 48, 76
"Less than" frequency tables, 31
"Less than" ogive, 32
Linear relationship:
    defined, 413, 463
    and sample coefficient of
      determination, 440
Location, measures of, 46
Losses:
    calculating expected, 173–74
    types of, 172–73
Lower confidence limit, 271
Lower-tailed test, defined, 307, 346

M

Magnitude of error, 420
Mann-Whitney *U* test, 484–88
    approaching problem using, 484–
      85
    calculating the *U* statistic, 486
    defined, 476, 484, 506
    special properties of, 488
    testing the hypothesis, 486–88
Marginal analysis, 205–206
    problem solving using, 210–212
Marginal gain (MG), 205
Marginal loss (ML), 206
Marginal probabilities:
    under statistical dependence, 142–
      43
    under statistical independence,
      130–31
Marginal probability:
    defined, 125, 152
    equation, 125, 152
Mean (*see also* Arithmetic mean;
      Geometric mean; Weighted
      mean):
    compared to median and mode,
      75–76
    computing for the binomial
      distribution, 182–83
    defined, 50, 76
    hypothesis testing of, 309–313,
      322–24, 325–37
    one-tailed tests of, 311–13, 324
    two-tailed tests of, 309–11
Mean of a binomial distribution,
    equation, 182, 215
Mean of the *r* statistic, equation, 492,
    507
Mean of sampling distribution of
    difference between sample
    means, 326
Mean of sampling distribution of the
    proportion of successes,
    equation, 277, 292

Mean of the U statistic, equation, 486, 507
Measures of central tendency, 46, 76
Measures of dispersion, 46, 76, 86, 457–59
Median, 64–68
  compared to mean and mode, 75–76
  defined, 64, 76
  equation, 64, 78
Median class, defined, 65–66, 76
Mendel's pea data, 367
Mesokurtic:
  curve, 48
  defined, 48, 76
Method of least squares, 418–22
Minimum probability equation, 207 217
  deriving, 206–209
Modal class, 70
Mode, 69–74
  compared to mean and median, 75–76
  defined, 69, 76
  in skewed distribution, 71
  in symmetrical distribution, 70
Mode of grouped data:
  equation, 72, 78
  graphical solution, 71–72
Monetary values, combining probabilities and, 172–74
"More than" frequency tables, 31
"More than" ogive, 32
Mu (μ), population mean, 50
Multicollinearity, defined, 457, 463
Multimodal distributions, 73
Multiple regression:
  computer and, 461–63, 464–65
  and correlation analysis, 451–60
  defined, 410, 463
  value of, 451–52
Multiple regression equation, 452–57
  symbols for, 452
  visualizing, 453–54
Multiple regression plan, 453–54
Multiplication rule:
  for joint, dependent events, 142
  for joint, independent events, 131–32
Mutually exclusive events:
  addition rule for, 126–27, 152
  defined, 117–18, 152

N

N, number of population elements, 50
n, number of sample observations, 50
Negative slope, 411
  and inverse relationship, 417
Nonparametric statistics:
  advantages and disadvantages, 476–77
  defined, 6
Nonparametric tests, defined, 476, 506
Nonrandom sampling (see Judgment sampling)

Normal curve:
  described, 191–93
  measuring area under, 193–95
  measuring distances under, equation, 196, 216
Normal distribution, 191–203 (see also Standard normal probability distribution)
  as approximation of binomial distribution, 201–3
  compared to t distribution, 281
  defined, 191–92, 215
  in hypothesis testing, 305–306
Normal probability distribution:
  characteristics of, 191–93
  shortcomings of, 201
Null hypothesis, defined, 301, 346
Numerical constant b for any given line, equation, 416, 466

O

Observed frequencies, 358–60
  calculating, 370–71
Obsolescence losses, 172–73
Odds, 116
Ogives, 31–34
  defined, 32, 36
$1 - \beta$, interpreting and computing values of, 314–16
One-sample run tests, 490–94
  basic idea of, 492
  defined, 476, 506
One-tailed tests:
  for difference between means, 330–333
  for difference between proportions, 341–43
  of hypothesis, 307–308, 346
  of means, 311–13, 324
  of proportions, 319–21
  using t distribution, 324
  of a variance, 392–93
  of two variances, 394–96
One-way analysis of variance, 385
Open-ended class, defined, 18, 36
Opportunity losses, 172–73
Optimum solution, minimizing expected losses, 174
Optimum stocking level, 209, 211–12

P

Paired data, sign test for, 478–81
Paired difference test, 336, 347
Paired samples, 333
Parameters:
  as characteristic of a population, 226
  defined, 50, 76, 226, 253
  significance of, in normal distribution, 192–93
Parametric statistics, 476
p-bar ($\bar{p}$), 277, 292
Peakedness, of a data set, 47–48
Percentage relative, 526
Percentiles, defined, 89, 107
Perfect correlation, 436–38, 498

Perfect inverse correlation, 499
Platykurtic:
  curve, 48
  defined, 48, 76
Point estimates, 264–66
  defined, 261, 291
Poisson distribution, 185–89
  as approximation of the binomial distribution, equation, 189, 216
  and binomial distribution, compared, 213–14
  calculating probabilities, 186–88
  characteristics of processes that produce a, 185
  defined, 185, 215
  equation, 186, 216
Population:
  compared to sample mean, 11, 226–27
  comparing parameters of two, 325
  defined, 5, 11, 36, 227
  function, 11–12
  sampling from normal, 240–43
Population arithmetic mean, equation, 51, 77
Population coefficient of variation, 104–105, 109
Population distribution, 238–39
Population mean:
  estimating, 260
  and sample mean, 240–42, 243, 268–69
  sample size for estimating, 286–88
  using sample mean to estimate, 248, 264
Population parameters:
  defined, 227
  estimating, 260
  making inferences about, 444–48
  probability of true, falling within the interval estimate, 268–69
Population proportion:
  estimating, 260
  and interval estimates, 277–79
  sample size for estimating, 288–90
  using sample proportion to estimate, 266
Population regression line:
  equations, 444. 445, 468
  slope of, 445–48
Population standard deviation, 95–97
  determining uncertainty attached to estimates of, 388–89
  equations, 96, 99, 108, 109
  and interval estimates, 267–68, 274–75, 291
  relationship to variance, 95–96
  using grouped data, 99–100, 109
  using the sample standard deviation to estimate, 264–66
Population standard score, equation, 98, 108
Population variance, 94–95
  equations, 95, 99, 108–109, 378, 401
  inferences about, 356
  inferences about a single, 388–93
  inferences about two, 394–98
  point estimate of, 264–66

Population variance (*cont.*)
  relationship to average absolute
    deviation, 94
  using grouped data, 99–100, 108–
    109
Positive slope, 411, 417
Posterior probabilities:
  calculating, 145–47
  defined, 145, 152
  example, based on three trials,
    148–49
  with inconsistent outcomes, 149–
    50
  with more information, 147–48
Power curve, in hypothesis testing,
    315–16, 347
Power of the hypothesis test, 315, 347
Precision, defined, 248, 253
Prediction intervals, approximate,
    430–33
Price index, defined, 526
Probabilities:
  combining with monetary values,
    172–74
  under conditions of statistical
    dependence, 138–43
  under conditions of statistical
    independence, 130–36
  level of, and hypothesis testing,
    302
  revising prior estimates of, 144–50
Probability:
  basic concepts in, 117–18
  defined, 117–152
  types of, 119–23
Probability distribution:
  choosing correct, 213–14
  compared to frequency
    distribution, 162, 164
  defined, 5, 162, 164, 215
  examples of, 162–64
  selecting correct, in hypothesis
    testing, 305–306
  testing appropriateness of a, 369–
    73
  types of, 164
Probability rules, 125–29
Probability sampling, defined, 227,
    253
Probability theory, 116
Probability tree:
  defined, 132, 152
  using, 132–35
Proportions:
  dependent and independent, 357
  hypothesis testing of, 317–21, 338–
    43
  sample size for estimating a, 288–
    90

Q

Quality control, use of dispersion
    measures in, 86
Quantity index, 526
Quartile, defined, 89, 107
Quartile deviation:
  defined, 90, 107
  equation, 90, 107

R

r, symbol for sample coefficient of
    correlation, 443
r statistic:
  defined, 492
  equations for mean and standard
    error of, 492, 493, 507
r^2, symbol for sample coefficient of
    determination, 436
Random digits, using table of, 230
Random disturbance e, 445
Random sampling, 228–33
  defined, 227, 253
Random variables, 166–69
  defined, 166, 215
Randomizing, in experimental
    design, 522
Randomness, concept of, 490–91
Range:
  of data set, 12, 46
  defined, 88, 107
  equation, 88, 107
  as limitation on extrapolation, 449–
    50
Rank correlation, 496–504
  advantage of using, 496, 503–504
  defined, 496, 506
Rank correlation coefficient, 497–99
  defined, 476, 496, 506
  equation, 497, 507
Rank sum tests (*see also* Mann-
    Whitney U test)
  defined, 484, 506
Raw data:
  defined, 13, 36
  examples of, 13–14
Real class limits, 24–25
Regions of significant differences,
    303
Regression, defined, 410, 463
Regression analysis, misuse of, 449–
    51
Regression line:
  estimation using, 415–33
  "fitting," mathematically, 418
  and hypothesis testing, 445–48
  relation of sample and population,
    444–45
Regression plane, 453–54
Rejection, and hypothesis testing, 300
Rejection region, 303
Relationships:
  between variables, types of, 410–12
  misuse of, in regression analysis,
    451
Relative dispersion, 104–105
Relative frequency distribution:
  characteristics of, 16–18
  defined, 16, 36
Relative frequency histogram, 28
Relative frequency of occurrence,
    120–22, 152
Relative frequency polygon, 30
Replacement, defined, 229–30
Representative sample, defined, 12,
    36
Revised probabilities (*see* Posterior
    probabilities)

Right-tailed test of hypothesis, 307–
    308
Risk, of rejection, 300
Rows, contingency table, 358
Run, defined, 491, 506

S

s_e, symbol for standard error of
    estimate, 426
Sample, defined, 5, 11, 12, 36, 226,
    253
Sample arithmetic mean:
  equation, 51, 77
  of grouped data, equations, 52, 54,
    77
Sample coefficient of correlation (r),
    442–44
  equation, 443, 468
Sample coefficient of determination
    (r^2):
  developing, 435–36
  equation, 436, 441, 467, 468
  explained by regression line, 440–
    41
  intuitive interpretation of, 436–40
  and linear relationship, 440
Sample mean:
  converting, to a z value, 242–43
  and population mean, 240–45,
    268–69
  precision of, 248–49
  using, to estimate the population
    mean, 264
Sample mean $\bar{x}$, equation, 242, 254
Sample median of grouped data,
    equation, 67, 68
Sample proportion, using to estimate
    the population proportion, 266
Sample sizes:
  for chi-square test, 367
  determining in estimation, 286–90
  as determining sampling precision,
    251
  diminishing returns with increase
    of, 249
  for precision specified, 287, 289
  relationship between standard
    error and, 248–51
  results of increasing, 245
Sample space, 117, 152
Sample standard deviation, 101–102
  equation, 101, 109
  using, to estimate the population
    standard deviation, 264–65,
    274–75, 291
Sample standard score, equation,
    101, 109
Sample statistics:
  conventional terminology, 237
  defined, 227
Sample variance, 101–102
  distribution, 388–89
  equation, 101, 109
Sampling:
  and confidence interval estimation,
    272
  from non-normal populations, 243–
    44

Sampling (*cont.*)
  from normal populations, 240–43
  in planning experiments, 519
  reasons for, 226
  relationship between sample size
    and standard error, 248–51
  types of, 227
Sampling distribution of the mean:
  defined, 235, 253
  deriving the, 238–40
  from non-normal populations, 243–45
  from normally distributed
    populations, 240–43
  properties of, 242
  use of, to exemplify distribution of
    any sample statistic, 240
Sampling distribution of the
  proportion, defined, 235
Sampling distribution of a statistic,
  240, 253
Sampling distributions:
  conceptual basis for, 238–40
  defined, 235
  describing, 235–36
  for difference between to
    population parameters, 326–28
  of difference between sample
    means, 325
  function of theoretical, 240
  of number of runs (*r* statistic), 492–93
  one-sample runs test, 492, 507
Sampling error, defined, 236, 253
Sampling fraction, defined, 251, 253
Scatter diagrams, 412–14
  defined, 412, 463, 466
Seasonal variation, 524, 525
Secular trend, 524, 525
S-shaped curves, 32
Sigma (Σ), "the sum of" symbol, 51
Sigma hat ($\hat{\sigma}$), 275, 284, 292
Significance level, 302–304, 347
Significant difference, regions of, 303
Sign test, defined, 476, 506
Sign test for paired data, 478–81
Simple random sampling, 229–30,
  233, 253
Skewness:
  defined, 47, 76
  measures of, 46–47
Slope, defined, 415, 466
Slope of best-fitting regression line,
  equation, 421, 466–67
Slope of population regression line,
  445–48
Slope of straight line, equation, 416,
  466
Squaring the error, 420–21
Standard deviation:
  calculating, using grouped data,
    99–100, 109
  characteristics of, 102
  compared to standard error of
    estimate, 426–27, 429
  defined, 95–96, 107
  equations, 96, 99, 101, 108, 109
  relationship to variance, 95–96

shortcomings of, 104
  uses of, 92, 97–99
Standard deviation of binomial
  distribution:
  equation, 182, 216
  and interval estimate, 277–78
Standard deviation from the range,
  estimating, 288
Standard error:
  defined, 236, 253
  proper use of term, 237
  relationship between sample size
    and, 248–51
Standard error of the difference
  between two means, equation,
  326, 347
Standard error of the difference
  between two proportions,
  equation, 338, 348
Standard error of estimate, 426–30
  defined, 426, 466
  equation, 427, 428, 467
  interpreting, 429–30
  measuring reliability of, 426
  using, to generate prediction
    intervals, 430–31
Standard error of estimate for
  multiple regression, equation,
  457, 458, 469
Standard error of the mean:
  defined, 236, 253
  equations, 241, 250, 254
  for finite populations, 250–51
  for infinite populations, 241–43
  and interval estimates, 267, 268–69
Standard error of the proportion:
  defined, 236
  equations, 278, 292
Standard error of the *r* statistic,
  equation, 493, 507
Standard error of r_s, equation, 501,
  507–508
Standard error of the regression
  coefficient, defined, 446, 466
Standard error of the regression
  coefficient *b*, equation, 446,
  468
Standard error of the *U* statistic,
  equation, 486, 507
Standard normal probability
  distribution:
  defined, 194–95, 215
  using, 210–212
Standard normal probability
  distribution table, 196–201
Standard score, 98–99
  defined, 98, 107
  equation, 99, 108
Stated class limits, 24–25, 37
Statistic, as characteristic of a
  sample, 226
Statistical deception, 513,–17
  averages, improper use of, 514
  biased samples, 513–14
  "big jump" assertions, 515–16
  comparisons without statistical
    significance, 514
  in graphs and pictures, 516–17
  means, improper use of, 514–15

Statistical decision theory, 5
Statistical dependence, 138, 152
Statistical independence, 130, 133,
  152
Statistical inference:
  as branch of statistics, 260
  defined, 233, 253
  simple random sampling, 233
Statistical sampling, defined, 5
Statistically dependent events,
  equations, 140, 142, 143, 153
Statistically independent events,
  equations, 131, 136, 143, 153
Statistics:
  conventional symbols used, 50
  defined, 50, 76, 226, 253
  improper use of (*see* Statistical
    deception)
  meanings and origin of word, 2
STATPACK, 462–63, 464–65
Straight line:
  equation, 415, 466
  "fitting," to a scatter diagram, 413
  using estimating equation for a,
    416–18
Strata, defined, 232, 253
Stratified sampling, 232, 253
  compared to cluster sampling, 233
Student's *t* distribution (*see t*
  distribution)
Subjective probability, 122–23, 152
Sufficient estimator, defined, 263,
  291
Summary statistics, defined, 46, 76
Symbols:
  for equations, 50
  for population parameters, 50, 100,
    227
  for sample statistics, 50, 100, 227
Symmetry, 46, 76
Systematic sampling, 231–32, 253

T

t distribution:
  characteristics of, 281
  compared to chi-square
    distribution, 361
  conditions for using, 280–81
  defined, 280–81, 291
  in hypothesis testing, 305–306,
    322–24
  for prediction intervals, 432–33
  and rank correlation coefficient,
    500
t distribution table, using, 282–85
Theoretical sampling distributions,
  function of, 240
Theory of runs, defined, 491, 506
Time, as a factor in sampling, 226
Time series, 523–25
  defined, 523
  kinds of variation in, 524–25
Time series analysis, 523
Total change, 300
Total error, using to determine best
  fit, 419
Trade-offs, significance level, 304

Trends, misuse of, in regression analysis, 450
Two-tailed tests:
  for difference between means, 328–30
  for difference between proportions, 338–41
  of hypothesis, 306, 347
  of means, 309–311, 322–24
  of proportions, 317–19
  of two variances, 396–98
  using t distribution, 322–24
  of a variance, 390–91
Type I error, defined, 305, 347
Type II error, defined, 305, 347

U

U statistic:
  defined, 486
  equation, 486, 488, 506–507
Unbiased estimator, defined, 262, 291
Uncertainty, dealing with, 298
Unconditional probability (*see* Marginal probability)
Unequal classes, problems with, 21
Unexplained deviation, 440
Unexplained variation, 440–41
Ungrouped data, 51
Upper confidence limit, 271
Upper-tailed test, defined, 307, 347

V

Value index, 526
Variability (*see* Dispersion)
Variables, types of relationships, 410–12
Variance, 92
  calculation of, using grouped data, 99–100, 108–109
  defined, 94, 107
  equations, 95, 99, 101, 108–109
  relationship to standard deviation, 95–96
Variance among sample means, equation, 377, 401
Variation, defined, 435
Variation of Y values around the regression line, equation, 435, 467
Variation of the Y values around their own mean, equation, 435, 467
Venn diagram, 125–26, 152
Vertical axis, in graphing, 28

W

Weighted mean, 58–60
  defined, 59, 76
  equation, 59, 77
Weighting of factors, inappropriate, 527–28

Within-column variance:
  defined, 379, 400
  equation, 379, 401

X

x-bar ($\bar{x}$), symbol for sample mean, 50

Y

$\hat{Y}$ approximate prediction intervals, 430–31
Y-hat ($\hat{Y}$), symbol for values of estimated points, 418
Y-intercept, defined, 415, 466
Y-intercept of best-fitting regression line, equation, 422, 467

Z

Zero correlation, 438–39
z table, compared to t table, 282–83
z value (*see also* Standard normal probability distribution table)
  converting sample mean to a, 242–43
  equation, 196, 216
  use of, 196

Pg 19. # 9,10,11,12,15
Pg 26-27 # 20-26
Pg 34 #30-34
Pg 48-49 #2,4
Pg 56-57 # 6,7,8,9,10